Biotechnology
The Renewable Frontier

Biotechnology
The Renewable Frontier

Edited by Daniel E. Koshland, Jr.

The American Association for the Advancement of Science

Library of Congress Cataloging-in-Publication Data
Biotechnology: The Renewable Frontier
Includes index.
1. Biotechnology. 2. Biology—Research.
I. Koshland, Daniel.
TP248.2.B546 1986 660'.6 86-3378
ISBN 0-87168-314-8
ISBN 0-87168-283-4 (pbk.)

This material originally appeared in ***Science***, the official journal of the American Association for the Advancement of Science.

Distributed to bookstores and overseas through Westview Press, Boulder, Colorado 80301

AAAS Publication No. 85-26

Printed in the United States of America

Other titles in this series of ***Science*** volumes include:
Biotechnology & Biological Frontiers
Edited by Philip H. Abelson

Neuroscience
Edited by Philip H. Abelson, Eleanore Butz, and Solomon H. Snyder

Astronomy & Astrophysics
Edited by Morton S. Roberts

AIDS: Papers from ***Science***, 1982–1985
Edited by Ruth Kulstad

Frontiers in the Chemical Sciences
Edited by William Spindel and Robert M. Simon

Contents

VII. PLANT SCIENCES

VIII. BEHAVIOR AND SENSORY PHENOMENA

Introduction

In Pursuit of the Renewable Frontier

Daniel E. Koshland, Jr.

The endearing feature of intellectual frontiers is that they are in endless supply. Explorers of continents fight their way through wildernesses until they arrive at the water's edge, and then sigh that there are no new mountains to conquer. Researchers, on the other hand, are part of an ever-expanding universe, inevitably creating new territories to explore as they complete the maps begun by the discoveries of the past.

No area of research illustrates this phenomenon more clearly than modern biology, which many believe is in its Golden Age. This book presents a collection of articles that contain illustrative—but certainly not exhaustive—areas that have great potential for the future.

The articles are grouped together under a number of sections, but simply organizing the articles was a revelation. Almost any of these papers could have been listed in another section equally well. For example, Sherie Morrison describes chimeric antibodies which could be listed under "New Techniques" or under "Immunology;" Smith, Duncan, and Moir's article could be listed under "Biotechnology" or under "New Techniques;" Goodenow, Vogel, and Linsk's article could be under "Immunology" or "Developmental Biology and Cancer." The list could go on, but it illustrates the extraordinary cross-disciplinary aspects of modern biology. Discoveries in one area almost inevitably generate ideas or techniques applicable to a widely different area.

The same interdisciplinary phenomenon applies to the line between a new biological frontier and biotechnology. We have listed two articles in the latter category because they deal directly with industry, but in effect all the articles could be listed under that one heading. "Safety Concerns and Genetic Engineering in Agriculture" is both a frontier in biology and of key interest in biotechnology. "Genetic Engineering of Novel Genomes of

Large DNA Viruses" is of great theoretical interest and involves new techniques, but it also may have vast practical applications. The line between basic and applied research becomes ever more fuzzy. The new frontiers in biology today thus become the frontiers of biotechnology tomorrow.

For many years biology seemed a study in pure science. Biologists were rarely consulted by industry, unlike their colleagues in chemistry, engineering, and law. Today all that has changed—and there is anguished ethical study of the dangers to academic research of the new practicality of biological research. The dangers are real, but no different from those found for other research areas at an earlier time. The finding that the discoveries in the modern biology laboratory are of great practical importance in industry, as they have been in medicine for many years, should only add interest and excitement to the frontiers of biology and biotechnology being uncovered each day.

As this book goes to press and we plan new volumes in other areas for the future, an editor cannot help but reflect on the personality of individuals who are satisfied by such an unceasing quest. Are we scientists just curious children who have never grown up? Are we the most idealistic of people, bravely confronting the ultimate challenges for the good of humankind? Or are we the most selfish of its citizens, who have discovered the ideal way of life: solving nature's crossword puzzles while being subsidized in our happiness? Whatever the answer, we are all, in the words of the poet, "emperors of the endless dark, even in seeking."

Biotechnology
The Renewable Frontier

Part I

New Techniques

1

Strategies and Applications of in Vitro Mutagenesis

David Botstein and David Shortle

Biochemists have begun to use mutations to probe the relationship between the structure and activity of proteins; cell biologists are using mutations to define the roles of particular proteins and protein assemblies in the cell; developmental biologists are using mutations to determine the logic and order of molecular events during differentiation and morphogenesis; and neurobiologists are beginning to turn to mutations to try to understand the way in which neural networks are formed and, eventually, how they function. These new and expanding applications of mutations in many disciplines of biology represent one of the most important consequences of the revolution in the life sciences that has resulted from the development of recombinant DNA technology. The objective of this review article is to sketch the outlines of the many mutagenesis strategies made possible by the availability of cloned genes. Our emphasis on general principles and applications has required that we gloss over many ideas and technical accomplishments in the field; such information, though, has been reviewed (*1*, *2*).

Before we address the question of how best to isolate or construct mutations, it is important to review the fundamental logic behind the use of gene mutations to analyze biological phenomena. The primary reason for isolating and characterizing a mutation is to assess its consequences, or in genetic terminology, its phenotype. In the ideal mutation experiment, two organisms are submitted to careful comparison, one being mutated at a single known site and the other lacking the mutation. Any observed difference between the two is then attributable to the mutation; by characterization

Science 229, 1193–1201 (20 September 1985)

of this difference, inferences can be made about the function of the corresponding wild-type gene, regulatory signal, or nucleotide pair. In other words, the properties of a normal function can be learned from the consequences of perturbing or eliminating a single gene or genetic element.

One approach for designing the ideal mutation experiment is to begin with a phenotype of interest (such as failure to complete mitosis at a given temperature) and search, after classical mutagenesis, for mutant organisms that exhibit this phenotype. The mutations obtained by this strategy would then be characterized by genetic mapping, followed by cloning and sequencing the wild-type and mutant genes. A second approach is to begin with the idea that a particular protein (such as tubulin) is likely to play an important role in mitosis. To establish this point genetically, it would be necessary to clone the gene, induce mutations in the cloned copy in a recombinant DNA molecule, and return these mutant genes to the organism in a way that would allow assessment of phenotype.

Both these approaches might lead to the same experiment, although with unequal likelihood. The first approach, in which the phenotype determines the selection of mutations for analysis, has the advantages that no detailed hypotheses are required to obtain material for study and that each new mutation has a high probability of contributing to an understanding of the phenomena underlying the phenotype. The major advantage of the second approach, in which the gene for a particular protein is cloned initially, is that very specific hypotheses, often arrived at by earlier biochemical or biological studies, can be put to definitive genetic tests.

Although, in principle, every gene that encodes information relevant to a given biological process should be found by the first approach if the "correct" phenotype has been used to search for mutations, the results, in practice, are different. For example, extensive screening for cell-division–cycle mutants in yeast failed to produce mutations in the β-tubulin gene, despite the fact that mutations in that gene isolated by the second approach (*3*) exhibited precisely the same cell-division–cycle phenotype used in the unsuccessful screening. Perhaps the most important feature of in vitro mutagenesis is the ability to efficiently and predictably introduce mutations into a gene of interest.

Classical in Vivo Mutagenesis

The only mutations available for study by the earliest geneticists were natural variants and occasional spontaneous mutations. The discoveries that organisms exposed to x-rays (*4*) and certain chemical compounds (*5*) yield much higher frequencies of mutant progeny led to a revolution in experimental genetics. Because these agents gave geneticists partial control over the process of mutagenesis, they could systematically study biological phenomena by collecting large numbers of different mutations displaying a characteristic phenotype, which could then be classified on the basis of complementation behavior, map position, and various other criteria. In this way the genetic loci relevant to the phenomena could be identified, counted, positioned on a genetic map, and then functional interactions between loci defined. With the classical studies of eye pigmentation mutations in *Drosophila melanogaster* (*6*) and of auxotrophic mutations in *Neurospora crassa* (*7*), it became apparent that the detailed phenotypic characterization of mutations could be extended to the level of individual molecules.

The power of classical in vivo mutagenesis to provide material for genetic and biochemical analysis is attested to by the progress made over the past few decades in the genetics of eukaryotes such as *D. melanogaster* and the molecular genetics of *Escherichia coli* and other microbes. Nevertheless, the requirement for a specific phenotype in order to identify rare mutations in a mixture of wild-type and irrelevant mutations imposed serious limits on the range of phenomena to which genetics could be applied. It was not feasible to study more than a few mutations in a single gene without the special circumstances of a unique phenotype that is easy to score plus an efficient genetic crossing system for weeding out secondary mutations in irrelevant genes caused by the general mutagenesis. Isolation of mutations in a gene of special biochemical interest by simple assay ("brute force," in the jargon of the geneticist) required not only heroic amounts of labor to screen thousands of mutagenized organisms but also major (even lucky) assumptions about the phenotype. A good example is the isolation of a mutation in DNA polymerase I of *E. coli* (*8*).

In bacteria, some improvements in mutagenesis were afforded by the concept of "localized mutagenesis" (*9*). New mutations could be limited to individual segments of a genome by mutagenizing transducing particles that carry only a small fragment of the bacterial chromosome, followed by generalized transduction with selection for a marker known to be closely linked to the genes targeted for mutagenesis. The development of specialized transducing phages and episomes carrying only a portion of the bacterial genome also allowed the mutagenesis of specific genes without exposure of the entire genome of the host cell to the action of a mutagen. Despite having some of the advantages of the new methods of in vitro mutagenesis, mutagenesis of these naturally occurring recombinants had only limited applicability because many genes of interest in *E. coli* could not be readily isolated on specialized transducing phage or small F′ episomes.

Transposon Mutagenesis

A second revolutionary development in mutagenesis came with the realization that insertion mutations could be induced in virtually any gene of interest by the appropriate manipulation of naturally occurring transposable elements. Although these mobile segments of DNA had been discovered many years earlier in maize and *Drosophila*, it was in bacteria that the advantages of insertion mutagenesis were first systematically exploited (*10*).

Unlike chemical or radiation mutagenesis, transposon insertion mutagenesis results in a single, unique physical alteration in the gene that has been mutated. Mutagenesis can frequently be carried out in such a way as to limit the number of transposons inserted to essentially one per genome. Most importantly, the insertion event can often be selected by means of drug resistance or phage immunity carried by the transposon. Thus, a population of organisms, each of which had a transposon insertion within a gene or intergenic segment, can easily be generated for phenotypic screening. Mutagenesis is therefore extremely efficient, the level of secondary mutations is very low, and most mutations lead to total inactivation of the gene. If the transposon inserts randomly, the probability of finding a mutation in a given gene is the fraction of the total genome that the gene of interest occupies. Thus, for bacteria, about 1 of 3000 to 5000 organisms with a single insertion are expected to be mutated in a given gene. Transposon mutagenesis can be applied to very small genomes, such as plasmids, and is a simple alternative to the use of deletion mutations for defining the extent of a gene of interest (or a small cluster of genes) after it has been cloned [for examples, see (*11*, *12*)].

Probably the single most important advantage of insertion mutations is that they contain an insertion of a known DNA element, the transposon. As a result, transposon mutagenesis can be used to isolate genes by first identifying an insertion mutation in or very near the gene of interest, on the basis of phenotype or genetic linkage, and then cloning a fragment of DNA from the mutant genome that harbors the nucleotide sequences of the transposon. This transposon-tagging technique has made possible the efficient identification and cloning of interesting genes in *Drosophila* (*13*) and also, by means of an RNA tumor virus as the transposon, in mice (*14*).

In Vitro Mutagenesis

Since the introduction of recombinant DNA methodology, genes can be removed from their normal environment in an intact genome and isolated as DNA fragments on cloning vectors. The availability of purified genes in vitro in microgram amounts has dramatically expanded the potential for inducing mutations. In the controlled environment of the test tube, it is now possible to alter, efficiently and systematically, the sequence of nucleotides in a segment of DNA. In the following sections the new methods of in vitro mutagenesis are divided into three broad categories: (i) methods that restructure segments of DNA, (ii) localized random mutagenesis, and (iii) oligonucleotide-directed mutagenesis. This classification emphasizes the practical aspects of each method's application.

The considerable increases in mutagenic efficiency and specificity attainable with the new methods, however, do exact a price. Because these methods are designed for use on isolated DNA molecules, a gene must almost always be removed from its normal genetic context—a unique locus on a large complex chromosome inside a living cell (or virus)—and inserted into the abnormal context of a small cloning vector propagated in *E. coli*. Unlike classical in vivo mutagenesis, in which all mutations are isolated in situ, in vitro mutagenesis invariably yields gene mutations out of their normal context. This is the most radical and most troublesome difference between the classical methods and the powerful in vitro methods.

For some applications, this change in genetic context is relatively unimportant. For other applications, though, inferences about a wild-type gene, on the basis of the phenotype of a mutation construction in vitro, can only be made after the mutant allele has been restored to its normal genetic context. In such situations, the genetic manipulations required to assess the consequences of a mutation in its proper context become the primary challenge to the molecular biologist. Therefore, the last two sections of this review article are devoted to a discussion of the variety of available solutions, some partial and some complete, to this important problem.

Restructuring of DNA Segments

After a gene of interest has been isolated, it is usually necessary to reduce to a minimum the size of the cloned DNA segment carrying the gene and to move it into a small, circular cloning vector. By a form of deletion analysis, extragenic flanking sequences are systematically eliminated from the initially cloned DNA segment and each new deletion is tested, with some assay of structure or function, to determine that the gene is still intact. In this way the ends or boundaries of the gene are roughly defined while, at the same time, smaller subclones are isolated that will simplify subsequent manipulations of the gene, such as DNA sequencing and mutagenesis.

Further reductions in size, which permit very precise definition of the functional boundaries at the 5′ and 3′ ends of the gene, are achieved by generating terminal deletions with one end point located outside of the gene and the other positioned progressively closer to the gene. Collections of deletions, which have been extensively used to identify such regulatory sequences as transcriptional promoters, can be readily constructed by using an exonuclease [either Bal 31 (*15*) or exonuclease III plus S1 nuclease (*16*)] to remove nucleotides starting from a unique site just outside the gene. This site is generated by restriction enzyme cleavage of a circular DNA molecule; in some cases, nuclease digestion can be confined to one of the two ends of the linear DNA molecule, leading to deletions that extend in a single direction (*17, 18*).

Insertion of a small synthetic oligonucleotide that encodes a unique restriction enzyme cleavage site at the end points of these deletions (*16*) or at positions randomly distributed across the cloned DNA segment (*19*) provides a readily available restriction site for use in further restructuring of the cloned DNA segment. These oligonucleotide insertions, termed linker mutations, are especially versatile for the modification of circular DNA molecules. For instance, cleavage at a linker insertion introduced near or within the gene provides a site at which additional nucleotides can be inserted or deleted. It may also provide ends by which the cloned DNA segment itself can be inserted into other types of DNA vectors for procedures such as nucleotide sequencing, production of large quantities of the gene product, and transformation of other types of cells. In addition, by joining restriction fragments isolated from pairs of linker insertion mutations of the appropriate size and position, mutations consisting of four to eight tightly clustered base substitutions can be generated, making it possible to efficiently "scan" a small region of DNA for regulatory sites (*20*).

When changes in the level of expression of a gene are difficult to monitor because of the lack of a convenient assay for the gene product, a common strategy is to construct a gene fusion in which the regulatory elements of the gene of interest control the expression of a gene product that can be readily quantitated. Typically, a second gene specifying an easily assayable enzyme, such as β-galactosidase (*21*) or chloramphenicol acetyltransferase (*22*), is joined downstream of the 5′ end of the first gene, either by fusing the two protein-encoding regions in frame to create a hybrid protein or by fusing the 5′ untranslated sequences of the two genes so that the hybrid gene encodes only the product of the second gene. This strategy, developed in *E. coli* before the availability of recombinant DNA methods (*23*), has been expanded and applied to the study of gene regulation in a large number of divergent organisms, including yeast (*24, 25*) and a number of higher eukaryotes (*26*).

Random Mutagenesis

Several of the mutagens used to randomly mutagenize organisms or viruses in vivo can also induce random point mutations in purified DNA molecules. Hydroxylamine and nitrous acid have been used to generate mutations in genes cloned into a plasmid or phage vector by simply reacting the entire recombinant DNA molecule with mutagen. Although this technically simple approach can be useful when coupled to a simple phenotypic selection or screen, the problem of a background of secondary mutations must be dealt with, either by reducing mutagenesis to a level where secondary mutations are rare or by using recombination to isolate the primary mutation free of any secondary mutations. Furthermore, as is typical of chemical mutagens in general, hydroxylamine and nitrous acid induce a limited number of base substitutions that, instead of being uniformly distributed across a nucleotide sequence, tend to cluster at a small number of "hot spots" (*27*). Although technically not an in vitro method of mutagenesis, brief passage of purified DNA through bacterial strains carrying various mutator activities can also be used to obtain random mutagenesis (*12, 28, 29*).

The potential problem of secondary mutations arising because of random mutagenesis can be avoided in two ways. First, transposon mutagenesis of a recombinant plasmid or phage produces single-insertion mutations that are often essentially random in distribution. These types of random mutations, which usually result in complete loss of function, can be isolated easily by propagating the cloned gene in an appropriate *E. coli* strain and then retransforming purified DNA with selection for the drug-resistance marker on the transposon (*12, 30*). A second, more general approach applicable to small circular DNA molecules is to introduce a single nick at a random site with deoxyribonuclease I plus ethidium bromide (*31, 32*). After conversion of the nick into a short single-stranded gap, deletions and insertions (*33*) or a variety of types of base-substitution mutations (as described in the next section) can be efficiently targeted to the sequence of nucleotides exposed in the single-stranded gap.

Localized Random Mutagenesis

In many applications involving cloned genes, random mutagenesis of the purified gene is still not a feasible approach for mutant isolation, usually because the assay for phenotype is too laborious to apply to the hundreds, if not thousands, of candidates. In response to the need for more direct routes to mutant isolation, in vitro methods have been developed that result in random mutations confined to small segments of large DNA molecules. To understand the principles behind these methods and to compare their relative advantages for a particular mutagenesis project, it is necessary to consider three properties: (i) site specificity, (ii) efficiency, and (iii) complexity.

Site specificity. All the methods discussed in this section can induce a variety of different point mutations localized within a defined interval of a DNA molecule. Exactly how a site or segment is targeted for mutagenesis and how large or small it can be are two of the principal variables that distinguish the different methods. Obviously, the optimal strategy for identifying the genetic information relevant to a particular phenotypic property is to target mutagenesis to the smallest interval in which all the information is likely to reside.

Efficiency. Every method of mutagenesis, whether targeted to the entire genome of a living organism or to a single nucleotide position in a purified DNA molecule, yields a mixture consisting of one or more mutant forms and wild type. As a measure of the accomplishment of mutation induction, efficiency can be defined as the fraction or percentage of all output DNA molecules (candidates) that do not have the starting wild-type sequence. In practical terms that are of the utmost importance, an increase in the efficiency of mutagenesis means that fewer candidates must be screened in order to find phenotypic mutations. Many of the new methods of in vitro mutagenesis have very high efficiencies and, consequently, new opportunities are presented for applying even very labor-intensive, "brute force" screens for mutants with new and rare phenotypes.

Complexity. In many situations, highly efficient mutagenesis is of value only when a large variety of different mutations is generated. Complexity represents a relative measure of the number of different types of nucleotide sequence changes that can be induced. With regard to base-substitution mutations, each of the four types of base pairs in DNA can substitute for each of the other three. Therefore, when mutagenesis is confined to one strand of DNA, there are 12 distinguishable types of base substitution. When both DNA strands are mutable, these 12 reduce to only six types. A low complexity, which is typical of most chemical mutagens used either in vivo or in vitro, translates into the induction of a subset of all possible substitutions at a subset of the different nucleotide positions in a segment of DNA.

One simple strategy for limiting the action of chemical mutagens to a specific DNA sequence is first to isolate a small restriction fragment containing the sequence and then, after in vitro reaction with the mutagen, to recombine the mutagenized fragment back into the DNA of origin (Fig. 1A). Nitrous acid, hydroxylamine, methoxyamine, and other mutagens have been successfully used to obtain localized random mutagenesis of acceptable efficiency, albeit of low complexity (*34–37*). One technical problem that frequently limits the applicability of this strategy is the requirement for efficient reconstruction of the original DNA molecule from the mutagenized fragment plus a larger untreated fragment. If instead of using DNA ligase to rejoin the correct fragments (which usually limits mutagenesis to those fragments generated with two single-cut restriction enzymes) the mutagenized fragment is reannealed to a single-stranded form of

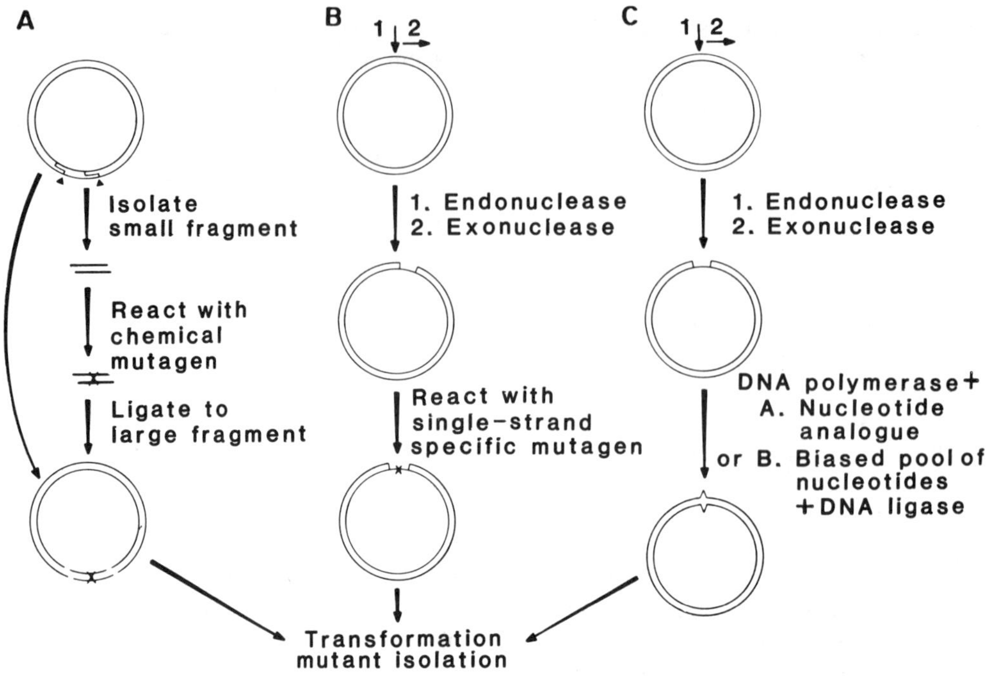

Fig. 1. Outline of three strategies for localized random mutagenesis. (A) Chemical mutagenesis of a restriction fragment isolated from a larger DNA molecule, followed by reassembly of complete molecules with DNA ligase. (B) Mutagenesis of a short single-stranded gap (or loop) in an otherwise duplex DNA molecule by means of a chemical mutagen that is specific for single-stranded DNA (for example, sodium bisulfite). (C) Nucleotide analogue incorporation, or nucleotide misincorporation, by DNA polymerase during synthesis from a primer-template.

the original molecule, mutations can be induced by in vitro synthesis of the second strand and DNA transformation (*38, 39*), in much the same way as with synthetic oligonucleotides (see below).

To increase both the complexity and efficiency of chemical mutagenesis of isolated DNA fragments, Meyers and Maniatis (*40*) have developed a system for in vitro DNA synthesis that generates mutations by the use of a damaged DNA strand as template. After inducing a very limited number of lesions in single-stranded DNA that destroy the base-pairing ability of one or more of the four types of bases, avian myeloblastosis virus (AMV) reverse transcriptase is used to synthesize a second, complementary strand. Apparently at sites in the template where there is a damaged base, or no base at all, this highly error-prone polymerase will insert one of the four nucleotides in a nonspecific manner, leading to a very large variety of types of base substitution. After recombining the fragment back into a specially constructed cloning vector and transforming into *E. coli*, restriction fragments containing single-point mutations can be physically separated from the original wild-type fragment by electrophoresis on urea-gradient gels at high temperature (*41*). Although it requires a considerable number of steps, this approach is particularly attractive in situations in which it is preferable to isolate and sequence mutant forms before determining their phenotypic consequences (the phenotypic screen is very labor-intensive) and in situations in which phenotypically silent mutations are of interest.

A second strategy for achieving site specificity is to modify the local structure of a sequence on a DNA molecule in such a way that the sequence becomes more sensitive to a mutagenic reaction than the flanking sequences. If such mutagen-sensitive target sites can be introduced without reducing the DNA molecule to fragments, the problem of reassembly is avoided altogether. The chemical reactivity of a particular sequence or segment of a duplex DNA molecule can be drastically modified by converting the nucleotides in the segment to be mutagenized into a single-stranded state. By the appropriate enzymatic manipulation of circular DNA molecules in vitro (*32*), each of the three types of single-stranded structures—gaps, deletion loops, and displacement loops—can be introduced at more or less precisely defined sites in duplex DNA molecules and then can be used as targets for localized mutagenesis (Fig. 1B).

For example, the simple inorganic salt sodium bisulfite efficiently induces GC to AT mutations by catalyzing the deamination of cytosine residues in single-stranded, but not double-stranded, DNA (*42*). Consequently, each of the cytosine residues on one strand of a segment of DNA can be made susceptible to mutation by exposing the sequence in a single-stranded gap or loop. The extent of deamination of such reactive cytosines can be carefully controlled by the concentration of bisulfite and the time of the reaction, giving in vitro mutagenesis that is both highly localized and efficient (*32, 43–46*). However, the only mutations induced are C to T substitutions (or G to A substitutions as seen from the opposite strand), a level of complexity too low for many types of mutational analysis.

DNA polymerases, somewhat surprisingly, are emerging as the most versatile of the in vitro mutagens for achieving localized random mutagenesis that is both efficient and high in complexity. As mentioned earlier, polymerases will, in some cases, insert nucleotides nonspecifically onto a growing DNA chain opposite sites of damage in the template strand. Errors in the choice of nucleotide incorporated from an undamaged template during in vitro synthesis can be increased, by many orders of magnitude over the in vivo error rate, either by using nucleotide analogues (*47, 48*) or simply by biasing the pool of nucleoside triphosphates available to the DNA polymerase (Fig. 1C).

For instance, when incubated with a DNA molecule containing a short single-stranded gap plus one α-thiophosphate nucleoside triphosphate, the large fragment of *E. coli* DNA polymerase I can be made to misincorporate a single nucleotide with an efficiency that approaches 100 percent (*49*). The normally very low frequency at which the polymerase makes mistakes is amplified in this in vitro reaction because only one of the four nucleotide substrates is present. Once the enzyme misincorporates a nucleotide, the normal correction mechanism involving the polymerase's intrinsic 3′ to 5′ nuclease activity is blocked by the sulfur atom present in the terminal phosphodiester bond. In a second reaction, the remainder of the single-stranded gap is correctly filled in by extension from the mispaired 3′OH-terminus. Eleven of the 12 types of base substitution have been induced at roughly comparable frequencies by means of this strategy of gap-misrepair mutagenesis with each of the four α-thiophosphate nucleotides (*50*).

A second method for enhancing the rate of errors made by polymerase during the repair of a short single-stranded gap is to leave out one of the four deoxynucleoside triphosphate substrates (*49*). The gap-filling reaction can be driven to completion by addition of DNA ligase plus adenosine triphosphate, which seals the resulting nick and prevents subsequent proofreading. Although this method gives base-substitution mutagenesis of high complexity (plus frequent insertions and deletions), it has the serious disadvantage of frequently inducing multiple, clustered mutations (*51*).

Alternatively, in vitro repair synthesis with a DNA polymerase lacking an editing 3′ to 5′ exonuclease can be used to efficiently misincorporate the conventional nucleoside triphosphates (*52*). After generating unique sites for priming by annealing a short oligonucleotide to circular single-stranded DNA, it is possible to use AMV reverse transcriptase to misincorporate nucleotides at downstream positions. Again, by proper choice of reaction conditions and time of incubation, a large variety of types of base-substitution mutation can be induced with a high efficiency (*53–55*).

The principal technical problem associated with the routine use of DNA polymerases as in vitro mutagens is the general problem of positioning the 3′OH-terminus, which determines the site for misincorporation, at any specified nucleotide or at sites uniformally distributed across any specified interval. As mentioned earlier, completely random mutagenesis of a small circular plasmid can be achieved by nicking with deoxyribonuclease I plus ethidium bromide followed by limited 3′ to 5′ exonucleolytic digestion (which further randomizes the 3′OH-terminus) to produce a short gap (*50*). To obtain mutagenesis localized to a defined sequence, it is sometimes possible to move the 3′OH-terminus to a more or less defined position from a unique site nearby by means of controlled digestion with a 3′ to 5′ exonuclease (*56*) or controlled nick translation with DNA polymerase I (*44*). In addition, modifications in the chemical structure (*57*), genetic information (*53*), or replication pathway (*58*) of the primer-template often permit further improvements in the overall efficiency of mutagenesis by selecting specifically against the template strand or for the strand synthesized in vitro.

Oligonucleotide-Directed Mutagenesis

With the methods of localized random mutagenesis, the desired product is a pool of many different mutations that can be searched, by use of a selection or a screen, for the subset of mutations that cause specific phenotypic changes. However, when only one or a few mutations of defined DNA sequence (genotype) are of special interest, searching through pools of mutations becomes a very inefficient strategy of mutant isolation. In these situations, synthetic oligonucleotides can be used to construct mutant alleles of a cloned gene with base substitutions, insertions, or deletions, either singly or in any combination (*1, 59, 60*).

The enormous versatility of oligonucleotide mutagenesis derives from the fact that the mutation (the part of a gene sequence that differs from the wild type) is actually synthesized de novo as a short oligonucleotide. Because the chemistry involved in coupling nucleotides imposes no constraints on the base sequence of the product oligonucleotides, there are no constraints on the types of mutation that can be induced by this method.

There have been several recent reviews of the technical advances in the chemistry of oligonucleotide synthesis (*59, 61, 62*), and the availability of commercial automated synthesizers has made the actual synthesis of mutant oligonucleotides the most routine and reliable step in the overall protocol. Therefore, this section concentrates on the somewhat more problematic subsequent steps in which the mutation encoded in the sequence of a synthetic oligonucleotide is introduced into a cloned gene.

In the first strategy developed for incorporating a short mutant oligonucleotide into a longer segment of DNA, DNA polymerase is used to copy wild-type sequences onto both ends of the oligonucleotide. The simplest way this can be done is to make a single-stranded oligonucleotide that is identical in sequence, except for the mutation to be induced, to a site on the wild-type gene (the target site for mutagenesis). On annealing the oligonucleotide to a single-stranded form of the wild-type gene, a short heteroduplex forms that can serve as a primer for extension by DNA polymerase with the wild-type sequence as a template. If the single-stranded template is circular, the DNA polymerase may extend the newly synthesized strand all the way around the template until it reaches the 5′ end of the oligonucleotide to form a single-stranded nick. Covalent closure of the nick by DNA ligase generates a duplex circular molecule that has a wild-type sequence in the template strand and a mutant sequence in the newly synthesized strand (Fig. 2A). On transformation into a cell in which the DNA molecule undergoes replication, the two genotypes segregate and subsequently can be isolated as pure clones.

Since the initial application of this approach to the site-specific mutagenesis of single-stranded phage φX174 (*63, 64*), a number of extensions have resulted in its general applicability to any gene cloned onto a circular vector. For example, the requirement that the target sequence be in a single-stranded state can be satisfied (i) by recloning the gene into an M13 single-stranded vector (*65*); (ii) or into a plasmid vector that contains an M13 replication origin [which leads to packaging of single-stranded plasmid DNA into virus particles after superinfection with M13 (*66*)]; (iii) by enzymatically degrading part or all of one strand of a duplex plasmid with exonuclease III (*67*); (iv) by forming a heteroduplex molecule with an appropriate single-stranded gap (*68*); or (v) by direct annealing of the oligonucleotide to duplex DNA to form a D-loop (*69*). Instead of requiring that the polymerase complete the entire strand to attach wild-type sequences to the 5′ end of the mutant oligonucleotide, a second oligonucleotide can be used to prime DNA synthesis from a site upstream of the mutant oligonucleotide, yielding significantly higher efficiencies of mutagenesis (*70, 71*).

Perhaps the most important innovation has been the use of the mutant oligonucleotide, after end labeling with ^{32}P, as a hybridization probe to physically screen for mutant alleles among the progeny molecules generated on replication of the mutagenized DNA (*67*). First,

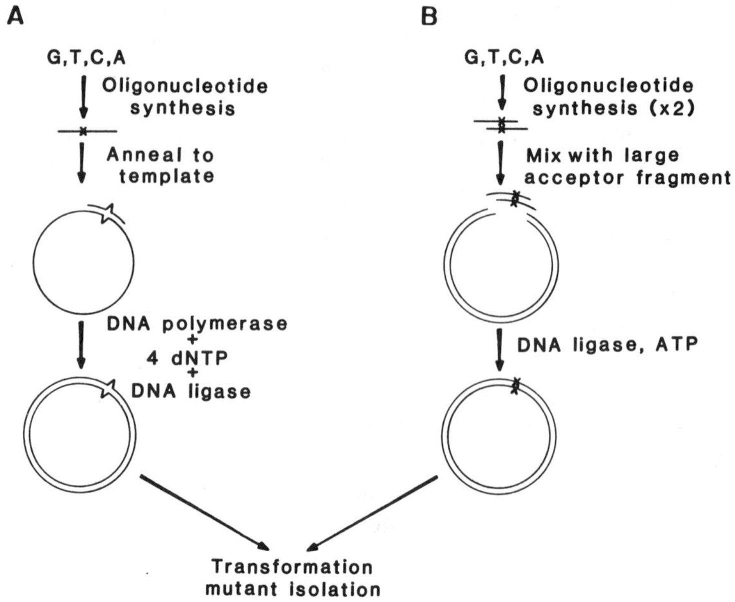

Fig. 2. Outline of two strategies for oligonucleotide-directed mutagenesis. (A) A single mutant oligonucleotide is synthesized and used as a primer for second-strand synthesis by DNA polymerase plus DNA ligase. (B) Two synthetic oligonucleotides (×2) of complementary sequence anneal to form a short duplex, which is joined to a large restriction fragment by DNA ligase to generate a complete duplex DNA molecule. ATP, adenosine triphosphate; dNTP, deoxynucleoside triphosphates.

the labeled oligonucleotide is hybridized with the DNA of candidate molecules that have been immobilized on nitrocellulose under conditions of very low stringency. Autoradiography is then performed after each in a series of successive washes at progressively higher stringency. This procedure can reliably discriminate between mutant and wild-type DNA molecules, either as plasmids in *E.coli* colonies (*72*) or as single-stranded M13 DNA in phage plaques (*65*). This technique permits the isolation of mutant alleles in the absence of a scoreable phenotype and solely on the basis of nucleotide sequence, even when the efficiency of mutagenesis is considerably less than 1 percent.

With the reduction in the time and expense of oligonucleotide synthesis, a second strategy requiring two unique oligonucleotides to incorporate a mutant sequence into an intact gene has become a feasible alternative to the oligonucleotide primer strategy. If a short DNA duplex containing a mutant sequence is formed by annealing two complementary oligonucleotides, it can be inserted into a much larger DNA molecule that contains all the remaining gene sequences, provided that the ends of both molecules have complementary single-stranded "sticky" ends. The DNA is circularized by joining of the two molecules with DNA ligase, and an intact, complete gene is formed that contains the mutant duplex segment (*73*) (Fig. 2B).

The synthetic mutant duplex can be made with any type of cohesive ends by appropriately extending the nucleotide sequence of each of the two synthetic oligonucleotides beyond the complementary region. However, to arrange for the appropriate sticky ends at precisely the position where the short mutant duplex is to be inserted almost invariably requires the introduction of unique restriction sites at the boundaries of the insertion site, either by the oligonucleotide primer method (*74, 75*) or through the total chemical synthesis of a gene with silent codon changes to create new restriction sites (*76*). Although this strategy requires considerably more oligonucleotide synthesis plus an initial partial or total redesign of the nucleotide sequence of the gene to be mutagenized, the efficiency of mutagenesis can approach 100 percent and the labor involved in recovering mutations in a double-stranded vector can be significantly less than with other procedures. Consequently, this strategy is likely to be most applicable for relatively small genes or DNA segments that are to be very intensively analyzed by site-specific mutagenesis.

Additional technical improvements and refinements are continuing to be made in both strategies of oligonucleotide-directed mutagenesis, particularly with regard to reducing the time and effort involved in assembling large collections of mutations of known sequence distributed over small intervals. At least two approaches have been described that permit the simultaneous synthesis of multiple oligonucleotides with different sequences. Small paper disks can be used as the support matrix for solid-phase synthesis; by simply sorting each disk (on which one individual chain is synthesized) into groups for condensation with A, G, C, or T before each coupling, as many as 100 different oligonucleotides can be synthesized in four reaction vessels (*77–79*). Alternatively, mixtures of oligonucleotides having the same nucleotide sequence, except for different base substitutions at one or more designated positions, can be synthesized in one reaction vessel by adding a mixture of several activated nucleotides to the coupling reaction when the appropriate point in the synthesis is reached (*74*). By coupling mixtures of nucleotides at several points in the synthesis, complex pools of oligonucleotides encoding different single- and multiple-base substitutions can be generated (*74*) for use in either the oligonucleotide primer or short duplex methods of mutagenesis.

The tactic of synthesizing a collection of mutant oligonucleotides to induce many different mutations in one experiment can be viewed as a form of localized random mutagenesis. It seems reasonable to expect that, in the future, oligonucleotide-directed mutagenesis will evolve from its current status as a reasonably efficient method for altering single nucleotide positions into a highly efficient method for saturation mutagenesis of DNA segments varying in length from 2 to perhaps as many as 50 base pairs. However, for mutagenesis of longer DNA segments, the methods of localized random mutagenesis discussed earlier are likely to continue to provide a more efficient means of generating pools of mutations. Thus, in many respects, these two different classes of methods for in vitro mutagenesis are distinctly complementary. Which strategy is optimal for a given mutational analysis depends primarily on the size of the target sequence that must be covered to extract all the genetic information relevant to the phenomenon under study.

Finally, all the methods of in vitro mutagenesis, including those that require synthetic oligonucleotides (*80*), can generate nucleotide sequence changes at sites other than those targeted for mutations. Therefore, even though a specific mutation of defined nucleotide sequence has been isolated and sequenced, an observed change in phenotype cannot immediately be attributed to the constructed mutation. Except for determining the entire nucleotide sequence of the gene or genetic element that has undergone mutagenesis, genetic mapping of the mutation is the only conclusive means of establishing the connection between a change in nucleotide sequence and its phenotypic consequences.

Determining Phenotype and the Context Problem

The methods described make possible the generation of a vast array of mutant forms of a cloned gene; DNA sequence technology makes possible the rapid determination and/or confirmation of the alterations in sequence, even before any attempt to assess phenotypic consequence. Up to this point, mutagenesis can be viewed as an exercise in nucleic acid biochemistry. To use mutations to draw inferences about underlying biological phenomena, the next step, and in almost all studies the most critical step, is the determination of the functional consequences of the mutations.

Although it is biochemically correct to define a gene mutation as a change in nucleotide sequence, from the point of view of genetic analysis it is more useful to define a mutation as a change in a unit of genetic information. As with all information, the meaning conveyed is dependent on the context in which it is expressed. Whether or not a mutation manifests itself by a change in phenotype can depend very significantly on the genetic context. For example, a recessive lethal mutation may lead to the immediate death of a haploid cell and yet be totally without observable effect when combined with a wild-type allele in a diploid cell. A mutant gene isolated after in vitro mutagenesis is, by definition, recovered in a foreign context. Before the functional consequences of such a mutation can be accurately determined and inferences correctly drawn about the wild-type gene, it may be necessary to partially, or perhaps totally, restore the mutant gene to the same genetic context in which the wild-type form occurs in nature.

The severity of this context problem varies with the system under study. For example, if biochemists are using in vitro mutagenesis to test hypotheses about the catalytic roles of certain amino acid residues in an enzyme of known three-dimensional structure, they will usually not be concerned that the organism in which the mutant gene was expressed or the transcriptional control sequences at its 5′ end will influence the conclusions concerning the properties of the purified protein. However, molecular biologists studying the details of expression of a mouse α-globin gene can draw no inferences at all from the level of gene transcription in *E. coli* from a cloned complementary DNA inserted into a λ phage vector. In this situation the ideal experiment would be to return the mutant mouse gene to its normal genetic context so that any alterations in phenotype could be confidently attributed to the mutation, rather than to possible artifacts arising from the abnormal environment in which gene has been placed.

To meet the challenge of studying mutations in their proper context, molecular biologists have devised a number of experimental solutions. One partial solution that permits study in a genetic context closer to that of the wild-type gene in nature is to examine the expression of a cloned gene, side by side with a collection of mutant forms, in a convenient eukaryotic cell. Genes can be injected into *Xenopus* oocytes and, in a number of cases, the transcriptional and translational behavior of mammalian genes in this context is comparable to the same processes in the cells of the parent mammal (*81*). Alternatively, it may be feasible to reintroduce the mutant genes into a cell line derived from the source organism or from a closely related species by DNA-mediated transformation and then to assay expression in either transiently or in stably transformed cells (*82*). In studies of this type with cultured cells, it is accepted practice to define "wild type" as the genotype and phenotype of the unmodified cloned gene under the same assay conditions. However, in the interpretation of phenotypes of mutant genes, it must be remembered that inferences with regard to the natural "wild type"—that is, the intact organism and the genetic context it provides for the wild-type gene—cannot be made without introducing untestable assumptions.

A complete solution to the context problem would require replacing the wild-type gene at its normal genetic locus in a germline cell (a cell that can give rise to an intact organism) with the mutant allele constructed by in vitro mutagenesis. As discussed in the next section, such allele replacements have become routine genetic manipulations in prokaryotes and several fungi. In a number of higher eukaryotes, cloned genes can be efficiently introduced into germline cells, albeit not at their normal chromosomal location. Microinjection of DNA into oocytes of mice (*82*) and *Caenorhabditis elegans* (*83*), for example, can lead to the stable germline inheritance of some of the injected DNA sequences. In *Drosophila* (*84*) and some dicotyledonous plants (*85*), stable inheritance is greatly facilitated by the presence of transposon sequences in the injected DNA molecules. Initial characterization of the expression of *Drosophila* genes integrated at different sites in the genome suggest that, in some cases, expression and regulation of the inserted gene is not greatly altered by the new and slightly different genetic context (*86–88*).

Allele Replacement

The idea of inducing a mutation into a small portion of a genome and using a cell's recombinational machinery to transfer the mutation onto the genome is not new. In *E. coli*, cellular genes carried on specialized transducing phages (or F′ episomes) can be mutagenized and transferred to an unmutagenized cell, where homologous recombination between the mutagenized copy and the chromosomal copy of the gene can result in the desired replacement of wild type by mutant. For this replacement of one allele by another to occur, the equivalent of two crossovers must take place, one on either side of the mutation.

To carry out a similar sort of allele replacement between a mutation constructed in vitro in a cloned gene copy on a circular plasmid and its normal genetic locus requires essentially the same sequence of events (Fig. 3A). The recombinant plasmid, which also carries a selectable genetic marker, is introduced into the host cell by DNA transformation. Recombination between gene sequences on the plasmid and the homologous sequences on the genome must occur. If

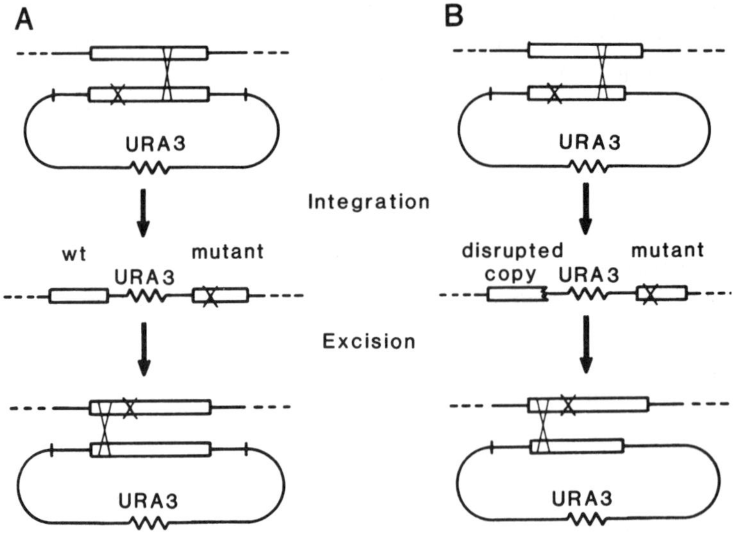

Fig. 3. Two strategies for recombining mutations constructed in vitro into the cellular gene. The *URA*3 gene is a selectable marker in the yeast *S. cerevisiae*. The small x indicates the mutant, the large X indicates the sites of recombination.

the only pathway for stable transformation of the genetic marker is through integration into the cell's genome (that is, the plasmid is unable to replicate autonomously in the host cell), the first crossover can be obtained by applying selection for the marker. Integration of the circular DNA results in a tandem duplication of the sequences of the cloned gene, a structure that is unstable to intramolecular recombination between the two repeats. One consequence of this instability is that the second crossover can usually be obtained by selecting, or in some cases simply by screening, for spontaneous loss of the selectable marker. When the two crossovers occur on opposite sides of the mutation, the end result is replacement of the wild-type gene with the mutant copy. Alternatively, both crossovers can be selected in a single event by transformation with a linear DNA fragment that includes a selectable marker closely linked to the mutation, a situation very similar to generalized transduction in bacteria.

Strategies involving circular and linear DNA have been successfully used to replace normal genes with mutant alleles made in vitro in a variety of bacteria [*Salmonella typhimurium* (*89*), *E. coli* (*90*), *Bacillus subtilis* (*91*)], as well as in the yeast *Saccharomyces cerevisiae* (*92*) and the filamentous fungus *Aspergillus nidulans* (*93*). In these bacteria and in yeast, homologous recombination appears to be the dominant mode for stable integration of selectable markers. In *Aspergillus*, however, many of the integrative events appear to involve nonhomologous recombination, thereby resulting in DNA insertions at improper sites in the genome (*93*). Regrettably, in *N. crassa* (*94*), as well as in nearly all higher eukaryotic cells (*95, 82*), integrative transformation by exogenous DNA molecules occurs predominantly via nonhomologous pathways, making replacement of alleles at the chromosomal locus problematic.

Two modified forms of allele replacement deserve brief mention. When mutations are induced in a cloned gene that has a deletion at one end, integration by homologous recombination results in a

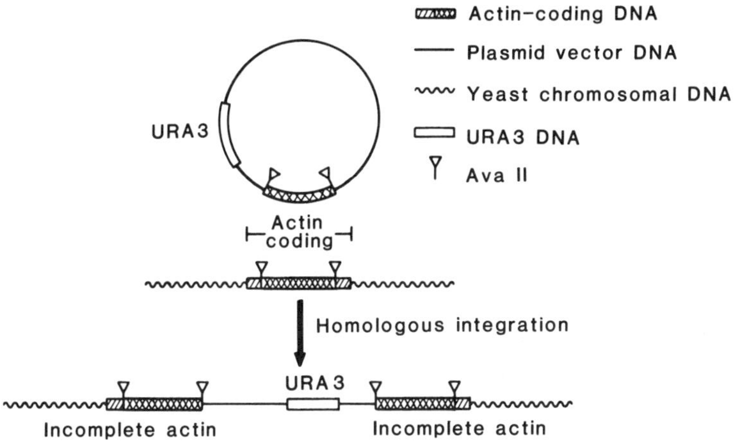

Fig. 4. Disruption of the yeast actin gene by integrative transformation (*99*).

tandem duplication in which only one of the two copies is intact (Fig. 3B). As a result, recessive mutations can be expressed in the transformed cell, there being no second, wild-type copy to mask the phenotypic consequences. This permits a selection or screen to be applied directly to the initial transformants (instead of transformants that subsequently excised the plasmid), thereby simplifying the identification of recessive mutations. This strategy has been used to isolate conditional lethal mutations in the yeast genes encoding actin (*96*), β-tubulin (*97*), and topoisomerase II (*98*).

Similarly, integration of a gene copy deleted at both ends (an internal DNA fragment) generates a tandem duplication in which neither of the two copies is intact (Fig. 4). This "integrative-disruption" technique has been used to determine the phenotype of null mutations in yeast (*3, 99*), *Aspergillus* (*92*), and several bacterial species (*100*). Alternatively, a gene can be disrupted by DNA transformation with a linear DNA fragment containing a selectable marker inserted within a cloned copy of the gene (*101, 102*), giving rise to a very stable null mutation.

In summary, methods for precise allele replacement are now routine for some microorganisms, making the ideal mutation experiment readily achievable. It is possible to clone a gene from these organisms, make a single mutation, replace the normal gene in its proper genetic context with the mutant allele, and then study the consequences, even if they are quite subtle or altogether unexpected. The ability to do such experiments in *S. cerevisiae* is a major reason for the current enthusiasm for studying gene regulation and basic problems of cell biology in this simple eukaryotic organism. It is hoped that methods for gene replacement will be developed for higher eukaryotes as well in the near future, at which time the genes and genetic regulatory elements involved in developmental pathways and neurological phenomena may become amenable to systematic analysis with gene mutations.

Conclusions

New in vitro methods for altering the nucleotide sequence of DNA molecules have emerged during the past few years and have greatly expanded the range of problems to which mutational analysis can be applied. To a considerable degree, the armamentarium the geneticist needs to collect gene mutations for studying phenomena at all levels of biological organization and complexity is now complete: Mutagenesis can be targeted to the entire genome, to individual genes and gene clusters, to structural and regulatory segments of genes, and to single nucleotide positions. The challenge for the future will not be in the isolation of greater and greater numbers of mutations, but rather in the detailed reading and interpretation of the story each mutation has to tell.

References and Notes

1. M. Smith, *Annu. Rev. Genet.*, in press.
2. D. Shortle, D. DiMaio, D. Nathans, *ibid.* **15**, 265 (1981).
3. J. H. Thomas, P. Novick, D. Botstein, in *Molecular Biology of the Cytoskeleton*, G. G. Borisy, D. W. Cleveland, D. B. Murphy, Eds. (Cold Spring Harbor Laboratory, Cold Spring Harbor, N.Y., 1984).
4. H. J. Muller, *Science* **66**, 84 (1927).
5. C. Auerbach and J. M. Robson, *Proc. R. Soc. Edinburgh B* **62**, 279 (1947).
6. G. W. Beadle and B. Ephrussi, *Genetics* **21**, 225 (1936).
7. G. W. Beadle and E. L. Tatum, *Proc. Natl. Acad. Sci. U.S.A.* **27**, 499 (1941).
8. P. deLucia and J. Cairns, *Nature (London)* **224**, 1164 (1969).
9. J. S. Hong and B. N. Ames, *Proc. Natl. Acad. Sci. U.S.A.* **68**, 3158 (1971).
10. N. Kleckner, J. Roth, D. Botstein, *J. Mol. Biol.* **116**, 125 (1977).
11. W. G. Shanabruch and G. C. Walker, *Mol. Gen. Genet.* **179**, 289 (1980).
12. G. B. Ruvkun and F. M. Ausubel, *Nature (London)* **289**, 85 (1981).
13. P. M. Bingham, R. Levis, G. M. Rubin, *Cell* **25**, 693 (1981).
14. M. Breindl, K. Harbers, R. Jaenisch, *ibid.* **38**, 9 (1984).
15. R. J. Legerski, J. L. Hodnett, H. B. Gray, *Nucleic Acids Res.* **5**, 1445 (1978).
16. S. Sakonju, D. Bogenhagen, D. D. Brown, *Cell* **19**, 13 and 27 (1980).
17. S. D. Putney, S. J. Benkovic, P. R. Schimmel, *Proc. Natl. Acad. Sci. U.S.A.* **78**, 7350 (1981).
18. S. Henikoff, *Gene* **28**, 351 (1984).
19. F. Heffron, M. So, B. J. McCarthy, *Proc. Natl. Acad. Sci. U.S.A.* **75**, 6012 (1978).
20. S. L. McKnight and R. Kingsbury, *Science* **217**, 316 (1982).
21. M. J. Casadaban, J. Chou, S. N. Cohen, *J. Bacteriol.* **143**, 971 (1980).
22. C. M. Gorman, L. F. Moffatt, B. H. Howard, *Mol. Cell. Biol.* **2**, 1044 (1982).
23. P. Bassford *et al.*, in *The Operon*, J. H. Miller and W. S. Reznikoff, Eds. (Cold Spring Harbor Laboratory, Cold Spring Harbor, N.Y., 1978).
24. M. Rose, M. J. Casadaban, D. Botstein, *Proc. Natl. Acad. Sci. U.S.A.* **78**, 2460 (1981).
25. L. Guarente and M. Ptashne, *ibid.*, p. 2199.
26. G. M. Weinstock, M. L. Berman, T. J. Silhavy, in *Expression of Cloned Genes in Prokaryotic and Eukaryotic Cells*, T. S. Papas *et al.*, Eds. (Elsevier/North-Holland, New York, 1983).
27. S. Benzer, *Proc. Natl. Acad. Sci. U.S.A.* **47**, 403 (1961).
28. E. C. Cox, *Annu. Rev. Genet.* **10**, 135 (1976).
29. P. Youderian, S. Bouvier, M. S. Susskind, *Cell* **30**, 843 (1982).
30. J. R. Broach, V. R. Guarascio, M. Jayaram, *ibid.* **29**, 227 (1982).
31. L. Greenfield, L. Simpson, D. Kaplan, *Biochim. Biophys. Acta* **407**, 365 (1975).
32. D. Shortle and D. Botstein, *Methods Enzymol.* **100**, 457 (1983).
33. D. Shortle, *Gene* **22**, 181 (1983).
34. D. Solnick, *Nature (London)* **291**, 508 (1981).
35. C. T. Chu, D. S. Parris, R. A. F. Dixon, F. E. Farber, P. A. Schaffer, *Virology* **98**, 168 (1979).
36. S. Busby, M. Irani, B. deCrombrugghe, *J. Mol. Biol.* **154**, 197 (1982).
37. J. T. Kadonaga and J. R. Knowles, *Nucleic Acids Res.* **13**, 1733 (1985).
38. W. E. Borrias, I. J. C. Wilschut, J. M. Vereijken, P. J. Weisbeek, G. A. van Arkel, *Virology* **70**, 195 (1976).
39. S. Hirose, K. Takeuchi, Y. Suzuki, *Proc. Natl. Acad. Sci. U.S.A.* **79**, 7258 (1982).
40. R. M. Meyers and T. Maniatis, *Science* **229**, 242 (1985).
41. S. G. Fischer and L. S. Lerman, *Proc. Natl. Acad. Sci. U.S.A.* **80**, 1579 (1983).
42. H. Hayatsu, *Prog. Nucleic Acids Res. Mol. Biol.* **16**, 75 (1976).
43. D. Shortle and D. Nathans, *Proc. Natl. Acad. Sci. U.S.A.* **75**, 2170 (1978).
44. D. DiMaio and D. Nathans, *J. Mol. Biol.* **140**, 129 (1980).
45. K. W. C. Peden and D. Nathans, *Proc. Natl. Acad. Sci. U.S.A.* **79**, 7214 (1982).
46. D. Kalderon, B. A. Oostra, B. K. Ely, A. S. Smith, *Nucleic Acids Res.* **17**, 5161 (1982).
47. W. Muller, H. Weber, F. Meyer, C. Weissmann, *J. Mol. Biol.* **124**, 343 (1978).
48. H. Weber *et al.*, in *The ICN-UCLA Symposia on Molecular and Cellular Biology* (Academic Press, New York, 1982), vol. 12.
49. D. Shortle, P. Grisafi, S. J. Benkovic, D. Botstein, *Proc. Natl. Acad. Sci. U.S.A.* **79**, 1588 (1982).
50. D. Shortle and B. Lin, *Genetics* **110**, 539 (1985).
51. K. M. Overbye and D. Botstein, unpublished observations.
52. R. A. Zakour and L. A. Loeb, *Nature (London)* **295**, 708 (1982).
53. C. Traboni, R. Cortese, C. Ciliberto, G. Cesareni, *Nucleic Acids Res.* **11**, 4229 (1983).

54. R. A. Zakour, E. A. James, L. A. Loeb, *ibid.* **12**, 6615 (1984).
55. J. J. Champoux, *J. Mol. Appl. Genet.* **2**, 454 (1984).
56. P. T. Englund, S. S. Price, P. H. Weigel, *Methods Enzymol.* **29**, 273 (1974).
57. T. A. Kunkel, *Proc. Natl. Acad. Sci. U.S.A.* **82**, 488 (1985).
58. P. Abarzua and K. J. Marians, *ibid.* **81**, 2030 (1984).
59. K. Itakura, J. J. Rossi, R. B. Wallace, *Annu. Rev. Biochem.* **53**, 323 (1984).
60. C. S. Craik, *Biotechniques* **3**, 12 (1985).
61. M. H. Caruthers, S. L. Beaucage, C. Becker, W. Efcavitch, E. F. Fisher, in *Genetic Engineering: Principles and Methods*, J. K. Setlow and A. Hollaender, Eds. (Plenum, New York, 1982), vol. 4.
62. E. Ohtsuka, M. Ikehara, D. Soll, *Nucleic Acids Res.* **10**, 653 (1982).
63. C. A. Hutchinson III *et al.*, *J. Biol. Chem.* **253**, 6551 (1978).
64. A. Razin, T. Hirose, K. Itakura, A. D. Riggs, *Proc. Natl. Acad. Sci. U.S.A.* **75**, 4268 (1978).
65. M. J. Zoller and M. Smith, *Nucleic Acids Res.* **10**, 6487 (1982).
66. L. Dente, G. Caesareni, P. Cortese, *ibid.* **11**, 1645 (1983); R. J. Zagursky and M. L. Berman, *Gene* **27**, 183 (1984); A. Levinson, D. Silver, B. Seed, *J. Mol. Appl. Genet.* **2**, 507 (1984).
67. R. B. Wallace *et al.*, *Science* **209**, 1396 (1980).
68. W. Kramer, V. Drutsa, H. W. Jansen, B. Kramer, M. Pflugfelder, H. J. Fritz, *Nucleic Acids Res.* **12**, 9441 (1984).
69. M. Schold, A. Colombero, A. A. Reyes, R. B. Wallace, *DNA* **3**, 469 (1984).
70. K. Norris, F. Norris, L. Christiansen, N. Fiil, *Nucleic Acids Res.* **11**, 5103 (1983).
71. M. J. Zoller and M. Smith, *DNA* **3**, 479 (1984).
72. R. B. Wallace, M. Schold, M. J. Johnson, P. Dembek, K. Itakura, *Nucleic Acids Res.* **9**, 3647 (1981).
73. K. M. Lo, S. S. Jones, N. R. Hackett, H. G. Khorana, *Proc. Natl. Acad. Sci. U.S.A.* **81**, 2285 (1984).
74. M. D. Matteucci and H. L. Heynecker, *Nucleic Acids Res.* **11**, 3113 (1983).
75. J. A. Wells, M. Vasser, D. B. Powers, *Gene*, **34**, 315 (1985).
76. K. P. Nambiar *et al.*, *Science* **223**, 1299 (1984).
77. R. Frank, W. Heikens, G. Heisterber-Montsis, H. Blocker, *Nucleic Acids Res.* **11**, 4366 (1983).
78. H. W. D. Mathes *et al.*, *EMBO J.* **3**, 801 (1984).
79. J. Ott and F. Eckstein, *Nucleic Acids Res.* **12**, 9137 (1984).
80. K. A. Osinga, A. M. van der Blick, G. van der Hoist, M. J. A. G. Koerkamp, H. F. Tabak, *ibid.* **11**, 3595 (1983).
81. J. B. Gurdon and D. A. Melton, *Annu. Rev. Genet.* **15**, 189 (1981).
82. R. Kucherlapati and A. I. Skoultchi, *Critical Rev. Biochem.* **16**, 349 (1984).
83. D. T. Stinchcomb, J. Shaw, D. Hirsh, in preparation.
84. A. C. Spradling and G. M. Rubin, *Science* **218**, 341 (1982).
85. K. A. Barton, A. N. Binns, A. J. M. Matzke, M. D. Chilton, *Cell* **32**, 1033 (1983).
86. S. B. Scholnick, B. A. Morgan, J. Hirsh, *ibid.* **34**, 37 (1983).
87. A. C. Spradling and G. M. Rubin, *ibid.*, p. 47.
88. D. A. Goldberg, J. W. Posakony, T. Maniatis, *ibid.*, p. 59.
89. N. I. Gutterson and D. E. Koshland, Jr., *Proc. Natl. Acad. Sci. U.S.A.* **80**, 489 (1983).
90. M. Jasin, L. Regan, P. R. Schimmel, *Cell* **36**, 1089 (1984).
91. P. Youngman *et al.*, *Science* **228**, 285 (1985).
92. S. Scherer and R. Davis, *Proc. Natl. Acad. Sci. U.S.A.* **76**, 4951 (1979).
93. R. Morris, personal communication.
94. C. Yanofsky, personal communication.
95. D. M. Robins, S. Ripley, A. S. Henderson, R. Axel, *Cell* **23**, 29 (1981).
96. D. Shortle, P. Novick, D. Botstein, *Proc. Natl. Acad. Sci. U.S.A.* **81**, 4889 (1984).
97. T. Huffaker and D. Botstein, unpublished observations.
98. C. Holm, T. Goto, J. C. Wang, D. Botstein, *Cell*, in press.
99. D. Shortle, J. E. Haber, D. Botstein, *Science* **217**, 371 (1982).
100. F. A. Ferrari, E. Ferrari, J. A. Hoch, *J. Bacteriol.* **152**, 780 (1982).
101. R. J. Rothstein, *Methods Enzymol.* **101**, 202 (1983).
102. S. C. Winans, S. J. Elledge, J. H. Krueger, G. C. Walker, *J Bacteriol.* **161**, 1219 (1985)
103. Supported by grants from the National Institutes of Health (D.B. and D.S.) and from the American Cancer Society (D.B.). We would like to thank M. Smith for making a copy of his review available before publication and our many colleagues, current and past, for helpful discussions.

2

Genetic Engineering of Novel Genomes of Large DNA Viruses

Bernard Roizman and Frank J. Jenkins

Viral genomes encode a vast variety of information. In addition to coding sequences for proteins, these genomes encode specific signals for the initiation and termination of transcription, the regulation of gene product abundance, and the modification of genome structure necessary for gene expression and genome replication. One central objective of studies on the molecular biology of viruses infecting human and animals is the elucidation of the functions encoded in the genomes of the viruses. A second objective, no less important, is the specific modification of the viral genomes for use as vaccines or as vectors of genes whose products provide protection against infectious disease.

Identification of nucleotide sequence function, unlike that of gene products, requires mutagenesis by nucleotide substitution, deletion, or insertion. In some instances the function of specific nucleotide sequences within viral genomes can be deduced from studies of isolated domains of the genome in vitro or in cells. However, the effects of the mutations observed in vitro and in cells must be confirmed by testing viral genomes in which these modifications have been introduced. Construction of novel genomes carrying insertions, deletions, or substitutions is therefore essential not only for the attenuation of viruses or the construction of virus vectors but also for analyses of function.

With small DNA viruses that can be cloned in their entirety (*1*), or DNA viruses that can be cleaved by various restriction endonucleases into a small number of fragments that can reassemble in the proper order with relatively good efficiency (*2, 3*), the construction of novel genomes carrying the desirable modifi-

Science 229, 1208–1214 (20 September 1985)

cations may be tedious but not impossible. This is not the case, however, for genomes of such large DNA viruses as poxviruses (200 kilobase pairs) or herpesviruses (120 to 250 kbp). The sheer size of the genomes renders the elegant techniques developed for genetic manipulation of papovaviruses (5 kbp) and adenoviruses (35 kbp) totally inappropriate.

In this article, we describe techniques designed for the genetic manipulation of large genomes such as those of poxviruses and herpesviruses. Because these techniques have been applied to solve specific questions regarding the function of domains of the genomes of the herpes simplex virus 1 (HSV-1) and because the properties of the novel genomes generated by these techniques are relevant for an assessment of the potential of these techniques, it is both convenient and desirable to begin with a brief description of these genomes and the information they encode.

The HSV-1 Genome: Structure and Information Content

The genome. The HSV-1 genome extracted from virions is linear, double-stranded DNA approximately 150 kbp in size (*4*). It is convenient to view the genome as consisting of two covalently linked components designated as L (long) and S (short) (Fig. 1). Both L and S components consist of largely unique sequences (U_L and U_S) flanked by inverted repeats (*5, 6*). The inverted repeat sequences of the L component are each 9 kbp in size and have been designated as *ab* and *b'a'*, whereas those of the S component are each 6.5 kbp in size and have been designated as *c'a'* and *ca* (*6*). The *a* sequence is the only one shared by both L and S components. It is approximately 200 to 500 bp in size, depending on the number of reiterations of small sequences contained in its domain (*7, 8*). Only one *a* sequence is located at the terminus of the S component, whereas from one to more than five may be present at the terminus of the L component and at the junction between the L and S components (*7, 9, 10*). The sequence arrangement of the HSV-1 genome may be represented by

$$a_nb\text{-}U_L\text{-}b'a'_mc'\text{-}U_S\text{-}ca$$

where the primes indicate sequence inversion and *n* and *m* are the number of copies of *a* sequences.

A remarkable property of the HSV-1 genome is that the L and S components can invert relative to each other. As a consequence, viral DNA extracted from virions or from infected cells consists of four equimolar populations differing solely with respect to the relative orientation of the two components (*11*).

To date, three origins have been mapped in the HSV-1 genome. Of these, two are located within the inverted repeats (*c* sequence) of the S component, whereas the third is located in the middle of the L component (*12–17*). It is not clear whether the three function together or sequentially and whether all three are essential. The terminal *a* sequence of the S component is the *cis*-acting site for packaging of the mature viral DNA into preformed capsids (*14, 18*).

The functions of the inverted repeats and of the inversions of the L and S components is but one of the problems approached experimentally by construction of novel, genetically engineered genomes.

Gene arrangement and regulation. By enumerating the relatively abundant novel proteins appearing in infected cells, it was originally estimated that the HSV-1 genome encodes approximately 50 polypeptides (*19*). Current estimates suggest that the number is closer to 70. Two observations are of particular significance. First, although herpesvirus genes carry transcriptional and translational signals similar to those of other DNA viruses infecting higher eukaryotic cells, the messenger RNA's arising from a vast majority of the genes are not spliced (*20*). Second, the information density is lower than that encoded in the genomes of smaller viruses, such as the papovaviruses.

The 50 genes specifying abundant products form coordinately regulated groups whose expression is sequentially ordered in a cascade fashion (*21, 22*). The five α genes, which make up the first group to be expressed, are operationally defined as the genes transcribed in the infected cells in the absence of de novo, viral protein synthesis. The β genes are operationally defined by their requirement for functional α gene products and by a total lack of sensitivity to inhibitors of viral DNA synthesis on their expression. The β genes are heterogeneous with respect to temporal patterns of their expression. The β_1 genes are generally expressed earlier than the β_2 genes, and in some cell lines they are difficult to differentiate from α genes. The onset of expression of β genes coincides with a decline in the rate of expression of α genes and irreversible shutoff of host protein synthesis. The γ genes are also heterogeneous. The γ_1 genes are expressed earlier and are differentiated from β genes solely by their requirement for viral DNA synthesis for maximal expression. In contrast, the γ_2 genes are expressed later in infection and stringently require viral DNA synthesis for their expression (*23*).

The α proteins are translocated into the nucleus. Ample evidence supports the view that the α4 gene product is a large multifunctional protein required for transcription of both β and γ genes (*24–26*). The products of the α22 and α27 genes play a role in the expression of γ genes (*27–29*), but their function—as well as the functions of the two other α genes (α0 and α47)—is less well understood than that of α4. The β genes identified to date appear to play a major role in the replication of viral DNA, in the shutoff of the expression of α genes [reputed to be a function of the major DNA-binding protein ($\beta_1$8) (*30*)], and in the induction by as yet unknown means of γ gene expression. The γ genes appear to be structural components of the virion. The mechanism of the shutoff of β gene expression by γ gene products is not understood, but it does not appear to be due to competition for transcriptional or translational factors (*22*). At least one γ gene product introduced into the cells during infection causes a transient, but not essential, shutoff of host macromolecular metabolism (*31, 32*). Another γ gene product abundantly represented in the virion appears to have a major role in the induction of α gene expression immediately after infection (*27, 33–38*).

A pattern of gene clustering has been observed, but it is not totally consistent. The α genes are located in or near the inverted repeats, and two genes (α0 and α4), located entirely within the repeats, are diploid (Fig. 1) (*24, 39, 40*). All α genes are transcribed from independent promoters that share AT-rich homologs required for the induction of these genes by the γ *trans*-inducing protein and GC-rich enhancer-like elements (*41*). Most of

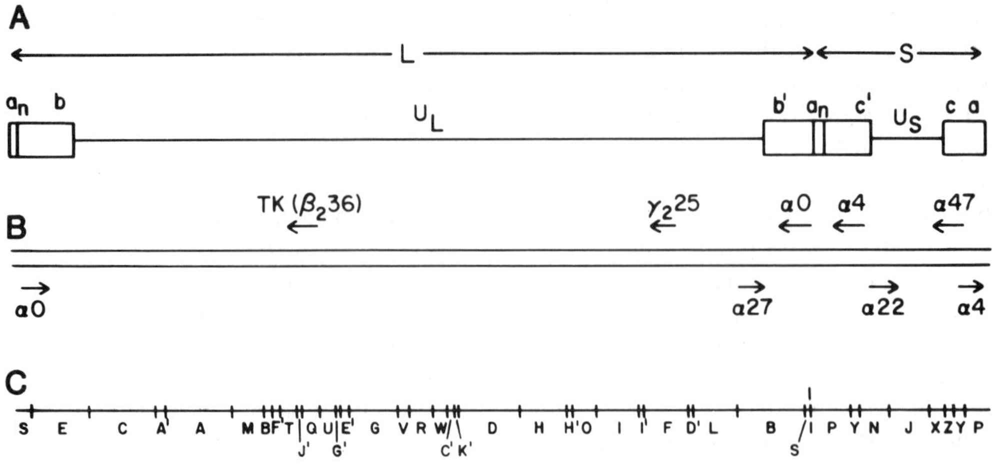

Fig. 1. Schematic of the HSV-1 genome. (A) Locations of unique sequences (U_L and U_S), terminal repeat sequences (a_nb and ca), and inverted internal repeat sequences ($b'a_nc'$). (B) Locations of certain α, β, and γ genes; the arrows represent direction of transcription and the numbers refer to the infected cell polypeptide (ICP) number of the corresponding gene. (C) Bam HI restriction endonuclease cleavage sites on the HSV-1 (F) genome.

the β and γ genes are dispersed throughout the genome, but small clusters of genes with related functions have been noted.

General strategy of virus multiplication. Following infection and multiplication at the portal of entry, the virus may infect nerve endings and ascend to the trigeminal and dorsal root ganglia and remain latent in neurons for the life of the host (*42*). Periodically, physical or emotional trauma or fever causes the latent virus to multiply, descend to or near the portal of entry, and initiate lesions apparent as vesicles and known as fever blisters.

Infection of cells is initiated by the fusion of the membrane covering the virus—the envelope—with the plasma membrane of the cell (*43*). The capsid is released into the cytoplasm and is transported to a nuclear pore, where the DNA is released into the nucleus (*44*). In the nucleus the viral DNA is transcribed by host RNA polymerase II (*45*) and ultimately specifies the α, β, and γ proteins described above. Viral DNA is replicated by a viral DNA polymerase (*46*).

Late in infection, the nascent viral DNA appears to be present in head-to-tail concatemers from which unit-length viral DNA is excised and packaged into capsids (*47, 48*). The capsids containing viral DNA acquire a new surface protein (*49, 50*), adhere to the underside of the nuclear membrane, and become enveloped by patches in the membrane containing viral membrane proteins (*51, 52*). The enveloped particles are transported to the cell surface through the endoplasmic reticulum, which is also modified by the insertion of the viral membrane proteins (*52*).

A characteristic of the larger DNA viruses that differentiates them from the smaller papovaviruses and adenoviruses is that they encode numerous enzymes involved in DNA metabolism. Among these enzymes are a DNA polymerase (*46*), a deoxynucleotide kinase (thymidine kinase) (*53*), ribonucleotide reductase (*54, 55*), deoxyribonuclease (DNase) (*46*), and uridine triphosphatase (UTPase) (*56, 57*). Not all of the gene products involved in viral DNA metabolism have been identified, but many do have counterparts in host cells. Because cells in culture express some gene products that can substitute for the products of viral genes, it has been possible, in some instances, to inactivate or delete the corresponding viral gene (*27, 58*). One objective of genetic engineering of novel HSV genomes is to identify viral genes whose function is similar to that of host genes and can be deleted.

Principles of Genetic Engineering of Novel HSV Genomes

Basic strategy. Deproteinized HSV DNA is infectious (*59–61*). Cotransfection of intact DNA with a molar excess of a mutagenized DNA fragment will result in progeny virus in which the mutagenized fragment recombines and replaces homologous DNA sequences (*62*). The recombination frequency varies, depending on the size and location of nonhomologous sequences and the size of homologous flanking sequences. For example, a 1-kbp insert located at the end of a 2-kbp DNA fragment is less likely to recombine than 5 kbp inserted in the middle of the same fragment. Recombinants can be isolated by screening for either DNA or expression of a gene contained in the insert (for example, β-galactosidase), but the work is tedious and time-consuming.

Another problem is that viruses modified by insertional mutagenesis may grow more slowly and yield less progeny than the wild-type parent. In consequence, the slower growing recombinants may be overgrown by the parent, wild-type virus. The isolation of desirable recombinants is greatly facilitated by the use of a selectable marker. Under selective pressure only the recombinants carrying the selectable marker are able to grow and are readily isolated by plaque purification.

Although several selectable markers are available (*63*), the one chosen by this laboratory is the HSV thymidine kinase (TK) gene. This gene was chosen because procedures are available both for and against the expression of the selectable marker, thereby permitting the selection of TK^- and TK^+ recombinant progeny (*12, 27*).

Properties of the HSV TK. Cells in culture have two pathways for the synthesis of thymidine monophosphate (TMP). The major pathway involves the conversion of deoxyuridine monophosphate to TMP by thymidylate synthetase. The second, or scavenger, pathway involves phosphorylation of thymidine by TK. Several herpesviruses and the poxviruses encode a TK. This enzyme appears to be important for normal virus growth in experimental animals (*64*), but it is not essential for growth in cell cultures because mutants lacking a functional TK gene have been known for many years (*58*). The host TK enzyme is also not essential for cell growth, provided that the primary pathway for TMP synthesis is not obstructed by metabolic inhibitors.

The HSV TK differs in two important respects from its cellular counterpart. First, the name is a misnomer because the viral enzyme phosphorylates deoxypurines and deoxypyrimidines whereas the host TK does not phosphorylate deoxycytidine or deoxypurines (*53*). The second and more important difference is that the substrate range of the viral TK for deoxynucleotide analogs is rather wide and includes analogs that bear little resemblance to deoxynucleotides, whereas the substrate range of the host TK is restricted to a small number of analogs (for example, bromodeoxyuridine) that closely resemble thymidine.

In consequence, analogs that are phosphorylated by the viral TK (for example, arabinosylthymine and acyclovir), but not by the host TK, will not affect uninfected TK^+ cells and will permit the multiplication of TK^- viruses but not that of TK^+ viruses in these cells. Arabinosylthymine and acyclovir can be used, therefore, to select TK^- viruses in both TK^+ and TK^- cell lines, but bromodeoxyuridine can be used for this purpose only in TK^- cells inasmuch as this analog is phosphorylated by the host TK.

Obstruction of the thymidylate synthetase pathway for TMP biosynthesis by such inhibitors as methotrexate or aminopterin is not deleterious in TK^+ cells but will result in the depletion of the thymidine triphosphate (TTP) pool and destruction of TK^- cells. Because of the depletion in the TTP pool, TK^- cells overlaid with medium containing methotrexate or aminopterin will permit the growth and plaque formation of TK^+ viruses but not of TK^- viruses. TK^+ viruses are therefore readily selected in TK^- cells overlaid in HAT medium, which contains hypoxanthine, aminopterin (or methotrexate), and thymidine (hypoxanthine and thymidine being used to compensate for other metabolic products whose synthesis is blocked by these drugs).

It should be pointed out that the selection of TK^- recombinants is generally more efficient than that of TK^+ recombinants. Although spontaneous TK^- mutants are readily obtainable, the frequency of such mutations in a cloned viral population is 10^{-4} to 10^{-5}—that is, lower than the recombination rate between viral DNA and a DNA fragment containing an insertion or deletion flanked by at least 2 kbp of DNA homologous to contiguous DNA sequences in the viral genome. Furthermore, in cells doubly infected with a TK^+ parent and a TK^- recombinant, both viruses are destroyed by the analog phosphorylated by the viral TK.

The TK^+ selection suffers from two problems. First, the uninfected TK^- cells survive only a few days in the presence of methotrexate or aminopterin, and debilitated TK^+ recombinants may not have sufficient time to spread and form plaques. Furthermore, in cells that are doubly infected with the TK^- parent and TK^+ recombinant, the TK^+ parent acts as a helper and both viruses grow.

Construction of novel genomes. Figure 2 illustrates three strategies used for construction of recombinant genomes based on the use of TK as a selectable marker. The first strategy (Fig. 2-1) involves the insertion of a sequence into the transcribed domain of the TK gene. The DNA fragment carrying the TK gene with the insert is then amplified by cloning in *Escherichia coli* and cotransfected with intact HSV wild-type DNA. TK^- recombinants carrying the insert are then selected from among the progeny of transfection as described above. This technique allows rapid selection of recombinant progeny, but it restricts the insertions to the domain of the TK gene. The preferred, but not unique, sites for insertional mutagenesis of the TK gene (Fig. 3) are the Bgl II site in the tran-

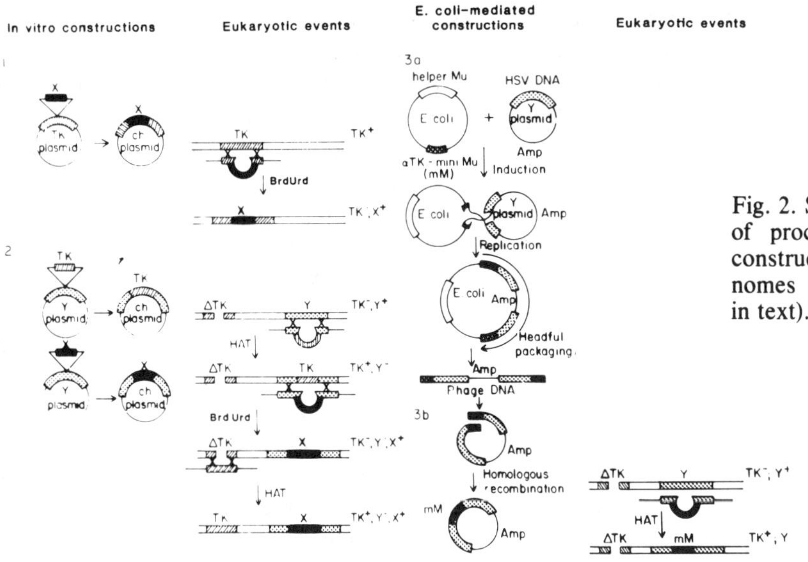

Fig. 2. Schematic outline of procedures used to construct novel HSV genomes (see explanation in text).

scribed noncoding sequences and the Sac I site in the coding sequences (*12, 27*).

The second strategy permits the construction of recombinants carrying insertions or deletions at specifically selected sites. In the first step of this procedure, sequences are deleted near one end of the domain of the TK gene. The DNA fragment containing the TK gene carrying the deletion is cotransfected with intact wild-type DNA, and TK$^-$ progeny carrying the deletion are then selected from among the progeny of transfection.

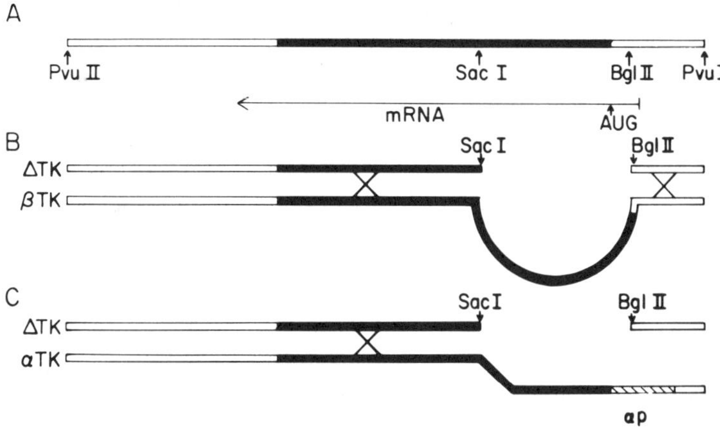

Fig. 3. (A) Schematic of Pvu II DNA fragment from HSV-1 containing the wild-type TK gene; the horizontal arrow represents the location of the TK messenger RNA transcript, and the solid line represents the coding region for the TK protein. (B) Homologous recombination between the deleted TK gene (ΔTK) in HSV-1(F) Δ305 (top line) and an intact wild-type TK gene (βTK) (bottom line). (C) Illustration of nonhomology between the 5′ end of the deleted TK gene and the 5′ end of the chimeric αTK gene used for insertional mutagenesis at sites other than the natural TK gene; αp refers to the location of the α promoter and regulatory region fused to the structural TK gene in the αTK chimeric gene.

The TK$^-$ recombinant constructed for these studies—HSV-1(F)Δ305—contains a 700-bp deletion extending downstream from the site of transcription initiation ($^+1$) beyond the Sac I site ($^+540$). Although this recombinant makes a truncated 19-kilodalton protein that reacts with antibody to TK (*65*), it is unable to convert thymidine to TMP.

In the second step (Fig. 2-2), the TK is inserted into a HSV DNA fragment at a specific site. In initial studies the wild-type TK gene was used (*27*). In subsequent studies, a chimeric gene consisting of the coding sequences of the TK gene fused to the promoter regulatory domain of α genes was used in order to reduce the probability that the TK gene will recombine at its natural location and to restore the integrity of the wild-type gene (*66, 67*).

Because the deletion in the natural TK gene extends both upstream and downstream from the Bgl II site, whereas the new promoter regulatory domain extends upstream from the Bgl II site (Fig. 3), the probability that the chimeric gene will restore TK activity at the natural site is greatly reduced; in fact, such recombinants have not been observed. Viral TK$^+$ recombinants carrying the TK insert at the desired site are selected by plating the progeny of transfection on TK$^-$ cells maintained in HAT medium. An appropriate deletion or insertion is made in the target sequence at the site of insertion of the TK gene to delete or insert a specific HSV or foreign gene in the domain of the target sequence. This DNA fragment is then amplified and cotransfected with intact DNA of the recombinant carrying the TK gene inserted in the target sequence. Viral TK$^-$ recombinants carrying the deletion or insertion in the target gene are then selected from among the progeny of transfection as described above.

When it is desirable to assess the effect of the deletion in the target sequence only, the native TK gene can be restored by cotransfecting the DNA of this recombinant with DNA fragments carrying the entire domain of the TK gene. The TK$^+$ progeny of this transfection should differ from the wild-type virus only with respect to the deletion or insertion in the target sequence (Fig. 2-2).

This strategy permits the selection of viable recombinants carrying insertions or deletions at specific sites. The major disadvantages are that the location of the target sequence within the cloned DNA fragment must be known. Furthermore, this sequence must contain a suitable restriction endonuclease cleavage site for the insertion of the TK gene. For example, if the objective is to inactivate by insertional mutagenesis a gene that overlaps in part with the domain of another gene either at its 3′ or 5′ terminus, a suitable site would be upstream or downstream from the overlapping regions. In some instances, no suitable sites are found or the precise domain of the target gene is not known.

The third strategy (Fig. 2-3, a and b), a modification of the second, involves the use of the mini-Mu derivative of the transducing bacteriophage Mu (*67, 68*). The mini-Mu prophage contains the left and right ends of the bacteriophage Mu plus the ner, A and B genes, a temperature-sensitive *c* repressor gene of Mu, and a selectable marker. The mini-Mu prophage is defective and requires a helper Mu for replication, but the presence of the left and right ends—plus the A and B genes of Mu—allow for transposition within an *E. coli* cell.

For use in insertional mutagenesis of

HSV genomes, a functional TK gene under the control of the HSV α4 gene promoter and regulatory region was inserted into the mini-Mu to yield an αTK mini-Mu (αTK-mM) (*67*). The αTK-mM was then used to lysogenize an *E. coli* strain containing a helper Mu prophage with a temperature-sensitive repressor *c* gene. Growth of the double lysogen *E. coli* strain at 30°C prevents replication and transposition of both the helper Mu and αTK-mM prophages.

The use of the αTK-mM system for insertional mutagenesis requires three steps. In the first step (Fig. 2-3a), the double lysogen *E. coli* strain is transfected with plasmid DNA containing the target HSV-1 DNA fragment. Induction by shift up to 42°C of the bacterial cells results in replication and transposition of both the helper Mu and αTK-mM. As a result of transposition, the αTK-mM will integrate into many different sites within the plasmid DNA and produce a cointegrate structure integrated into the *E. coli* DNA consisting of the plasmid DNA flanked by single copies of the αTK-mM.

In the case of the cointegrate structures, packaging begins with the leftmost copy of the αTK-mM and proceeds through the plasmid DNA and into the second copy of the αTK-mM. If the cointegrate structure is less than 38 kbp, then flanking *E. coli* DNA sequences will be packaged. Cointegrate structures greater than 38 kbp result in deletions of the second αTK-mM copy. The αTK-mM is 9.6 kbp in size. Therefore, recombinant plasmids 18 kbp or more in length will cause deletion of part or all of the second copy of the αTK-mM. As is evident below, at least a portion of the second copy of the αTK-mM is essential, and therefore the recombinant plasmid must be less than 28 kbp in size.

The lysates of the double lysogen transfected with plasmids containing HSV DNA and induced by shift up to 42°C should contain phages with defective genomes consisting of the recombinant plasmid DNA's linearized at random sites and flanked by direct copies of the αTK-mM. The second step (Fig. 2-3b) involves infection of an *E. coli* Rec(A^+) strain lysogenized by a helper Mu prophage. The presence of the helper Mu prevents the cointegrate structure from transposing to the *E. coli* DNA. Homologous recombination between the flanking copies of the αTK-mM results in the production of plasmid molecules in which the αTK-mM is randomly inserted into the recombinant plasmid. *E. coli* carrying the plasmid molecules can then be readily grown by selection for the antibiotic resistance marker contained in the αTK-mM. The isolated plasmid DNA consists of a pool of plasmid molecules containing an αTK-mM inserted at many different sites within the HSV DNA insert and nonessential vector sequences.

In the third step (Fig. 2-3c), the amplified αTK-mM plasmids are cotransfected with intact TK^- HSV DNA, and the progeny of the transfection are then plated on TK^- cells under HAT medium. The TK^+ progeny carrying the αTK-mM insert may be heterogeneous inasmuch as the αTK-mM may be expected to insert at any site within the target sequences of HSV DNA that is not essential for growth in the cells in which the selection is done.

The αTK-mM system offers several advantages. Foremost, it obviates the need for inserting the selectable marker at a specific and often inaccessible site. The production of recombinant plasmids containing randomly inserted αTK-mM is rapid—less than 48 hours. The αTK-mM is an excellent probe for sequences and genes nonessential for growth in the cells used for the selection of recombinant viruses. Furthermore, as is the case of the inserted TK gene, the αTK-mM can be replaced by recombination with an insertion or deletion in the target sequence.

Application of Genetic Engineering to Construction of Novel HSV Genomes

The techniques described above have been used to insert HSV and foreign sequences into HSV genomes and to probe HSV-1 DNA for domains not essential for growth in cell culture. Among the central issues to be considered are the capacity of the genome as a vector for HSV or foreign genes, stability of the recombinants, and privileged sites characterized by spontaneous rearrangements.

Minimum and maximum sizes of HSV genomes that can be packaged. Current studies by Frenkel and associates (*14, 18*) and others (*69*) indicate that packaging of the genome requires the presence of terminal *a* sequences and is not by a simple headful mechanism. The capsids are capable of packaging small head-to-tail concatemers of defective genomes (*18*), but capsids containing concatemers significantly smaller than the wild-type genome are not enveloped. One hypothesis that could explain this observation is that capsids containing standard-length DNA become modified and bind an additional protein (*50*) and that capsids packaging less than a minimum-size DNA do not become modified and fail to bind this protein.

The minimum size of the packaged genome is not known but the smallest nondefective genome packaged efficiently is that of the I358 recombinant in which a 2-kbp sequence containing the TK gene replaced 15 kbp of DNA that included a portion of the unique sequences and nearly the entire internal inverted repeat sequence (*70*). The largest genomes packaged to date have included an αTK-mM 9.6 kbp in size (*67*). Taken together, these results indicate that the 2-kbp sequence containing the TK gene in an I358 recombinant could be replaced by an insert at least 24 kbp in size.

Sites and sequences not essential for growth in cell culture. The sites and sequences shown to date to be nonessential for viral replication include (Fig. 1) (i) various portions of the domain of the TK gene (*27*), (ii) the sequences located between the 3′ terminus of the α27 and α0 genes (*67, 70*), (iii) the sequences (approximately 15 kbp) located between the 3′ terminus of the α27 gene and the promoter-regulatory domain of the α4 gene and constituting unique sequences of the L component and nearly the entire internal inverted repeat sequence $b'a_mc'$ (*67, 70*), (iv) the sequences located between the terminal *a* sequence and the 3′ terminus of the α4 gene at both ends of the S component (*66*), (v) the sequences flanking the origins of DNA synthesis of the S component (*71*) and portions of the domains of the α22 gene (*27*), and (vi) the sequences located between the terminal *a* sequence and the glycoprotein D gene including glycoprotein E and the α47 gene (*88*).

The recombinants carrying inserts or deletions in these regions of HSV DNA vary in their capacity to grow. The recombinants R325 and R328 carrying 500- and 100-bp deletions in the coding sequences of the α22 gene grow as well as the wild-type parent in Vero and HEp-2 cells but poorly in human embryonic lung cells and in various rodent cell lines. Studies on these recombinants suggest that a host factor is able to complement the deleted gene in Vero and HEp-2 cell lines but not in the restrictive cells (*29*). The I358 recombinant described above yields approximately one-tenth as much virus as the wild type in Vero and HEp-2 cells and considerably less virus in rodent cell lines or human embryonic cell lines. A characteristic of this virus is that its DNA is frozen in one arrangement; that is, the L and S components do not invert relative to each other.

The insertion of the TK gene, the selectable marker, into appropriate sites of target sequences has been the rate-limiting step in the analyses of HSV DNA for sequences and genes not essential for virus growth in the cells used for selection of recombinant genomes. The development of the αTK-mM system for insertional mutagenesis will considerably facilitate these studies.

Insertion of HSV and foreign sequences into HSV genome. The sequences and genes inserted into HSV-1 DNA to date include (i) HSV DNA sequences inserted for the purpose of identifying putative *cis*-acting sites; (ii) HSV genes inserted for the purpose of identifying their function by complementing the authentic, mutagenized copy; and (iii) foreign genes inserted to determine whether the HSV genome can serve as a vector for the expression of foreign genes.

Specifically, to identify the promoter-regulatory domains of the α genes, a fragment containing the capping site and upstream sequences of both copies of the α4 gene was inserted into the Bgl II site (nucleotide +50) of the TK gene in the proper transcriptional orientation. In the resulting recombinants, the chimeric TK gene was regulated as an α gene (*27*). Similar insertion of a fragment carrying a putative γ_2 promoter-regulatory sequence converted the natural βTK into a γ_2 gene (*72*). Insertion of DNA fragments containing the *a* sequence into the Bgl II site of the TK gene led to the identification of this sequence as the site-specific recombination site for the inversion of the L and S components (*12*).

The HSV genes specifically inserted into the genome were designed to complement mutations in the native gene (*73*) or to express a mutated gene (*74*). The foreign genes inserted into the genome were the *S* gene of hepatitis B virus (*75*), chick ovalbumin gene (*76*), and the EBNA 1 and EBNA 2 genes of Epstein-Barr virus (*77*). The processing of products of foreign genes inserted into HSV-1 vectors has not been studied extensively; normally, secreted proteins are secreted from cells infected with HSV-1 vectors carrying their genes (*75*).

Requirements for expression of foreign genes. Host genes and genes of other viruses linked to their natural promoters are expressed poorly if at all in HSV vectors (*78*). When linked to promoter-regulatory domains of HSV genes, the foreign genes are expressed and regulated as the authentic HSV genes whose promoter was "borrowed" (*75*). HSV-2 genes are expressed in HSV-1 vectors under their own promoters, and it is conceivable that genes of other herpesviruses may also be expressed in HSV-1 vectors under their own promoters.

Stability of genetically engineered

HSV-1 genomes. Several major types of rearrangement in genetically engineered strains have been noted. These include the following:

1) Cell-specific restrictions. An example of this phenomenon is recombinant R316 constructed by the insertion of Bam HI N fragment of HSV-1 (Fig. 1) into the Bgl II site of the TK gene (Fig. 3) and converting the natural (β) TK into an α-regulated gene (*27*). This recombinant is stable in Vero cells; in various other cell lines and in experimental animals, mutants carrying various size deletions in the inserted Bam HI N fragment emerge and rapidly overgrow the R316 recombinant (*29*). No explanation for this bizarre, cell-specific restriction has emerged so far.

2) Equalization of genetically related duplicated genes. Insertion of a gene closely related but nonidentical to that resident in the HSV genome may result in the accumulation of identical recombinant sequences as a consequence of genetic recombination and segregation (*79*).

3) Insert-dependent defective genomes. Insertion of a second copy of a sequence in the same orientation relative to the first copy may result in the generation of head-to-tail reiterations of novel subsets of defective HSV genomes in which all sequences between the two copies of the reiterated sequence had been deleted. Defective genomes require an origin of DNA synthesis for amplification and a terminal *a* sequence for packaging by factors supplied by the helper virus. In addition, the length of the subset must be such that an integral number of tandem reiterations fall within the minimum- and maximum-size DNA that is both packaged and enveloped (*80*). The insertion of a second copy of sequences in the inverted orientation relative to the native copy has resulted, in some instances, in the inversion of the sequences flanked by the inverted repeats (*12, 79*). Whether a particular set of inverted repeats will cause inversions is totally unpredictable and may reflect recombinational "hot spots" within the inserted sequence.

4) Privileged sites. These sites are defined as most likely to result in major rearrangements of the HSV genome. Thus, insertion of sequences at or near the termini of the inverted repeats may result in the spontaneous deletion of most of the inverted repeat sequence $b'a_n'c'$ (Fig. 1) (*67, 70*). To date, insertions inside the inverted repeats have not caused such deletions (*66, 81*). Insertions within one of the inverted repeat sequences of the S component (*c'a'* or *ca*) (Fig. 1) result in the duplication of the insertion in the other repeat (*66*). The duplication is almost obligatory if the insert is close to the *a* sequence but diminishes in frequency with distance.

Construction of Novel Genomes of Other Large DNA Viruses

The strategies for construction of novel genomes described in this article have been used for the insertion and expression of foreign genes in vaccinia virus. Like HSV, vaccinia virus encodes a TK gene, and both the vaccinia and HSV TK genes have been used to insert foreign genes linked to vaccinia virus promoters (*82–86*). The extensive experience obtained with vaccinia virus in the course of its use as a vaccine for the eradication of smallpox has been cited as the basis for its use as a vector of vaccines for the prevention of a wide variety of infectious diseases. The genomic capacity of vaccinia virus to carry foreign genes is at least equal to that of HSV.

Conclusions

Various techniques based on the use of selectable markers are now available for the analysis of the function of specific domains of large DNA genomes and for the construction of vectors capable of delivering vaccines for the prevention of infectious diseases in humans and animals.

References and Notes

1. K. W. C. Peden, J. M. Pipas, S. Pearson-White, D. Nathans, *Science* **209**, 1392 (1980).
2. Q. S. Kapoor and G. Chinnadurai, *Proc. Natl. Acad. Sci. U.S.A.* **78**, 2184 (1981).
3. J. E. Mertz, J. Carbon, M. Herzberg, R. W. Davis, P. Berg, *Cold Spring Harbor Symp. Quant. Biol.* **39**, 69 (1974).
4. E. D. Kieff, S. L. Bachenheimer, B. Roizman, *J. Virol* **8**, 125 (1971).
5. P. Sheldrick and N. Berthelot, *Cold Spring Harbor Symp. Quant. Biol.* **39**, 667 (1974).
6. S. Wadsworth, R. J. Jacob, B. Roizman, *J. Virol.* **15**, 1487 (1975).
7. E. S. Mocarski and B. Roizman, *Proc. Natl. Acad. Sci. U.S.A.* **78**, 7047 (1981).
8. A. J. Davison and N. M. Wilkie, *J. Gen. Virol.* **55**, 315 (1981).
9. M. M. Wagner and W. C. Summers, *J. Virol.* **27**, 374 (1978).
10. H. Locker and N. Frenkel, *ibid.* **32**, 424 (1979).
11. G. S. Hayward, R. J. Jacob, S. C. Wadsworth, B. Roizman, *Proc. Natl. Acad. Sci. U.S.A.* **72**, 4243 (1975).
12. E. S. Mocarski and B. Roizman, *Cell* **31**, 89 (1982).
13. N. Frenkel, H. Locker, W. Batterson, G. Hayward, B. Roizman, *J. Virol.* **20**, 527 (1976).
14. D. A. Vlazny and N. Frenkel, *Proc. Natl. Acad. Sci. U.S.A.* **78**, 742 (1981).
15. N. D. Stow, *EMBO J.* **1**, 863 (1982).
16. ——— and E. C. McMonagle, *Virology* **130**, 427 (1983).
17. S. K. Weller *et al.*, *Mol. Cell. Biol.* **5**, 930 (1985).
18. D. A. Vlazny, A. Kwong, N. Frenkel, *Proc. Natl. Acad. Sci. U.S.A.* **79**, 1423 (1982).
19. R. W. Honess and B. Roizman, *J. Virol.* **12**, 1346 (1973).
20. E. K. Wagner, in *The Herpesviruses*, B. Roizman, Ed. (Plenum, New York, 1985), vol. 3, pp. 45–104.
21. R. W. Honess and B. Roizman, *J. Virol.* **14**, 8 (1974).
22. ———, *Proc. Natl. Acad. Sci. U.S.A.* **72**, 1276 (1975).
23. A. J. Conley, D. M. Knipe, P. C. Jones, B. Roizman, *J. Virol.* **37**, 191 (1981).
24. D. M. Knipe, W. T. Ruyechan, B. Roizman, I. W. Halliburton, *Proc. Natl. Acad. Sci. U.S.A.* **75**, 3896 (1978).
25. D. M. Knipe, W. T. Ruyechan, R. W. Honess, B. Roizman, *ibid.* **76**, 4534 (1979).
26. R. A. F. Dixon and P. A. Schaffer, *J. Virol.* **36**, 189 (1980).
27. L. E. Post and B. Roizman, *Cell* **25**, 227 (1981).
28. W. R. Sacks, C. C. Greene, D. P. Aschman, P. A. Schaffer, *J. Virol.* **55**, 796 (1985).
29. A. E. Sears, I. W. Halliburton, B. Meignier, S. Silver, B. Roizman, *J. Virol.* **55**, 338 (1985).
30. P. J. Godowski and D. M. Knipe, *ibid.* **47**, 478 (1983).
31. M. L. Fenwick and M. J. Walker, *J. Gen. Virol.* **41**, 37 (1978).
32. G. S. Read and N. Frenkel, *J. Virol.* **46**, 498 (1983).
33. S. Mackem and B. Roizman, *ibid.* **43**, 1015 (1982).
34. ———, *Proc. Natl. Acad. Sci. U.S.A.* **79**, 4917 (1982).
35. ———, *J. Virol.* **44**, 939 (1982).
36. W. Batterson and B. Roizman, *ibid.* **46**, 371 (1983).
37. M. E. M. Campbell, J. W. Palfreyman, C. M. Preston, *J. Mol. Biol.* **180**, 1 (1984).
38. P. E. Pellett, J. L. C. McKnight, F. J. Jenkins, B. Roizman, *Proc. Natl. Acad. Sci. U.S.A.* **82**, 5870 (1985).
39. R. J. Watson, C. M. Preston, B. Clements, *J. Virol.* **31**, 42 (1979).
40. S. Mackem and B. Roizman, *Proc. Natl. Acad. Sci. U.S.A.* **77**, 7122 (1980).
41. T. M. Kristie and B. Roizman, *ibid.* **81**, 4065 (1984).
42. T. J. Hill, in *The Herpesviruses*, B. Roizman, Ed. (Plenum, New York, 1985), vol. 3, pp. 175–240.
43. C. Morgan, H. M. Rose, B. Mednis, *J. Virol.* **2**, 507 (1968).
44. W. Batterson, D. Furlong, B. Roizman, *ibid.* **45**, 397 (1983).
45. F. Constanzo, G. Campadelli-Fiume, L. Foa-Tomas, E. Cassai, *ibid.* **21**, 996 (1977).
46. H. M. Keir and E. Gold, *Biochim. Biophys. Acta* **72**, 263 (1963).
47. R. J. Jacob, L. S. Morse, B. Roizman, *J. Virol.* **29**, 448 (1979).
48. T. Ben-Porat and S. Tokazewski, *Virology* **79**, 292 (1977).
49. W. Gibson and B. Roizman, *J. Virol.* **10**, 1044 (1972).
50. D. K. Braun, B. Roizman, L. Pereira, *ibid.* **49**, 142 (1984).
51. R. W. Darlington and L. H. Moss III, *Prog. Med. Virol.* **11**, 16 (1969).
52. J. Schwartz and B. Roizman, *J. Virol.* **4**, 879 (1969).
53. S. Kit and D. R. Dubbs, *Virology* **26**, 16 (1965).
54. G. H. Cohen, *J. Virol.* **9**, 408 (1972).
55. D. Huszar and S. Bacchetti, *ibid.* **37**, 580 (1981).
56. F. Wohlrab and B. Francke, *Proc. Natl. Acad. Sci. U.S.A.* **80**, 100 (1980).
57. V. G. Preston and F. B. Fisher, *Virology* **138**, 58 (1984).
58. D. R. Dubbs and S. Kit, *ibid.* **22**, 493 (1964).

59. D. Lando and M. L. Rhyiner, *C. R. Acad. Sci.* **269**, 527 (1969).
60. F. L. Graham, G. Velihaisen, N. M. Wilkie, *Nature (London) New Biol.* **245**, 265 (1973).
61. P. Sheldrick *et al.*, *Proc. Natl. Acad. Sci. U.S.A.* **70**, 3621 (1973).
62. W. T. Ruyechan, L. S. Morse, D. M. Knipe, B. Roizman, *J. Virol.* **29**, 677 (1979).
63. J. Hubenthal-Voss and B. Roizman, unpublished observations.
64. R. B. Tenser and M. E. Dunston, *Virology* **99**, 417 (1979).
65. M. Ackermann, M. Sarmiento-Batterson, B. Roizman, in preparation.
66. J. Hubenthal-Voss and B. Roizman, *J. Virol.* **54**, 509 (1985).
67. F. J. Jenkins, M. J. Casadaban, B. Roizman, *Proc. Natl. Acad. Sci. U.S.A.* **82**, 4773 (1985).
68. B. A. Castilho, P. Olfson, M. J. Casadaban, *J. Bacteriol.* **158**, 488 (1984).
69. N. D. Stow, E. C. McMonagle, A. J. Davison, *Nucleic Acids Res.* **11**, 8205 (1983).
70. K. L. Poffenberger, E. Tabares, B. Roizman, *Proc. Natl. Acad. Sci. U.S.A.* **80**, 2690 (1983).
71. P. Mauromara-Nazos, J. Hubenthal-Voss, B. Roizman, unpublished observations.
72. S. Silver and B. Roizman, *Mol. Cell. Biol.* **5**, 518 (1985).
73. G. T.-Y. Lee, K. L. Pogue-Geile, L. Pereira, P. G. Spear, *Proc. Natl. Acad. Sci. U.S.A.* **79**, 6612 (1982).
74. M. G. Gibson and P. G. Spear, *J. Virol.* **48**, 396 (1983).
75. M.-F. Shih, M. Arsenakis, P. Tiollais, B. Roizman, *Proc. Natl. Acad. Sci. U.S.A.* **81**, 5867 (1984).
76. M. Arsenakis, K. L. Poffenberger, B. Roizman, in preparation.
77. M. Hummel *et al.*, *Virology*, in press.
78. C. Tackney, G. Cachianes, S. Silverstein, *J. Virol.* **52**, 606 (1984).
79. K. L. Pogue-Geile, G. T.-Y. Lee, P. G. Spear, *ibid.* **53**, 456 (1985).
80. K. L. Poffenberger and B. Roizman, *ibid.* **53**, 587 (1985).
81. F. J. Jenkins and B. Roizman, unpublished observations.
82. M. Mackett, G. L. Smith, B. Moss, *Proc. Natl. Acad. Sci. U.S.A.* **79**, 7415 (1982).
83. G. L. Smith, M. Mackett, B. Moss, *Nature (London)* **302**, 490 (1983).
84. G. L. Smith, B. R. Murphy, B. Moss, *Proc. Natl. Acad. Sci. U.S.A.* **80**, 7155 (1983).
85. M. Mackett, G. L. Smith, B. Moss, *J. Virol.* **49**, 857 (1984).
86. S. Gillard, D. Spehner, P. Drillien, *ibid.* **53**, 316 (1985).
87. Supported in part by grants from the National Cancer Institute (CA08494 and CA19264), United States Public Health Service, American Cancer Society (MV2T), and Institut Merieux. F.J.J. is a postdoctoral trainee (CA09241) of the National Cancer Institute.
88. R. Longnecker and B. Roizman, submitted for publication.

3

Heterologous Protein Secretion from Yeast

Robert A. Smith, Margaret J. Duncan, and Donald T. Moir

The yeast *Saccharomyces cerevisiae* has recently gained popularity as a host for heterologous gene expression and protein secretion. This interest is due both to the ease and favorable economics of yeast fermentation, developed over years of industrial experience, and to the rapid progress made in the molecular genetics of the organism. The ability to introduce exogenous DNA into the yeast genome, coupled with advancements in molecular cloning techniques and understanding of the genetics and physiology of protein secretion from yeast cells (*1*), has led to the development of strains that produce and secrete such proteins as α-interferon (*2*), epidermal growth factor (*3*), and β-endorphin (*4*).

In this article, we describe the development of yeast strains that secrete calf prochymosin, the inactive precursor of chymosin (also known as rennin), which is used in the manufacture of cheese. Obvious advantages to be gained from the secretion of prochymosin, as opposed to its production in the cytoplasm of microorganisms, include the ease of continuous fermentation and processing and the relatively high initial purity of the desired protein product. However, we are interested in the secretion of prochymosin primarily because secreted prochymosin is fully activable to chymosin, unlike cytoplasmically produced prochymosin, which is largely insoluble and unactivable (*5*). Other naturally secreted mammalian proteins, such as bovine growth hormone (*6*), human tissue plasminogen activator (*7*), and human γ-interferon (*8*), are also insoluble and inactive when produced in the cytoplasm of microorganisms. Secretion, therefore,

Science 229, 1219–1224 (20 September 1985)

should play a major role in future industrial fermentation processes.

A secreted protein follows a specific pathway in the cell. Transit through this pathway may or may not occur as efficiently for heterologous proteins as for homologous ones. In this article, we discuss some of the parameters affecting the efficiency of secretion from yeast. Using calf prochymosin as a model for yeast secretion of heterologous proteins, we present results on the effect of promoter strength, gene dosage (both as plasmid copies and as integrated copies), secretion signal, and host mutations on the amount of protein production and secretion. The results indicate that heterologous protein secretion from yeast becomes saturated at a surprisingly low level. However, we show that the saturation level can be increased significantly by changes in the secretion signal and the mode of maintenance of the transcriptional unit, as well as by mutations in the host genome.

Secreted Prochymosin Fully Activable to Chymosin

Prochymosin has been a prime target for production in microorganisms because of its industrial utility and limited availability. Indeed, strains of both yeast and *Escherichia coli* that produce calf prochymosin cytoplasmically have been described (*9, 10*). A striking characteristic of the prochymosin produced by all of these strains is that it is largely insoluble and unactivable. It can be isolated only after solubilization with protein denaturants. Even after removal of the denaturant and refolding of the protein, incubation of this prochymosin at low *p*H converts less than 25 percent to active chymosin (*11*). Native calf prochymosin is autocatalytically activated to chymosin at low *p*H with nearly 100 percent efficiency by removal of 42 amino acid residues from the amino terminus (*12*).

The structure of insoluble prochymosin remains obscure because it can be examined only after denaturation. Nevertheless, it is clear that one or more of the three disulfide bonds present in the native molecule do not form, or form incorrectly, in the insoluble product found in yeast and *E. coli*. In polyacrylamide gels containing sodium dodecyl sulfate, the electrophoretic mobilities of prochymosin derived from calves and prochymosin produced in yeast cells are identical after reduction with dithiothreitol. Without prior reduction, however, the yeast-produced material migrates more slowly than the calf prochymosin. The slower migration is consistent with a more extended structure for the protein made in yeast, which may result because conditions inside cells are too reducing to permit disulfide bond formation. The fact that most proteins containing disulfide bonds are extracellular suggests that the secretion pathway may favor disulfide bond formation. The inactivity of intracellular prochymosin from yeast and *E. coli* is not surprising because the protein is normally made with a 16-residue amino-terminal secretion signal (*13*) and is secreted into the fourth stomach of the calf.

We achieved secretion of prochymosin from yeast cells by use of the secretion signal from yeast invertase, and we found that the solubility and activability of prochymosin are then identical to those of native calf prochymosin (*5*). A segment of DNA containing the *SUC2* (invertase) transcriptional promoter region, the coding region for the amino-terminal secretion signal, and codons for the first 11 amino acid residues of mature invertase were fused in translational reading frame with a prochymosin-encoding DNA fragment and incorporated into a plasmid vector (Fig. 1A) (*5, 14*). In yeast the expression product from this construct is an invertase-prochymosin fusion protein that after removal of the signal peptide during secretion is processed at low *p*H in the same way that natural calf zymogen is processed to yield active chymosin (Fig. 1B and lane g in Fig. 2).

Prochymosin production by yeast strains containing this plasmid, and by others in which the *SUC*2 promoter and secretion signal have been replaced by different promoters or secretion signals (Fig. 1A), is shown in Table 1. A comparison of strains with similar total prochymosin expression levels (experiments 1 and 4, 2 and 5, and 3 and 6 in Table 1; activable enzyme levels from wild-type cells) shows that the addition of the invertase secretion signal results

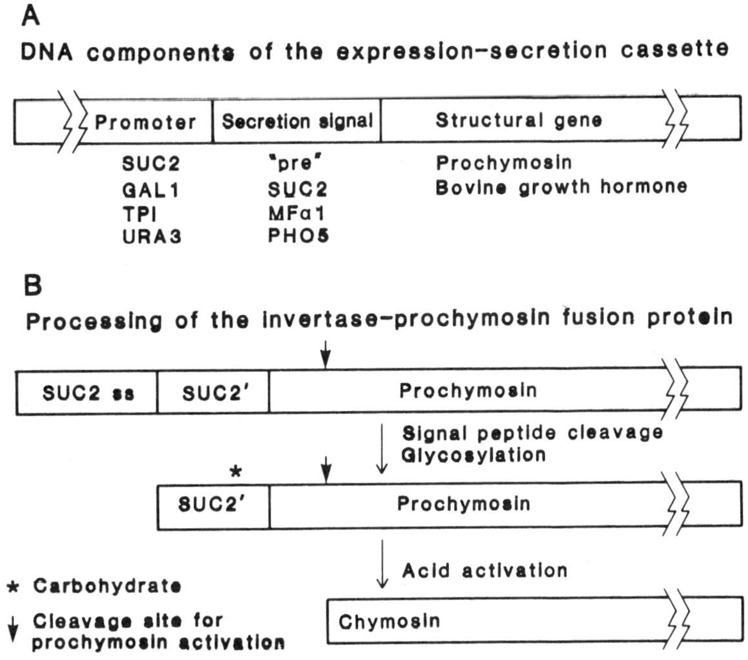

Fig. 1. (A) DNA components of the expression-secretion cassette. Several autonomously replicating yeast plasmids [see pCGS40 in (*9*)] were constructed in which three elements—the promoter, secretion-signal coding region, and structural gene—were varied. The *SUC*2 (*24*) and MFα1 (*25*) secretion signals also include regions of the structural genes preceding convenient restriction endonuclease access sites (Ava II and Hind III, respectively). The synthesis of the *PHO*5 signal sequence encoding DNA was based on the published sequence (*26*). The designation "pre" refers to the DNA encoding the native secretion signal present on prochymosin or BGH complementary DNA. Access to the prochymosin encoding portion of the complementary DNA was obtained as previously described (*9*). The Met-Met-Ala-Ala initiating codons of pre-BGH were replaced with Met to yield a slightly truncated pre-BGH gene. The promoters are approximately 1-kb DNA segments found upstream from the following genes: *TPI* (*27*), with a 3′ end at −18 relative to the ATG translation initiation codon; *GAL*1, as previously described (*9*); *SUC*2, from plasmid pRB58 (*28*); *URA3*, from plasmid pRB71 (*29*), with a 3′ end at +12 relative to the ATG. (B) Processing of the invertase-prochymosin fusion protein. The initial translation product is a fusion of the *SUC*2 secretion signal and 11 amino acid residues of the invertase structural gene to prochymosin [see (*5*) for details]. At least four processing steps are observed: (i) the secretion-signal peptide is removed, (ii) a single core carbohydrate unit is added, (iii) the outer mannose chains are added, and (iv) the secreted fusion protein is activated to chymosin by a low *p*H incubation.

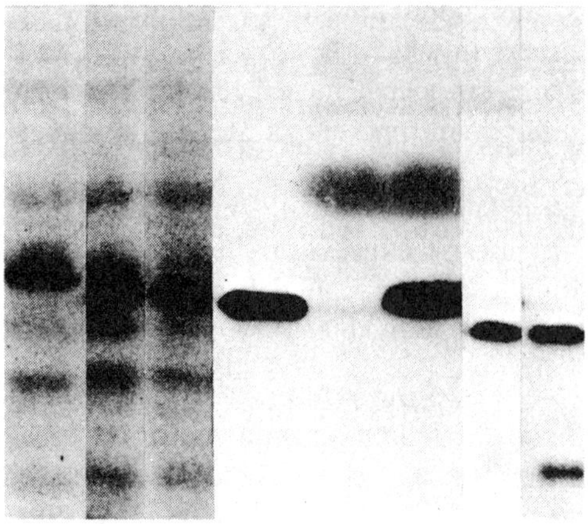

Fig. 2. Processing of invertase-prochymosin in the yeast secretion pathway. Internal invertase-prochymosin in crude extracts (100 μg of protein) or secreted invertase-prochymosin in culture broth (2 ml concentrated by ultrafiltration) from yeast cultures (2×10^7 cell/ml) was subjected to SDS-polyacrylamide gel electrophoresis and transferred to nitrocellulose. Chymosin-related protein species were detected by immunoblot (*30*), for which rabbit antiserum against purified calf chymosin was used. Control immunoblots performed on yeast cell extracts (300 μg protein per lane) containing no prochymosin revealed no significant cross-reactivity of the antiserum with yeast proteins in the chymosin or prochymosin region of the gel [lane B of figure 5 in (*9*)]. Internal invertase-prochymosin made from a *SUC*2-promoted construct (experiment 5 in Table 1) in cells derepressed by growth for 3 hours in YNB (*31*) containing 0.05 percent glucose and 3 percent glycerol is shown untreated (lane a) and treated (lane b) with endoglycosidase H (*16*). The protein produced under derepressed conditions in the presence of tunicamycin (20 μg/ml) is shown in lane c. Secreted invertase-prochymosin produced from a *TPI*-promoted construct (experiment 6 in Table 1) in cells grown in succinate-buffered YNB (*31*) is shown before (lane e) and after (lane f) treatment with endoglycosidase H. The secreted product (160 ng of protein) of a strain carrying the *SUC*2-promoted construct derepressed on YNB agar plates (*31*) is shown in lane g. Both YNB and succinate-buffered YNB liquid media contained bovine serum albumin (25 μg/ml) to stabilize secreted chymosin. The diffusely stained region above the prochymosin band in lanes e and f is bovine serum albumin, which, at this heavy loading, shows some cross-reactivity with the antiserum. Calf prochymosin (lane d, 34 ng) and calf chymosin (lane h, 25 ng) are shown as markers.

in the production of secreted activable prochymosin, in addition to an amount of internal activable prochymosin similar to that produced from constructs without the secretion signal. Most of the antigen produced, either with or without a secretion signal, remains internal and cannot be activated efficiently to chymosin (experiments 1 to 6 in Table 1; internal activable enzyme levels in wild-type cells). All of the secreted prochymosin, however, can be activated to chymosin (see below). Other secretion signals, both from yeast and from the native gene, are less effective than the invertase signal (experiments 7 to 9 in Table 1; secreted levels from wild-type cells).

Processing and Glycosylation of Secreted Prochymosin

Processing of the invertase-prochymosin fusion protein during its production and secretion is illustrated schematically in Fig. 1B and by gel electrophoretic analysis in Fig. 2. In wild-type cells, most of the fusion protein is internal. This fusion protein (lane a in Fig. 2) is glycosylated because cell extracts treated with endoglycosidase H (lane b in Fig. 2) contain a smaller protein, as do extracts from cells treated with tunicamycin (lane c in Fig. 2). Incubations with endoglycosidase H for varying lengths of time indicate that only one glycosylation site is used even though there are three potential sites for asparagine-linked glycosylation in this fusion protein—one in the invertase portion of the molecule and two in the chymosin portion [amino acid residues 310 and 349 (*13*)]. Activation of the invertase-prochymosin fusion protein to chymosin at low *p*H removes all of the carbohydrate, demonstrating that the sugar moieties must be present on the glycosylation site in the invertase portion of the fusion protein (lane g in Fig. 2). It is likely that certain asparagines within Asn-x-Thr/Ser sequences of a protein are not glycosylated because they are inacessible to the glycosylation enzymes (*15*). The fact that identical sequences in chymosin fail to be glycosylated in both the calf and yeast cells suggests that folding of the protein chain during secretion is similar in the two organisms.

The fusion protein, which does not contain carbohydrate because of treatment with endoglycosidase H or production in cells treated with tunicamycin, is only slightly larger than native calf prochymosin (lane d in Fig. 2). Compared to the much greater difference in mobility between prochymosin and chymosin (lanes d and h in Fig. 2), reflecting a difference of 42 amino acid residues, the slight mobility difference between deglycosylated invertase-prochymosin and native calf prochymosin (lanes c and d in Fig. 2) is consistent with the loss of the 19 amino acid residues of the invertase secretion signal and the retention of the 11 amino acid residues of mature invertase. Because the entire invertase secretion signal and mature invertase junction region are present on the initial product, it is likely that processing of the secretion signal has occurred at the normal junction. However, we have not determined the precise processing point of the secretion signal.

The primary secreted product appears to be heavily glycosylated, but is not visible in lane e of Fig. 2. When an equivalent amount of secreted material is first treated with endoglycosidase H, an intense band is visible in the region of the gel corresponding to the unglycosylated invertase-prochymosin fusion protein (lane f of Fig. 2). When more untreated secreted material is loaded on the gel, a broad region of antigenic activity can be seen in the high molecular weight portion of the gel. This heterogeneous electrophoretic mobility is analogous to that seen with secreted invertase (*16*) and is probably due to varying degrees of mannose outer chain addition. These additional mannose residues may interfere with antibody binding or electrophoretic transfer to nitrocellulose. Internal invertase-prochymosin appears to contain only core oligosaccharide because it migrates as a distinct band that is not altered in antigenic activity by endoglycosidase H treatment (lanes a and b in Fig. 2).

All of the activated secreted invertase-prochymosin exhibits an electrophoretic mobility identical to that of natural calf chymosin (lanes g and h in Fig. 2). No additional invertase-prochymosin is apparent after endoglycosidase H treatment, and no carbohydrate is present on the chymosin portion of the molecule. The amount of chymosin predicted from the milk-clotting activity [assuming a specific activity of 100 unit/mg; (*17*)] is consistent with the amount of secreted antigen detected by immunoblot. Therefore, all secreted prochymosin can be converted to active chymosin. This result is striking in comparison with the extremely poor activability of internal

Table 1. Heterologous protein production and secretion. Extrachromosomal 2-μm-based plasmid constructions described in Fig. 1 were introduced into one or more of the following yeast host strains: CGY80 (*MATa his3 ura3-52 leu2-3,112 trpl-289*), CGY150 (*MATα leu2-3 ura3-52*), CGY339 (*MATα his4-29 ura3-52 pep4-3*), CGY434 (*MATα his4-519 ura3-50 pep4-3*), and supersecreting mutants derived from CGY339 by ethyl methane sulfonate mutagenesis and backcrossed to wild-type laboratory strains. Constructions on YIp5 vectors (*32*) were integrated at three different yeast loci by means of sequence homology present on the vector; multiple integrants were obtained by genetic crosses. Transformants were grown in YNB (*31*) with appropriate amino acid supplements to a density of about 2×10^7 cells per milliliter and harvested by centrifugation. Cells were lysed with glass beads, and extracts were prepared by treatment with trichloroacetic acid according to method B (*9*). The total antigen level (both inside and outside the cells) was measured by immunoblot (*30*) with appropriate dilutions of samples and standards (purified from the native source); the total antigen level varied with different promoters and signal sequences but was independent of the host strain in every case examined. Levels of total antigen and activable enzyme are expressed as milligrams or micrograms per gram of soluble yeast protein, which was determined by means of a commercial reagent (Bio-Rad). Chymosin milk-clotting activity was determined after activation of prochymosin by incubation in 40 m*M* lactic acid and neutralization with NaOH (*9*). Activity measurements were converted to micrograms of activable enzyme assuming a specific activity for chymosin of 100 units per milligram (*17*). Internal activity was measured after activation of cell extracts prepared in 50 m*M* tris-HCl [method A in (*9*)]. Secreted protein was isolated by growing cells on an agar plate containing YNB with appropriate amino acid supplements for 3 days at 30°C, removing cells, and washing them with 50 m*M* sodium phosphate, *p*H 5.8. The activity was measured in the wash-buffer supernatant. Acid activation was unnecessary because the poorly buffered YNB medium reaches sufficiently low *p*H to activate all prochymosin secreted. This assay is convenient but underestimates the level of secretion compared with that observed in liquid culture by a factor of about 2 because some product diffuses into the agar layer beneath the cells. Abbreviations: P, plasmid; I, integrant; 4×I, four-copy integrant; pC, prochymosin; BGH, bovine growth hormone; *SUC2*(33) indicates an in-frame addition (33 bp, in this case) of a yeast gene (*SUC2*, in this case) between the indicated secretion-signal coding region and prochymosin; *SSCX* denotes a partially dominant supersecreting mutant whose complementation group has not been established.

Exp.	Promoter	Secretion signal	Gene	Vector	Total antigen produced (mg/g)	Level of activable enzyme [μg/g (percent of antigen)]					
						Internal	Secreted				
						Wild type	Wild type	*ssc1*	*ssc2*	*ssc1ssc2*	*SSCX*
1	*GAL1*		pC	P	2.5	10 (0.4)					
2	*URA3*		*URA3*(12)pC	P	0.7	4 (0.5)					
3	*TPI*		pC	P	2.0	15 (0.8)					
4	*GAL1*	*SUC2*	*SUC2*(33)pC	P	3.0	18 (0.6)	27 (0.9)	65 (2.2)	141 (7.9)	238 (7.9)	177 (5.9)
5	*SUC2*	*SUC2*	*SUC2*(33)pC	P	0.5	<2 (0.5)	21 (4.2)	50 (10)		256 (51)	
6	*TPI*	*SUC2*	*SUC2*(33)pC	P	2.5	18 (0.7)	23 (0.9)	264 (10)	274 (11)	675 (27)	126 (5.0)
7	*GAL1*	pre	pC	P	2.5		<6 (0.2)	117 (4.7)	46 (1.8)	275 (11)	
8	*MFα1*	*MFα1*	*MFα1*(267)pC	P	2.0		11 (0.6)		68 (3.4)		43 (2.2)
9	*GAL1*	*PHO5*	pC	P	3.0		<6 (0.2)	268 (8.9)	129 (4.3)	390 (13)	45 (1.5)
10	*TPI*	*SUC2*	*SUC2*(33)pC	I	0.7		82 (12)	246 (35)	264 (38)		
11	*TPI*	*SUC2*	*SUC2*(33)pC	4×I	3.0		240 (8.0)	1382 (45)	1224 (40)		
12	*GAL1*	pre	BGH	P	3.0		3 (0.1)*	36 (1.2)*	4 (0.2)*	40 (1.6)*	

*Level of secreted antigen, with percentage of total antigen in parentheses, as determined by immunoblot (*30*).

prochymosin, made both with and without a secretion signal sequence (experiments 3 and 6 in Table 1; internal levels from wild-type cells).

Limitations on the Efficiency of Prochymosin Secretion

These results demonstrate that secretion of prochymosin overcomes the insolubility and inactivity associated with the cytoplasmically produced protein. However, they also show that less than one-tenth of the invertase-prochymosin made by the cells under these conditions is secreted. Replacement of the *SUC2* promoter with the stronger promoter triose phosphate isomerase (*TPI*) increases the expression level to 0.25 percent of total cell protein, but does not significantly increase the amount secreted (experiments 5 and 6 in Table 1; levels secreted from wild-type cells).

Two observations indicate that there is a limitation in the secretion pathway itself. First, protein secreted into the medium contains both core oligosaccharide and the outer-chain carbohydrate added in the Golgi apparatus of the cell. By contrast, most of the intracellular fusion protein contains only the core oligosaccharide added in the endoplasmic reticulum. Because very little of the fully glycosylated protein accumulates inside, it would appear that it is transported rapidly to the outside of the cell. These observations suggest that translocation of core oligosaccharide–containing protein from the endoplasmic reticulum to the Golgi, transit of protein through the Golgi, or the addition of outer-chain carbohydrate in this organelle may be a rate-limiting step in the secretion of invertase-prochymosin.

Second, indirect immunofluorescence studies indicate that a substantial fraction of the core glycosylated invertase-prochymosin within the cell is located in the cell vacuole (*14*). A rate-limiting Golgi-associated step in the secretion pathway might cause excess protein to be diverted to the vacuole. Alternatively, proteins that are "foreign" to yeast may contain insufficient sequence information to ensure their efficient secretion; thus, the vacuole may be a repository for protein that cannot be sorted properly.

In summary, prochymosin-secreting yeast cells appear to direct all of the prochymosin they produce to the secretion pathway, but they secrete only a small fraction of it. In a well-characterized pathway, the rate-limiting step could be accelerated by overproduction of the enzyme catalyzing it. In this case, however, the secretion pathway is complex, our understanding preliminary, and the actual rate-limiting step ill-defined.

Supersecreting Mutants

To identify mutant yeast strains with elevated secretion levels, we chose to take a general mutagenesis approach coupled with a suitable screen for the desired phenotype. We recently succeeded in isolating a number of mutant yeast strains that are capable of secreting a larger proportion of the prochymosin they produce (*18*). We have called them "supersecreting" mutants, and where it is clear that the supersecreting phenotype is due to a single gene mutation, we have assigned the designation *ssc* (for supersecretion) to that mutation.

The isolation of supersecreting strains of yeast was greatly facilitated by the development of a rapid screening assay that allows estimation of the relative amount of chymosin secreted by individual yeast colonies. This assay relies upon the fact that a yeast colony which secretes prochymosin leaves a "footprint" of secreted material on the surface of the nutrient agar plate on which it grows. This secreted material, which is activated by the low *p*H of the medium, can be assayed by overlaying the surface of the plate with a mixture of milk and molten agarose after the yeast colonies have been removed. The speed with which opaque regions of clotted milk form, and their size and intensity, are related to the amount of chymosin secreted by individual colonies. Thus, colonies that secrete more prochymosin can be distinguished from colonies that secrete less.

This assay was used to identify 39 supersecreting strains of yeast from a collection of approximately 120,000 mutagenized colonies. All of the mutant strains produce the same total amount of prochymosin as the parent strain, but some secrete as much as eight or ten times the wild-type amount. [A genetic analysis of these strains will be described in more detail elsewhere (*19*).] Both partially dominant and recessive mutations were obtained. The recessive mutations have been analyzed extensively, and they behave as simple Mendelian traits through several backcrosses. Complementation analysis of the recessive mutations indicates that mutations in at least four genes can lead to a supersecreting phenotype, but mutations in two genes in particular—designated *SSC*1 and *SSC*2—are the strongest and most easily manipulated alleles.

To determine the additivity of the supersecretion effects, we constructed haploid yeast strains that contain both *ssc*1 and *ssc*2 mutations. Such doubly mutant strains secrete prochymosin more efficiently than either *ssc*1 or *ssc*2 single mutants (experiment 6 in Table 1). Other strains with particularly high secretion efficiencies could also be constructed by combining supersecreting mutations from other complementation groups. These results are important because they imply that each of a number of mutational changes can have significant independent, additive effects on the intracellular distribution of a protein such as prochymosin.

Supersecreting Mutations Are Effective for Other Proteins and Signals

We have tested the general utility of supersecreting strains in a number of ways. Clearly, their effects extend beyond the features of the expression-secretion vector used in their isolation. For example, secretion of prochymosin from vectors having other promoter and secretion-signal sequences is also more efficient in supersecreting strains (experiments 7 to 9 in Table 1). The increased secretion resulting from supersecreting mutations occurs in constructions integrated into the yeast genome, as well as those carried on plasmid vectors (experiments 10 and 11 in Table 1). Also, the dramatic effects of the supersecreting mutations are not limited to laboratory shake-flask cultures; rather, the yield of secreted invertase-prochymosin is proportional to cell number up to commercially useful cell densities of a least 25 g/liter (dry weight), resulting in secretion into the growth medium of 20 mg per liter of activable prochymosin, or about 80 to 85 percent of the prochymosin produced by the cells (*20*). Finally, some mutant strains selected solely on the basis of increased prochymosin secretion also have an increased secretion capacity for at least one other foreign protein. We introduced an expression vector for the secreted precursor of bovine growth hormone (pre-BGH) into supersecreting strains and found that the *ssc*1-1 mutation increases the secreted yield at least tenfold, and *ssc*2-1 exhibits a detectable but smaller effect (experiment 12 in Table 1).

Chromosomal Integration Improves Secretion Efficiency

Perhaps one of the most surprising observations we have made is that the efficiency with which a yeast strain secretes prochymosin can be affected by the manner in which the prochymosin gene is maintained in the cell. In particular, when the gene for prochymosin fused to a yeast promoter and secretion signal sequence is integrated into a yeast chromosome, a considerably greater proportion of the prochymosin is secreted than when the same promoter–signal sequence–prochymosin construction is introduced on a multicopy plasmid vector. Under conditions in which plasmid constructions result in a secretion efficiency of 1 to 2 percent, integrants typically secrete 8 to 10 percent of the prochymosin they produce (experiments 6, 10, and 11, in Table 1; levels secreted from wild-type cells).

Multicopy plasmids might be expected to display a lower secretion efficiency because of their higher copy number and correspondingly higher gene expression level and larger reservoir of unsecretable prochymosin. However, this hypothesis is clearly not correct because we have constructed strains with several integrated copies of invertase-prochymosin fusion genes. Such strains produce approximately the same absolute amount of prochymosin as strains containing the same construction on a multicopy plasmid, but they secrete at least four times as much (experiments 6 and 11 in Table 1).

Thus, chromosomal integration appears to be the method of choice for introducing foreign genes into yeast—especially if secretion of the products is desired. Integration results in improved secretion efficiency for prochymosin and presumably also for other gene products. The intrinsic genetic stability of integrated constructions is well established (*21*). Unlike most plasmid constructions, these chromosomally integrated constructions do not require selective pressure for maintenance, thus removing restrictions on the type of growth medium that can be used and allowing growth of cells in continuous culture for long periods. Finally, the major anticipated shortcoming of the integrated constructions—namely, their relatively low copy number—does not diminish the usefulness of integration as a gene maintenance strategy because, in practice, many copies of the transcriptional unit are only marginally more productive than a few copies (*14*).

Discussion

We have shown that secretion of prochymosin is critical for obtaining soluble activable enzyme (*5*). The failure to form disulfide bonds, or their incorrect formation, appears to be characteristic of prochymosin produced in the cytoplasm, both in yeast (as we have shown) and in *E. coli* (*22*). Secretion of a fully activable proenzyme eliminates many of the processing steps required to isolate activable protein from the cytoplasm. Unfolding in denaturants and refolding, which is frequently inefficient in vitro, are not required. However, yeast cells have a limited capacity for prochymosin secretion. For example, a fivefold increase in production of prochymosin through use of a stronger promoter results in secretion of only one-tenth more prochymosin (experiments 5 and 6 in Table 1).

We have identified effective methods for reducing the limitations on secretion of prochymosin. The nature of the secretion signal is important because the native calf signal is only about one-fourth as efficient as the yeast signal from invertase. However, even with the invertase signal sequence, less than 5 percent of the prochymosin produced is secreted. Since secretion of prochymosin can be blocked either by the absence of additional signals required for yeast secretion or by the presence of signals that shunt it to intracellular compartments, we decided to alter the host cell function by mutation. This approach is effective not only with prochymosin but also with bovine growth hormone (BGH).

The simplest model to account for the supersecreting mutations is that they open new bypass routes around a single rate-limiting step in the secretion of heterologous proteins. This model is also consistent with two other characteristics of the mutants—additivity and protein specificity. Clearly, if the bypass pathway is still rate-limiting, additional alternative routes will improve the overall rate. At the same time, it is reasonable to assume that certain proteins will not flow through certain bypass pathways. The fact that BGH secretion is improved dramatically by the *ssc*1-1 mutations but only marginally by the *ssc*2-1 mutation could be an example of differential flow through the alternative routes.

At least two groups have reported isolation of mutants with altered localization of homologous yeast vacuolar proteins (*23*). However, the effects of those mutations on heterologous proteins have not been reported.

Finally, increased efficiency of both prochymosin gene expression and protein secretion has been observed from a surprising manipulation—integration of the transcriptional unit into the genome. The effect is significant and reproducible, but the mechanism is unclear; it may reflect differences in either the rate or the level of expression from plasmid-borne and integrated copies of the gene. We find that prochymosin production from plasmids does not respond in a linear fashion to increases in transcriptional unit copy number because the expression level from plasmids at more than 100 copies per cell is not 100 times but only about 10 times the level from a single-copy integrant (*14*). Transcriptional or translational limitations for some promoters may prevent a linear response to copy number. Whatever the explanation, the efficiency of expression per gene copy is much higher at low copy numbers.

The supersecreting mutants increase the secretion level for prochymosin produced from both integrated and plasmid-borne transcriptional units; however, they do not alter the nonlinear response between plasmid-borne gene dosage and expression. Furthermore, secretion of prochymosin made from plasmid copies of the gene appears to be saturated at a lower level than secretion of prochymosin made from integrated copies, even in *ssc*1 and *ssc*2 host strains. One possible explanation for this saturation is that gene expression from plasmids may be more synchronous than that from multiple, integrated gene copies dispersed throughout the genome. Thus, pulses of expression delivering saturating amounts of protein to the secretion pathway would overwhelm not only the rate-limiting step but also the putative bypass steps opened by the supersecreting mutations, causing a logjam effect.

In conclusion, secretion is necessary for production of activable prochymosin. By combining the benefits of yeast secretion signals, multiple integrated transcriptional units, and specific mutations of the host genome, we have increased secretion of prochymosin into the growth medium at least 80-fold, thereby allowing production of fully activable prochymosin at levels of 20 mg per liter of culture medium. The existence of techniques to improve yeast secretion of heterologous proteins should aid in the production of other mammalian secreted proteins in their native active conformations.

References and Notes

1. R. Schekman and P. Novick, in *The Molecular Biology of the Yeast Saccharomyces: Metabolism and Gene Expression*, J. N. Strathern, E. W. Jones, J. R. Broach, Eds. (Cold Spring Harbor Laboratory, Cold Spring Harbor, N.Y., 1982), pp. 361–393.
2. A. Singh, J. M. Lugovoy, W. J. Kohr, L. J. Perry, *Nucleic Acids Res.* **12**, 8927 (1984).
3. A. J. Brake *et al.*, *Proc. Natl. Acad. Sci. U.S.A.* **81**, 4642 (1984).
4. G. A. Bitter, K. K. Chen, A. R. Banks, P. H. Lai, *ibid.*, p. 5330.
5. D. T. Moir, J. Mao, M. J. Duncan, R. A. Smith, T. Kohno, in *Developments in Industrial Microbiology*, L. Underkofler, Ed. (Society for Industrial Microbiology, 1985), vol. 26, pp. 75–85.
6. R. G. Schoner, L. F. Ellis, B. E. Schoner, *Biotechnology* **3**, 151 (1985).
7. D. Pennica *et al.*, *Nature (London)* **301**, 214 (1983).
8. G. Simons, E. Remaut, B. Allet, R. Devos, W. Fiers, *Gene* **28**, 55 (1984).
9. C. G. Goff *et al.*, *ibid.* **27**, 35 (1984).
10. J. S. Emtage *et al.*, *Proc. Natl. Acad. Sci. U.S,A.* **80**, 3671 (1983).
11. F. A. O. Marston *et al.*, *Biotechnology* **2**, 800 (1984).
12. V. B. Pedersen and B. Foltmann, *Eur. J. Biochem.* **55**, 95 (1975).
13. D. Moir *et al.*, *Gene* **19**, 127 (1982).
14. M. J. Duncan and K. Stashenko, in preparation.
15. D. D. Pless and W. J. Lennarz, *Proc. Natl. Acad. Sci. U.S.A.* **74**, 134 (1977).
16. R. B. Trimble and F. Maley, *J. Biol. Chem.* **252**, 4409 (1977).
17. B. Foltmann, *Methods Enzymol.* **19**, 421 (1970).
18. R. A. Smith and T. Gill, *J. Cell. Biochem.* (suppl. **9C**), 157 (1985).
19. ———, in preparation.
20. R. J. Summers and S. Roof, personal communication.
21. J. B. Hicks, A. Hinnen, G. R. Fink, *Cold Spring Harbor Symp. Quant. Biol.* **43**, 1305 (1979).
22. J. M. Schoemaker, A. H. Brasnett, F. A. O. Marston, *EMBO J.* **4**, 775 (1985).
23. S. D. Emr, V. A. Bankaitis, J. M. Garrett, M. G. Douglas, in *Protein Transport and Secretion*, M.-J. Gething, Ed. (Cold Spring Harbor Laboratory, Cold Spring Harbor, N.Y., 1985), p. 184; J. H. Rothman *et al.*, *ibid.*, p. 190. Complementation tests show that the *vpt3* and *vpt5* mutants of Emr are different from *ssc1* and *ssc2*.
24. M. Carlson, R. Taussig, S. Kustu, D. Botstein, *Mol. Cell. Biol.* **3**, 439 (1983).
25. J. Kurjan and I. Herskowitz, *Cell* **30**, 933 (1982).
26. K. Arima *et al.*, *Nucleic Acids Res.* **11**, 1657 (1983).
27. T. Alber and G. Kawasaki, *J. Mol. Appl. Genet.* **1**, 419 (1982).
28. M. Carlson and D. Botstein, *Cell* **28**, 145 (1982).
29. M. Rose and D. Botstein, *J. Mol. Biol.* **170**, 883 (1983).
30. H. Towbin, T. Staehelin, J. Gordon, *Proc. Natl. Acad. Sci. U.S.A.* **76**, 4350 (1979).
31. YNB contained, per liter: 7 g of yeast nitrogen base (Difco), 20 g of glucose or galactose, and 20 g of agar (when used in plates). Detection of active chymosin required the addition of bovine serum albumin (25 μg/ml) to liquid medium but not to agar plates. Acid activation was unnecessary, except when succinate-buffered YNB (YNB buffered at *p*H 6 with 10 g of succinic acid and 6 g of NaOH per liter) was used, in that yeast metabolism reduces the *p*H of the medium to levels sufficient to activate all the prochymosin produced.
32. K. Struhl, D. T. Stinchcomb, S. Scherer, R. W. Davis, *Proc. Natl. Acad. Sci. U.S.A.* **76**, 1035 (1979).

33. We thank T. Gill and K. Stashenko whose work was vital to the success of this project. We are grateful for key contributions from the following people at Collaborative Research, Inc.: J. Mao, T. Kohno, E. Yamasaki, M. E. Rhinehart, R. Knowlton, V. Brown, S. Porteous, and C. Stillman. We thank R. J. Summers and S. Roof of Dow Chemical Company for unpublished yeast fermentation results. Purified native calf prochymosin and rabbit antiserum directed against calf chymosin were gifts from R. Goltz of Dow Chemical Company. We thank G. Fink and G. Vovis for critically reading the manuscript. This work was partially supported by a contract with Dow Chemical Company.

4

The Genetic Linkage Map of the Human X Chromosome

Dennis Drayna and Ray White

The X chromosome has been studied more intensively than any other human chromosome, because its unique inheritance allows identification of recessive X-linked traits. A linkage map provides an essential framework for understanding and predicting chromosome behavior at the molecular genetic level. Because of the importance of the X chromosome for the study of human genetics, we have devoted our efforts to the construction of a genetic linkage map for this chromosome. The map is based on meiotic recombination frequencies rather than on physical localization. A widely applicable genetic map of this chromosome will be useful in the diagnosis and prediction of many X-linked genetic diseases.

The primary genetic defect is unknown for many X-linked diseases. Some of these disorders, such as Duchenne muscular dystrophy and fragile X-linked mental retardation, are genetic diseases of major importance. A genetic trait can be mapped by linkage without a detailed knowledge of the biochemical defect causing the disease. Specifically, even in the absence of a direct test for a specific mutant gene, it is possible to diagnose or predict a disease if ubiquitous closely linked markers are available. For such markers, we have relied on cloned DNA segments that reveal polymorphism in the length of the fragments produced by cleavage of genomic DNA with certain bacterial endonucleases (restriction enzymes). These restriction fragment length polymorphisms (RFLP's) can be detected in human DNA by hybridization of a single-copy DNA probe to a Southern blot of total genomic DNA digested with restriction enzymes (*1*). RFLP's have several important advantages as genetic markers.

Science 230, 753–758 (15 November 1985)

They can be found in virtually all regions of the genome, and they act as codominant systems, providing complete genotypic information at their respective loci. Furthermore, a high degree of polymorphism can often be developed in RFLP's, and this makes them useful as genetic markers for most families.

To construct the map, we determined the genetic linkage relationships among a number of X chromosome RFLP markers in a series of normal families, because a normal population is the best source of families with pedigree structures optimal for efficient accumulation of linkage data. These markers serve to identify loci of genes associated with disease (OTC, HPRT, FVIII, and FIX) as well as arbitrary loci defined solely by DNA. The DNA marker loci used in our study together with their characteristics and sources are listed in Table 1.

The X chromosome map resulting from our linkage studies (Fig. 1) has allowed us to examine the meiotic behavior of X chromosomes as a population and to improve the resolution of linkages between DNA markers and X-linked diseases.

Table 1. DNA markers on human chromosome X.

Marker	HGM8 workshop symbol	Physical location	Polymorphism	Allele frequency	Reference
Xg^a		Xp22.3	Protein polymorphism	0.35/0.65	(*26*)
dic56	DXS143	Xp22.3	Bcl I	0.45/0.55	(*25*)
pD2	DXS43	Xp21-p22	Pvu II	0.29/0.71	(*19*)
RC8	DXS9	Xp21	Taq I	0.15/0.85	(*3*)
p99-6	DXS41	Xp21-p22	Pst I	0.29/0.71	(*19*)
754	DXS84	Xp21	Pst I	0.40/0.60	(*24*)
OTCase	OTC	Xp21	Msp I	0.30/0.25/0.25/0.20	
			Bam HI	0.35/0.65	(*23*)
L1.28	DXS7	Xp1	Taq I	0.35/0.65	(*22*)
p58	DXS14	Xcen-p21	Msp I	0.35/0.65	(*19*)
p8	DXS1	Xcen-q13	Taq I	0.12/0.88	(*19*)
DXYS1	DXYS1	Xq12	Taq I	0.48/0.52	(*21*)
p19-2	DXS3	Xq13-q22	Msp I	0.20/0.80	(*19*)
S21	DXS17	Xq22	Taq I	0.35/0.65	(*17*)
			Msp I	0.10/0.90	
22–23	DXS11	Xq24-Xqter	Taq I	0.16/0.84	(*19*)
HPRT	HPRT	Xq26	Bam HI	0.70/0.22/0.08	(*20*)
43–15	DXS42	Xq24-qter	Bgl II	0.19/0.81	(*19*)
52A	DXS51	Xq27	Taq I	0.50/0.50	(*17*)
F IX	F9	Xq27–28	Taq I	0.30/0.70	(*18*)
DX13	DXS15	Xq28	Bgl II	0.35/0.65	(*17*)
F VIII	F8	Xq28	Bcl I	0.30/0.70	(*10*)
St14	DXS52	Xq28	Taq I	0.36/0.20/0.15/ 0.12/0.11/0.05/0.01	(*16*)
			Msp I	0.35/0.35/0.19/0.11	

Building the Map

Two types of information contribute to the construction of a genetic map of the X chromosome: localization of the DNA markers by physical methods, and determination of the various genetic linkage relationships among them. The physical locations of these markers were approximated by in situ hybridization to metaphase chromosomes (*2*) or by hybridization to the DNA of a panel of hybrid cell lines, each containing a different portion of the human X chromosome (*3*). In some instances, both techniques were used.

We determined the genetic linkage relationships among 21 DNA markers (Table 1) by examining DNA from 38 normal families. These families are characterized by the presence of all four grandparents and by large sibships, averaging nine children each (*4*). We gathered complete genotypic information for each marker in each family in which the marker was informative—that is, the mother was heterozygous. Because the maternal grandfather in each family passes an exact copy of his only X chromosome to his daughter, the genotype of the maternal grandfather explicitly determines the disposition of alleles on the two maternal chromosomes (that is, their haplotype phase). When phase is known, the recombination frequency between two marker loci can be estimated by counting the recombinant and nonrecombinant chromosomes.

In the construction of the linkage map from our genotypic data we considered two features, genetic distance and genetic order. The distinction between these two parameters is important, for while gene distances are expected to change as the data set enlarges, it would be useful

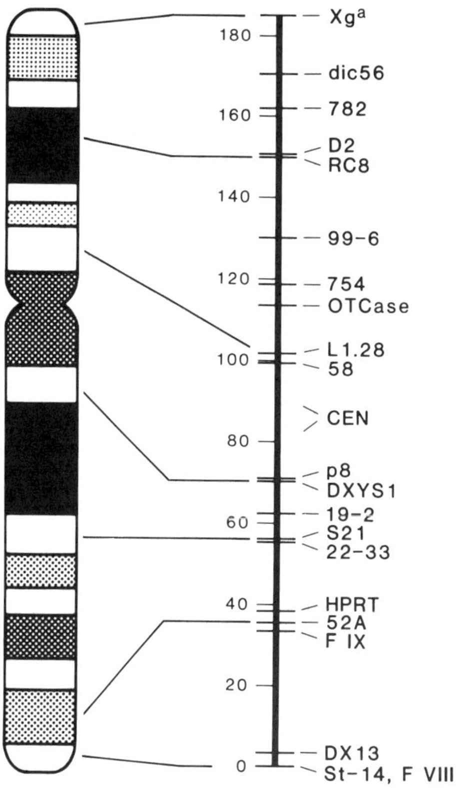

Fig. 1. Linkage map of the human X chromosome. Genetic distances are represented in recombination units, and physical locations of selected markers are indicated on the ideogram.

for the gene order to remain largely invariant (*5*).

Gene distances. We obtained a first estimate of genetic linkage distance between loci by examining the recombination frequency between pairs of marker loci. The number of meioses generating X chromosomes in which recombination could be scored between any two marker loci ranged from 0 to 88, with a mean of 26.5. Listed in Table 2 are the recombination fractions observed between each marker locus and the four marker loci that emerged as the closest adjacent loci

Table 2. Linkage data between DNA markers on the human X chromosome. N.I., not informative.

Marker		Two-factor crosses	Recombinants per informative meioses	Recombination fraction	Most likely gene order	Log_{10} of ratio of likelihood of most likely order to next most likely order
St14	×	F VIII	0/57	0.0 ± .048	No recombinants between St14 and FVIII; order not determined	
		DX13	3/54	.055 ± .021		
		F IX	17/57	.285 ± .059		
		52A	31/88	.352 ± .051		
		43-15	23/53	.434 ± .068		
FVIII	×	DX13	0/27	0.0 ± .074	(St14, FVIII) DX13 FIX	0.14
		52A	6/27	.222 ± .080		
		43-14	12/26	.461 ± .096		
		HPRT	6/20	.300 ± .102		
DX13	×	FIX	13/33	0.393 ± .085	52A FIX DX13 St14	32.14
		52A	20/69	.290 ± .055		
		43-15	6/19	.316 ± .107		
		HPRT	5/13	.385 ± .135		
FIX	×	52A	1/55	0.018 ± .017	43-15 52A FIX DX13	0.43
		43-15	0/7	0.0 ± .125		
		HPRT	1/13	.077 ± .074		
		22-33	1/7	.143 ± .132		
52A	×	43-15	3/35	0.086 ± .047	HPRT 43-15 52A FIX	3.49
		HPRT	1/20	.050 ± .049		
		22-23	4/22	.182 ± .082		
		S21	7/29	.241 ± .079		
43-15	×	HPRT	1/25	0.040 ± .039	22-33 HPRT 43-15 52A	0.26
		22-33	3/14	.214 ± .110		
		S21	2/16	.125 ± .083		
		19-2	3/8	.375 ± .171		
HPRT	×	22-33	4/14	0.285 ± .121	S21 22-23 HPRT 43-15	8.64
		S21	2/7	.285 ± .171		
		19-2	2/11	.182 ± .116		
		DXYS1	4/7	≧.50 ± .189		
22-33	×	S21	0/7	0.0 ± .125	19-2 S21 22-23 HPRT	0.01
		19-2	0/8	0.0 ± .112		
		DXYS1	5/15	.300 ± .118		
		p8	N.I.			
S21	×	19-2	1/17	0.059 ± .057	DXYS1 19-2 S21 22-23	0.66
		DXYS1	4/32	.125 ± .058		
		p8	N.I.			
		58	1/10	.100 ± .095		

19-2	×	DXYS1	1/23	0.044 ± .043	p8 DXYS1 19-2 S21	0.91
		p8	N.I.			
		58	1/19	.053 ± .051		
		L1.28	6/16	.375 ± .121		
DXYS1	×	p8	N.I.		L1.28 58 p8 DXYS1	Order based on physical location
		58	0/16	0.0 ± .070		
		L1.28	8/15	≧.50 ± .130		
		OTCase	19/73	.261 ± .051		
p8	×	58	N.I.		58 p8 DXYS1 19-2	Order based on physical location
		L1.28	4/8	0.50 ± .176		
		OTCase	6/16	.375 ± .121		
		754	3/8	.375 ± .171		
58	×	L1.28	N.I.		OTC L1.28 58 p8	Order based on physical location
		OTCase	1/9	0.111 ± .105		
		754	3/10	.300 ± .145		
		99-6	5/9	≧.50 ± .130		
L1.28	×	OTCase	4/22	0.182 ± .082	754 OTC L1.28 58	1.41
		754	4/32	.125 ± .058		
		99-6	3/6	≧.50 ± .204		
		RC8	7/20	.350 ± .107		
OTCase	×	754	4/88	0.045 ± .022	99-6 754 OTC L1.28	0.84
		99-6	12/67	.179 ± .047		
		RC8	9/27	.333 ± .098		
		D2	16/40	.400 ± .077		
754	×	99-6	3/37	0.081 ± .045	RC8 99-6 754 OTC	5.30
		RC8	5/19	.263 ± .101		
		D2	18/62	.290 ± .098		
		782	9/33	.273 ± .077		
99-6	×	RC8	5/34	0.147 ± .061	D2 RC8 99-6 754	2.86
		D2	4/17	.235 ± .103		
		782	9/30	.300 ± .084		
		dic56	8/23	.348 ± .065		
RC8	×	D2	0/7	0.0 ± .125	782 D2 RC8 99-6	0.00
		782	2/12	.167 ± .087		
		dic56	2/12	.167 ± .087		
		Xg^a	5/14	.357 ± .128		
D2	×	782	3/16	0.187 ± .074	dic56 782 D2 RC8	0.50
		dic56	2/17	.118 ± .075		
		Xg^a	N.I.			
782	×	dic56	2/18	0.111 ± .073	Xg^a dic56 782 D2	3.71
		Xg^a	3/8	.375 ± .092		
dic56	×	Xg^a	0/6	0.0 ± .140		

in the final map. Examining pairwise cross data alone, we can often estimate a consistent marker order. However, the standard errors associated with each linkage distance are frequently such that several different orders are possible.

Gene order. To determine gene order, we began by considering the physical evidence for localization of our 21 DNA markers. As shown in Table 1, the physical order of many markers along the chromosome is known with sufficient resolution that order is left uncertain for only a subset of markers in certain regions. On the long arm, for example, the physical order is cen-DXYS1-S21-HPRT-52A-DX13 (*2*); this establishes a framework that can be used to order other loci on the long arm.

When physical methods were insufficient for resolving marker order, we relied on genetic evidence derived from multifactor crosses. These can be scored in progeny chromosomes from mothers simultaneously informative for three or more linked markers. Multifactor crosses are particularly powerful for determining gene order because they provide an internal accounting of crossover events among the loci being examined. If, for example, the order of three loci is A-B-C, then a crossover between A and B will reveal recombination between A and B and between A and C, but not between B and C, neglecting double exchanges. Likewise, a single crossover between B and C will show recombination between B and C and between A and C, but not between A and B (Fig. 2). The likelihood of the order A-B-C over the alternative orders can then be quantified on the basis that chromosome classes requiring double exchanges will be the least frequently observed. This logic can be extended to more than three markers, and confidence in gene order can be quantified in the same fashion.

Likelihood Calculations

The support for a gene order can be quantified by the method of maximum likelihood; the relative likelihood of each order is calculated at the most likely recombination fractions between the markers (*6*). These overall likelihoods can then be compared to determine the relative support for one gene order over another. Computerized systems for multilocus likelihood calculations, recently developed by Lathrop *et al.* (*7*), greatly facilitate these analyses. We have made extensive use of the linkage analysis program ILINK for this purpose (*7*). The ILINK program iteratively calculates the most likely recombination fraction for several loci considered jointly, taking into account information available from both pairwise and multifactor crosses. We have analyzed each successive set of four loci with ILINK. The program was run serially, each time specifying a different gene order consistent with known physical locations. This analysis produced a series of likelihoods, one for each marker order. ILINK analysis also provided the most likely genetic distances among the four loci; in consequence, we were able to build the map with the most likely parameters for both distance and order, based on markers taken four at a time.

Examination of markers four at a time allows the map locations for one group of four loci to be conditional on the information from the previous, overlapping group of four loci. For example, in a given set of four loci, the most likely order and distances between loci 1, 2, 3,

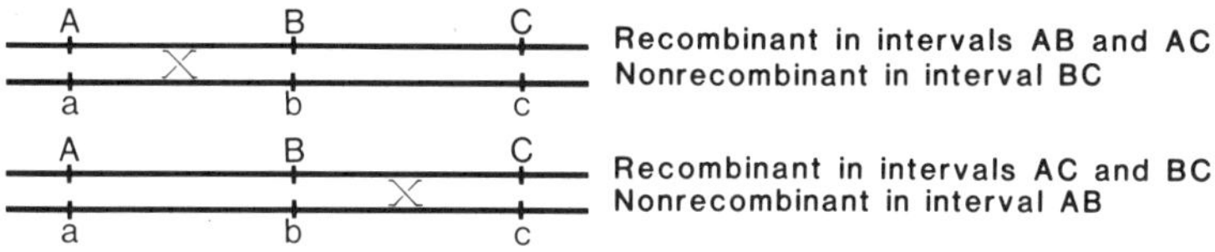

Fig. 2. Logic for definition of gene order when multifactor crosses are used. Gene orders other than A-B-C will require double exchanges to produce chromosomes that display this distribution of recombination and nonrecombination among the three markers.

and 4 were determined. The resulting distance between loci 2 and 3 was then fixed and used to determine the most likely order and distances for loci 2, 3, 4, and 5. The map produced by this technique integrates the information on marker order and marker distance for all 21 loci. The most distal and highly polymorphic locus, St14, was chosen as the zero point of the map.

For each set of four marker loci, the results of multifactor analysis with ILINK are listed in Table 2, which gives the most likely gene order within each successive group of four loci. Table 2 also lists the likelihood of that order when compared to the second most likely order. The second most likely order was usually that of the middle two loci inverted. In some cases, pairwise cross information on one interval was limited; either there were no informative meioses (such as in the case of 58 × L1.28), or we observed no recombinants in a small number of meioses (as in D2 × RC8). In these cases, the genetic evidence for relative order had to be based solely on two-factor cross information with neighboring markers. This relatively weak method of determining gene order results in limited support for a given order, and the order shown in Fig. 1 is in some instances based solely on physical location, as noted in Table 2. In these cases, the confidence in the gene order is presented with respect to the second most likely order that is consistent with the physical locations of those markers. Where the $\log_{10}$ of the odds for a unique order is less than 3, we must consider the most likely order as only tentative. However, this method does provide evidence as to where the map is strong and where it is weak.

In regions of the map in which genetic evidence for order was limited, loci were analyzed five at a time with ILINK to extract the most information possible from the genotypic data available. For example, the position of marker 754 was determined by analyzing 754 with RC8, 99-6, OTCase, and L1.28, and again with 99-6, OTCase, L1.28, and 58. The final map reveals that the human X chromosome is at least 185 recombination units in length; the short arm displays roughly the same genetic length as the long arm.

The results obtained in 38 normal Utah kindreds are generally consistent within the set. However, one instance represented striking inconsistency between families in the interval DX13 to St14. In seven families, no recombinants were observed in 47 meioses. In the eighth family, these two markers displayed three recombinants in seven meioses. A recombination fraction of 0 obtained in 47 meioses would place these two markers within 5.6 recombination units of each other (95 percent confidence interval). The probability of observing three recombinants out of seven if the true recombination fraction was 0.056 is less than 2.0×10^{-4}. It is difficult, however, to determine whether this inconsistency is the result of statistical fluctuation or of some fundamental heterogeneity.

Distribution of Crossover Events

The current data set contains 285 X chromosomes in the third generation—that is, chromosomes in which recombination can be scored. The average X chromosome in this set is informative for 7.3 markers, with a fraction of this set (105 chromosomes) informative for eight or more markers evenly distributed throughout the length of the chromosome. When this subset of well-marked chromosomes was examined for the number and distribution of recombination breakpoints, we found that chromosomes can undergo zero to four recombination events per meiosis, as illustrated in Fig. 3. At this level of resolution, the X chromosomes displayed a mean of 1.5 recombination events per chromosome, with 14 percent of the X chromosomes showing no recombination, 34 percent showing a single exchange, 35 percent showing two exchanges, 12 percent showing three exchanges, and 1 percent showing four exchanges. The distribution of crossover events shows deviation from the Poisson expectation which is significant at a level of 0.05, according to the χ^2 test. The deviation indicates that the number of recombination events on the X chromosome may be nonrandom.

Mapping Genetic Diseases

The current distribution of RFLP markers on the X chromosome is such that any X-linked disease locus will map within ten recombination units of at least two markers. Because of this proximity of markers, we should be able to locate disease loci whose ultimate genetic defect is unknown. Toward this end, we have studied linkage relationships among a number of DNA markers and several X-linked diseases.

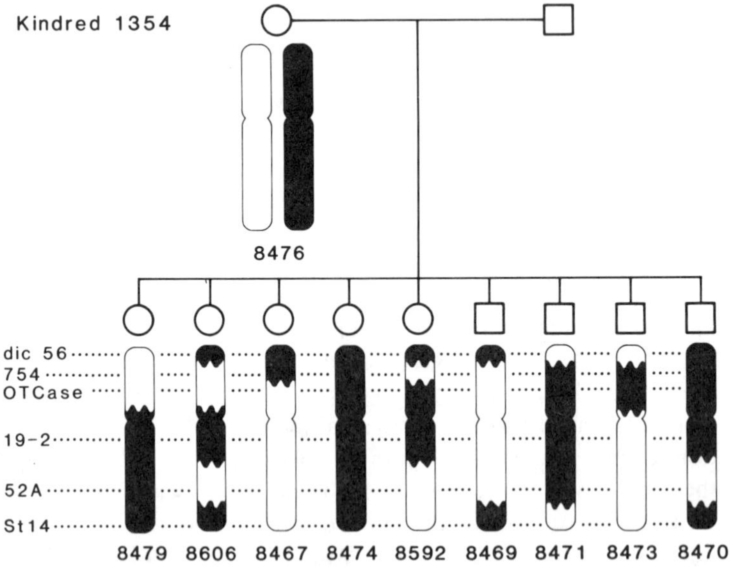

Fig. 3. Distribution of recombination breakpoints among X chromosomes in one family. The contribution of each maternal X chromosome toward the constitution of the X chromosomes in her progeny is shown by shading. Since the precise locations of recombination breakpoints between any two loci in the progeny X chromosomes are not known, the breakpoints are indicated by wavy lines.

For example, we sampled six families in Utah with Duchenne muscular dystrophy (DMD). Each family contained three or four generations. These six families provided information on the linkages between DMD and the markers 99-6, 754, OTCase, and 58. The most likely linkage distances and marker order, calculated with ILINK, indicate that the DMD locus lies between the markers 754 and 99-6, being 15 ± 6 (standard error of the mean) recombination units proximal to 99-6 and 21 ± 11 units distal to 754. Findings were similar in a recent collaborative study (*8*).

Disease families have also been studied for linkage between DNA markers and several other X-linked diseases, including X-linked retinitis pigmentosa (*9*), hemophilia A (*10*), X-linked retinoschisis (*11*), and choroideremia (*12*).

Using the Normal Map to Refine Disease Linkages

The linkage results obtained for the diseases Duchenne muscular dystrophy, hemophilia A, and X-linked retinitis pigmentosa indicate that a reasonable level of certainty of genetic location (and thus certainty of predictive value of the DNA markers) is possible if large numbers of families are available to build linkage data. Most X-linked diseases, however, are much less common than DMD, hemophilia A, or retinitis pigmentosa, and frequently occur in only one family or a few families (*13*). How can such diseases be accurately mapped and predicted?

We suggest that RFLP linkage data gathered in normal families can greatly improve the quality of linkage information in cases of rare genetic diseases. When only a few informative meioses are available, the linkage distances from RFLP markers to a disease locus will be characterized by a large variance. This variance can be constrained, however, by the relatively large amount of information available on the linkage distance between the RFLP markers themselves, obtained from normal families.

An example of the usefulness of the normal map is provided by our analysis of the limited amount of linkage information we obtained in the six Utah DMD kindreds. These families were genotyped at the loci RC8, D2, 99-6, 754, L1.28, and 58. The number of meioses informative for DMD and for each of the DNA markers ranged from none (for RC8 and L1.28) to 11 for 99-6; very little information on linkage between DNA markers was obtained. From these families alone, information on the pairwise linkage distances between each marker and DMD (calculated with LINKMAP) was very weak, as shown in Fig. 4A; the pairwise LOD scores for linkage to DMD did not exceed +0.40 for any DNA marker.

By contrast, the linkage database gathered on the same DNA markers in normal families is substantial. We took advantage of the information in the larger database by examining the genetic linkages between the DMD locus and the four closest loci—namely, RC8 and 99-6 distal, and 754 and OTCase proximal—pooling the DMD and normal family data. We analyzed the entire data set with ILINK to determine the most likely genetic distances between the loci taken five at a time. With normal and DMD data sets combined, the most likely position of the DMD locus was established between 99-6 and 754, with the distance from 99-6 to DMD now estimated at 9.1 ± 3.6 units, and the distance from DMD to 754 at 12.9 ± 3.5 units (Fig.

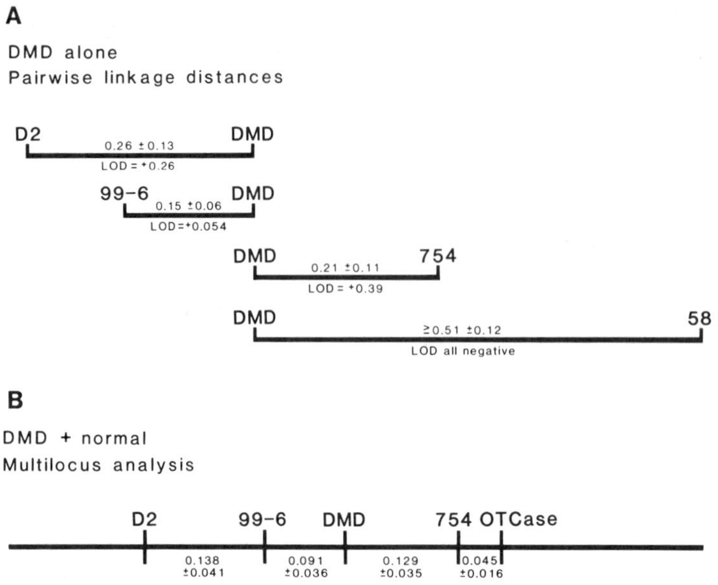

Fig. 4. Refinement of estimates of gene distance by combining the normal map with linkage information on DMD disease families. The numbers on the distance map represent the recombination fractions with standard errors. Placement of DNA markers proximal or distal to the DMD locus was based on physical mapping (*25*).

4B). This example demonstrates the potential usefulness of combining in a multifactor analysis the normal map with data from disease families in the effort to map rare genetic diseases accurately.

Discussion

A genetic map represents a description of an important part of the fundamental biology of an organism. In addition, it serves as a tool for the experimental study, and ultimately for experimental manipulation, of that organism. While technical and ethical considerations make experimental modification of the human genome problematic, the genetic map of the human X chromosome is of immediate value in the study and diagnosis of human X-linked genetic diseases.

The data set on which the X chromosome linkage map is based is sufficiently developed to allow efficient mapping of new markers. The average X chromosome in our set is currently informative for more than seven markers, and it will be possible to add new markers rapidly by observing their segregation in X chromosomes that contain defined crossover breakpoints. The genotypic data derived from this study, as well as DNA samples from family members, will be made available through the Centre d'Etude du Polymorphisme Humain in Paris, to assist in an international collaboration to construct a linkage map of the human genome.

Finally, the genetic map will provide information important to the task of spanning large regions of the chromosome by physical methods now being developed (*14*, *15*). The defined loci will provide essential reference points for studies that seek to "walk" along the X chromosome.

References and Notes

1. D. Botstein, R. White, M. Skolnick, R. Davis, *Am. J. Hum. Genet.* **32**, 314 (1980).
2. D. Hartley, K. Davies, D. Drayna, R. White, R. Williamson, *Nucleic Acids Res.* **12**, 5277 (1984).
3. J. Murray *et al.*, *Nature (London)* **300**, 69 (1982).

4. R. White *et al.*, *ibid.* **313**, 101 (1985).
5. C. Bridges and T. Morgan, publication of the Carnegie Institution of Washington, Washington D.C. (1923).
6. A. Edwards, *Likelihood: An Account of the Statistical Concept of Likelihood and its Application to Scientific Inferences* (Cambridge Univ. Press, Cambridge, England, 1972).
7. G. Lathrop, J. Lalouel, C. Julier, J. Ott, *Proc. Natl. Acad. Sci. U.S.A.* **81**, 3443 (1984).
8. E. Bakker *et al.*, *Lancet* **1985-I**, 655 (1985).
9. S. Bhattacharya *et al.*, *Nature (London)* **309**, 253 (1984).
10. J. Gitschier, D. Drayna, E. Tuddenham, R. White, R. Lawn, *ibid.* **314**, 738 (1985).
11. P. Wieacker *et. al.*, *Hum. Genet.* **64**, 143 (1983).
12. R. Nussbaum, personal communication.
13. V. McKusick, *Mendelian Inheritance in Man* (University Park Press, Baltimore, Md., ed. 6, 1984).
14. D. Schwartz and C. Cantor, *Cell* **37**, 67 (1984).
15. F. Collins and S. Weissman, *Proc. Natl. Acad. Sci. U.S.A.* **81**, 6812 (1984).
16. I. Oberle, D. Drayna, G. Camerino, R. White, J. Mandel, *ibid.* **82**, 2824 (1985).
17. D. Drayna *et al.*, *ibid.* **81**, 2836 (1984).
18. G. Camerino *et al.*, *ibid.*, p. 498.
19. J. Aldridge *et al.*, *Am. J. Hum. Genet.* **36**, 546 (1984).
20. R. Nussbaum, W. Crowder, W. Nyhan, C. T. Caskey, *Proc. Natl. Acad. Sci. U.S.A.* **80**, 4035 (1983).
21. D. Page *et al.*, *ibid.* **79**, 5352 (1982).
22. K. Davies *et al.*, *Nucleic Acids Res.* **11**, 2302 (1983).
23. R. Rozen, J. Fox W. Fenton, A. Horwich, L. Rosenberg, *Nature (London)* **313**, 815 (1985).
24. B. Bakker and P. Pearson, personal communication.
25. L. Kunkel, personal communication.
26. R. Race and R. Sanger, *Blood Groups in Man* (Blackwell, Oxford, ed. 6, 1975), p. 610.
27. Supported by a postdoctoral fellowship from the Muscular Dystrophy Association (to D.D.) and by a Howard Hughes Medical Institute investigator award (to R.W.). We thank investigators who generously agreed to share X chromosome DNA probes before publication, including L. Kunkel and J. Aldridge, P. Pearson and B. Bakker, and J.-L. Mandel and I. Oberle. We also thank F. Ziter for his role in gathering and sampling muscular dystrophy families and M. Hoff for excellent technical assistance. We thank P. Callahan and T. Elsner for help with linkage analysis and B. Ogden for help in sampling muscular dystrophy families.

5

Multiple Mechanisms of Protein Insertion Into and Across Membranes

William T. Wickner and Harvey F. Lodish

Each of the 20 membrane-limited compartments of a mammalian cell contains a particular set of proteins that enables it to carry out its specific functions. The accurate and swift delivery of each protein to its correct compartment is an important step in gene expression. Except for the few proteins made within mitochondria and chloroplasts, protein synthesis begins with the formation of polysomes in the cytoplasm. An important concept that has guided recent work is that protein localization is initiated by binding to a specific receptor on an intracellular membrane. For different organelles, binding occurs either while the protein is still growing on the ribosome or after it is completed. Subsequent translocation of the protein into or across the membrane requires an input of energy. In different cases this is provided either by a transmembrane electrochemical potential or by the folding of the protein during or after its translocation through the membrane. In many cases there is one or more "maturation" steps involving covalent modifications or folding on the opposite membrane surface.

Different integral membrane proteins—those bound to the phospholipid bilayer by hydrophobic interactions—can have distinct asymmetric structures (Fig. 1). A protein may span the membrane once or several times, with the NH_2- and COOH-terminus on either side of the bilayer. Our ideas of membrane assembly necessarily rest on our knowledge of the complexities of the structures of these proteins. Surprisingly different answers have emerged for various membrane proteins, organelles, and organisms, confounding early global hypotheses (*1–3*) and raising the need for better

Science 230, 400–407 (25 October 1985)

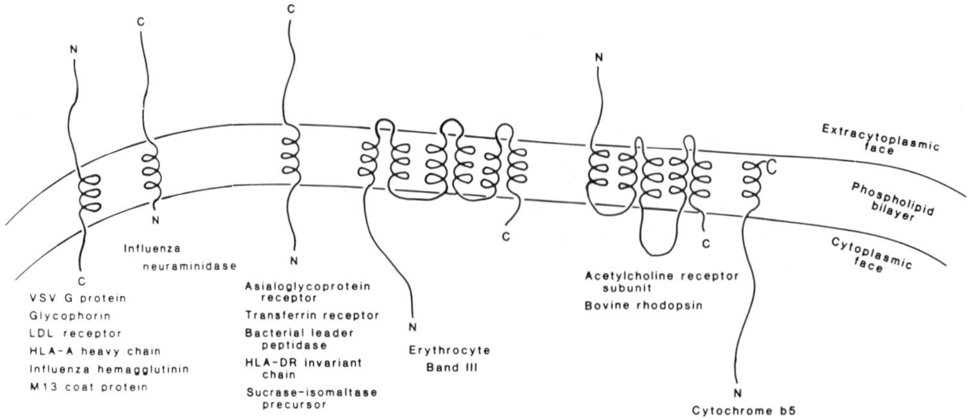

Fig. 1. Topologies of integral membrane proteins. Segments of the chain within the bilayer are depicted as helices. Extramembrane regions are drawn as lines, and no attempt is made to depict the folding of these segments of the proteins. References for the amino acid sequences and transmembrane topologies are: VSV G protein (*103*); glycophorin (*104*); LDL (low-density lipoprotein) receptor (*105*); HLA-A heavy chain (*106*); influenza hemagglutinin (*107*); M13 coat protein (*108*); influenza neuraminidase (*109*); asialoglycoprotein receptor (*28, 111*); transferrin receptor (*112*); bacterial leader peptidase (*68*); HLA-DR invariant chain (*27*); acetylcholine receptor subunit (*41*); cytochrome b5 (*113*); erythrocyte Band III (*37, 114*). Although only six membrane-spanning regions are drawn, recent data on the sequence of the entire Band III messenger RNA (*37*) indicates that there are 12 stretches of hydrophobic residues of length sufficient to span the membrane. Initial work suggested that the precursor of sucrase-isomaltase spanned the plasma membrane twice, with both the NH_2- and COOH-termini remaining exoplasmic (*110*). However, the complete complementary DNA sequence, as well as other data, indicate only a single membrane-spanning segment, with the NH_2-terminus facing the cytoplasm (*115*). N, NH_2-terminus; C, COOH-terminus.

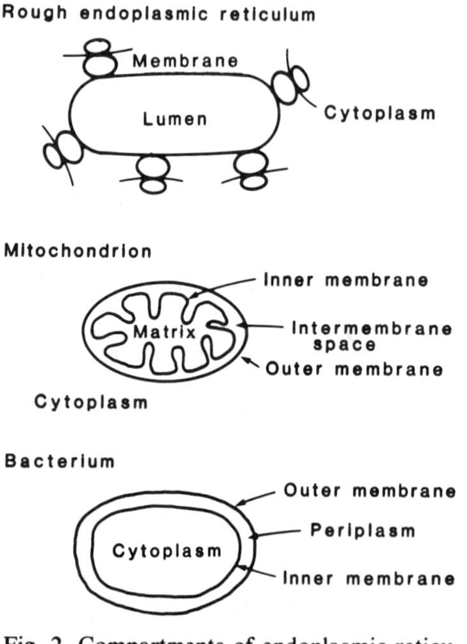

Fig. 2. Compartments of endoplasmic reticulum, mitochondria, and bacteria.

definition of fundamental mechanisms. In this article we focus on protein assembly into the endoplasmic reticulum, mitochondria, and the bacterial cell surface (Fig. 2). We do not discuss other assembly-competent organelles such as chloroplasts, peroxisomes, and the nuclei, nor will we include organelles (such as Golgi or lysosomes) that are derived by membrane fission and fusion.

Secretory Proteins and the Endoplasmic Reticulum

Classic studies by Palade and his colleagues (*4*) showed that secretory proteins are first found within the lumen of the rough endoplasmic reticulum (RER). Sealed fragments of the isolated organelle (rough microsomes) carry polysomes, which are highly enriched for secretory proteins and capable of chain completion in vitro (*5–7*). Nascent chains, emerging from the ribosome, cross the membrane (*8, 9*). In 1972, Milstein *et al.* (*10*) showed that a cell-free reaction, programmed by myeloma messenger RNA, synthesized a larger form of immunoglobulin light chains than was finally secreted from the cell. They proposed that this extra piece might be NH_2-terminal and serve as a "signal" to direct secretion. Microsomes from these myeloma cells made mature-length protein, demonstrating that the signal sequence was cleaved during translocation into the RER. In 1975, Blobel and Dobberstein (*1*) showed that polysome-free dog pancreas microsomes would sequester newly made precursor of the immunoglobulin κ light chain and proteolytically process it to its mature molecular weight if the microsomes were present during protein synthesis. "Signal sequences" were discovered at the NH_2-termini of many nascent secretory proteins (*11*). They consist of 16–26 residues and have a polar, basic NH_2-terminus and a central, apolar domain (*12, 13*).

Rothman and Lodish (*14*) used a synchronized cell-free synthesis of vesicular stomatitis virus (VSV) G protein to show that translocation requires membranes early in nascent chain growth. Walter and Blobel (*15*) and Meyer *et al.* (*16*) isolated two receptor proteins, the signal recognition particle (SRP) and the docking protein (DP), which coordinate the synthesis of nascent pre-secretory proteins (with an NH_2-terminal signal peptide) with their insertion into microsomes (Fig. 3A). SRP is a complex of six polypeptides and 7*S* RNA (*17*). It binds to polysomes making pre-secretory or pre-membrane proteins and causes arrest of chain growth after approximately 80 residues. This is only a few more than the 40 residues buried within the ribosome plus the 25 (approximately) that comprise the average signal peptide. The SRP-polysome complex binds to DP, a 72,000-dalton integral membrane protein of the RER (*16*). Upon binding, the SRP is released and the polysome resumes chain elongation (Fig. 3A). The growing polypeptide chain passes through the endoplasmic reticulum membrane and into the lumen, where the signal peptide is cleaved and core glycosylation occurs.

In contrast to our present understanding of SRP-DP, less is known about the mechanism of polypeptide translocation through the apolar center of the endoplasmic reticular membrane. It is not known whether translocation is directly through lipid (*2*), as shown in Fig. 3, or through a proteinaceous pore. Unlike translocation in mitochondria and bacteria, translocation into the RER does not require a transmembrane electrochemical potential. We propose that binding of the complex of SRP and nascent chain to the DP on the endoplasmic reticulum membrane would cause the membrane insertion of the signal sequence and the segment of amino acids adjacent to its COOH-terminus as a helical hairpin. SRP and DP would be dislodged as part of their recycling. Signal sequence cleavage of secretory proteins and many transmembrane glycoproteins occurs on the extracytoplasmic face of the membrane (Fig. 3A). This model incorporates features proposed previously by several investigators. Receptors to target nascent proteins as well as signal and "stop-transfer" or "membrane-anchor" sequences, and the idea that the information for the orientation of each membrane protein is contained in discrete, short sequences of the polypeptide chain

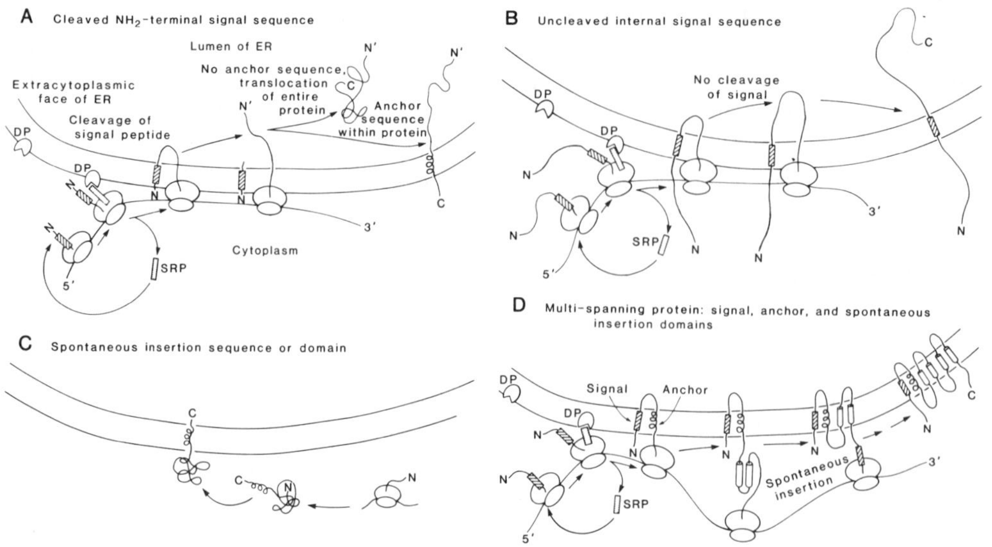

Fig. 3. A model for cotranslational insertion of membrane and secreted proteins into or through the endoplasmic reticulum membrane. (A) Secreted proteins and VSV G protein. Binding of the complex SRP and nascent chain to the DP on the endoplasmic membrane would cause the insertion into the endoplasmic reticulum membrane of the signal sequence (hatched box) and the segment of amino acids adjacent to its COOH-terminus as a helical hairpin. After the nascent chain is cleaved, continued extrusion across the endoplasmic reticulum membrane would generate a secretory protein or, if there is an anchor (or stop-transfer) sequence, a single-spanning transmembrane protein with the same conformation as the VSV G protein. (B) Asialoglycoprotein receptor. If the signal sequence is uncleaved and there is no anchor sequence, continued growth and translocation of the nascent chain across the endoplasmic reticulum membrane would generate a protein with its NH_2-terminus facing the cytoplasm and its COOH-terminus in the lumen of the RER. (C) The synthesis of proteins such as cytochrome b5 occurs on cytoplasmic polysomes. The completed protein then inserts spontaneously into the RER, without mediation of SRP or DP, by means of an insertion sequence or domain. (D) Multi-spanning membrane proteins. The first helical hairpin could result from a combination of signal and anchor sequences; subsequent helices could fold against each other, forming a domain that would insert spontaneously as the peptide grows in the cytoplasm. The cylinders represent a possibly alternative structure of a membrane-spanning α-helix.

that act independently, are from the signal hypothesis. (*1, 3, 18*). The membrane trigger hypothesis (*2*) emphasized assembly information in the mature sequence, the importance of protein folding, and direct, hydrophobic interaction between the polypeptide chain and the hydrocarbon core of the membrane bilayer. The importance of the energetics of transfer of polypeptide domains between water and hydrocarbon was emphasized in the direct transfer model (*19*) and the helical hairpin hypothesis (*20*). The loop models (*21, 22*) and the helical hairpin hypothesis proposed that pairs of helices insert into the membrane.

Synthesis of Membrane Proteins on the Endoplasmic Reticulum

Biosynthesis of single-spanning membrane proteins whose NH_2-termini face the extracytoplasmic face is similar to that of secretory proteins (Fig. 3A). These membrane proteins contain a sequence of about 20 hydrophobic amino acids that anchors the growing polypeptide in the phospholipid bilayer. For example, Yost *et al.* (*23*) and Guan and

Rose (*24*) have shown that the apolar sequences at the COOH-terminus of immunoglobulin M (IgM) heavy chain and the VSV G protein stop translocation across the endoplasmic reticulum. These long stretches of hydrophobic amino acids could bind so tightly to the fatty acid core of the membrane that continued extrusion of the protein would be blocked; thus it would be considered a stop-transfer sequence. As the ribosomes continued elongation of the nascent chain, the newly added residues would remain on the cytoplasmic face. However, in some proteins long stretches of hydrophobic amino acids are translocated entirely across the bilayer (*25*); recognition of a stop-transfer sequence might also require an appropriate "receptor" protein in the endoplasmic reticulum.

Membrane proteins such as erythrocyte Band III (*26*), the histocompatibility antigen (HLA)-DR-associated invariant chain (*27*) and the asialoglycoprotein receptor (*28*), as well as a few secreted proteins such as ovalbumin (*29*) are synthesized on the endoplasmic reticulum, yet are not cleaved during insertion. This led to the concept of an internal, uncleaved signal sequence (*3, 30*), namely a sequence that could be recognized by the same proteins [except the leader (signal) peptidase] and would perform all the same functions as its cleaved counterparts. Studies with ovalbumin (*26, 29*) indicate that it has an uncleaved signal sequence near its NH_2-terminus and have illustrated the difficulty in experimentally establishing the location of such an uncleaved signal sequence. One criterion is that such sequences be required for the insertion of a protein into microsomes (*31, 32*). Bos *et al.* (*33*) have provided the clearest demonstration of an uncleaved signal.

Uncleaved signal peptides, together with anchor sequences and other internal membrane insertion domains, can generate the complex topologies of different types of membrane proteins. For example, the bulk of the influenza virus neuraminidase is extracytoplasmic and is anchored to the membrane by a hydrophobic sequence at its NH_2-terminus. Neuraminidase does not undergo endoproteolytic cleavage during its biosynthesis, and the NH_2-terminal hydrophobic domain functions both as a signal and a membrane anchor sequence. Genetic fusion of this NH_2-terminal domain to a different viral glycoprotein that had its NH_2-terminal (cleaved) signal deleted restored translocation of this glycoprotein (*33*). This suggests that neuraminidase has an NH_2-terminal uncleaved signal sequence, and that all signal sequences, whether or not cleaved, function similarly in initiating translocation across the endoplasmic reticulum membrane. Similar studies (*34*) on the asialoglycoprotein receptor showed that the membrane spanning segment of 21 hydrophobic amino acids also functions as a signal sequence; SRP is required for its insertion into the RER. In Fig. 3B, we show how insertion of this protein might occur; as for secretory proteins, the signal peptide and the adjacent segment of the protein are postulated to insert as a helical hairpin (*20*).

Proteins such as cytochrome b5 are anchored to the membrane by apolar residues at the COOH-terminus, are synthesized on free polysomes, and probably insert into the endoplasmic reticulum membrane posttranslationally (Fig. 3C). SRP is not required for insertion of cytochrome b5 (*35*).

As proposed by Blobel (*3*), multi-spanning proteins such as sucrase-isomaltase and Band III could achieve their final topology by a succession of internal signal sequences and membrane-anchoring stop-transfer sequences. Hydrophobic side groups of the amino acids project outward to interact with the apolar fatty acyl core of the bilayer. However, many transmembrane sequences in multi-spanning proteins such as bacteriorhodopsin (*36*), Band III (*37*), or the acetylcholine receptor (*38*) have polar residues, which are not seen in the intramembrane segments of single-spanning proteins like VSV G protein, the asialoglycoprotein receptor, or M13 coat protein. These membrane-spanning helices may be amphipathic, with charged or polar residues confined to one face of the helix. Polar faces of several adjacent sequences could, in the mature protein, form a polar "pore" or "channel" through the membrane (*36, 37, 39–41*). These amphipathic helices may not be hydrophobic enough to function as simple signal or stop-transfer sequences. The proteins might utilize only one signal sequence, that would catalyze insertion of the most NH_2-terminal helical hairpin of the nascent chain into the endoplasmic reticulum membrane, and one membrane anchor. As the nascent chain continued to grow in the cytoplasm, domains with hydrophobic surfaces could form and insert spontaneously into the membrane (*2*) without involvement of SRP and DP (Fig. 3C). Several helices could associate with each other to shield their polar surfaces and form an "insertion domain" that could spontaneously insert into the phospholipid bilayer, presenting only an apolar face to the fatty acyl side chains of the phospholipids. Recombinant DNA techniques should allow the construction of novel membrane proteins that will test these concepts.

Demonstration of the close coupling between protein synthesis and membrane translocation in the endoplasmic reticulum has come from studies in a crude reticulocyte or wheat germ cell-free translation reaction supplemented with dog pancreas microsomes. Mutants, specific drugs, and in vivo studies, perhaps in a microorganism such as yeast, will be important to confirm or modify the current picture and to help in dissecting the crucial membrane translocation step.

Mitochondria

Mitochondria differ from the endoplasmic reticulum in almost every aspect of their biogenesis (*41, 42*). Except for the few proteins encoded by the mitochondrial DNA (*43*), all mitochondrial proteins are specified by nuclear genes. Each is synthesized in the cytoplasm and imported to one of the four mitochondrial compartments; outer membrane, intermembrane space, inner membrane, or matrix (Fig. 2). Isolated yeast and *Neurospora* mitochondria specifically imported only mitochondrial proteins from a cell-free translation reaction (*44*). Each of the proteins examined went to its correct compartment, and the uptake was just as efficient when the mitochondria were added posttranslationally to the protein synthesis reaction as when they were present throughout the reaction (*45*). In vivo pulse-chase studies in *Neurospora* mycelia and yeast (*46, 47*) have shown that mitochondrial proteins pass through a cytoplasmic pool prior to binding to the organelle. Isolated mitochondria have polysomes on their outer surfaces, and these polysomes are highly enriched in nascent mitochondrial preproteins (*48*). However, Suissa and

Schatz (*49*) showed that these represent only a small fraction of the polysomes for any given protein and that the proportion is governed simply by the rate of protein synthesis. Protein uptake into the matrix, inner membrane, or (except for cytochrome c) into the intermembrane space requires an electrochemical potential across the inner membrane (*46, 50*).

Many mitochondrial proteins are synthesized with a transient NH_2-terminal leader peptide, while others are made and imported without cleavage (*51*). Mitochondrial leader peptides (*52*) are basic and have a different sequence pattern from pre-secretory proteins. Removal of mitochondrial leader peptides is catalyzed by a soluble matrix protease that has a specificity distinct from its counterpart in the endoplasmic reticulum or in *Escherichia coli* (*53*) (Fig. 4). In some cases, for proteins of the outer surface of the inner membrane or of the intermembrane space, the pre-sequence is removed by two successive proteolytic cleavages, one of which is catalyzed by the matrix protease (Fig. 4) (*50, 54*). Precursors of these proteins are thought to span the inner membrane transiently, with their NH_2-termini facing the matrix space. This model would explain the puzzling requirement for an electrochemical potential across the inner membrane for translocation of several proteins ultimately located in the intermembrane space. As shown in Fig. 4, the second cleavage of these proteins is thought to occur at the outer surface of the inner membrane (*52*).

Mitochondrial pre-proteins use several outer membrane receptors for import (*52, 57*). Cytochrome c, a protein of the intermembrane space, is made as a precursor (apocytochrome) without the heme group and without a cleaved presequence. Uptake requires heme addition. Microgram amounts of apocytochrome (but not holocytochrome) block uptake of radioactive apocytochrome c, but not uptake of inner membrane or matrix proteins, suggesting the involvement of a specific receptor (*58*). A chimeric protein with 350 amino acids of the β subunit of yeast F1 adenosine triphosphatase at its NH_2-terminus and a large portion of β-galactosidase at its COOH-terminus can be inserted into yeast mitochondria. This is also true for a gene fusion of the NH_2-terminal 53 amino acids of the precursor to subunit IV of cytochrome oxidase and a different cytosolic enzyme, demonstrating that part of the NH_2-terminus is sufficient to target a protein to the mitochondrion (*59*). Recently, several laboratories have reported that a soluble, cytosolic protein fraction is also necessary for import (*60*) (Fig. 4).

The outer membrane of mitochondria is distinct from the three other compartments in its biogenesis. Outer membrane proteins do not have cleaved leader sequences (*61*), and their membrane insertion does not require an electrochemical potential across the inner membrane. All of the information for targeting and anchoring the 70,000 molecular weight outer membrane protein is contained within the NH_2-terminal 41 amino acids (*62*). Like the proteins of the internal mitochondrial compartments, assembly into the outer membrane is not coupled to translation.

Several fundamental questions of mitochondrial biogenesis are unanswered. What is the role of the electrical potential? Does import into the matrix involve separate steps of traversing the inner and outer membrane or, as illus-

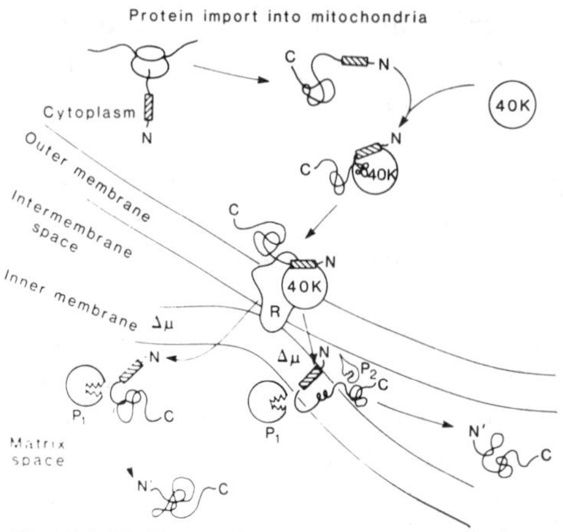

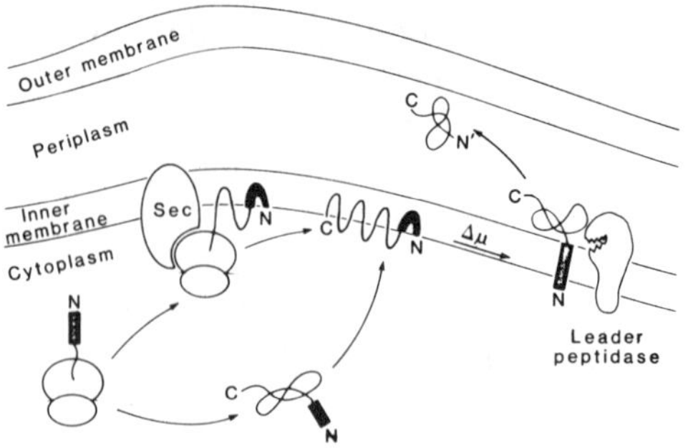

Fig. 4 (left). Protein import into mitochondria. The import of a matrix protein and an intermembrane protein with cleaved leader sequences is depicted. The function of the cytoplasmic import element, a 40-kilodalton protein (*60*), is not known. Proteins of the intermembrane space may partially insert across the inner membrane and undergo two-step proteolysis. R represents the outer membrane receptors for pre-mitochondrial proteins. P_1 and P_2 are proteases that remove the mitochondrial leader peptide. Fig. 5 (right). Assembly of the bacterial cell surface. The leader peptide is indicated by a hatched rectangle. As discussed in the text, many (but not all) pre-proteins require functional *sec* genes for export. Proteins reach critical molecular weight (*78*) or full length prior to beginning potential-dependent translocation.

trated in Fig. 4, does this occur at adhesion zones between the two membranes as previously suggested (*41, 42*)? What is the number of outer membrane receptors, how do they catalyze translocation, and what other proteins are needed? Genetic identification of elements needed for import (*63*) may be vital to answering this question.

Bacterial Cell Surface

The cell surface of Gram-negative bacteria consists of three layers: the plasma membrane, an aqueous periplasm, and the outer membrane (Fig. 2). The mechanisms of membrane assembly and protein secretion in bacteria include those found in both mitochondria and RER. As in import of proteins into mitochondria, translocation of bacterial proteins across the plasma membrane requires a transmembrane electrochemical potential and, in different cases, can occur cotranslationally or after translation is complete. However, leader sequences of bacterial membrane and of exported proteins closely resemble those of the eukaryotic RER in structure (*12, 64*). Protein products of the bacterial *sec* (secretion) genes may, in some respects, have a function similar to that of the RER SRP. There is selective arrest of the synthesis of secretory proteins in *sec*C mutants (*65*), even though translocation occurs late in translation or posttranslationally (as discussed below).

All known proteins of the periplasm and outer membrane, and at least several of the inner membrane, are made with NH_2-terminal leader (signal) sequences (*66*). The majority of inner membrane proteins are made without a cleaved leader sequence (*67, 68*). Bacterial leader sequences are processed by a membrane-bound leader peptidase whose active site is on the periplasmic face of the plasma membrane (*68*) and whose substrate specificity is identical to that of the RER enzyme (*69*). However, unlike the

RER, efficient bacterial protein export requires the membrane electrochemical potential (*70*). In this regard, bacterial export resembles the mitochondrial import process. Bacteria and mitochondria are also similar in the relative timing of protein synthesis and protein translocation across a membrane.

There has, until recently, been some confusion over whether bacterial proteins must begin crossing the plasma membrane early in their synthesis in order to be exported, as is seen in the RER, or whether their export is not coupled to polypeptide chain growth, similar to posttranslational import by mitochondria. Much of this confusion is semantic. "Cotranslational" refers to any event that occurs before the end of translation. However, it has been confused with an obligate coupled extrusion of the nascent chain through the bilayer as it emerges from the ribosome, as occurs in the eukaryotic RER. Such extrusion is apparently not seen in bacterial protein export, yet there are important cotranslational events, such as folding and interaction with *sec*-encoded proteins, which are necessary for later translocation.

Bacterial membrane fractions are enriched in polysomes coding for certain pre-secretory proteins (*71*). Davis and co-workers found that nascent chains of alkaline phosphatase could be labeled by a membrane-impermeant radioactive reagent added to intact cells or spheroplasts (*72*). M13 procoat, synthesized in a cell-free reaction (*73*), was shown to assemble efficiently into plasma membrane vesicles present during the synthesis of the protein (*74*). However, other experiments have shown that bacterial protein export normally does not occur early in the growth of the polypeptide chain. Ito *et al.* (*75*) observed that the pulse-labeling of the periplasmic or outer membrane proteins in intact cells was dramatically delayed relative to the labeling of cytoplasmic or inner membrane proteins. The export of β-lactamase to the periplasm is entirely posttranslational (*76*), as is the insertion of M13 procoat protein into the plasma membrane in vivo (*77*). With the discovery that a membrane potential is needed for export, it was possible to experimentally separate the synthesis of both pro-OmpA and M13 procoat protein from their translocation (*70*).

These apparently contradictory observations, membrane-bound polysomes and membrane-spanning nascent chains on the one hand and two-step export on the other, were resolved by the experiments of Randall (*78*). She demonstrated that nascent chains grow to at least 80 percent of their final size before they begin translocation through the plasma membrane. This critical chain length is different for each exported protein. Translocation does not then begin synchronously at the critical molecular weight, but, instead, a stochastic "race" ensues between polypeptide chain growth and protein translocation. Thus ribose-binding protein and pre–β-lactamase are exported entirely posttranslationally, while only a fraction of the other protein species are exported cotranslationally. The remainder are completed as full-length pre-proteins within the cell, then are translocated entirely posttranslationally (*79*). These data are consistent with membrane-bound polysomes and with membrane-spanning nascent chains, yet clearly show that translocation is not strictly coupled to polypeptide chain elongation in bacteria.

Our current concept of protein export in bacteria is shown in Fig. 5. As with uptake of proteins into mitochondria, more than one receptor system targets different proteins to the bacterial plasma membrane. Mutations in any of several *sec* genes abolish export of a number of inner membrane, periplasmic, and outer membrane proteins but do not affect export of other proteins to these compartments (*80*). Since nascent pre-secretory proteins are not extruded through the bacterial membrane as they emerge from the ribosome, as in the RER, the *sec* proteins may serve functions distinct from those of SRP and DP. The *sec* proteins might stabilize certain pre-proteins until they can begin translocation. Alternatively, as illustrated in Fig. 5, they may bring specific proteins to the membrane early in their synthesis, analogously to SRP and DP. In bacteria, this may not lead to immediate translocation but may allow the polypeptide to grow at the interface between the aqueous cytoplasm and the apolar membrane.

Steps of Protein Translocation

Protein translocation in bacteria, mitochondria, and RER are summarized in Table 1. There are clearly no universal themes, or even completely consistent groupings of export themes. For example, bacterial pre-proteins have leader sequences like those of the endoplasmic reticulum but require a potential, as is seen for mitochondrial import. A new framework is needed for coherent organization of our knowledge of protein insertion into and across membranes. We suggest that the common features of protein translocation are its three necessary steps, the association of the protein with receptors on the correct membrane, the translocation through the membrane, and covalent modifications and folding on the opposite membrane surface. Individual proteins have evolved to use different combinations of the translocation themes to accomplish each step in export.

Protein binding to receptors on the correct membrane is essential to provide accurate protein sorting within the cell. In eukaryotic cells, recognition is mediated by specific soluble and organelle-bound elements. Bacteria, in which all exported proteins initially cross the same membrane, may only require sorting after translocation is complete. In addition to sorting, the binding step may stabilize

Table 1. Characteristics of the three steps

Assembly system	Recognition step		
	Signal sequence	Soluble receptor	Membrane receptor
Endoplasmic reticulum	RER/bacterial type	Signal recognition particle	Docking protein
Mitochondria			
Outer membrane	None		
Inner compartments	Distinct structure	40K protein	Distinct receptors in OM
Bacteria			
Inner membrane	Usually none	*sec*C*	*sec*Y*
Outer compartments	RER-bacterial type	*sec*C*	*sec*Y*

**Sec* gene function is required for approximately half of the exported proteins.

pre-proteins against denaturation or possibly folding into a "dead-end" structure. This may be one function of the bacterial *sec* genes and of the mitochondrial cytoplasmic assembly component. This may even be viewed as a role of the SRP, which prevents elongation of the nascent chain unless DP is present. Receptors must facilitate translocation by either stabilizing the protein, catalyzing its refolding into a competent conformation, or transferring it to other elements that catalyze translocation.

Leader peptides of bacteria and RER (*12, 13*) average approximately 23 residues in length and have three characteristic domains. The NH_2-terminal domain is short (1–5 residues), basic, and polar. The central domain is nonpolar and contains a "core" of 4–8 strongly hydrophobic residues. Mutations that alter the charge at the NH_2-terminus (*81*) or introduce charged residues into the apolar domain (*82*) strongly inhibit protein export. The third domain begins with a helix-breaking residue (usually proline) and has small residues, characteristically glycine or alanine, at positions −3 and −1 relative to the cleavage site. These residues are not essential for translocation but serve as a leader peptidase recognition site (*76, 83*). Despite these conserved features, there is no true conservation of sequence.

The leader peptide is clearly essential for export (*84, 85*). It is not clear whether the leader is sufficient to specify export, that is, whether the match between leader sequence and the mature protein is critical, or whether part of the information for export lies in the sequence of the rest of the protein. Despite the similarities of the leader sequences, the answers to this question may be different for bacterial protein export and for protein secretion into the lumen of the RER. In bacteria, the fact that proteins grow to 80 percent of their final molecular weight before beginning translocation across the plasma membrane (*78*) suggests that there is information late in the protein sequence that is needed for secretion. This idea is supported by genetic studies; a fusion protein of the *lam*B leader joined to β-galactosidase is not secreted across the plasma membrane (*86*), while a fusion product that contains virtually the entire *lam*B protein is efficiently secreted (*87*). A fusion of the β-lactamase leader sequence to a foreign cytoplasmic protein (*85*) produced a chimeric protein that also failed to be secreted. Lipopro-

of protein insertion into membranes.

Translocation step		Maturation step	
Membrane potential	Translation-coupled	Proteolysis	Other
None	Yes	Signal peptidase	Core glycosylation
None	No	No	None
Required	No	One- or two-step matrix protease	Addition of prosthetic group (heme)
Required	No	Leader peptidase (ER-like)	
Required	No	Leader peptidase and lipoprotein signal peptidase	Coupling to peptidoglycan

tein was still secreted when its leader sequence was replaced by that of OmpF (*88*); however, replacement of the rat pre-proinsulin leader sequence with the leader sequence from pre–β-lactamase inhibited its secretion from *E. coli* (*89*). Mutations have been described in the mature portions of bacterial prolipoprotein (*90*), M13 procoat (*91*), and pre-maltose binding protein (*92*) that affect their export. On balance, it seems that the leader sequence is necessary, but not sufficient, for bacterial protein export.

In contrast, Yost *et al.* (*23*) have shown that the leader sequence of pre–β-lactamase, fused to the membrane-spanning segment of IgM and the COOH-terminal portion of globin, will direct the insertion of the hybrid protein into dog pancreas microsomes. A chimeric protein consisting of the NH_2-terminus of IgM heavy chain or of the asialoglycoprotein receptor (containing the internal signal) and the COOH-terminus of globin is completely translocated across the endoplasmic reticulum (*23*, *34*). This suggests that insertion into the RER does not require information from the mature protein sequence. This difference between bacterial secretion and that in the RER is in accord with the differences in coupling between translation and translocation in these two systems.

The translocation step is not as well understood. Eukaryotic ribosomes appear to form a very tight junction with the RER membrane, and pre-secretory proteins may never contact the cytoplasm or fold prior to translocation. Thus, translocation of secretory proteins may be insensitive to the exact sequence of the polypeptide. In contrast, stop-transfer and insertion sequences are important information in the mature region of membrane proteins. Proper folding of soluble, secreted proteins that are synthesized on the RER may be essential for completion of translocation. Immediately after completion of synthesis, ribosomes release the nascent chain and dissociate into subunits. This leaves the COOH-terminal 25–35 amino acids (those formerly embedded in the large ribosomal subunit) exposed on the cytosolic face, and approximately 20 amino acids spanning the endoplasmic reticulum membrane. Since translocation of these last 45–55 residues cannot be coupled to chain elongation, some other process must provide a driving force. This could be the folding of the rest of the chain on the lumenal side of the membrane. This may be responsible for the SRP-independent translocation of short proteins such as the 70-residue precursor of bee venom mellitin (*93*) across microsomal membranes. The translation of this protein is virtually complete by the time the entire signal sequence has emerged from the ribosome; its assembly into the microsome must therefore be essentially posttranslational.

What are the energetics of translocation and what is the role of the electrical potential? While answers to these questions must await further experiments, several facts are noteworthy. Bacterial pre-proteins are exported in a direction from the negative to positive with respect to the transmembrane electrochemical potential, while the opposite is true for the import of mitochondrial pre-proteins. If the mechanism underlying these requirements is the same, then it becomes difficult to envision a simple electrophoretic model. A mutant of M13 procoat has been described which, while unaltered in net charge in the translocated region, displays dramatically less dependence on the electrochemical potential for export (*94*). Bakker and Randall (*95*) showed that the chemical portion of the potential can substitute for the electrical component in driving bacterial export. This also casts doubt on simple electrophoretic models. Other possible roles for the potential include affecting the lipid structure, governing the concentration of other critical solutes, or even driving a protein-proton transport system. The well-studied voltage-dependent translocation of diphtheria toxin (*96*), mellitin (*97*), and asialoglycoprotein receptor (*98*) across lipid bilayers may be analogous to the translocation of preproteins; translocation of premitochondrial proteins and bacterial pre-secretory proteins need to be assayed in these systems.

When translated in a cell-free extract, the human erythrocyte glucose transporter can insert into the endoplasmic reticulum entirely posttranslationally (*99*). This glycoprotein probably spans the membrane as 12 α-helixes (*100*) and bears a single *N*-linked oligosaccharide. The observation that its insertion into the endoplasmic reticulum membrane and its glycosylation requires SRP, but not concommitant translation, indicates that the binding and translocation steps need not be obligatorily coupled even in the RER. Its insertion thus resembles that of bacterial membrane proteins, except that there is no obvious requirement for a membrane potential.

The maturation step, which follows translocation, may be essential to the operation of additional sorting steps. Import of apocytochrome c across the mitochondrial outer membrane requires heme addition in the intermembrane space. When heme addition is blocked, the apocytochrome c remains bound on the outer mitochondrial surface (*55*). Explanations other than the reversibility of the insertion steps in the absence of maturation are, of course possible. For example, proteins catalyzing translocation might also need to donate pre-proteins to the appropriate maturation enzyme in order to catalyze another translocation event.

Prospectus

Further progress will depend on: (i) genetic and in vivo studies to define the physiological pathways and provide strains that are optimal for biochemical analysis, (ii) development of specific drugs to interrupt the pathway, reveal intermediates, and assist studies of enzymology, and (iii) analysis of cell-free reactions that are amenable to fractionation and reconstitution from their purified components. Bacterial export has benefited from intensive genetic study, while this approach is only beginning (in yeast) to be used in investigations of translocation into the RER and for mitochondrial biogenesis. Cloning has allowed the isolation of substantial quantities of the bacterial leader peptidase (*101*); it will allow preparation of large quantities of other catalysts of protein translocation in the near future. Mutants have been isolated to test the functions of different domains of pre-secretory and mitochondrial proteins. Pulse-labeling of a microorganism such as yeast may allow detection of predicted complexes, such as cytoplasmic SRP-polysomes, and may reveal new intermediates.

Cell-free translocation reactions (*102*) may provide assays for the products of (*sec*) and protein localization (*prl*) genes (*64*) and allow the study of the role of the electrical potential. Submitochondrial translocation across isolated inner or outer membrane has not yet been reported, nor has a soluble detergent extract of RER been reconstituted to yield a translocation-competent liposome. Each of these cell-free reactions is the focus of intensive research and will provide further insights into the molecular mechanisms of translocation.

References and Notes

1. G. Blobel and B. Dobberstein, *J. Cell Biol.* **67**, 835 (1975).
2. W. Wickner, *Annu. Rev. Biochem.* **48**, 23 (1979).
3. G. Blobel, *Proc. Natl. Acad. Sci. U.S.A.* **77**, 1496 (1980).
4. G. Palade, *Science* **189**, 347 (1975).
5. S. J. Hicks, J. W. Drysdale, H. N. Munro, *ibid.* **164**, 584 (1969).
6. M. C. Ganozo and C. A. Williams, *Proc. Natl. Acad. Sci. U.S.A.* **75**, 2737 (1969).
7. C. M. Redman, *J. Biol. Chem.* **244**, 4308 (1969).
8. ——— and D. D. Sabatini, *Proc. Natl. Acad. Sci. U.S.A.* **56**, 608 (1966).
9. D. D. Sabatini and G. Blobel, *J. Cell Biol.* **45**, 146 (1970).
10. C. Milstein, G. G. Brownlee, T. M. Harrison, M. D. Mathews, *Nature (London) New Biol.* **239**, 117 (1972).
11. I. Schechter, *Proc. Natl. Acad. Sci. U.S.A.* **70**, 2256 (1973); B. Kemper *et al.*, *ibid.* **71**, 3731 (1974).
12. G. von Heijne, *Eur. J. Biochem.* **133**, 17 (1983).
13. D. Perlman and H. O. Halvorson, *J. Mol. Biol.* **167**, 391 (1983).
14. J. E. Rothman and H. F. Lodish, *Nature (London)* **269**, 775 (1977).
15. P. Walter and G. Blobel, *J. Cell Biol.* **91**, 557 (1981).
16. D. I. Meyer, E. Krause, B. Dobberstein, *Nature (London)* **297**, 647 (1982).
17. P. Walter and G. Blobel, *ibid.* **299**, 691 (1982).
18. M. Sjostrom *et al.*, *FEBS Lett.* **148**, 321 (1982).
19. G. von Heigne and C. Blomberg, *Eur. J. Biochem.* **97**, 175 (1979).
20. D. M. Engelman and T. A. Steitz, *Cell* **23**, 411 (1981).
21. D. F. Steiner *et al.*, in *Proinsulin, Insulin, and C-peptide*, S. Baba, T. Kaneko, N. Yanaihara, Eds. (Excerpta Medica, Amsterdam, 1979), pp. 9–19.
22. S. Halegoua, M. Inouye, in *Bacterial Outer Membranes*, M. Inouye, Ed. (Wiley, New York, 1979), pp. 67–114.
23. C. S. Yost, J. Hedgepeth, V. R. Lingappa, *Cell* **34**, 759 (1983).
24. J.-L. Guan and J. K. Rose, *ibid.* **37**, 779 (1984).
25. M. J. Gething, J. M. White, M. D. Waterfield, *Proc. Natl. Acad. Sci. U.S.A.* **75**, 2737 (1978).
26. W. A. Braell and H. F. Lodish, *Cell* **28**, 23 (1982).
27. M. Strubin, B. Mach, E. O. Long, *EMBO J.* **3**, 869 (1984).
28. E. C. Holland, J. Leung, K. Drickamer, *Proc. Natl. Acad. Sci. U.S.A.* **81**, 7338 (1984); M. Speiss, A. L. Schwartz, H. F. Lodish, *J. Biol. Chem.* **260**, 1979 (1985).
29. R. D. Palmiter, J. Gagnon, K. A. Walsh, *Proc. Natl. Acad. Sci. U.S.A.* **75**, 94 (1978); V. R. Lingappa, J. R. Lingappa, G. Blobel, *Nature (London)* **281**, 117 (1979); R. L. Meek, K. A. Walsh, R. D. Palmiter, *J. Biol. Chem.* **257**, 12245 (1982).
30. D. D. Sabatini, G. Kreibich, T. Morimoto, M. Adesnick, *J. Cell Biol.* **92**, 1 (1982).
31. D. A. Brown and R. D. Simoni, *Proc. Natl. Acad. Sci. U.S.A.* **81**, 1674 (1984).
32. M. Sakaguchi, K. Mihara, R. Sato, *ibid.*, p. 3361.
33. T. J. Bos, A. R. Davis, D. P. Nayak, *ibid.*, p. 2337.
34. M. Spiess and H. F. Lodish, unpublished data.
35. D. J. Anderson, K. E. Mostov, G. Blobel, *Proc. Natl. Acad. Sci. U.S.A.* **80**, 7249 (1983).
36. D. Engelman, A. Goldman, T. Steitz, *Methods Enzymol.* **88**, 81 (1982).
37. R. Kopito and H. F. Lodish, *Nature (London)* **316**, 234 (1985).
38. J. Finer-Moore and R. M. Stroud, *Proc. Natl. Acad. Sci. U.S.A.* **81**, 155 (1984).
39. D. M. Engelman and G. Zaccai, *ibid.* **77**, 5894 (1980).
40. C. Tanford, *The Hydrophobic Effect* (Wiley, New York, ed. 2, 1980).
41. W. Neupert and G. Schatz, *Trends Biol. Sci.* **6**, 1 (1981).
42. G. Schatz and R. A. Butow, *Cell* **32**, 316 (1983).
43. F. Cabral *et al.*, *J. Biol. Chem.* **253**, 297 (1978); G. Attardi *et al.*, in *International Cell Biology 1980–1981*, H. G. Schweiger, Ed. (Springer-Verlag, Berlin, 1981), pp. 225–238; P. Brost, *ibid.*, pp. 239–249.
44. N. Nelson and G. Schatz, in *Membrane Bioenergetics*, C. P. Lee, G. Schatz, L. Ernster, Eds. (Addison-Wesley, New York, 1979), pp. 133–152; M. A. Harmey, G. Hallermayer, W. Neupert, in *Genetics and Biogenesis of Chloroplasts and Mitochondria* (Elsevier/North-Holland, Amsterdam), (1976); M. A. Harmey, G. Hallermayer, H. Korb, W. Neupert, *Eur. J. Biochem.* **81**, 533 (1977).
45. M. Schleyer and W. Neupert, *J. Biol. Chem.* **259**, 3487 (1984).
46. G. Hallermayer and W. Neupert, in *Genetics and Biogenesis of Chloroplasts and Mitochondria*, Th. Buchler, Ed. (Elsevier/North-Holland, Amsterdam, 1976).
47. G. Hallermayer, R. Zimmermann, W. Neupert, *Eur. J. Biochem.* **81**, 523 (1977); G. A. Reid and G. Schatz, *J. Biol. Chem.* **257**, 13062 (1982).
48. I. Z. Ades and R. A. Butow, *J. Biol. Chem.* **255**, 9918 (1980).
49. M. Suissa and G. Schatz, *ibid.* **257**, 13048 (1982).
50. N. Nelson and G. Schatz, *Proc. Natl. Acad. Sci. U.S.A.* **76**, 4365 (1979); G. Daum, S. M. Gasser, G. Schatz, *J. Biol. Chem.* **257**, 13075 (1982); M. Teintze, M. Slaughter, H. Weiss, W. Neupert, *ibid.*, p. 10364; G. A. Reid and G. Schatz, *ibid.*, p. 10364.
51. R. Zimmermann and W. Neupert, *Eur. J. Biochem.* **109**, 217 (1980).

52. A. Viebrook, A. Perz, W. Sebald, *EMBO J.* **1**, 565 (1982); J. Kaput, S. Soltz, G. Blobel, *J. Biol. Chem.* **257**, 15054 (1982); A. L. Horwitz *et al.*, *Science* **224**, 1068 (1984).
53. P. Boehni, S. Gasser, C. Leaver, G. Schatz, in *The Organization and Expression of the Mitochondrial Genome*, A. M. Kroon and C. Saccone, Eds. (Elsevier/North-Holland, Amsterdam, 1980).
54. A. Ohashi, J. Gibson, I. Gregor, G. Schatz, *J. Biol. Chem.* **257**, 13042 (1982).
55. B. Hennig and W. Neupert, *Eur. J. Biochem.* **121**, 203 (1981).
56. R. Zimmermann, B. Hennig, W. Neupert, *ibid.* **116**, 455 (1981).
57. C. Zwizinski, M. Schleyer, W. Neupert, *J. Biol. Chem.* **258**, 4071 (1983).
58. H. Korb and W. Neupert, *Eur. J. Biochem.* **91**, 609 (1978).
59. M. G. Douglas, B. L. Geller, S. D. Emr, *Proc. Natl. Acad. Sci. U.S.A.* **81**, 3984 (1984); E. C. Hurt, B. Pesold-Hurt, G. Schatz, *EMBO J.* **3**, 3149 (1984).
60. S. Miura, M. Mori, M. Tatibana, *J. Biol. Chem.* **258**, 6671 (1983); C. Argan, C. J. Lusty, G. C. Shore, *ibid.*, p. 6667; S. Ohta, G. Schatz, *EMBO J.* **3**, 651 (1984).
61. K. Mihara, G. Blobel, R. Sato, *Proc. Natl. Acad. Sci. U.S.A.* **79**, 7102 (1982); T. Hase, H. Riezman, K. Suda, G. Schatz, *EMBO J.* **2**, 2169 (1983).
62. T. Hase, U. Müller, H. Riezman, G. Schatz, *EMBO J.* **3**, 3157 (1985).
63. M. P. Yaffe, G. Schatz, *Proc. Natl. Acad. Sci. U.S.A.* **81**, 4819 (1984).
64. S. Michaelis and J. Beckwith, *Annu. Rev. Microbiol.* **36**, 435 (1982).
65. S. Ferro-Novick, M. Honm, J. Beckwith, *Cell* **38**, 211 (1984).
66. M. Inouye, S. Halegoua, *Crit. Rev. Biochem.* **7**, 339 (1980).
67. R. Ehring, K. Beyreuther, J. K. Wright, P. Overath, *Nature (London)* **283**, 537 (1980); M. I. Poulis, D. C. Sham, H. D. Campbell, I. G. Young, *Biochemistry* **20**, 4178 (1981); W. S. A. Brusilow *et al.*, *J. Biol. Chem.* **256**, 3141 (1981); E. Santos, H. King, I. G. Young, H. R. Kaback, *Biochemistry* **21**, 2085 (1982); H. Yazyu *et al.*, *J. Biol. Chem.* **259**, 4320 (1984).
68. P. B. Wolfe, W. Wickner, J. M. Goodman, *J. Biol. Chem.* **258**, 12073 (1983).
69. C. Watts, W. Wickner, R. Zimmermann, *Proc. Natl. Acad. Sci. U.S.A.* **80**, 2809 (1983).
70. T. Date, J. M. Goodman, W. Wickner, *ibid.* **77**, 4669 (1980); T. Date, C. Zwizinski, S. Ludmerer, W. Wickner, *ibid.*, p. 827; R. Zimmermann and W. Wickner, *J. Biol. Chem.* **258**, 3920 (1983); H. G. Enequist *et al.*, *Eur. J. Biochem.* **116**, 227 (1981); C. J. Daniels, D. G. Bole, S. C. Quay, D. L. Oxender, *Proc. Natl. Acad. Sci. U.S.A.* **78**, 5396 (1981).
71. R. Cancedda and M. J. Schlesinger, *J. Bacteriol.* **117**, 290 (1974); L. L. Randall and S. J. S. Hardy, *Eur. J. Biochem.* **75**, 43 (1977).
72. W. P. Smith, P.-C. Tai, B. D. Davis, *Biochemistry* **18**, 198 (1979).
73. P. Model and N. Zinder, *J. Mol. Biol.* **83**, 231 (1974).
74. N. C. Chang, P. Model, G. Blobel, *Proc. Natl. Acad. Sci. U.S.A.* **76**, 1251 (1979).
75. K. Ito, T. Sato, T. Yura, *Cell* **11**, 551 (1977).
76. D. Koshland and D. Botstein, *ibid.* **20**, 749 (1980); *ibid.* **30**, 893 (1982).
77. K. Ito, G. Mandel, W. Wickner, *Proc. Natl. Acad. Sci. U.S.A.* **76**, 1199 (1979); K. Ito, T. Date, W. Wickner, *J. Biol. Chem.* **255**, 2123 (1980).
78. L. L. Randall, *Cell* **33**, 231 (1983).
79. L.-G. Josefsson and L. L. Randall, *ibid.* **25**, 151 (1981).
80. D. B. Oliver and J. Beckwith, *ibid.*, p. 765.
81. G. P. Vlasuk *et al.*, *J. Biol. Chem.* **258**, 7141 (1983).
82. T. J. Silhavy, S. A. Benson, S. D. Emr, *Microbiol. Rev.* **47**, 313 (1983).
83. A. Kuhn, R. Dierstein, W. Wickner, in preparation.
84. D. Perlson and H. O. Halvorson, *Cell* **25**, 525 (1981); M. Carlson, R. Taussig, S. Kustu, D. Botstein, *Mol. Cell. Biol.* **3**, 439 (1983).
85. J. T. Kadonaga *et al.*, *J. Biol. Chem.* **259**, 2149 (1984).
86. F. Moreno *et al.*, *Nature (London)* **286**, 356 (1980).
87. S. A. Benson and T. J. Silhavy, *Cell* **32**, 1325 (1983).
88. F. Yu, H. Furukawa, K. Kakamura, S. Mizushima, *J. Biol. Chem.* **259**, 6013 (1984).
89. S. Stahl and A. Gautier, personal communication.
90. C. Z. Giam, S. Hayashi, H. C. Wu, *J. Biol. Chem.* **259**, 56011 (1984).
91. A. Kuhn, W. Wickner, G. Kreil, in preparation.
92. V. A. Bankaitis, B. A. Rasmussen, P. J. Bassford, Jr., *Cell* **37**, 243 (1984).
93. R. Zimmermann and C. Mollay, personal communication.
94. R. Zimmermann, C. Watts, W. Wickner, *J. Biol. Chem.* **257**, 6529 (1982).
95. E. P. Bakker and L. L. Randall, *EMBO J.* **3**, 895 (1984).
96. J. J. Donovan, M. I. Simon, M. Montal, *Nature (London)* **298**, 669 (1982).
97. M. T. Tosteson and D. C. Tosteson, *Biophys. J.* **36**, 109 (1981); C. Kempf *et al.*, *J. Biol. Chem.* **257**, 2469 (1982).
98. R. Blumenthal, R. D. Klausner, J. N. Weinstein, *Nature (London)* **288**, 333 (1980).
99. M. Mueckler and H. Lodish, unpublished data.
100. M. Mueckler *et al.*, *Science* **229**, 941 (1985).
101. P. B. Wolfe, P. Silver, W. Wickner, *J. Biol. Chem.* **257**, 7898 (1983).
102. L. Chen, D. Rhoads, P. C. Tai, *J. Bacteriol.* **24**, 1252 (1960); M. Muller and G. Blobel, *Proc. Natl. Acad. Sci. U.S.A.* **81**, 7737 (1984).
103. J. K. Rose *et al.*, *Proc. Natl. Acad. Sci. U.S.A.* **77**, 3884 (1980).
104. V. T. Marchesi, H. Furthmayr, M. Tomita, *Annu. Rev. Biochem.* **45**, 667 (1976); A. H. Ross *et al.*, *J. Biol. Chem.* **257**, 4152 (1982).
105. D. W. Russel *et al.*, *Cell* **37**, 577 (1984).
106. H. L. Ploegh, H. T. Orr, J. L. Strominger, *ibid.* **24**, 287 (1981).
107. W. Min-Jou *et al.*, *ibid.* **19**, 683 (1980).
108. W. Wickner, *Proc. Natl. Acad. Sci. U.S.A.* **73**, 1159 (1976); I. Ohkawa and R. E. Webster, *J. Biol. Chem.* **256**, 9951 (1981).
109. S. Fields, G. Winter, G. C. Brownlee, *Nature (London)* **290**, 213 (1981); J. Blok *et al.*, *Virology* **119**, 109 (1982).

110. M. Spiess, J. Brunner, G. Semenza, *J. Biol. Chem.* **257**, 2370 (1982).
111. K. Drickamer, J. F. Mamon, G. Bins, J. O. Leung, *ibid.* **259**, 770 (1984).
112. C. Schneider, M. J. Owen, D. Banville, J. G. Williams, *Nature (London)* **311**, 675 (1984).
113. H. G. Enoch, P. J. Fleming, P. Stritmatter, *J. Biol. Chem.* **254**, 6483 (1979); Y. Takagaki, R. Radhakrishnan, K. W. A. Wirtz, H. G. Khorana, *ibid.* **258**, 9136 (1983).
114. I. G. Macara and L. C. Cantley, in *Cell Membranes, Methods and Reviews* (Plenum, New York, 1982) vol. 1, pp. 41–87.
115. W. Hunziker, M. Spiess, G. Semenza, H. F. Lodish, unpublished data.

Part II

Immunology

6

Transfectomas Provide Novel Chimeric Antibodies

Sherie L. Morrison

Nearly 100 years ago it was shown that the antibodies circulating in the serum provided the basis for the immune response. With the development of serum electrophoresis it was possible to demonstrate that these antibodies were protein molecules (*1*). Many years of research have revealed that antibodies are responsible for specific protection against bacterial and viral diseases and are involved in normal and disease-related immune reactions, including inflammation, autoimmunity, graft rejection, and idiotype-mediated network regulation. Antibodies have also proven to be invaluable, exquisitely sensitive reagents for the location, identification, and quantification of antigens in many different assay systems.

The advent of hybridoma technology in 1975 made it possible to obtain antibodies of a defined specificity in large quantities (*2*). However, limitations persisted, and it was not always possible to generate antibodies with the precise specificity desired or with the appropriate combination of specificity and effector functions. In addition, hybridomas cannot be produced with equal ease from all species; in particular, human hybridoma antibodies have been difficult to obtain. Thus, an exciting recent advance has been the development of transfectomas, in which a combination of recombinant DNA techniques and gene transfection can be used to create novel, chimeric immunoglobulin molecules.

Structural Basis of Antibody Function

Two functions are characteristic of every antibody molecule: (i) specific binding to an antigenic determinant, and

Science 229, 1202–1207 (20 September 1985)

(ii) participation in effector functions, such as binding and activation of complement, stimulation of phagocytosis by macrophages, and triggering of granule release by mast cells. The specific binding of antigen by antibodies is determined by the structure of the variable regions of both heavy (V_H) and light (V_L) chains (see Fig. 1). The effector functions are determined by the structure of the constant region (C) of the heavy chains. Immunoglobulins (Ig's) with different constant regions—that is, Ig's of differing isotypes (in the human IgM, IgD, IgG_1, IgG_2, IgG_3, IgG_4, IgA_1, IgA_2, and IgE)—therefore exhibit different biologic properties. Antibodies are glycoproteins, and the presence of carbohydrate on the antibody molecule is essential for some of the effector functions of antibodies (*3*). Cleavage with the enzyme papain splits the Ig molecule into an Fab region that can bind antigen and into an Fc region containing the effector functions (Fig. 1).

Both the heavy and light chains are encoded by multiple DNA segments. A functional Ig gene is generated only after somatic rearrangement of distinct DNA segments (*4*). In the functional Ig gene, intervening sequences separate the hydrophobic leader sequence from the variable region and also the variable-region gene segment from the constant region; in addition, intervening sequences separate the different domains of the constant region in heavy chains so that each functional region of the heavy chain con-

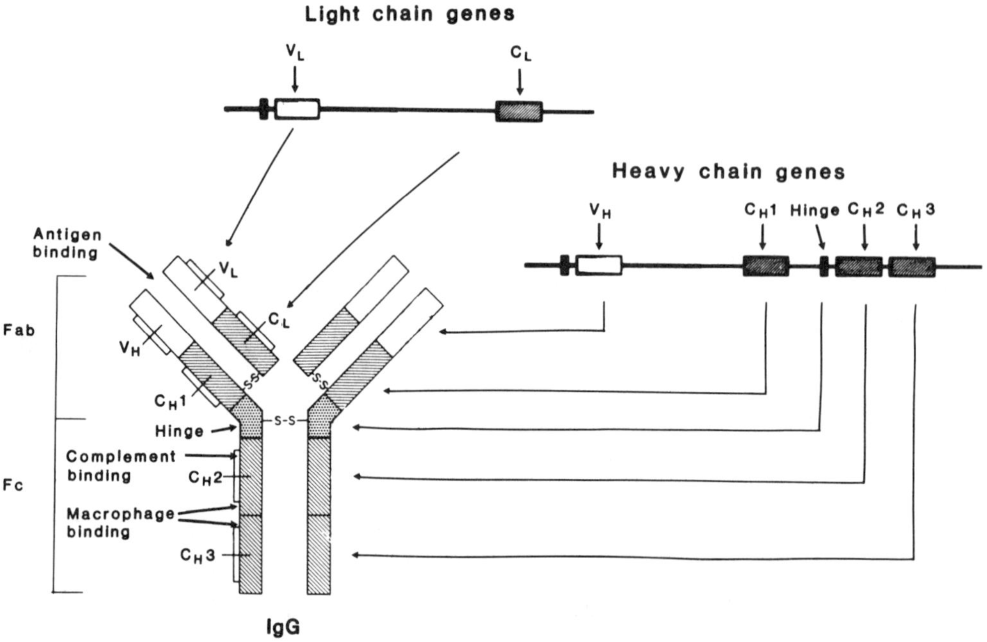

Fig. 1. Structure of an Ig molecule and the genes that encode it. The regions of the molecule that participate in antigen binding (Fab) or different effector functions (Fc) are indicated. Arrows indicate the correspondence between the DNA segments and the different domains of the Ig polypeptide chain that they encode. The hydrophobic leader sequence of both heavy and light chains is removed immediately after synthesis and is not present in the mature immunoglobulin molecule.

stant region is on a separate exon (Fig. 1).

In the past, structural and functional studies of Ig molecules were complicated by the heterogeneity of the molecules. Because antibodies present in an antiserum are a mixture of specificities, antisera of defined specificities were difficult to obtain in large quantities. These problems were solved to a certain degree when it was realized that the disease multiple myeloma represents a monoclonal proliferation of plasma cells and that large quantities of homogeneous antibodies could be obtained. The majority of what is known about the structure of antibody molecules and the organization of Ig genes has come from the study of these myeloma proteins and cells. Unfortunately, only a few of the myeloma proteins can be demonstrated to bind known antigens, so their usefulness as specific reagents is severely limited.

Monoclonal antibodies of defined specificity became available when Köhler and Milstein demonstrated that cultured mouse myeloma cells could be fused to spleen cells from immunized mice (*2*). The hybrid cells (hybridomas) grow continuously in culture (property acquired from the myeloma cell) and continue to produce in large quantities the Ig that had been synthesized by the spleen cell partner. These antibody molecules are monoclonal; hence, they have an advantage over classical antisera in that their combining sites are homogeneous. However, the fact that the Ig produced by one hybridoma is of only one isotype can be a problem if the isotype expressed does not have the effector function, such as complement fixation or binding of staphylococcal protein A, required for a certain assay.

The large quantities of chemically defined antibodies produced by hybridomas have been invaluable tools for immunologists, but they do have certain limitations. Although hybridoma proteins that bind defined antigen can be identified, the affinity of the binding cannot be determined by the researcher. Furthermore, the isotype of the antibody is determined by the constant region gene being used by the normal cell at the time of fusion.

An additional problem associated with hybridomas is their species limitations. While it is relatively easy to produce mouse or rat monoclonal antibodies, it is difficult to produce human monoclonal antibodies, which would be desirable for certain studies and in vivo use. Many human hybridomas produce only limited quantities of Ig and are unstable in their production.

The isolation of somatic mutants of hybridoma cells has provided investigators with additional capabilities. Cell lines with both decreased and increased affinity for antigen can be isolated (*5*). In addition, it is possible to isolate somatic mutants with alterations in their Fc regions (*6*). These mutations may be either structural changes in the Fc region or may represent isotype switch mutants. Switch variants have permitted an assessment of the contribution of the hinge region to the segmental flexibility of Ig molecules (*7*).

Ig Gene Transfection into Lymphoid Cells

Gene transfection provides a method for making novel Ig molecules (Fig. 2). Ig genes, either wild-type or altered in vitro, can be transfected and expressed (*8, 9*). The transfected cells (transfectomas) grow continuously in culture and produce the Ig specified by the transfected gene. This approach has the advantage over the isolation of mutants that alteration can be predetermined and does not depend on the chance occurrence of the desired changes.

The expression of transfected genes can be studied transiently (a few days after transfection) or stable transfectants in which the transfected gene has integrated into the chromosome can be isolated. If the objective is to produce a new protein, stable transfectants need to be isolated to provide a continuous source of the protein. Because only a small percentage (10^{-3} to 10^{-6}) of the cells exposed to foreign DNA go on to become stably transformed, selective techniques are required that permit the isolation of rare, stably transformed cells from among the many nontransformed cells. Initial experiments used the thymidine kinase gene from herpes simplex virus as the selective marker (*10*). However, this required that the recipient cells be deficient in endogenous thymidine kinase (TK^-); consequently, only a few available cell lines lacking thymidine kinase could be used as recipients.

To overcome these limitations, vectors with dominant selectable markers have been developed (*11–13*). Dominant-acting genetic markers produce a selectable change in the phenotype of normal cells; such markers have been produced by placing bacterial genes within mammalian transcription units. The most commonly used vectors have been the pSV2 vectors (*11*), in which the selectable marker is placed under the control of the SV40 early promoter; SV40 sequences provide splice signals and a polyadenylation site.

Two selectable bacterial genes have been used: (i) the xanthine-guanine phosphoribosyltransferase gene (*gpt*) (*12*), and (ii) the phosphotransferase gene from Tn5 (designated *neo*) (*13*). Selection with *gpt* is based on the fact that the enzyme encoded by this gene can use xanthine as a substrate for purine nucleotide synthesis, whereas the analogous endogenous enzyme cannot. Thus, for cells provided with xanthine and in which the conversion of inosine monophosphate to xanthine monophosphate is blocked by mycophenolic acid (*14*), only cells expressing the bacterial gene can survive. The product of the *neo* gene inactivates the antibiotic G418 (*15, 16*), which interferes with the function of 80*S* ribosomes and blocks protein synthesis in eukaryotic cells. The two selection procedures depend on two entirely different mechanisms; therefore, they can be used simultaneously to select for the expression of genes introduced on two different DNA segments. Alternatively, they can be used to select for the expression of different genes introduced sequentially into cells.

The recipient cell type of choice for the production of Ig molecules would appear to be myeloma cells. Myelomas represent malignancies of plasma cells, and they are capable of producing large quantities of Ig. Expression of transfected heavy chain genes by myelomas approaches the level seen for the endogenous myeloma protein (*17, 18*). Expression of light chain genes after transfection has been more of a problem, and frequently it is only 5 to 10 percent of that seen in myeloma cells expressing the same gene (*19*). However, it has been possible to identify transfectants in which the synthesis of both the light and heavy chain of the functional Ig molecule has been directed by a transfected gene (*20–22*).

Several methods exist for introducing DNA into eukaryotic cells. Calcium phosphate precipitation is routinely used to introduce DNA into many cell types (*10, 23, 24*). However, this technique results in a low frequency of recovery of transfectants from myeloma cells (*9*). A more efficient way of introducing DNA into lymphoid cells is by protoplast fusion (*9, 24*). In this method, lysozyme is used to remove the bacterial cell walls from *Escherichia coli* bearing the plasmids of interest (*9, 24*), and the resulting spheroplasts are fused with myeloma cells by means of polyethylene glycol. After protoplast fusion, stable transfectants have been isolated at frequencies ranging from 10^{-4} to 10^{-3} (*9, 17, 20*). Recent results have suggested that electroporation may also be an efficient method for introducing DNA into lymphoid cells (*25*). For electroporation, a high-voltage pulse is applied to cells in suspension in the presence of DNA. Stable transfectants have been isolated from many different cell types at frequencies as high as 3×10^{-4}.

Variations in transfectability have been observed among the different myeloma cell lines and among different clones of the same myeloma cell line. The cause of the variations in unknown. The myeloma J558L, which produces λ light chains and no heavy chains, is among the best recipient cell lines. With the appropriate vectors, it can be transfected with efficiencies approaching 10^{-3} (*9, 17, 26, 27*). However, J558L has the disadvantage that it continues to synthesize its own λ light chain; it has been impossible to isolate nonproducing variants of J558L. Although some nonproducing myelomas, such as SP2/0, have been transfected, the transfection frequencies are 10^{-1} to 10^{-2} of that seen with J558L.

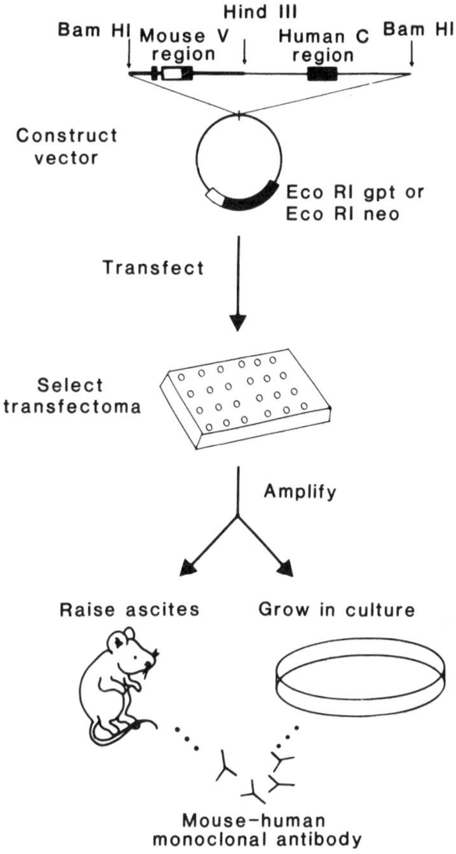

Fig. 2. Steps in production of chimeric Ig molecules. First, a chimeric gene containing the sequences of interest is ligated into an expression vector containing a selectable marker. Then, this vector is transfected into the appropriate recipient cell line, and stable transfectants synthesizing the genes of interest are identified; if vectors are used with independent selectable markers, transfectants synthesizing more than one novel chain can be isolated. Finally, the chimeric proteins are isolated either from tissue culture or from the ascitic fluid of tumor-bearing mice.

Expression of Transfected Genes

It is necessary to prove that the proteins synthesized after gene transfection are the same as those synthesized in myeloma cell lines. When the MOPC-41 κ light chain was introduced into a cell line that had been transformed with Abelson murine leukemia virus, the cells directed the synthesis of a κ chain that could assemble with the endogenous γ_{2b} heavy chain to produce molecules containing two heavy and two light chains (H_2L_2) (*8*). The light chain produced after the transfection of the S107A κ chain was identical, by two-dimensional gel analysis, to the S107A myeloma κ chain (*9*). The S107A light chain is not secreted by light chain–producing variants of this myeloma; a transfected S107A light chain was also not secreted when introduced into a myeloma that did not produce any heavy chain. However, when the S107A κ chain was introduced into a hybridoma cell line that synthesized a heavy chain, it assembled with the endogenous heavy chain to produce H_2L_2 molecules; these H_2L_2 molecules were secreted.

When genes isolated from a trinitrophenyl (TNP)-binding hybridoma were used the faithful expression of transfected genes with the retention of the original antigen-binding specificity was seen (*20, 28*). Transfection of the light chain from this hybridoma into a cell line that synthesized the TNP-specific heavy chain, but that had lost the ability to synthesize the TNP-specific light chain, resulted in the synthesis of a complete H_2L_2 molecule with the ability to bind antigen. This TNP-specific IgM was secreted by the transfected cells, and the hemagglutination titer of antibody to TNP (anti-TNP) in some transformants was comparable to that of the parental anti-TNP–producing hybridoma. Conversely, by transferring a μ heavy chain gene from an anti-TNP–specific myeloma into a light chain–producing variant of that myeloma, it was also possible to express a pentameric IgM molecule with anti-TNP activity. Simultaneous transfer of specific light and heavy chains into a nonproducing myeloma also resulted in the production of pentameric, antigen-binding antibody, although in this case the amount of antibody synthesized was less than that seen in the parental myeloma.

Gene transfection can be used to identify tissue-specific regulatory elements. A segment of DNA located between J_4 and the switch region in the major intervening sequence of a heavy chain gene is necessary for the expression of heavy chain genes transfected in lymphoid cells of the B lineage (*17*); this region does not influence the expression of Ig genes transfected into L cells (*29*). The location of this DNA segment 5′ of the switch site guarantees that the segment will remain associated with the expressed variable region gene during isotype switching and is consistent with the segment having a regulatory role. However, cell lines have been reported in which deletion of this segment does not abolish heavy chain synthesis (*30*). Therefore, a question remains as to relationship between the sequences needed for the expression of transfected genes and those needed for the expression of endogenous genes.

Controlling regions have been identified in the major intervening sequence of κ light chains. A mouse sequence lying approximately 600 base pairs 5′ of C_κ enhanced production of κ chain genes transfected into lymphoid cells (*19, 31*). Like the heavy chain controlling region, it does not function in nonlymphoid cells. This sequence corresponds to a DNA segment (*32*) that is highly conserved among rabbit, mouse, and man and that becomes sensitive to deoxyribonuclease when Ig is expressed (*33*). In addition, a second DNA segment from the 5′ half of the κ intervening sequence has been identified that facilitates expression of κ genes transfected into mouse myeloma but not into hamster lymphoma cells (*19, 34*).

Production of Novel Ig Molecules by Gene Transfection

With the demonstration that transfected Ig genes faithfully exhibited the expected properties, it became possible to produce novel Ig molecules. These novel Ig's can be divided into two categories: (i) those synthesizing wild-type Ig chains but in a combination not normally expressed and (ii) those synthesizing molecules with a novel gene structure that has been produced by recombinant DNA techniques (Table 1).

Novel combinations of wild-type heavy and light chains. In vitro reassembly of Ig heavy and light chain polypeptides has been used to create novel heavy-light chain combinations and to study chain interactions (*35–37*). This method, however, requires that the original interchain disulfide bonds in the Ig be broken and that heavy and light chains be separated in the presence of denaturing agents. As a result, only a small percentage of the chains renature to produce functional molecules (*36*). On the other hand, chain assembly between transfected gene products occurs within the cell by normal pathways of assembly. Because the assembled molecules have never been subjected to a denaturing step, their structure and function should more accurately reflect those of native molecules.

Experiments either with cell lines transformed by Abelson murine leukemia virus or with hybridomas have indicated that assembly can occur between the transfected and the endogenous Ig chains (Fig. 3) (*8, 9*), and this has been found to be generally true. When the γ_{2b} heavy chain of the myeloma MPC-11 was transfected into the λ myeloma J558, it was expressed (Fig. 3, lane 3), assembled with the endogenous λ chain, and secreted as H_2L_2 molecules (Fig. 3, lane 4). These novel molecules, in which the MPC-11 heavy chains are associated with λ light chains, have been useful in exploring the relative contribution of V_H and V_L to idiotypic determinants (*38*).

Novel combinations of variable and constant regions. By means of standard recombinant DNA techniques, it is possible to attach any variable region to any heavy chain constant region. The existence of the large intervening sequence between the two regions (see Fig. 1) makes such joining easier. Because the reading frame does not have to be maintained and the intervening sequence will be removed by splicing, many different joints will lead to functional molecules. This approach can be used to produce isotype switch variants of mouse hybridoma proteins.

Another family of molecules can be created in which the variable regions from the heavy and light chains of a mouse myeloma are joined to human constant regions (Fig. 2). These molecules have the antigen-binding specificity of the mouse hybridoma, but they should exhibit the effector function associated with the human constant regions

Table 1. Creation of novel molecules by gene transfection.

Genes used for transfection	Objective	Potential uses
	Wild-type Ig chains	
Wild-type heavy or light chains (or both)	To create new heavy and light chain combinations, both within a species and between species	Explore the nature of heavy and light chain interactions
		Define the relative contributions of each chain to idiotype expression and antigen-binding specificity
		Determine if all chain combinations and thus combinational diversity are possible
	Novel Ig chains	
V_H attached to new C_H; V_L attached to new C_L	To create chains with the same variable region associated with a new isotype	Produce isotype switch variants within one species
		Create cross-species chimerics, with a variable region from one species and a constant region from another
V_H attached to C_L; V_L attached to C_H	To create chains with a variable region attached to an isotype with which it is never associated in vivo	Evaluate contributions of constant regions to functions
		Create molecules with altered effector functions
V_H or V_L attached to a non-Ig sequence	To create fusion proteins	Attach antibody specificity to enzymes for use in assays
		Isolate non-Ig proteins by antigen columns
		Specifically deliver toxic agents

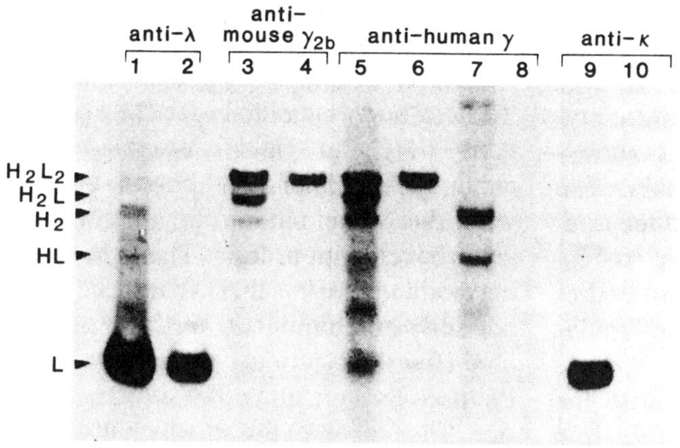

Fig. 3. Synthesis of Ig chains in transfectomas. Cells were labeled with [^{14}C]valine, [^{14}C]threonine, and [^{14}C]leucine for 3 hours, and cytoplasmic lysates and secreted material were immunoprecipitated with the antibodies shown at the top of the figure. (Lanes 1 and 2) J558L; (lanes 3 and 4) J558L containing a transfected MPC-11 γ_{2b} heavy chain; (lanes 5 and 6) J558L containing a chimeric mouse-human heavy chain; (lanes 7 and 8) nonproducing myeloma cells containing a chimeric mouse-human heavy chain; and (lanes 9 and 10) V_HC_κ in light chain–producing variant of anti-phosphocholine myeloma S107. Lanes 1, 3, 5, 7, and 9 represent cytoplasmic extracts; lanes 2, 4, 6, 8, and 10 represent secreted material.

and should be less antigenic in humans than are totally mouse antibodies.

Initially, the variable region from the heavy chain of the anti-phosphocholine myeloma S107 was joined to either human γ_1 or γ_2 heavy chain (*21*). After transfection into J558L, the chimeric gene was expressed (Fig. 3, lane 5), the chimeric heavy chain was assembled with the endogenous λ chain, and H_2L_2 molecules were secreted. The chimeric heavy chain was also expressed by myeloma cells not synthesizing any light chains (nonproducing) (Fig. 3, lane 7); however, these heavy chains were not secreted (Fig. 3, lane 8).

Subsequently, the antigen-specific light chain of S107 was joined to human C_κ and ligated into the pSV2neo vector, and transfectants synthesizing both chimeric heavy and light chains were identified (*21*). The chimeric heavy and light chains of the expected molecular weight were assembled and secreted as H_2L_2 molecules. Comparison of the apparent molecular weights of heavy chains synthesized in the presence and absence of tunicamycin showed that the mouse myeloma cell attaches *N*-linked carbohydrate to the human heavy chain. The chimeric H_2L_2 molecules bound the antigen phosphocholine and were recognized both by monoclonal antibodies specific for human κ chain antibodies and by monoclonal antibodies specific for the idiotype prepared against the mouse myeloma protein. The cells from the transfectoma produced an ascites after being injected into a mouse, and the chimeric proteins were found to be present in the ascitic fluid.

Chimeric human-mouse molecules have also been produced in which the variable regions of the heavy and light chains from an anti-TNP mouse myeloma were linked to human μ and κ genes, respectively, and were transfected into myeloma cells (*22*). The chimeric genes were expressed and assembled such that pentameric IgM was present in the secretions; however, both the chimeric μ and κ chains migrated more slowly than was expected on gels containing sodium dodecyl sulfate. There was no significant difference in affinity for TNP between the mouse and the chimeric IgM's. The chimeric IgM, however, was about one-fourth as effective as the mouse IgM in hemolysis of TNP-coupled sheep red blood cells and showed a displaced binding curve in hapten inhibition assays. The molecular basis for these apparent binding differences remains unclear.

Another chimeric molecule has been produced in which a heavy chain with the variable region of a mouse antibody to 4-hydroxy-3-nitro-phenacetyl (NP) was joined to human ϵ heavy chain (*27*). This fused segment was transfected into the J558L cell line, which synthesizes a mouse λ light chain identical to that used by mouse antibody to NP. The J558L λ chain assembled with the chimeric heavy chain to produce H_2L_2 molecules, which were secreted and, like human ϵ, were heavily glycosylated. The chimeric IgE bound NP efficiently and, when the chimeric antibodies were bound to the appropriate receptors on human basophils, antigen was able to stimulate a dose-dependent release of histamine.

Variable regions can also be placed on constant regions with which they would not normally be associated. When the V_H from an anti-azophenylarsonate hybridoma was joined to mouse C_κ and introduced into a mouse myeloma cell, a 25,000-dalton chimeric protein was produced (*39*). When the chimeric V_H-C_κ gene was transfected into a light chain–producing variant of the anti-azophenylarsonate hybridoma from which the gene was isolated, the V_H-C_κ protein was synthesized and assembled with the endogenous V_L-C_κ to yield a secreted light chain heterodimer that bound antigen with the same affinity as the original H_2L_2 hybridoma protein. The light chain heterodimer represents, in essence, an antibody-combining site with no associated effector functions, as light chains by themselves have no known effector function. In a second case in which V_H from the S107 myeloma was joined to Cκ and introduced in a light chain–producing variant of S107, a 25,000-dalton chimeric protein was produced, but heterodimers were not secreted (Fig. 3, lanes 9 and 10).

The V_H from a human myeloma can be joined to mouse C_κ (*40*). Although this human-mouse chimeric protein was produced, it did not assemble with the endogenous λ light chain of the recipient mouse myeloma and remained cytoplasmic. These experiments demonstrated that the human heavy chain promoter can function in mouse myelomas. Additional studies have shown that the human light chain promoter also can function in mouse myeloma cells (*25*). It is thus possible to consider using gene transfection techniques for gene rescue; that is, genes could be cloned from human myelomas or lymphoblastoid lines where they are expressed at low levels and then introduced by gene transfection into mouse myeloma lines. These initial studies suggest that such transfected genes will effectively function to produce high levels of protein.

Gene transfection has also been used to create cell lines synthesizing $F(ab')_2$-like antibody (*26*). [$F(ab')_2$ is two Fab's joined by disulfide bonds (see Fig. 1)]. A chimeric gene was constructed in which the variable region of a mouse NP-binding heavy chain was joined to the C_H1 and hinge regions of mouse γ_{2b}. The fifth exon from the δ heavy chain provided translation termination and polyadenylation signals at the end of the heavy chain.

This chimeric gene was transfected into λ_1-producing J558L cells; the λ_1 light chain associated with the NP-specific heavy chains to produce an antigen-binding protein that could be isolated from culture supernatants of transfected cells. The protein had a molecular weight of approximately 110,000 daltons and, on reduction, yielded one band that comigrated with λ light chain and several other bands of higher molecular weight; the predominant species was approximately 31,000 daltons in size. It thus appeared that an $F(ab')_2$-like antibody to NP was synthesized and secreted in large quantities.

Chimeric molecules with Ig sequences joined to non-Ig sequences. Finally, chimeric molecules can be produced in which Ig and non-Ig sequences are joined. The gene for *Staphylococcus aureus* nuclease was inserted into the C_H2 exon of a mouse γ_2 heavy chain specific for NP, and the construction was transfected into J558L cells (*26*). The chimeric heavy chain was produced and assembled with the λ light chain to form an NP-binding protein. Molecules of the appropriate size to be H_2L_2 were isolated from the secretions; these molecules bound the antigen NP and also had nuclease activity that, like the activity of authentic *S. aureus* nuclease, was dependent on Ca^{2+} but not Mg^{2+} ions. On a molar basis, the catalytic activity of the constructed nuclease was about 10 percent of that of authentic *S. aureus* nuclease. Similarly, in another study, the C_H2 and C_H3 domains of the heavy chain were replaced with the third exon of c-*myc*. After transfection into J558L cells, a secreted protein was produced that retained its ability to bind antigen and that was recognized by a monoclonal antibody to c-*myc*. These molecules show the feasibility of making unique combinations of antibody and protein functions that are secreted in large quantities by myeloma cells and retain their ability to bind antigen; hence, they can be rapidly purified on antigen columns.

Expression of Ig Genes in Nonlymphoid Cells

Expression of Ig genes and synthesis of Ig molecules in bacteria provide an alternative to expression in lymphoid cells. Bacteria have been used to make fragments of ϵ heavy chain that were at least partially biologically active in that they bound to the IgE receptor on basophils (*41–43*). However, heavy chains, light chains, and light chain fragments made in bacteria became insoluble components of inclusion bodies (*42–44*) rather than functionally intact molecules. Assembly did not occur even when efforts were made to promote heavy and light chain assembly by including them within the same bacterium on two different plasmids. Antigen-binding Ig molecules that bound antigen could be produced only by in vitro assembly and at very low yields (*44*). In addition, bacteria have the inherent limitation that they do not glycosylate proteins; therefore, any biologic function of Ig's that depends on carbohydrate would be missing.

Yeast can glycosylate proteins and hence may be preferred over bacteria for Ig expression. Recently, the synthesis in yeast of both λ light chains and μ heavy chains has been reported (*45*); a portion of the μ chains was *N*-glycosylated. Both the λ and μ chains were secreted, and, when μ and λ chains were expressed together, some assembly into antibody molecules that bound antigen occurred. However, the specific activity of antibody molecules isolated from a soluble extract of yeast cells was only about 0.5 percent, indicating that the efficiency of heavy and light chain assembly into functional molecules was low.

Future Perspectives and Applications

The initial experiments demonstrated the feasibility of producing novel Ig molecules by gene transfection. Myeloma cells appear to be the best recipient for the Ig genes because the transfected Ig genes are faithfully transcribed, translated, and glycosylated in them. The antibody components are synthesized in large amounts and can be assembled either with endogenous Ig chains or with other transfected chains to produce functional molecules. Nonlymphoid cells synthesize only small amounts of Ig, and both bacteria and yeast have proved to be inefficient in assembling functional Ig molecules.

Chimeric Ig molecules should provide a new family of reagents of wide potential application. Changing the Fc portion of the Ig can alter its ability to bind staphylococcal protein A, to be multivalent, or to fix complement and, therefore, would affect the usefulness of the Ig for many in vitro assays. Molecules with increased or decreased binding affinity can be useful in creating detection assays that have different levels of sensitivity and stringency.

The ability to produce antibody-protein chimeras permits the construction of a molecule with the Fab of an Ig molecule covalently bound to a non-Ig protein. When the Fab is bound to an enzyme, the binding of the Fab to its antigen can be detected by addition of the appropriate substrate. If the Fab is bound to a toxin, it can potentially be used as a specific cytotoxic agent. In addition, chimeric Ig molecules can function to facilitate protein isolation in that the non-Ig protein attached to a functional antibody-combining site can be purified on an antigen column.

In vitro mutagenesis may provide an additional way to generate molecules of altered function. It should be possible to generate molecules lacking only one specific effector function, with the remainder of the molecule intact; to eliminate or generate glycosylation sites; and through small changes in the variable region, to increase or decrease affinity for antigen. In turn, analysis of the exact structure of the mutated region is expected to reveal those areas of the Ig molecules required for effector function or antigen binding.

Human antibodies would clearly be optimal for use in man, but it has been difficult to produce human hybridoma that synthesize large quantities of antibodies with the desired specificities. Mouse antibodies have been used in trial studies for the treatment of certain human diseases. However, on continued use, they frequently elicit an immune response that renders them noneffective (*46*). Chimeric molecules may help solve this problem because molecules in which only the variable region is of nonhuman origin should be much less antigenic than completely foreign molecules. Chimeric molecules may prove useful in antibody-mediated cancer therapy and in the treatment of certain autoimmune diseases (*47*).

Even in the diagnosis or treatment of diseases in which chronic exposure to the treatment agent is not necessary (resulting in fewer anticipated problems with the agent's antigenicity), the use of chimeric molecules would appear to be preferable to the use of totally foreign Ig. Such chimeric antibodies are potentially useful, for example, for the treatment of immunosuppressed individuals who are subject to fatal viremias and of nonimmunized individuals exposed to tetanus who would be treated with horse antitoxin. Use of chimeric molecules also provides a choice of Ig's with the desired effector functions; alternatively, if no effector function is wanted, use of light chain heterodimers of IgA or Fab-type molecules should permit antigen clearing without complement activation. The application of these molecules to many clinical problems should be quite rewarding.

References and Notes

1. A. Tiselius and E. A. Kabat, *J. Exp. Med.* **69**, 119 (1939).
2. G. Köhler and C. Milstein, *Nature (London)* **256**, 495 (1975).
3. M. Nose and H. Wigzell, *Proc. Natl. Acad. Sci. U.S.A.* **80**, 6632 (1983).
4. S. Tonegawa, *Nature (London)* **302**, 575 (1983).
5. W. D. Cook and M. D. Scharff, *Proc. Natl. Acad. Sci. U.S.A.* **74**, 5687 (1977); W. D. Cook, S. Rudikoff, A. Giusti, M. D. Scharff, *ibid.* **79**, 1240 (1982).
6. J.-L. Preud'homme, B. K. Birshtein, M. D. Scharff, *Proc. Natl. Acad. Sci. U.S.A.* **72**, 1427 (1975); B. Liesegang, A. Radbruch, K. Rajewsky, *ibid.* **75**, 3901 (1978); D. E. Yelton and M. D. Scharff, *J. Exp. Med.* **156**, 1131 (1982); C. A. Muller and K. Rajewsky, *J. Immunol.* **131**, 877 (1983).
7. V. T. Oi *et al.*, *Nature (London)* **307**, 136 (1984).
8. D. Rice and D. Baltimore, *Proc. Natl. Acad. Sci. U.S.A.* **79**, 7862 (1982).
9. V. T. Oi, S. L. Morrison, L. A. Herzenberg, P. Berg, *ibid.* **80**, 825 (1983).
10. M. Wigler *et al.*, *Cell* **11**, 223 (1977); A. Pellicer, M. Wigler, R. Axel, S. Silverstein, *ibid.* **14**, 133 (1978).
11. R. C. Mulligan and P. Berg, *Science* **209**, 1422 (1980).
12. R. C. Mulligan and P. Berg, *Proc. Natl. Acad. Sci. U.S.A.* **78**, 2072 (1981).
13. P. J. Southern and P. Berg, *J. Mol. Appl. Genet.* **1**, 327 (1982).
14. T. J. Franklin and J. M. Cook, *Biochem. J.* **113**, 515 (1969).
15. J. Davies and A. Jiminez, *Am. J. Trop. Med. Hyg.* **29** (5) (suppl.), 1089 (1980).
16. J. Davies and D. I. Smith, *Annu. Rev. Microbiol.* **32**, 469 (1978).
17. S. D. Gillies, S. L. Morrison, V. T. Oi, S. Tonegawa, *Cell* **33**, 717 (1983).
18. S. L. Morrison, L. A. Wims, B. Kobrin, V. T. Oi, *Mt. Sinai J. Med.*, in press.
19. S. L. Morrison and V. T. Oi, in *Transfer and Expression of Eukaryotic Genes*, H. Ginsberg and H. Vogel, Eds. (Academic Press, New York, 1984), p. 93.
20. A. Ochi, R. G. Hawley, M. Shulman, N. Hozumi, *Nature (London)* **302**, 340 (1983).
21. S. L. Morrison, M. J. Johnson, L. A. Herzenberg, V. T. Oi, *Proc. Natl. Acad. Sci. U.S.A.* **81**, 6851 (1984).
22. G. L. Boulianne, N. Hozumi, M. J. Shulman, *Nature (London)* **312**, 643 (1984).
23. F. L. Graham and A. J. van der Erb, *Virology* **52**, 456 (1973); G. Chu and P. A. Sharp, *Gene* **13**, 197 (1981).
24. R. M. Sandri-Goldin, A. L. Goldin, M. Levine, J. C. Glorioso, *Mol. Cell. Biol.* **1**, 743 (1981).
25. H. Potter, L. Weir, P. Leder, *Proc. Natl. Acad. Sci. U.S.A.* **81**, 7161 (1984).
26. M. S. Neuberger, G. T. Williams, R. O. Fox, *Nature (London)* **312**, 604 (1984).
27. M. S. Neuberger *et al.*, *ibid.* **314**, 268 (1985).
28. A. Ochi *et al.*, *Proc. Natl. Acad. Sci. U.S.A.* **80**, 6351 (1983).
29. S. D. Gillies and S. Tonegawa, *Nucleic Acids Res.* **11**, 7981 (1983).
30. M. Wabl and P. D. Burrows, *Proc. Natl. Acad. Sci. U.S.A.* **81**, 2452 (1984); D. M. Zeller and L. A. Eckhardt, *ibid.* **82**, 508 (1985).
31. C. Queen and D. Baltimore, *Cell* **33**, 741 (1983).
32. L. Emorine, M. Kuehl, L. Weir, P. Leder, E. E. Max, *Nature (London)* **304**, 447 (1983).
33. T. G. Parslow and D. K. Granner, *ibid.* **299**, 449 (1982); S.-Y. Chung, V. Folsom, J. Wooley, *Proc. Natl. Acad. Sci. U.S.A.* **80**, 2427 (1983); T. G. Parslow and D. K. Granner, *Nucleic Acids Res.* **11**, 4775 (1983).
34. S. L. Morrison, K. P. Sun, V. T. Oi, unpublished observations.
35. B. N. Majula, C. P. J. Glaudemans, E. B. Mushinski, M. Potter, *Proc. Natl. Acad. Sci. U.S.A.* **73**, 932 (1976).
36. D. M. Kranz and E. W. Voss, Jr., *ibid.* **78**, 5807 (1981).
37. C. Horne, M. Klein, I. Polidonlis, K. J. Dorrington, *J. Immunol.* **129**, 660 (1982).
38. C. Victor-Kobrin and C. Bona, unpublished observations.
39. J. Sharon *et al.*, *Nature (London)* **309**, 364 (1984).
40. L. K. Tan, V. T. Oi, S. L. Morrison, *J. Immunol.*, in press.
41. T. Kurokawa *et al.*, *Nucleic Acids Res.* **11**, 3077 (1983).
42. J. Kenten, B. Helm, T. Ishizaka, P. Cattini, H. Gould, *Proc. Natl. Acad. Sci. U.S.A.* **81**, 2955 (1984).

43. F.-T. Lui, K. A. Albrandt, C. G. Bry, T. Ishizaka, *ibid.*, p. 5369.
44. M. A. Boss, J. H. Kenten, C. R. Wood, J. S. Emtage, *Nucleic Acids Res.* **12**, 3791 (1984).
45. C. R. Wood *et al.*, *Nature (London)* **314**, 446 (1985).
46. R. L. Levy and R. A. Miller, *Annu. Rev. Med.* **34**, 107 (1983).
47. M. K. Waldor *et al.*, *Science* **227**, 415 (1985).
48. I would like to acknowledge the invaluable contribution of V. T. Oi to the collaborative experiments on the production of chimeric human-mouse antibody molecules, of L. Wims for assistance in performing many experiments discussed here, and of B. Newman for assistance with and critical comments on the manuscript. Supported by grants from the National Institutes of Health (CA 16858, CA 22736, CA 13696, and AI 19042) and the American Cancer Society (IMS-360) and a research career development award (AI 00408).

7

Histocompatibility Antigens on Murine Tumors

Robert S. Goodenow, Julie M. Vogel, and Richard L. Linsk

The H-2 Antigens of the Major Histocompatibility Complex

The major histocompatibility complex (MHC) of the mouse consists of three classes of genes that encode cell surface and secreted products involved in immune regulation and function (*1*). Within the MHC, the class I genes represent a large multigene family of distinct, but related, sequences encoding several types of products. These products include the transplantation antigens, which were first identified on the basis of their role as the targets of graft rejection. The class II genes, encoding the immune response–associated or Ia antigens, are involved in the regulation of antibody responses to certain foreign antigens. The class III genes encode several components of the complement cascade.

There are approximately 32 class I genes in the mouse, which map either to the H-2 or Tla regions of the MHC (*1*). Each inbred mouse strain expresses a distinct set of MHC antigens (the haplotype). The products of the H-2 loci include the highly polymorphic histocompatability antigens H-2K, D, and L. On the other hand, the Qa and Tl genes (which map to the Tla region) are much less polymorphic among different strains of mice. Although the products of some of these genes appear to be expressed on certain lymphoid subsets and in the liver, their function has not been determined (*2, 3*).

The H-2 antigens are cell surface glycoproteins that function as targets directing the attack of cytolytic T lymphocytes (CTL's) against virally infected or neoplastic cells (Fig. 1). Although the precise molecular basis of T-cell recognition

Science 230, 777–783 (15 November 1985)

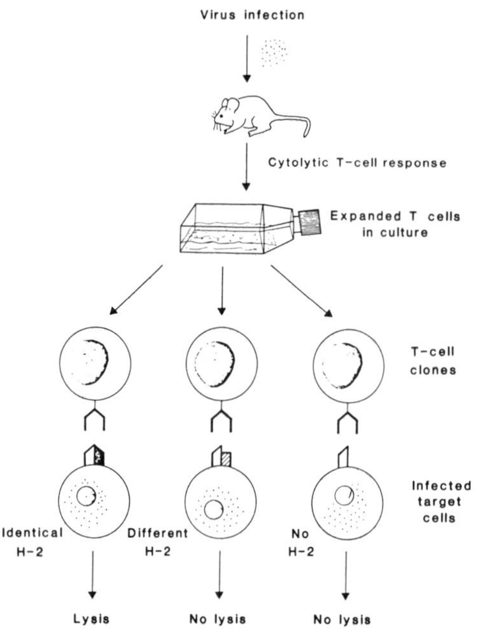

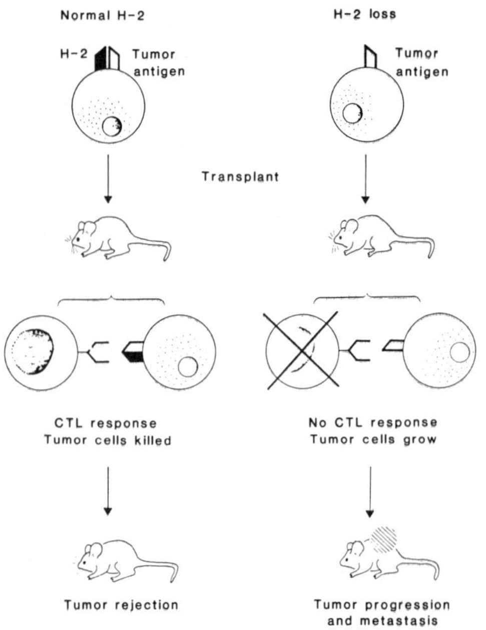

Fig. 1 (left). H-2–restricted T-cell recognition. Some of the major features characteristic of H-2–dependent CTL recognition of virally infected cells are shown. (Far left) CTL's generated during viral infection (shown expanded in culture) are able to lyse cells that express the viral antigen in association with the appropriate H-2 antigen. In the absence of H-2 (far right) or viral antigen, CTL recognition is blocked. This demonstrates the dual recognition of self plus nonself by cytolytic T cells. In the presence of an inappropriate H-2 molecule (for example, on target cells derived from a mouse strain different from that in which the CTL's are generated) the same cytolytic T cells are unable to recognize or lyse the target cell even in the presence of the viral antigen (center). Thus, the H-2 requirement is both absolute and specific. The T-cell receptor is represented as a single unit to depict the general features of H-2–restricted T-cell killing. Recent evidence suggests that the receptor may possess recognition sites for both self and foreign target antigens (*6*). Fig. 2 (right). Rejection of transplanted tumors based on class I–dependent T-cell responses. Transformed cells expressing a tumor-specific target antigen in association with the appropriate H-2 molecule on the cell surface can elicit a specific CTL-mediated immune response in transplanted recipients. In certain cases, the CTL's recognize the transplanted tumor cells as foreign, leading to the destruction of the transplanted tumor by the recipient mice (left). Such a CTL response, however, would not be generated when recipient mice are immunized with tumor cells that have lost the expression of the appropriate H-2 product (right). Although the tumor-specific target antigen is expressed, it would not be sufficient for CTL recognition. The transplanted tumor cells, left unchecked by the immune system of the recipient mouse, would be able to grow and progress to metastasis.

remains unknown, CTL function exhibits an absolute requirement for associated recognition between foreign antigens and self MHC components (*4–6*). As the products of specific class I genes are required for the restricted recognition of different viral antigens, the H-2 molecules are often referred to as restriction or control elements for CTL function. In addition, some CTL's can recognize non-self class I antigens as distinct targets and this phenomenon—known as allorecognition—is the cellular basis of the transplant rejection that occurs between histoincompatible strains of mice (*7*). The properties of CTL's contrast with those of natural killer (NK) cells. NK cells comprise a distinct subset of lymphoid cells able to kill certain tumors without preimmunization or MHC restriction. NK cells are believed to be important in the surveillance and elimination of cancer (*4, 8*).

Many tumors appear to express cell surface molecules that can be recognized in association with self H-2 as foreign and become the object of an immune response (*4*). Thus, it is possible that the host immune response against a nascent neoplasm has a significant effect on tumor survival and metastasis. Although it is difficult to estimate to what extent such immune surveillance is responsible for the elimination of incipient tumors in healthy animals, recent evidence suggests that the level of the immune response against transplanted tumors is a major factor in determining tumor growth and metastatic behavior (*9–13*). Therefore, any process perturbing immune recognition, for example modulation of class I expression (*14*), might facilitate the escape of these target cells from immune destruction in the host (Fig. 2).

The class I alterations that have been observed on tumors can be grouped into the following three classes: (i) the loss or quantitative attenuation of the expression of specific class I antigens; (ii) the enhanced expression of H-2 or the activation of unexpressed class I loci mapping to the Qa/Tla region of the MHC; and (iii) the generation of novel class I antigens by mutation or recombination. While these phenomena are not observed in all tumor systems, tumor-associated alterations of MHC components may be one of several decisive factors determining the course of certain neoplastic diseases. The immunobiology associated with these phenomena has been recently reviewed (*4*). We will focus our discussion on the insights afforded by the recent application of molecular technology to the analysis of class I expression on certain metastatic tumors.

Attenuated Expression of Class I Products on Metastatic Tumors

Tumors, especially those with virally or chemically induced etiologies, frequently exhibit an altered profile of class I products on their surface (*4, 15*). For example, the virally induced AKR T-cell leukemia line K36.16 expresses normal levels of H-2D antigen but negligible H-2K, as compared to normal lymphocytes (*14*). Unlike related AKR tumors that express H-2K and are rejected in immunocompetent, histocompatible mice, the K36.16 cell line is resistant to H-2K–restricted killing by T cells and readily progresses when transplanted into normal hosts (*14*). Presumably, the H-2K antigen, but not the H-2D antigen, is capable of acting as a restriction element for CTL recognition of the Gross/AKR murine leukemia viral antigen expressed on these cells, such that reduced H-2K expression allows the cells to escape immune destruction. To establish whether attenuation of the products of the H-2K locus was responsible for the altered growth behavior, Festenstein and coworkers (*9*) introduced the normal AKR H-2K gene into K36.16 cells. Clones of the transfected tumor cells expressing the transferred H-2K sequences at high levels were rejected when transplanted into normal, histocompatible mice. The same cells were able to progress to tumors when the hosts had been immunosuppressed by moderate doses of radiation. Thus, the attenuated H-2K expression on these cells correlates directly with their ability to avoid immune rejection and grow in immunocompetent hosts.

In the case of K36.16, H-2–restricted CTL killing may not be sufficient to account for the immunologic behavior and metastatic potential of the H-2K transfectants. For example, mice preimmunized with H-2K–positive K36.16 transfectants were later able to reject a second challenge with the original H-2K–negative K36.16 cells (*9*). Although H-2K expression appears to be required to initiate an adequate immune response for the rejection of the tumor, mice primed with tumor cells that express H-

2K are able to reject a subsequent challenge by H-2K–negative cells. Thus, this response probably includes additional humoral as well as nonspecific components, such as NK cells, capable of eliminating the parental CTL-resistant tumor.

The relationship between class I expression and the metastatic behavior of tumors has also been examined by Katzav, Segal, Feldman, and their colleagues in studies of variants of the methylcholanthrene (MCA)-induced sarcoma T10 (*12*). Like many MCA- and virally-induced sarcomas, T10 expresses a tumor-associated antigen (TAA). The TAA's expressed on these tumors are antigenically diverse and have been characterized on the basis of their ability to immunize histocompatible mice against the tumor of origin (*16, 17*). These molecules appear to be unrelated to the class I antigens but can be recognized in the context of self-MHC as the target of a CTL response against the tumor (*18*).

T10 is derived from an F1 hybrid of two distinct inbred mouse strains and should express the allelic products of both parental H-2K and H-2D loci. The genotype of this tumor thus facilitated the analysis of the effects of the expression of the individual H-2 antigens on the metastatic progress of the tumor in hosts of the same F1 origin. Although a highly metastatic variant of this tumor, IE7, and a moderately metastatic variant, IC9, express H-2D–encoded class I products, neither type of cell expresses either H-2K allele (*19*). Hammerling and his colleagues (*20*) introduced each parental H-2K allele separately into IE7 and IC9 and found that expression of either of the parental H-2K antigens severely reduced the metastatic phenotype of the transplanted cells. The H-2K products appeared to function as classical CTL restriction elements in this system: CTL's generated against T10 cells that expressed one of the parental alleles could only recognize target cells expressing that same antigen. These studies suggest that a CTL response is responsible for the rejection of the H-2K–expressing T10 transfectants, at least in the context of the transplantation assay, and that H-2K loss allows the tumor to escape immune destruction. Such conclusions are only possible given the ability to specifically manipulate the profile of class I expression of a tumor by transfection with cloned class I genes (*21*).

The role of class I expression in the immunosurveillance of tumors has recently been highlighted by the analysis of several adenovirus-transformed cell lines having different metastatic potentials. Several strains of adenovirus (for example, Ad2 and Ad5) can transform rodent cells in culture, but these cells are unable to progress to tumors when transplanted into histocompatible hosts—presumably because of immune rejection. However, cells transformed with the highly oncogenic adenovirus strain Ad12 are able to grow and metastasize (*22*). One difference between Ad5- and Ad12-transformed rat cells that could explain the oncogenicity of Ad12 is the apparent absence of class I antigens on the Ad12 transformants (*23*). Low levels of class I transcripts were detected in these cells. These results have been confirmed with adenovirus-transformed mouse cells (*24*). H-2 expression was low or absent on the surface of the Ad12 transformants, as determined by means of monoclonal antibodies to H-2. However, residual levels of class I transcripts are present.

Although it is possible that such a reduction in class I expression could have a significant impact on the efficiency of CTL recognition, it is not clear that evasion of T-cell recognition is an adequate explanation for the oncogenic character of the Ad12 transformants. The oncogenic potential of hamster and rat cells transformed with the different adenovirus strains appears to correlate more meaningfully with their resistance to killing by macrophages and NK cells than with their resistance to CTL lysis (*25, 26*). In fact, the various sets of adenovirus transformants do not appear to differ significantly in their sensitivity to killing by either alloreactive CTL's or, more relevantly, CTL's generated against adenovirus-transformed tumors (*25, 27*). It is not clear, however, whether NK function alone is sufficient for the rejection of an incipient tumor, since nude mice (which are CTL deficient but which express high levels of NK activity) are highly susceptible to tumorigenesis by cells transformed with all strains of adenovirus (*28*).

Whatever the immune component(s) responsible for the rejection of adenovirus-induced tumors, recent transfection experiments suggest that the level of class I expression on these tumors determines, in part, whether they will be oncogenic (*29*). In these experiments, an H-2L gene from the BALB/c mouse strain was transfected into Ad12-transformed cells of the C57BL/6 strain. The transfectants were much less oncogenic when transplanted into BALB/c mice than was the parental Ad12 tumor. Thorough biochemical and immunological characterization of these transfectants is necessary, however, before it will be possible to conclude that these cells are rejected by the same immune components that are responsible for the rejection of the nononcogenic Ad2-transformed cells in histocompatible mice. It is possible that tumor rejection in this case might be the result of phenomena more directly comparable to graft rejection and unrelated to the differences between Ad2- and Ad12-transformed cells. The C57BL/6 cells from which the tumor was derived undoubtedly express several polymorphic surface proteins that may be sufficient to elicit graft rejection even between H-2–compatible mice.

Molecular Mechanisms Involved in Attenuation of Class I Expression

Since alterations in class I expression appear to be involved in the escape from immune surveillance and the growth of some tumors, it is of interest to determine whether these tumor-associated alterations are the direct result of some mechanism related to cellular transformation. Given the variety of altered class I profiles that have been observed in tumors of diverse etiologies, a single unifying mechanism would be difficult to discern. Furthermore, only a fraction of all tumors exhibit any obvious changes in class I expression. Although it is likely that specific molecular mechanisms will be implicated in the class I alterations observed on certain classes of tumors—particularly those of lymphoid origin or viral etiology—we believe that most class I alterations arise from random events and are immortalized and amplified in the course of neoplastic progression. Just as the genotype of neoplastic cells in culture is extremely unstable, tumor cells in vivo probably progress via the sequential activation, mutation, and inactivation of a variety of gene products (*30–32*). Therefore, any molecular event in a tumor variant that results in the loss of the class I restriction element required for CTL recognition of the tumor cell could provide a selective advantage, leading to clonal expansion of this variant and fixation in the terminal phenotype of the metastasis.

Obviously, a number of molecular mechanisms might be responsible for the attenuation of class I expression seen on tumors of certain etiologies. These could include deletion, mutation, or recombination of class I sequences; transcriptional inactivation; defects in posttranslational processing; or even loss of β_2-microglobulin, a small protein associated with class I products on the cell surface (*10, 33–36*).

Activation or loss of specific transcriptional regulators might also be relevant, although there is, as yet, no molecular description of the factors that might be involved. One clue to the identity of these factors might be derived from the studies of the adenovirus system. Ad12-transformed cells express low levels of class I transcripts. However, expression can be restored by treatment with γ-interferon, an intercellular messenger that acts as a potent activator of class I expression (*24, 37*). Furthermore, repression of class I transcription in Ad12-transformed cells does not affect the expression of transfected class I genes (*29*). Surprisingly, Ad12 infection, in contrast to cellular transformation, results in a transient 10- to 15-fold increase in the expression of class I messenger RNA (*38*).

Tanaka *et al.* have shown that changes in the levels of DNA methylation surrounding class I genes might also be relevant (*39*). The activation of H-2K expression that occurs in differentiating F9 embryonal carcinoma cells was associated with an increase in the methylation state of the H-2K gene, as detected by a specific 3′ H-2K probe. These findings contrast with most other systems in which gene activation is correlated with a decrease in the state of DNA methylation (*40*). However, as the authors note, the behavior of the H-2K gene appears unique even among the class I genes of the F9 cells, most of which appear to become relatively hypomethylated during differentiation. The developmental changes in the methylation patterns of these genes are probably complex. Such changes in methylation patterns are generally believed to represent a response to changes in gene expression mediated by specific factors (for example, steroid hormone receptors) (*41, 42*). Since methylation patterns tend to be clonally maintained (*40, 43*) and DNA hypermethylation is capable of inactivating the expression of specific genes in other systems (*44*), it is possible that changes in the methylation pattern of specific class I genes might be responsible for the observed differences in class I expression on some tumors and their derivatives.

Methylation has been directly implicated in the regulation of class I expression in at least one tumor. Olsson and Forchhammer (*45*) have shown that both metastatic and nonmetastatic phenotypes could be imparted to subclones of the Lewis lung carcinoma by treatment with 5-azacytidine, a drug believed to specifically inhibit the enzymes responsible for the maintenance of DNA methylation patterns. The change in metastatic potential of the cells was associated with a change in expression of a particular class I molecule (*46*).

Enhancement and Activation of Class I Sequences in Tumors

While various studies point to a clear relationship between loss of class I products and tumor growth and progression, in other tumor systems just the opposite relationship has been observed. The metastatic potential of certain variants of the H-2K–negative T10 sarcoma, for example, appears to be directly related to the expression of a particular H-2D gene product (*37*). Although transfectants expressing H-2K were nononcogenic (*20*), analysis of the H-2 expression of metastatic and nonmetastatic derivatives of the H-2K–negative parental tumor revealed that, in each case, metastatic potential in histocompatible hosts was correlated with the expression of the H-2D antigen from one of the parental strains (*47*). The expression of this antigen was enhanced several fold with respect to normal spleen cells (*47*). Expression of this antigen appeared to be causally related to the invasiveness of these cells since the metastatic potential could be reversed by selection of variants that had lost only one of the parental H-2D products (*47*). Moreover, H-2D–positive T10 subclones were more metastatic than variants that were completely H-2–negative (*47*). However, expression of the other H-2D parental allele tended to make the T10 variants less metastatic. Similar observations have been reported for other systems (*48*).

The expression of the H-2D antigen does not enhance the metastatic potential of all tumors. Generally, H-2D appears to be important in the resistance to the tumorigenic effects of murine retroviral infection. The H-2D molecule is the restriction element for CTL-mediated killing of BALB/c cells that have been transformed by certain retroviruses, including radiation leukemia virus and Friend murine leukemia virus (*4*). Therefore, it is surprising that H-2D can mediate such distinct immunological effects on derivatives of the T10 sarcoma. The H-2D expressed on T10 might function, not as a conventional CTL restriction element, but in yet another capacity. For example, the expression of H-2D, in association with tumor-associated target antigens, might exert a regulatory effect on the specific immune response against the tumor. This might occur through interactions with suppressive immune components such as suppressor T cells (*49*) or by induction of CTL's capable of recognizing T-cell receptor structures on CTL's specific for the tumor (*50*). Such specific immune suppression could explain the lack of immunogenicity associated with these subclones (*51*). Alternatively, the apparent H-2D enhancement might actually represent the activation of another class I molecule, antigenically related to but distinct from H-2D, since the recognition of Qa-Tla class I gene products by monoclonal antibodies specific for H-2 antigens has been reported (*52*). This is particularly important since class I activation has been observed on a large number of thymic leukemias (*2*), on rat cells transformed with polyoma virus (*53*), on some retrovirally transformed mouse cells (*54*), and on mouse fibroblasts transformed in vitro with simian virus 40 (*55*).

The de novo activation of a normally silent, presumably intact class I locus has been implicated in the enhanced

invasiveness of several tumors, including the 5-azacytidine–induced metastatic variant of the Lewis lung carcinoma (*45, 56*). A monoclonal antibody generated against this variant may recognize a novel class I antigen, perhaps encoded within the Qa/Tla region of the MHC, which is specifically activated in this metastatic subclone (*45*). This product might exert an effect on the immune response to the tumor similar to that postulated for the H-2D antigen on T10. In addition, this category of altered H-2 expression may also affect the invasiveness of these cells through a nonimmunological mechanism. The H-2D products might be required for the functional assembly of membrane protein complexes essential for tumor localization and implantation (*57*).

Molecular Mechanisms Responsible for Class I Activation

In each of the systems described above, the tumors exhibiting altered class I expression were derived from rare variants and selected by passage in vivo. To gain insight into the molecular mechanisms underlying enhanced or activated class I expression it might be necessary to study systems that operate independently of immune selection. BALB/c 3T3 cells, which normally express reduced H-2, have enhanced levels of class I transcripts after transformation in vitro with SV40 (*55*). Although Rigby and co-workers originally identified a complementary (cDNA) clone derived from SV40-transformed 3T3 cells as the product of an activated Tla gene (*55*), further analysis revealed that this clone was, in fact, derived from H-2D (*58*). However, SV40-transformed 3T3 cells do express de novo–activated class I transcripts in addition to enhanced H-2D. Robinson, Hunt, and Hood have isolated a cDNA clone from the metastatic SVT2 tumor and have shown by DNA sequence analysis that this cDNA is derived from the Tla class I gene, TL10 (*56*).

Analysis of SV40-transformed 3T3 cells reveals that a variety of class I genes, as identified by specific oligonucleotide probes, can be activated in these cells (59). However, no consistent pattern of activation has been reported. Expression of these antigens is dependent upon the function of SV40 large T antigen (60), a viral regulatory protein required for maintenance of cellular transformation. Many of the class I genes activated by SV40 appear to contain repetitive elements known as B2 repeats. Since transcription of these sequences is enhanced in SV40-transformed cells (*61*) it is possible that B2 repeats are involved in the regulation of class I transcription by SV40.

Although DNA sequence analysis of one of the Tla genes activated in SV40-transformed 3T3 cells indicates that the transcripts could encode functional molecules (*45*), it is necessary to demonstrate that the transcripts give rise to class I molecules before their function can be discerned. The generation of specific serological probes to peptides predicted from these gene sequences should aid in the identification and characterization of such products and of their role in tumorigenesis. Since enhanced or activated class I expression in transformed cells might affect the immune response to such cells, it is also vital to identify the immunological factors underlying the phenomena. This includes not only biochemical characterization of the products of the activated class I genes but also the dissection of those immunological components (such as suppressor T cells) interacting with the tumor cells.

Expression of Unique Class I Products on Tumors

There have been many reports in the literature of tumor-specific expression of histocompatibility antigens alien to the strain from which the neoplasm was derived (*4, 15*). Such reports generally stemmed from investigations of the anomalous growth of tumors in transplanted recipients, either in their rejection in syngeneic hosts or progression in histoincompatible strains of mice. At issue was whether expression of these novel antigens involved alteration of classical H-2 structures, perhaps through mutational or recombinational events, or the activation of normally silent and intact class I loci. Early attempts to identify these tumor-specific antigens were often complicated by a number of problems including the use of complex (and occasionally contaminated) antisera to characterize the unique antigens, the lack of biochemical characterization of the products involved, and the use of tumors of uncertain origin. For these reasons, many of the early reports describing the identification of novel class I antigens on tumors actually concerned the identification of antigens that were not class I, or that were the normal H-2 antigens of the strain from which the tumor was derived. Nonetheless, the concept of tumor-specific class I antigens remains intriguing, particularly in view of the possible functions of the MHC in tumorigenesis.

Structural alteration of a CTL restriction element might allow a tumor variant expressing only a fraction of its H-2 genes to evade the immune response against the parental tumor (*4*). Novel class I products might also act to suppress the immune response against the tumor, perhaps by mechanisms comparable to those discussed above for the effect of elevated levels of H-2D on the T10 sarcoma. Cell-surface expression of novel class I products might be a by-product of the expression of a secreted form of class I antigen that had been generated by alternative RNA splicing events, such as that reported in liver (*3, 62*). Secreted class I antigens have been proposed to function in the regulation of CTL activity (*63*). It could conceivably be advantageous in evolutionary terms for normal tissue to express novel class I antigens when transformed, given the efficiency with which CTL's identify and eliminate cells expressing foreign MHC.

Schreiber and his colleagues have extensively analyzed an ultraviolet (UV)-induced fibrosarcoma (designated 1591) to determine the basis of its extreme immunogenicity after transplantation into immunocompetent, histocompatible mice (*64*). Specific monoclonal antibodies generated against this C3H tumor identified some of the immunogenic molecules on 1591 as class I antigens (*65*). The novel antigens are similar to, but distinct from, the class I products normally expressed by C3H and other strains of mice on the basis of serological studies, tryptic peptide analysis, and two-dimensional gel analysis (*66*). Goodenow and co-workers have cloned the genes encoding these novel antigens and demonstrated that 1591 expresses, in addition to the normal complement of H-2K and H-2D, at least three novel class I antigens not normally expressed on C3H tissue (*67*), one of which appears to be derived from the H-2K gene (*68*). Analysis of mouse fibroblast lines transfected with the novel genes indicates that these unique class I antigens function as the targets of many of the CTL's generated against the tumor in normal mice (*69*). Furthermore, loss of the novel class I molecules from variants of 1591 correlates with their immunogenicity (*65*). Thus, these molecules appear to limit tumor progression in a transplant situation.

Southern blot and sequence analysis of the class I genes of 1591 and C3H tissue suggests that these novel antigens are generated by multiple recombination events among the endogenous class I genes of C3H (for example, H-2K) that might have occurred during the course of tumorigenesis (*67*). Recombination of class I sequences, as distinct from gene activation, might therefore be responsible for some of the tumor-specific transplantation antigens reported. Since recombination rather than simple mutation is believed to be the primary mechanism responsible for generating the enormous polymorphism seen among the H-2 antigens of different strains of mice (*1*), it is reasonable to implicate recombination in the generation of novel class I antigens. Events that could best be explained by recombination have also been observed by Martin *et al.* in their studies of the structural alterations of a variant class I antigen expressed on a C3Hf adenocarcinoma (*70*). C3Hf is a strain of mouse derived from C3H by mutational alteration of the H-2K gene. The alterations detected on the tumor antigen appear to represent a partial reversion of these original meiotic mutational events (*71*). Thus, somatic recombination of class I genes in tumors may be mechanistically comparable to events that occur normally in the germline (*72–74*).

Ordinarily, the expression of such strongly immunogenic molecules would be expected to cause rejection of a nascent tumor and prevent progression to metastasis. However, since up to 70 percent of all (UV-induced fibrosarcomas that progress in their original hosts are as immunogenic as 1591 as transplants, this is clearly not the case (*75*). The immunobiological basis for tumor susceptibility may involve both immunogenic and suppressive elements. This stems from the following observations: (i) UV-irradiated mice cannot reject UV-induced tumors, but continue to have the ability to limit the growth of other non–UV-induced tumors. (ii) Normal mice surgically linked to UV-irradiated mice lose the capacity to resist UV-induced tumors. This suggests that the UV treatment may elicit the production of suppressive immune components capable of specifically eliminating the T-cell response against UV-induced tumors. Moreover, this suppressor population is specific, distinguishing between transplants derived from UV-induced and, for example, MCA-induced tumors. In fact, these cells even appear to be able to distinguish between tumors induced by specific wavelengths of UV-irradiation (*75*). One possible explanation is that UV light induces a common, stably inherited antigenic change in all exposed cells that mediates suppression. Tumors arising from irradiated cells would then express this common antigen and be protected from immune rejection (*75*). Although it is unclear what relationship, if any, exists between the novel class I antigens of 1591 and UV-specific suppression, it is possible that these molecules were important to the progression of this tumor in the original irradiated host.

Discussion

Neoplastic progression can be thought of as the evolution of the tumor through stages of sequential selection and adaptation, stages which can now be defined by means of molecular probes (*30, 31*). Neoplastic development must be influenced by the selection of tumor variants best suited to growth, expansion, and metastasis. As a result, the interactions between the host and the tumor undoubtedly play a dynamic role in the pathology of the disease. Evidence for these interactions can be inferred from the phenotype of the tumor. Just as the elevated levels of plasminogen activator detected within metastatic clones of some tumors (*76*) might reflect the need for proteolytic activity at some time during the metastatic process, the altered class I expression frequently observed on metastatic tumors might imply that evasion of a CTL response is a common event during neoplastic progression.

Immune recognition could obviously have a dramatic effect on the evolution of many tumors through selection of tumor subsets capable of avoiding rejection. Careful analysis of the immunological behavior of the terminal metastatic cells should provide some insight into the nature of the selective pressures that had acted upon the parental tumor population. Unfortunately, most studies rely upon transplantation as the primary measure or indication of the immunogenicity and metastatic potential of a variant tumor, and the transplantation assay might not accurately reflect the unique immunological environment facing a variant cell in a nascent tumor. However, within the context of the transplantation assay, it is clear that modulation of class I expression can be considered a major factor in the ability of some tumors to evade immune destruction. It will be necessary to confirm that the phenomena defined by transplantation are relevant to the process of tumorigenesis.

The results discussed in this review have been derived from studies in rodent systems. It will be essential to determine whether any of the phenomena which we have described also apply to human tumors. However, the results obtained for the murine neoplasms are relatively unambiguous, since these studies take advantage of inbred mouse strains expressing well-characterized complements of MHC genes and products. In contrast, comparable human studies may be obscured by the complexities inherent in the human equivalent of the MHC, the HLA system. Furthermore, it may be impossible to make any meaningful judgments about the immunogenicity of human tumors without the aid of appropriate genetically characterized transplant hosts. The example of the highly immunogenic UV-induced murine tumors has shown that "if the only measure of antigenicity available were the immune response induced by the tumor in its primary host, we would conclude, incorrectly, that the tumors were not antigenic. This is, of course, precisely the situation with human cancers" (*75*). Despite the difficulties of studying human cancer, epidemiological evidence with respect to immunosuppressed patients indicates that the immune system plays an essential role, at least in the preven-

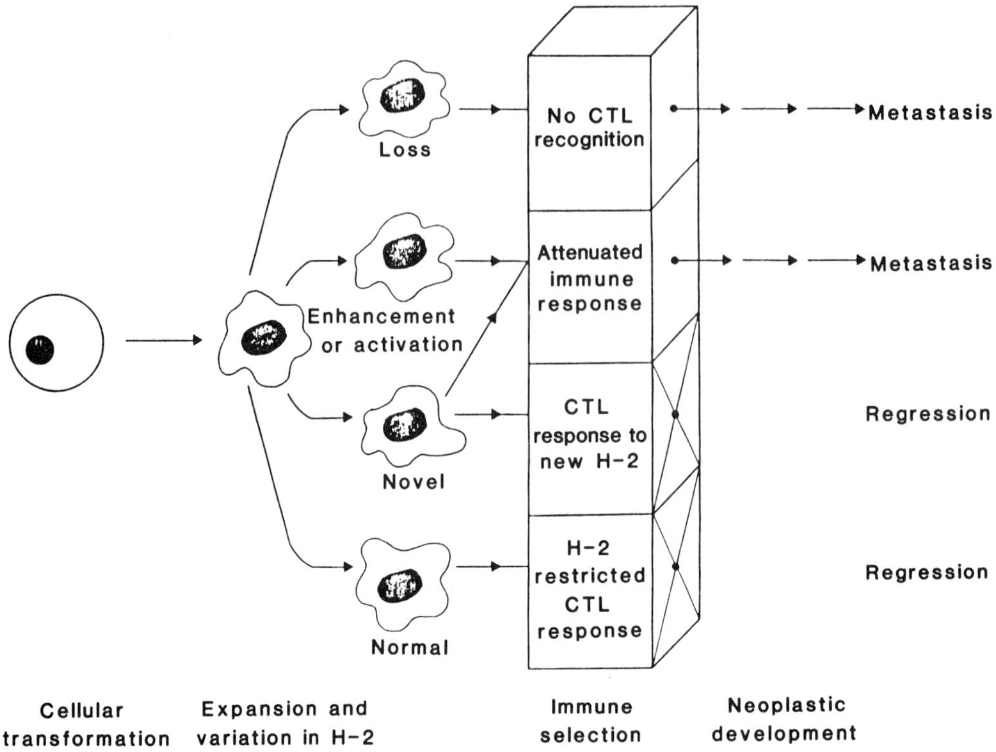

Fig. 3. Class I–mediated immune surveillance during tumorigenesis. The immunological fate of transformed cells displaying various H-2 expression profiles is depicted. Within the mouse, a cell may become transformed, allowing it to grow and expand beyond the limit permitted by the normal cellular regulatory processes. During expansion, the transformed cell can undergo changes in the expression of a number of genes, including those encoding class I antigens. The animal's immune system may or may not produce a CTL response to the altered class I phenotype of the tumor cell. Loss of expression of the appropriate H-2 antigen would allow the tumor cell to evade CTL recognition. Enhanced or activated class I expression may result in a reduced ability for the immune system to respond to the target antigens on the transformed cell. In both cases, such cells may undergo additional stages of selection, perhaps mediated by humoral or other cellular (for example, natural killer) responses, before reaching the final stage of metastasis. Novel class I antigens expressed on tumor cells may elicit a CTL response, in which case these transformed cells would be lysed and the tumor would regress. Similarly, a class I–restricted CTL response to a tumor cell expressing an unaltered H-2 profile would block neoplastic progression.

tion of those human tumors with a viral etiology, such as Burkitt's lymphoma (*31*) and papilloma (*77*).

A number of issues remain unresolved concerning the extent to which alterations in class I antigen expression are responsible for the progression of neoplastic cells in the murine system. Foremost among these is a meaningful estimation of the generality of these phenomena in diverse tumor systems. Many metastatic tumors show no apparent alterations in the level or nature of the class I products expressed (*4*). However, without the appropriate antibody probes to rule out the possibility of expression of class I antigens with subtle structural alterations, anomalous expression may remain undetected. The discovery of serologically undefined class I products on some metastatic tumor variants emphasizes this possibility (*45*). However, class I anomalies might not be expected to be associated with tumors of all etiologies, since immune selection is probably only one of many factors involved in neoplastic development.

It is equally important to identify the molecular mechanisms responsible for the alterations in class I expression that occur during tumor progression. Since the events that modify class I expression in tumor variants probably occur only rarely in the tumor populations, their characterization may be extremely difficult.

The number of different relationships described between H-2 and the growth behavior of certain tumors points to the potential complexity of the interactions between a disseminating tumor and the immune system (Fig. 3). A tumor expressing a foreign antigen in association with an appropriate class I restriction element could be eliminated by the immune system of a healthy animal. Although any alteration in the class I expression of the tumor could profoundly affect the efficacy with which the immune system recognizes a malignant cell, the oncogenic outcome of any specific alteration will depend upon the complex interactions of immune regulation. Additional studies may elucidate the multiplicity and generality of class I-dependent immunoselection during tumorigenesis. This will require the use of cloned genes, defined populations of lymphoid effector cells, and specific antibodies capable of identifying target molecules or distinguishing discrete differences in the expression of class I antigens. Future experiments, including tumor induction in transgenic mice that express class I mutations, are needed to describe some of the relevant molecular interactions between the T-cell network and the H-2 molecules and should facilitate the careful dissection of the molecular bases for MHC-dependent immune-responsiveness in tumorigenesis.

References and Notes

1. L. Hood, M. Steinmetz, B. Malissen, *Annu. Rev. Immunol.* **1**, 529 (1983).
2. L. J. Old, *Cancer Res.* **41**, 361 (1981).
3. J.-L. Lalanne *et al.*, *Cell* **41**, 469 (1985)
4. P. C. Doherty, B. B. Knowles, P. J. Wettstein, *Adv. Cancer Res.* **42**, 1 (1984).
5. R. M. Zinkernagel and P. C. Doherty, *Adv. Immunol.* **27**, 51 (1979).
6. B. Pernus and R. Axel, *Cell* **41**, 13 (1985).
7. T. J. Braciale, M. E. Andrew, V. L. Braciale, *J. Exp. Med.* **153**, 1371 (1981).
8. N. Hanna and R. C. Burton, *J. Immunol.* **127**, 1754 (1981).
9. K. Hui, F. Grosveld, H. Festenstein, *Nature (London)* **311**, 750 (1984).
10. M. Imamura and W. J. Martin, *J. Immunol.* **129**, 877 (1982).
11. L. R. Gooding, *ibid.*, p. 1306.
12. P. De Baetselier, S. Katzav, E. Gorelik, M. Feldman, S. Segal, *Nature (London)* **288**, 179 (1980).
13. H. Festenstein and W. Schmidt, *Immunol. Rev.* **60**, 85 (1981).
14. W. Schmidt and H. Festenstein, *Immunogenetics* **16**, 257 (1982).
15. G. Parmiani *et al.*, *ibid.* **9**, 1 (1979).
16. G. Parmiani, *Transplant Proc.* XII, 50 (1980).
17. G. C. DuBois, E. Appella, L. W. Law, *Int. J. Cancer* **34**, 561 (1984).
18. L. Åhrlund-Richter, E. Klein, F. Merino, *Immunogenetics* **15**, 53 (1982).
19. S. Katzav *et al.*, *Transplant. Proc.* XV, 162 (1983).
20. R. Wallich *et al.*, *Nature (London)* **315**, 301 (1985).
21. R. S. Goodenow, M. McMillan, A. Örn, M. Nicolson, N. Davidson, J. A. Frelinger, L. Hood, *Science* **215**, 677 (1982).
22. G. Huebner, in *Perspectives in Virology*, M. Pollard, Ed. (Academic Press, New York, 1967), vol. 5, p. 147.
23. P. I. Schrier, R. Bernards, R. T. M. J. Vaessen, A. Houweling, A. J. van der Erb, *Nature (London)* **305**, 771 (1983).
24. K. B. Eager *et al.*, *Proc. Natl. Acad. Sci. U.S.A.* **82**, 5525 (1985).
25. K. Raška, Jr., and P. H. Gallimore, *Virology* **123**, 8 (1982).

26. A. M. Lewis, Jr., and J. L. Cook, *Science* **227**, 15 (1985).
27. G. H. Mellow *et al.*, *Virology* **134**, 460 (1984).
28. R. B. Herberman and H. T. Holden, *Adv. Cancer Res.* **27**, 305 (1978).
29. K. Tanaka, K. J. Isselbacher, G. Khoury, G. Jay, *Science* **228**, 26 (1985).
30. P. C. Nowell, *ibid.* **194**, 23 (1976).
31. G. Klein and E. Klein, *Nature (London)* **315**, 190 (1985).
32. G. Poste and I. J. Fidler, *ibid.* **283**, 139 (1980).
33. P. Gladstone, L. Fueresz, D. Pious, *Proc. Natl. Acad. Sci. U.S.A.* **79**, 1235 (1982).
34. B. Holtkamp, M. Cramer, K. Rajewsky, *EMBO J.* **2**, 1943 (1983).
35. H.-G. Burghert and S. Kvist, *Cell* **41**, 987 (1985).
36. F. Rosa *et al.*, *EMBO J.* **2**, 239 (1983).
37. D. Wallach, M. Fellows, M. Revel, *Nature (London)* **299**, 833 (1982).
38. A. Rosenthal *et al.*, *ibid.* **315**, 579 (1985).
39. K. Tanaka, E. Appela, G. Jay, *Cell* **35**, 457 (1983).
40. W. Doerfler, *Annu. Rev. Biochem.* **52**, 97 (1983).
41. J. B. E. Burch and H. Weintraub, *Cell* **33**, 65 (1983).
42. R. M. Grainger *et al.*, *Nature (London)* **306**, 88 (1983).
43. R. Stein, Y. Gruenbaum, Y. Pollack, A. Razin, H. Cedar, *Proc. Natl. Acad. Sci. U.S.A.* **79**, 61 (1982).
44. M. Busslinger, J. Hurst, R. A. Flavell, *Cell* **34**, 197 (1983).
45. L. Olsson and J. Forchhammer, *Proc. Natl. Acad. Sci. U.S.A.* **81**, 3389 (1984).
46. L. Olsson, *Fed. Proc. Fed. Am. Soc. Exp. Biol.* **44**, 5414 (1985).
47. S. Katzav, P. De Baetselier, B. Tartakovsky, M. Feldman, S. Segal, *J. Natl. Cancer Inst.* **71**, 317 (1983); S. Katzav, S. Segal, M. Feldman, *Int. J. Cancer* **33**, 407 (1984).
48. L. Eisenbach, S. Segal, M. Feldman, *Int. J. Cancer* **32**, 113 (1983).
49. B. Payelle, G. Lespinats, S. Tlouzeau, *ibid.* **34**, 95 (1984).
50. P. M. Flood, M. L. Kripke, D. A. Rowley, H. Schreiber, *Proc. Natl. Acad. Sci. U.S.A.* **77**, 2209 (1980).
51. S. Katzav, S. Segal, M. Feldman, *J. Natl. Cancer Inst.* **75**, 307 (1985).
52. S. O. Sharrow, L. Flaherty, D. H. Sachs, *J. Exp. Med.* **159**, 21 (1984).
53. B. Majello, G. La Mantia, A. Simeone, E. Boncinelli, L. Lania, *Nature (London)* **314**, 457 (1985).
54. D. C. Flyer, S. J. Burakoff, D. V. Faller, *J. Immunol.* **135**, 2287 (1985).
55. P. M. Brickell, D. S. Latchman, D. Murphy, K. Willison, P. W. J. Rigby, *Nature (London)* **306**, 756 (1983).
56. R. Robinson, S. Hunt, L. Hood, in preparation.
57. M. Simonsen *et al.*, *Prog. Allergy* **36**, 151 (1985).
58. P. M. Brickell, D. S. Latchman, D. Murphy, K. Willison, P. W. J. Rigby, *Nature (London)* **316**, 162 (1985).
59. R. Robinson and S. Hunt, in preparation.
60. R. Robinson, personal communication.
61. K. Singh, M. Carey, S. Saragosti, M. Botchan, *Nature (London)* **314**, 553 (1984).
62. M. Kress, Y. Barra, J. G. Seidman, G. Khoury, G. Jay, *Science* **226**, 974 (1984).
63. M. Kress, D. Cosman, G. Khoury, G. Jay, *Cell* **34**, 189 (1983).
64. R. D. Wortzel, C. Philipps, H. Schreiber, *Nature (London)* **304**, 165 (1983).
65. C. Philipps *et al.*, *Proc. Natl. Acad. Sci. U.S.A.* **82**, 5140 (1985).
66. M. McMillan, K. D. Lewis, D. M. Rovner, *ibid.*, p. 5485.
67. H. Stauss *et al.*, in preparation.
68. S. Watts, in preparation.
69. J. Forman, personal communication.
70. J. Martin *et al.*, *J. Immunogenetics* **5**, 255 (1978).
71. P. Callahan *et al.*, *ibid.* **130**, 471 (1983).
72. E. H. Weiss *et al.*, *Nature (London)* **301**, 671 (1983).
73. A. L. Mellor *et al.*, *ibid.* **306**, 792 (1983).
74. H. Sun, R. S. Goodenow, M. McMillan, L. Hood, *J. Exp. Med.*, in press.
75. M. Kripke, *Adv. Cancer Res.* **34**, 69 (1981).
76. L. Eisenbach, S. Segal, M. Feldman, *J. Natl. Cancer Inst.* **74**, 77 (1985).
77. H. Kirchner, *Immunol. Today* **5**, 272 (1984).
78. We thank J. Allison, M. Koshland, P. St. Lawrence, and M. Blanar for their helpful comments and review of the manuscript, and H. Kanera for technical assistance. Supported in part by NIH grant CA-37099A-02 and NIH Graduate Student Training Grant GM 07127-11 (R.L. and J.V).

8

Intrinsic and Extrinsic Factors in Protein Antigenic Structure

Jay A. Berzofsky

Interest in defining and predicting antigenic sites of proteins stems, in part, from the recent advances in the production of synthetic vaccines (*1*, *2*) and from the development of techniques for the cloning and sequencing of genes. Even in the absence of a known gene product, synthetic peptides can be made on the basis of the DNA sequence, and antibodies to these peptides can be used to isolate and characterize the unknown gene product (*1*). This interest goes back to Landsteiner (*3*), who studied antigen structure before anything was known of the structure of an antibody. A fundamental problem is that, experimentally, an antigenic site (also called an antigenic determinant or epitope) can *only* be defined by examining the products of the immune response (antibodies or lymphocytes) of a particular animal or person (the host) that has been exposed to thc antigen. Thus, for example, the portion of a protein bound by a specific antibody molecule can be called the antigenic site recognized by that antibody. The fundamental question is, can that site be called an antigenic site in its own right, or only with respect to the particular antibody which binds that site, or with respect to a host that can make such antibodies.

One viewpoint is that certain parts of proteins are inherently antigenic sites and that this property would be intrinsic to the nature of the protein molecule and independent of the host to be immunized (*4*). On the basis of this concept, attempts have been made to define the properties of certain protein substructures which might make them inherently antigenic. To design synthetic vaccines or prepare antibodies to proteins that have not been isolated and whose pri-

Science 229, 932–940 (6 September 1985)

mary sequence is known only from the DNA sequence, it is important to be able to predict which stretch of primary sequence is most likely to elicit antibodies that will also bind to the native protein.

The alternative viewpoint is that virtually any accessible part of a protein is potentially an antigenic site, and that the choice of sites that elicit an immune response in a particular case depends largely on the bias of the immune system of a specific host (*5*). For instance, for mammalian proteins injected into a mammalian host, self-tolerance to the homologous host protein may strongly influence the result. Even in the case of a nonmammalian protein for which there is no host counterpart (such as viral, bacterial, or invertebrate parasite antigens), the regulatory mechanisms of the host, such as immune response genes linked to the major histocompatibility complex of transplantation antigens (*6*), or networks of idiotype–anti-idiotype interactions (*7*), will determine the outcome of an immunization. Such regulatory mechanisms may also involve the interplay of helper and suppressor T lymphocytes, each with its own repertoire of antigenic specificities (*6, 8*). If this viewpoint is correct, the ability to rationally design synthetic vaccines or antibody probes would be limited and would require a more empirical approach, although comparisons of the antigen with homologous host proteins would provide a rational starting point. However, from a more optimistic point of view, if any accessible part of a protein is a potential antigenic site, the likelihood of being able to synthesize a useful site in the absence of much secondary or tertiary structural information about a protein is greatly increased.

The purpose of this review is to examine the evidence for each of these hypotheses, and to distinguish between properties intrinsic to the antigen and factors dependent on the host that determine the epitope specificity of an immune response. Although protein antigenic sites recognized by T cells have recently been characterized (*9, 10*), this review will be limited to the regulation of antibody specificity.

First, we must consider the semantic problem of immunogenicity versus antigenicity. Immunogenicity refers to the ability to elicit an immune response (antibody or T cell) when used to immunize an animal, whereas antigenicity refers to the ability to be recognized by the product of a previous immune response, either antibody or T cell. Although these properties often coincide, a few examples may illustrate the distinction. A hapten is defined as a small molecule (such as dinitrophenol) that is not immunogenic alone but, when attached to an immunogenic protein (called the carrier), will elicit antibodies that can bind to the free hapten. Thus, the free hapten is antigenic without being immunogenic. In another situation, although hen lysozyme is not immunogenic in a strain of mice such as C57BL/10 for reasons of T-cell suppression or Ir gene control (or both), when these mice are immunized with a fragment of hen lysozyme or with certain other lysozymes, the mice can still make antibodies that will bind cross-reactively to hen lysozyme (*8*). In this case, the sites on hen lysozyme bound by these antibodies are antigenic without being immunogenic in this strain. However, by definition, anything that is immunogenic must also be antigenic.

In principle, antigenic sites can be divided into two structural categories (*5*). A segmental (continuous) site exists wholly within a continuous segment of the amino acid sequence. An assembled topographic site consists of amino acid residues far apart in the primary sequence but brought together in the surface topography of the native protein by the way it folds in three dimensions (*5, 11, 12*). These categories are the outgrowths of the earlier categorizations of sequential and conformational sites (*13*), respectively. However, it has become clear that even segmental sites bind with highest affinity in a preferred conformation (*14, 15*). Furthermore, peptides have been synthesized either with the correct order of L-amino acids or with the reverse order of D-amino acids, which should result in approximately the same arrangement and order of amino acid side chains but a different peptide backbone configuration (*1*). Antibodies to the former did not bind the latter, an indication that recognition involves more than just a linear sequence of side chains. Finally, knowing that an antibody combining site consists of a three-dimensional array of amino acids with a defined surface contour, the prediction can be made that among various possible conformations of a polypeptide sequence, some would fit better into the antibody combining site than others and thus bind with higher affinity. In other words, the three-dimensional conformation of the antibody combining site defines an antigen conformation most complementary to itself. In this sense, all antigenic determinants must be conformational (*5, 15*).

The type of antigenic site observed depends in part on the probe or method used to investigate such sites. Early investigators of polyclonal antisera to sperm whale myoglobin used synthetic peptides or proteolytic peptides as probes and detected only segmental sites (*4, 16*). It was claimed that all of the antigenic activity of the protein could be accounted for by five short segments of the sequence consisting of six to seven residues each (a total of only about 20 to 25 percent of the sequence) (*4*), no matter which species of animal was immunized (*4*). However, with the advent of monoclonal antibodies (*17*), studies could be carried out that would not have been interpretable with heterogeneous serum antibodies. By use of native myoglobins with known sequence differences, rather than short peptides, it was found that most of the monoclonal antibodies studied reacted with assembled topographic sites, not segmental ones (*11, 12*). Moreover, these were mostly not in the region of the five segments delimited with synthetic peptides. How frequently do antibodies to assembled topographic sites occur in polyclonal antisera made against native myoglobin? A minimum estimate was obtained by exhaustive depletion of antibodies that could bind to any of the three large cyanogen-bromide cleavage fragments that together span the entire sequence of myoglobin (*18*). For each serum, from three different species (goat, sheep, and rabbit), there were always 28 to 40 percent of the antibodies remaining that could bind with high affinity to native myoglobin but which failed to bind to any of the peptide fragments in a radioimmunoassay. Even conformation-dependent antibodies to segmental sites should have been removed on the affinity columns because very low affinity binding to peptides is boosted by attachment of one of the reactants to a solid phase (*19*) and by the multiple plating efficiency of the column. Thus, more than a third and perhaps a majority of antibodies made by immunizing with native protein react with assembled topographic sites.

Similar results were found for a large number of monoclonal and polyclonal antibodies to hen lysozyme (*20*). A number of assembled topographic antigenic sites as well as some that are segmental but conformation-dependent have been defined by careful mapping, involving sequence variants (*21*), substrate or enzyme inhibitor competition, competitive binding of clusters of monoclonal antibodies, and even x-ray crystallography. Also, of the monoclonal antibodies to influenza hemagglutinin, influenza neuraminidase (*22*), and tobacco mosaic virus coat protein (*23*), many bind to assembled topographic sites.

The importance of physical contiguity can be understood by examining a space-filling model of a native globular protein such as myoglobin (Fig. 1). In contrast to a stick-figure model in which the polypeptide backbone is easy to trace, the space-filling model presents a continuous surface of abutting atoms, obscures the helices, and makes it difficult to distinguish between sequentially distant residues that are adjacent on the surface because of protein folding and those that are adjacent by being neighbors in the primary sequence. An antibody that binds to a native protein must interact with a surface more like that of the space-filling model. The probability that all of the antigenic contact residues (those that fit within the antibody combining site and contribute to the affinity of binding) happen to come from the same continuous segment of polypeptide chain should be extremely low.

A second way in which the type of probe used influences the repertoire of antibodies detected is illustrated by the use of synthetic peptides of varying length. In addition to the five segmental sites defined by short peptides (*4*), other segmental sites were identified when larger peptides of myoglobin were used (*24*) (see Fig. 1). Similarly, in the case of tobacco mosaic virus protein, synthetic peptides 13 to 20 residues in length were able to detect antibodies to regions of the molecule previously thought to be nonantigenic because of lack of antibody binding to hexapeptides within these segments (*25*). The ability to detect additional antibodies with longer peptides may be due to the involvement of more than six or seven sequential residues within the site or to greater stabilization of certain secondary structures in the longer peptides. Although peptides of 15 to 20 residues are still too short to stabilize an α helix in water, significant helix formation was detectable, in the case of the myoglobin peptides 25 to 55 and 72 to 88, when they were studied by circular dichroism in trifluoroethanol (*24*). An additional caveat regarding peptide length has come from the studies of short myoglobin synthetic peptides. The binding of shorter peptides of two to seven residues was dominated by nonspecific influences of charge and hydrophobicity, so that antibodies to staphylococcal nuclease would bind to these myoglobin peptides as well as antibodies made against myoglobin. Only when the peptide length was closer to 15 to 20 residues was biologically meaningful binding detected (*24*).

When all of these approaches are combined, it is generally found that virtually the entire accessible surface of the protein can be immunogenic in one animal or another. In addition to myoglobin, this is true for lysozyme (*20, 26*), bovine and human serum albumin (*5, 27*), tobacco mosaic virus protein (*23, 25*), and influenza neuraminidase (*22*). Even for influenza hemagglutinin, (*22*), for which

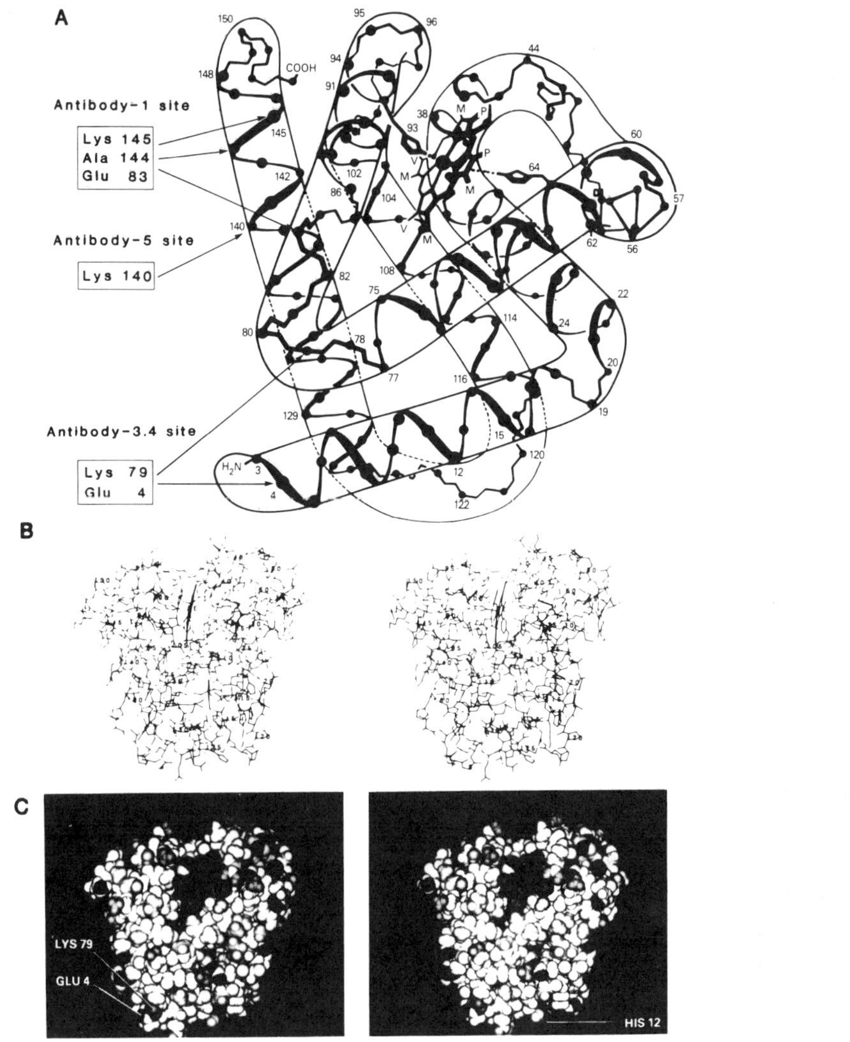

Fig. 1. Three ways of viewing the same face of sperm whale myoglobin, illustrating antigenic sites. (A) A line drawing representing the three-dimensional structure of the α-carbon backbone, modified from Dickerson (*73*). The numbers indicate the amino acid residue positions. M, V, and P are the methyl, vinyl, and propionyl groups of the heme. Side chains are omitted except for two histidines (64 and 93) involved with the heme. The residues involved in binding three conformation-specific monoclonal antibodies (*11*) are shown. Sites of antibodies 1 and 3.4 are assembled topographic sites. Although only one residue in the site has been defined, the binding of antibody 5 is also highly conformation-dependent. Other monoclonal antibodies, to beef myoglobin, bind assembled topographic sites that include residues 74, 87, and 142 and residues 34, 53, and 113, as described by East *et al.* (*11*). Five regions of segmental sites include residues 15 to 29, 56 to 69, 70 to 76, 139 to 146, and 147 to 153 as described by Crumpton and Wilkinson (*16*) or residues 15 to 22, 56 to 62, 94 to 99, 113 to 119, and 145 to 151 as described by Atassi (*4*). Longer peptides outside the latter five regions, which bind a significant fraction of antibodies (*24*), include residues 25 to 55 and 72 to 88 of beef myoglobin. (B) Computer-generated stereo view of the same face of sperm whale myoglobin, showing the amino acid side chains in place (*74*) based on the x-ray crystallographic coordinates of Takano (*75*). (C) Computer-generated stereo space-filling model of sperm whale myoglobin (*74*) and modified from Berzofsky *et al.* (*11*). Stereopairs may be viewed in three dimensions with a stereo viewer. The carboxyl oxygens are shaded darkest, followed by the heme and aromatic carbons, aliphatic side-chain carbons, noncarboxylic oxygens, primary amino groups, and other nitrogens. The backbone and side chains of nonaliphatic residues, except for functional groups, are shown in white. Residues 4, 12, and 79 in the assembled topographic site of antibody 3.4 are indicated. As one progresses from (A) through (B) to (C), it becomes progressively harder to discern the backbone helices, and what is seen is a surface similar to that seen by antibody.

only four discrete clusters of antigenic sites had been detected originally, studies with larger panels of antibodies showed the originally discrete regions of antigenicity to overlap as a continuum.

A continuum of antigenic sites over virtually the entire surface of many proteins was also indicated by pairwise comparisons among proteins of evolutionarily related species, such as lysozymes, myoglobins, ribonucleases, cytochromes c, azurins, and albumins. These studies demonstrated, by effects on cross-reactivity, that about 80 percent of the amino acid substitutions, scattered around the protein surface, were detectable by antibodies (*5, 26*).

All of these results lead to the formulation of a "multideterminant-regulatory hypothesis" (*5*). This hypothesis suggests that the surface of a protein is a continuum of potential antigenic sites, and that the selection of sites that elicit antibody production in a given animal depends on a number of regulatory influences in the host. Before discussing these extrinsic factors, we must examine the factors intrinsic to the protein antigen that have been considered as possibly predictive of antigenic sites.

Intrinsic Properties of the Antigen Used as Predictors of Antigenic Sites

A number of approaches have been taken to search for sites that are intrinsically antigenic and for principles that would predict such sites. Notwithstanding the evidence that the entire surface of a protein may be antigenic, these principles may be useful in studying relative antigenic potency, in analyzing the forces involved in antigen-antibody interactions, and in selecting optimal sequences to use as synthetic immunogens.

Accessibility. As a minimum requirement, antigenic sites should be accessible on the surface (*28*) in order to be detected by antibodies binding to the native protein. For example, studies of monoclonal antibodies to influenza neuraminidase showed the surface of the protein heads to be a continuum of overlapping antigenic sites (*22*), but no antibodies were found binding to the stalks. This apparent paradox was explained when it was noted that the protein of the stalks was almost entirely covered by carbohydrate and thus not accessible (not to say that carbohydrate cannot also be immunogenic).

Although accessibility is necessary, it is not known whether accessibility of a site on the native protein surface is sufficient for antibody binding. Also, accessibility is useful for predicting antigenic sites only for proteins whose three-dimensional structure is known from x-ray crystallography. It would be more broadly useful to have a method of predicting sites from primary sequence.

Hydrophilicity. Water-soluble globular proteins, to be stable in the aqueous environment, fold in the native conformation so as to bury hydrophobic residues in the interior and expose mostly hydrophilic ones on the surface. Hopp and Woods (*29*) suggested that possible antigenic sites might be predicted from primary sequence data by determining the most hydrophilic segments of the sequence. They averaged hydrophilicity parameters for overlapping segments of six amino acid residues, assigning to residue 1 the average for residues 1 to 6, to residue 2 the average for residues 2 to 7, and so forth. By analyzing 12 proteins for which antigenic sites had been reported, they found that the segment of greatest hydrophilicity for the entire protein was invariably in one of the known antigenic sites. However, other secondary peaks of hydrophilicity did not correlate very well with known antigenic sites, so that hydrophilicity alone was not sufficiently predictive. Nevertheless, they were able to predict an antigenic site of hepatitis B surface antigen and to verify its existence experimentally.

Fraga (*30*) extended this approach by considering semiempirical "recognition factors," which would reflect the likelihood that hydrophilic amino acids might interact with one another and thus be able to be buried in the protein despite their hydrophilicity. Combining these factors with the Hopp-Woods procedure produced a reasonable correlation between predicted antigenic regions and sites reported to bind antibodies, at least for the limited database of sites considered for each protein. However, in most of these studies of intrinsic factors, the correlation between antigenic sites and the intrinsic property in question is qualitative, not quantitative. It would be useful to define overlaps and statistical significance more quantitatively (*10*).

The advantage of hydrophilicity as a predictor of antigenic sites is that it requires only primary sequence information, not x-ray diffraction crystal structures. However, if the success of the hydrophilicity method is due solely to its correlation with surface exposure, then it is the surface exposure, rather than the hydrophilicity of the residues involved, that is important for antigenicity. Disproportionate increases in antibody binding have been observed during sequential peptide synthesis after addition of lysine residues (*24*), consistent with a major role of these hydrophilic residues in binding. However, such charged residues also contribute greatly to nonspecific binding (*24*). As further support for the importance of charge in antigen-antibody interactions, the classic study of Sela and Mozes and co-workers (*13, 31*) showed that antibodies specific for basic antigens tended to be acidic, whereas those specific for acidic antigens tended to be more basic. As most of the antibody charge variability is in the hypervariable segments of the variable region, which contribute to the combining site, this result suggests that antigen-antibody charge complementarity is frequently important for binding. On the other hand, a significant fraction of surface residues can be nonpolar (*28*), and several groups have described the importance of hydrophobic and especially aromatic residues in antigenicity (*13, 24, 32*). For several nonantibody protein-protein interactions, it has been estimated that most of the bonding energy derives from hydrophobic interactions (that is, exclusion of water over a large surface area of contact, with resultant increase in solvent entropy) (*33*). However, as this contribution to the energy does not contribute much to specificity, the specificity of interaction between two proteins (the selectivity with which binding occurs) depends more on the complementarity of the surfaces for hydrogen bonding and van der Waals contacts (*33*), and on more polar interactions.

In conclusion, although hydrophilicity, as a predictor of surface exposure, may have some value in predicting potential antigenic sites on proteins, it is necessary to be cautious about inferring any mechanistic implication for the biophysics of antigen-antibody interactions.

Mobility. Several groups have proposed that segmental mobility of the polypeptide backbone in portions of protein molecules may contribute to the antigenicity of these sites (*34, 35*). In contrast to hydrophilicity, this parameter may be more important for its theoretical implications in protein dynamics and protein-protein interactions than for its practical predictive value. The studies so far have depended on temperature factors or B values from x-ray diffraction data, which indicate the degree of atomic motion within the crystal structure. Such values can be obtained only from highly refined x-ray crystallographic data, available for only a handful of proteins. Although useful information may also be obtainable from temperature-dependent nuclear magnetic resonance studies (*36*), these, too, are difficult and not readily available for most proteins. Certainly, the method is not applicable to cases in which only the primary sequence is known, although exon-intron boundaries may be indicative of at least some mobile regions (*35*).

Second, for methodological reasons, mobility has been examined only for segmental antigenic sites. Therefore, the large fraction of sites that are of the assembled topographic type are excluded from consideration and would be very difficult to analyze or predict by this approach.

The issue of mobility has been studied from two very different approaches, the binding of antibodies against peptides and that of antibodies against proteins. These must be considered separately. First, Lerner and colleagues (*1, 37*) were surprised by the high frequency with which antibodies to synthetic peptides too short to have much native structure could nevertheless bind to the intact (although not always fully native) protein. If only a small fraction of the peptide molecules used for immunization were folded in a native-like conformation at any given time, a much smaller proportion of the antibodies might be expected to cross-react with the native protein. Yet, of 38 monoclonal antibodies to six peptides from four proteins, 24 reacted with the intact protein either in solution (where it should be native) or on nitrocellulose sheets (where it would likely not be native, but would also not be as random in conformation as the peptide) (*37*). One explanation considered (*37*) was that the protein structure was flexible enough to fit itself into an antibody combining site specific for a more unfolded conformation of the immunizing peptide (an induced fit hypothesis) (*38*) (see Fig. 2). Tainer *et al.* (*35*), using antisera against 12 overlapping peptides (of about ten residues each) from myohemerythrin, found an extremely good correlation between mobility of a segment of the native protein and the ability of antibodies to the corresponding peptide to bind the native protein in solution. Of course, surface segments of polypeptide may be more mobile than internal ones, but the correlation was not just with surface accessibility. Two regions of low mobility were as exposed as several of the more mobile regions, and yet bound much less to the antisera to the corresponding peptide. It was concluded that atomic mobility of a segment of the native protein was a critical factor in determining the ability of antibodies against short peptides corresponding to that segment to bind to the native protein.

This conclusion is a statement about the specificity and cross-reactivity of antibodies to peptides, not a statement about the immunogenicity of sites on the native protein. As such, it is compatible

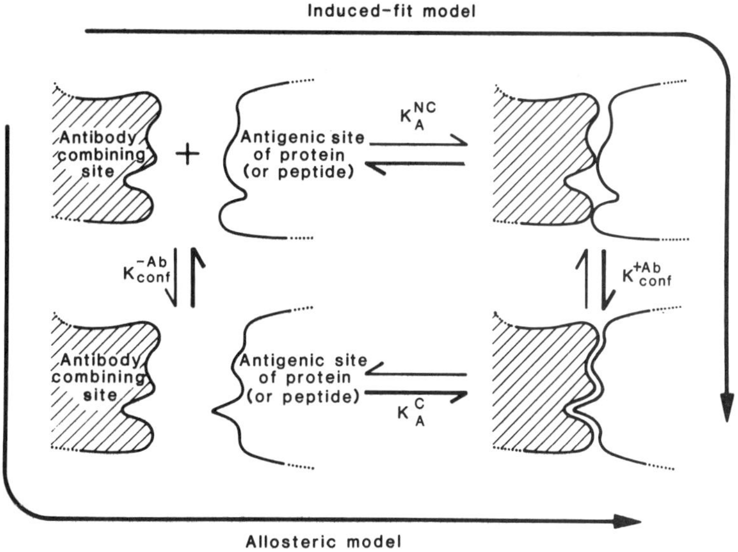

Fig. 2. Schematic comparison of the induced fit (*38*) (top and right) and allosteric (*40*) (left and bottom) models of the equilibriums between an antibody and a protein or peptide antigen whose predominant conformation (top left) does not make a best fit with the antibody combining site. The antigen can achieve greater complementarity by undergoing a conformation change (vertical equilibriums) (*14, 15, 38, 40*). K_A is the association constant for the bimolecular interaction with the antigen in the noncomplementary (NC) or complementary (C) forms. K_{conf} is the equilibrium constant for the unimolecular conformational equilibrium in the absence (−Ab) or presence (+Ab) of antibody. Bold arrows indicate the predominant direction of the reactions. In the induced fit model, an initial weak binding is stabilized by a conformational change in the antigen induced by the antibody. In the allosteric model, an unfavorable conformational equilibrium is "pulled" toward the complementary form by trapping of the complementary molecules in a tight complex with the antibody. A more complicated version, which may be more realistic (*45*), would allow the antibody also to undergo a conformational change.

with much other data (*14*). This approach may be useful in the selection of peptides to elicit antibodies cross-reactive with the native protein, but it does not predict which sites of the native molecule will elicit antibodies when the native protein is used as an immunogen. However, for the purpose of making site-specific antibodies as probes of the native molecule or for production of synthetic vaccines, it is sufficient to be able to select useful peptides, rather than to make predictions about immunization with the native protein.

The concept is also useful as it applies to the mechanisms of antigen-antibody interaction and, more broadly, to protein-protein interactions and the molecular dynamics of proteins. Studies of proteins in solution, such as by nuclear magnetic resonance, paint a picture of proteins that is much more mobile and fluctuating than that envisioned from the structure in crystals (*36, 39*). This mobility enables an antibody specific for an unfolded peptide conformation to bind to the intact protein, but at an energy cost. Studies measuring the affinity of antipeptide antibodies for binding to the native protein have found it to be lower than the affinity for the peptide itself by two to three orders of magnitude (*14*). This energy may be required to induce the optimal conformation complementary to the antibody combining site, as in the induced fit model (Fig. 2). Alternatively, it may be the entropic energy of locking in one conformation out of the many that are in equilibrium in solution before the antibody binds. This may be viewed as a type of allosteric model (*40*) (Fig. 2). Either way, since these mechanisms are two paths between the same end points (Fig. 2), the net change in free energy must be the same by the laws of thermodynamics.

Thus, antibodies may be useful for determining conformation (*41*) and dynamic mobility of proteins (*12*) and also for inducing (or stabilizing) conformational changes in a protein. A graphic example was the demonstration by Crumpton (*42*) that antibodies made against apomyoglobin (the form with the heme prosthetic group removed) reacted with native myoglobin to induce or stabilize a conformation in which the heme was squeezed out of the protein; brown myoglobin, precipitated with the antibodies, formed a white precipitate. More recently, monoclonal antibodies to native myoglobin have been found to induce or stabilize a conformation that favors the low-spin over the high-spin electronic state of the ferric heme iron, as measured by optical spectroscopy (*12*). Also, antibodies that bind the oxyhemoglobin conformation with higher affinity than the deoxy conformation enhance the oxygen affinity of human hemoglobin (*43*). Thus, antibodies may act as allosteric effectors and so alter the function of their target antigens.

A second approach to the question of mobility of antigenic structures was to examine the mobility of sites bound by antibodies raised against the native protein. Westhof *et al.* (*34*) compared the atomic mobility of segments of the polypeptide backbone of tobacco mosaic virus protein (as determined from x-ray crystallography) with the known antigenic sites (as determined from antibodies raised against the native protein and mapped to particular segments by binding to synthetic peptides). Only segmental sites could be examined by this approach. Nevertheless, in a plot of mobility (temperature factor) versus sequence, seven of seven sites corresponded to local maxima of mobility, and six of the seven corresponded to major peaks. Moreover, the correlation with mobility seemed better than with surface accessibility, as three hexapeptides that did not bind antibodies corresponded to regions that were exposed but not more than average in mobility. However, subsequently longer peptides corresponding to regions of low mobility were found to bind to antibodies from several antisera against the native protein (*25*). The longer peptides span a broad stretch of low mobility from about residues 115 to 150, to which antibodies had not been detected with the hexapeptides. Thus, the exposed regions of low mobility in the native protein were immunogenic, but the antibodies specific for these sites did not cross-react with the short hexapeptides. Similarly, a good correlation was found between reported segmental antigenic sites for myoglobin and lysozyme and the segmental mobility of these proteins (*34*). However, for myoglobin, seven of the nine sites examined were based on short peptides six to eight residues long (*4*). Thus, with this limited database, antibodies cross-reactive to short peptides were the major category of antibodies studied. Combining the two studies, these investigators have concluded that the entire surface of the tobacco mosaic virus protein is antigenic, and suggest that no single criterion can be used to distinguish nonantigenic from antigenic regions (*25*). As noted earlier, the sites found depend considerably on the probes used in the investigation. Mobility in this case, as for the antipeptide antibodies, is important in determining cross-reactivity rather than immunogenicity. Antibodies to more mobile regions of native proteins are more likely to cross-react with short peptides than are antibodies to less mobile regions, but both are produced on immunization with the native protein. Indeed, if native molecules bearing mutational changes are used as probes, instead of fragments, another category of antibodies is frequently found. These antibodies react with assembled topographic sites that depend strongly on the maintenance of the local tertiary structure (*5*). Thus, we cannot conclude that mobile segments are inherently more immunogenic or antigenic in the native protein than less mobile segments, although it is possible that the relative frequency of antibodies to different regions is affected by their mobility.

Nevertheless, the studies of mobility of antigenic sites provide important insights. In order to use peptides on columns to fractionate antisera against the native protein, it is important to choose peptides corresponding to more mobile regions or to use longer peptides, and to know which antibodies may be missed in such a procedure. Secondly, as discussed by Westhof *et al.* (*34*), antigenic flexibility may contribute to the ability of the immune system to provide a defense against such an enormous diversity of antigens. The clonal selection theory (*44*), for which an enormous body of evidence has been amassed, states that the specificity of antibodies on B lymphocytes is determined genetically before antigen enters the system. The antigen activates and expands those clones of B cells that bear antibodies capable of binding the antigen. However, it is unlikely that there will be preformed antibodies on B cells that can bind every possible conformation of every polypeptide sequence. If the protein antigens are flexible, they are more likely to find some antibody for which they can achieve a satisfactory (induced) fit (*34, 35*). This possibility expands the potential repertoire of antibodies, but at the cost of reduced affinity.

Finally, the implications of the mobility hypothesis for the biophysics of antigen-antibody and, in general, protein-protein interactions may be of great importance. Neither the antigen nor the antibody combining site can be viewed as a static structure. X-ray diffraction of a series of fluorescent ligands interacting with the combining site of an immunoglobulin light-chain dimer showed that both side chains and polypeptide backbone of the antibody combining site move to accommodate different ligands (*45*). Thus, structural flexibility of both the protein antigen and the antibody may be valuable to achieve optimum complementarity.

Factors Extrinsic to Antigen Influencing Immunogenicity of Specific Sites

Although intrinsic factors may determine the repertoire of potential antigenic sites in a protein antigen, only a subset of these sites will elicit antibodies when any given person or animal is immunized. For example, two peptides reported to bind antibodies made against sperm whale myoglobin in rabbits, goats, and mice (*4*) did not bind any detectable antimyoglobin antibodies made in two other goats, a sheep, and several high-responder strains of mice (*46*). Thus, factors in the host being immunized, as distinct from any structural features inherent in the antigen, may be of paramount importance in determining the outcome of any immunization. The discussion here will be limited to a summary of the impact of these factors on antibody specificity.

Tolerance to self. Self-tolerance is one of the fundamental properties of the immune system (*47*). The ability to break tolerance by experimental manipulation and the appearance of autoantibodies in certain pathologic states indicates that the potential structural gene repertoire is present in the genome to make antibodies that will react with self. Nevertheless, under normal circumstances, antibodies to host proteins are not made. Therefore, when the immunogen is a mammalian or avian protein homologous to a protein of the host, antibodies will be made primarily to those sites that differ from those of the host protein (*5*). For example, when rabbits were immunized with guanaco, mouse, or horse cytochrome c, the subpopulations of antibodies detected could mostly be accounted for by binding to the sites at which rabbit cytochrome c differs from the immunogen (*48*).

A second example is that of antibodies to beef myoglobin raised in various species (*49*). The antibodies raised in rabbits, dogs, and chickens bound almost equally well to bovine or sheep myoglobin, whereas the antibodies raised in sheep distinguished strongly between beef and sheep myoglobin, which differ at only 6 of 153 residues. Thus, the sheep's self-tolerance led to antibodies specific for sites at which beef and sheep myoglobin differ, whereas most of the antibodies raised in the other species bound sites shared by beef and sheep myoglobins. Differences between the immunogen and homologous host proteins are more important than anything inherent in the structure of the immunogen in determining the outcome of an immune response, at least for antigens which have homologues in the host. The same should apply to monoclonal antibodies as well, although an occasional autoantibody may be isolated.

Immune response (Ir) *genes.* Ir genes are genes that regulate the ability of an individual to make an immune response to a specific antigen (*6*). Antigen specificity is a key part of the definition, as the gene responsible for a broad-spectrum immunodeficiency disease would not be called an Ir gene. Most of the immunoregulatory genes that have proved to be antigen-specific, and thus qualify as Ir genes, are part of the major complex of genes encoding transplantation antigens, known as the major histocompatibility complex (MHC). Structural mutations have definitively shown that the Ir genes are actually the structural genes for MHC antigens (*6*). Their mechanism of action appears to be involved with the way in which T lymphocytes are activated, not by free antigen in solution, but only by a combination of the antigen with an appropriate MHC antigen on the surface of another cell. The Ir genes thus determine which T lymphocytes are activated, although it is still debated whether they accomplish this through a direct, specific interaction with the antigen or through an effect on the available repertoire of T cells.

In the case of antibody responses, Ir genes act indirectly, by influencing the level and specificity of helper T lymphocytes, which are required for B cells to be activated to make antibodies to protein antigens. Nevertheless, probably via this indirect route, Ir genes influence the site specificity of antibodies raised to protein antigens (for example, see Fig. 3). Mice of strains B10 and B10.A are genetically identical except for their MHC genes. Although B10 is initially a lower responder, after three immunizations with staphylococcal nuclease both strains make equivalent levels of antinuclease antibodies (*50*). However, only the B10.A sera contain antibodies that bind fragment 99–149; the B10 sera do not detectably bind this fragment (*50*). A second example is the case of antibody responses against sperm whale myoglobin in three strains of mice, B10.D2, B10.A, and B10.BR, which are again congenic (identical except for their MHC genes). B10.D2 mice are high responders to sperm whale myoglobin and make antibodies to a number of sites, including antibodies which bind to fragment 132–153 (*51*). B10.BR mice are low responders and make little total antibody, of which almost none binds to fragment 132–153. The B10.A mice are intermediate in total magnitude of response, but the antibodies to myoglobin that are made contain no more antibodies binding to fragment 132–153 than do those of the low responder B10.BR mice (*51*) (Fig. 3). Thus, Ir genes control the site specificity of antibodies produced, not just the total magnitude of the response. These congenic strains of mice do not differ in their myoglobin structural genes, and they have no protein homologous to staphylococcal nuclease. Therefore, the differences in antibodies produced are not due to tolerance to homologous host proteins. The first case of this type described was for antibodies to a synthetic random polymer of three amino acids

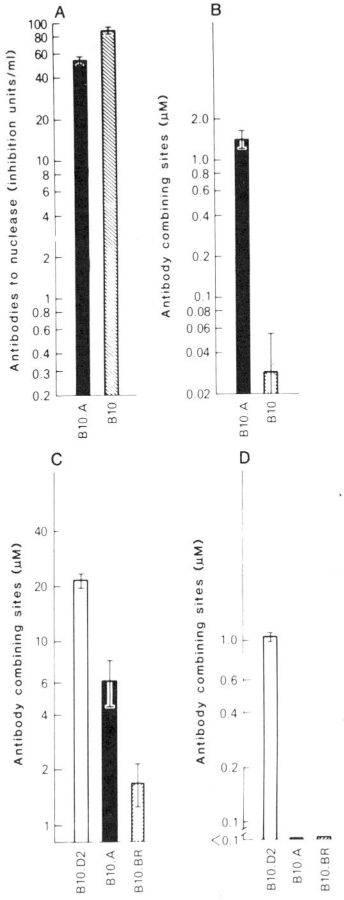

Fig. 3. Effect of Ir genes on antibody specificity (*50, 51*). (A and B) Antibodies made by immunizing three times with native staphylococcal nuclease and testing the same sera for binding to either native nuclease (A) or fragment 99–149 of nuclease (B). The congenic strains of mice, B10.A and B10, differ only in their major histocompatibility genes, including Ir genes. Although the total level of antinuclease antibodies is comparable, only the B10.A sera contain antibodies binding fragment 99–149. (C and D) Antibodies made by immunizing with native sperm whale myoglobin and testing the same sera for binding to native myoglobin (C) or the cyanogen-bromide cleavage fragment 132–153 (D) by radiobinding assay. Means and geometric standard error of the means are shown.

(*52*). Thus, Ir genes are a second host factor that strongly influences which sites on a protein are immunogenic in a given animal. In an outbred species such as man, every individual will have a different complement of Ir genes. Other examples of potential clinical relevance include Ir gene control of the specificity of antibodies to hepatitis B surface antigen (*53*) and to peptides of myelin basic protein (*54*). In this context, the frequency of low responders to a peptide vaccine bearing a single antigenic site would be expected to be greater than the frequency of low responders to a whole protein with many sites, each under different Ir gene control (*51*). Perhaps mixtures of peptides could circumvent this potential problem.

Helper and suppressor T-lymphocyte specificity and T-B reciprocity. B lymphocytes specific for protein antigens generally require "help" from a T lymphocyte specific for the same antigen molecule (although not necessarily the same site) in order to be activated to secrete antibodies. Ir genes act on the magnitude and specificity of T-cell help, but to explain their action on antibody specificity, it is necessary to postulate that the specificity of T cells elicited influences the specificity of antibodies produced (*55*). Because Ir gene products that influence T-cell specificity are expressed on B lymphocytes, and the resulting T-cell specificity appears to lead to selective activation of a subset of B lymphocytes, this postulated reciprocal effect of T and B lymphocytes on each other's specificity has been termed "T-B reciprocity" (*55*). Evidence supporting this notion is still largely indirect but has been found for a number of antigens (*6, 50, 51, 56*).

There are also several cases in which Ir genetic low responsiveness to an antigen is accompanied by active suppression of the response by suppressor T lymphocytes (*6, 8, 57*). Such suppressor T cells have been found, in at least one case, to act selectively on a subset of helper T cells specific for certain antigenic sites (*58*). If suppressor cells influence the specificity of the helper population and this, in turn, selects the B cell repertoire to be activated, suppressor T cells might influence antibody specificity as well. However, a direct demonstration is lacking.

The effect of T-cell helper specificity on the specificity of antibodies produced need not be limited to Ir genes. Anything else which affects T-cell specificity could influence the antibody response. One example is antigen processing. For T lymphocytes to be activated by antigen "presented" on the surface of another cell rather than free in solution, the antigen must first be proteolytically digested or processed by the presenting cell (*59*). The purpose of this processing may be to unfold the protein and expose sites that are necessary for interaction with the MHC antigen or the plasma membrane of the presenting cell (*60*). Amino acid substitutions that affect processing can affect the T-cell response. For instance, Shastri *et al.* (*61*) studied mouse T-cell clones that responded to a tryptic fragment of residues 74 to 96 from either chicken or ring-necked pheasant lysozyme, which share this part of their sequence. Yet, when native lysozymes were used, only that from pheasant would stimulate the clones. Cleaving a small segment from both the amino and carboxyl termini of the lysozymes made the two lysozymes equivalent in stimulatory activity. Thus, a difference far from the antigenic site seemed to influence the proteolytic processing of the chicken lysozyme so as to make the antigenic site seen by the T-cell clone unavailable. If differences in antigen processing lead to differences in helper T-cell specificity, they may also affect antibody specificity.

Processing differences among B cells, related to their antibody specificity, has also been suggested as a possible mechanism for T-B reciprocity (*55*). In contrast to other cells that present antigen to T cells, such as macrophages or dendritic cells, B cells bind specific antigen via the combining site of their surface immunoglobulin (*62*). If the B-cell uptake of antigen is via receptor-mediated endocytosis, an antigen-antibody complex (rather than free antigen) may be processed. Each B cell, with a different surface immunoglobulin specific for a different antigenic site, would be processing a different complex, with a different part of the antigen molecule sterically protected from proteolysis. Thus, each B cell may preferentially present a different set of fragments and so be able to be helped preferentially by T cells specific for those fragments (*55*). This hypothesized selective processing is an appealing way to explain the phenomenon of T-B reciprocity without requiring that the T-cell receptor and B-cell immunoglobulin bind the antigen at the same time (*55*).

The requirements for helper T-cell recognition also have important implications for peptide vaccine development. In order to ensure an anamnestic response when the recipient is challenged with whole antigen or virus, the peptide must contain not only a site that can induce antibody but also a site that will elicit helper T cells. This point has recently been demonstrated experimentally for peptide antigens from tobacco mosaic virus protein (*63*).

Idiotype networks. The idiotype of an antibody, first described in 1963 (*64*), is that part which is antigenically unique, as determined by a second set of antibodies made against the first one. Idiotypes, therefore, are in the variable region, mostly in or near the combining site. As the same combining site structure may determine both the antigen specificity and the idiotype of an antibody, there should be some correlation between specificity and idiotype, although not a perfect one. Jerne (*7*) proposed that a system of antibodies and other antibodies against them (anti-idiotypes) could act as a self-contained homeostatic network. The role of antigen was to perturb this internal network. Many studies have supported the existence of such networks (*7*). These networks may involve idiotype-specific helper T cells as well as antibodies. If networks regulate idiotype, they must influence the prevalence of different combining sites and thus antibody specificity. This regulatory mechanism may even be subject to nongenetic inheritance. Rubinstein *et al.* (*65*) injected newborn mice with A48, a monoclonal antibody to levan with a rare idiotype not normally found in antilevan antisera. When these mice were immunized with bacterial levan 4 weeks later, the antibody response was dominated by the normally rare idiotype of A48. The predominance of this rare idiotype could be transferred to other mice with helper T cells from the treated mice. Thus, the prevalence at birth of this idiotype induced complementary idiotype-specific helper cells which altered the spectrum of idiotypes produced on immunization later in life. Since immunoglobulin G crosses the placenta, Rubinstein *et al.* (*65*) suggested that the idiotype history of the mother could be passed on to her offspring, a form of nongenetic inheritance that may have some protective advantage.

The idiotype expressed in a given immune response can also be influenced by MHC-linked Ir genes (*66*). Thus, there is an interface at which these two major regulatory mechanisms can interact. The net effect is that antigen specificity of the antibody response depends on both the genetics and the immune history of the host, and even on the immune history of the previous generation.

Structural gene repertoire. The diversity of the potential antibody repertoire depends on the combinatorial joining of structural genes for several elements which make up the antibody combining site (V_H, D, and J_H on the heavy chain and V_L and J_L on the light chain). There is a further diversity generated by combinatorial pairing of heavy and light chains and by somatic mutation (*67*). Loh *et al.* (*68*) showed that the V_H that dominates the antibody response to the 4-hydroxy-3-nitrophenyl acetyl group (NP) in one strain of mice (BALB/c) comes from a completely different family of V_H genes from that which dominates the response to NP in another strain of mice, C57BL/6. Thus, there appears to have been enough genetic divergence in V_H between mouse strains to influence the specificity of antibody responses to the same small antigen.

Stochastic effects and clonal preemption. The antibody response to *p*-azobenzene arsonate (ABA) in A/J mice is dominated by a major idiotype (*69*). However, the repertoire of B-cell clones available prior to immunization is quite diverse, and the dominance of a particular idiotype in this response is due to selective forces during the immune response itself, such as affinity of B cells for antigen (*70*). Later in the immune response, clones lacking the dominant idiotype may arise by somatic mutation with affinities higher than that of the dominant idiotype. However, it is supposed that once the clones with dominant idiotype have gained a head start at expansion, they preempt the response, perhaps by consuming antigen, so that the other clones cannot catch up. An earlier study of the same ABA response supports the idea that clones with a head start can preempt the response (*71*). When ABA-specific B cells expressing idiotypes other than the dominant one were transferred into naive A/J mice prior to immunization, these clones now dominated the ABA response, and the major idiotype that dominates in unmanipulated A/J mice was suppressed (*71*). Thus, in addition to all the regulatory mechanisms that influence antibody specificity, there is a stochastic element as well that can result in animal to animal variability. In each animal, the clone that gets the antigen first may preempt the response. However, there are cases, such as the response to the hapten 4-hydroxy-3-nitrophenyl acetyl, in which an idiotype that predominates early in the response becomes negligible after repeated immunizations, when other idiotypes take over (*72*). Thus, other regulatory mechanisms can override clonal preemption. Indeed, all of the host regulatory mechanisms act in concert to determine the ultimate magnitude and specificity of the observed antibody response.

Conclusions

In summary, the weight of the evidence is that virtually the entire accessible surface of a protein may be antigenic. Intrinsic features such as hydrophilicity of stretches of amino acids help predict accessibility from primary sequence. Mobility of a polypeptide segment of a protein, in the few cases in which mobility data can be obtained, is useful for predicting which segmental sites of a protein are likely to elicit antibodies that cross-react with the corresponding peptide or are likely to cross-react with antibodies made against a synthetic peptide of like sequence. Flexibility of an antigenic site also has important implications for the dynamics of antigen-antibody and protein-protein interactions in general. However, mobility probably does not enhance immunogenicity per se, and these studies cannot be applied to the large number of antibodies that bind assembled topographic sites—that is, sites comprising amino acid residues far apart in the primary sequence but brought together on the surface by the folding of the protein in its native conformation. Of all the potential antigenic sites of a protein, only a subset will be immunogenic in any individual host. Thus, host factors (extrinsic to the antigen molecule) are critical to immunogenicity of a potential antigenic site. These host regulatory factors include self-tolerance, immune response genes, specificity of helper and suppressor T cells, antigen processing, idiotype networks, the host's structural antibody gene repertoire, and clonal preemption. While these extrinsic factors make it difficult to predict accurately the structure of individual antigenic sites from a knowledge of the protein structure alone, it is still possible to make worthwhile selections of peptide segments likely to be useful as probes or as synthetic vaccines. Indeed, one of the powers of the synthetic immunogen approach is that peptides may be immunogenic in a given host even when the corresponding site in the native protein is not (*1, 8, 61*). Thus, synthetic peptide immunogens may circumvent some extrinsic factors.

References and Notes

1. R. A. Lerner, *Nature (London)* **299**, 592 (1982); *Adv. Immunol.* **36**, 1 (1984).
2. R. Arnon, in *Impact of Protein Chemistry on the Biomedical Sciences*, A. N. Schechter, A. Dean, R. F. Goldberger, Eds. (Academic Press, Orlando, Fla., 1984), p. 187.
3. K. Landsteiner, *The Specificity of Serological Reactions* (Thomas, Springfield, Ill., 1936).
4. M. Z. Atassi, *Immunochemistry* **12**, 423 (1975); S. S. Twining, H. Lehmann, M. Z. Atassi, *Biochem. J.* **191**, 681 (1980); S. S. Twining, C. S. David, M. Z. Atassi, *Mol. Immunol.* **18**, 447 (1981).
5. D. C. Benjamin *et al.*, *Annu. Rev. Immunol.* **2**, 67 (1984).
6. B. Benacerraf and R. N. Germain, *Immunol. Rev.* **38**, 71 (1978); A. S. Rosenthal, *ibid.* **40**, 136 (1978); J. A. Berzofsky, in *Biological Regulation and Development*, R. F. Goldberger, Ed. (Plenum, New York, 1980), vol. 2, pp. 467–594; R. H. Schwartz, in *Ia Antigens*, vol. 1, *Mice*, S. Ferrone and C. S. David, Eds. (CRC Press, Boca Raton, Fla., 1982), pp. 161–218; J. A. Berzofsky, in *The Antigens*, M. Sela, Ed. (Academic Press, New York, in press).
7. N. K. Jerne, *Ann. Immunol. (Paris)* **125C**, 373 (1974); C. A. Bona and B. Pernis, in *Fundamental Immunology*, W. E. Paul, Ed. (Raven, New York, 1984), pp. 577–592.
8. E. E. Sercarz *et al.*, *Immunol. Rev.* **39**, 108 (1978); J. W. Goodman and E. E. Sercarz, *Annu. Rev. Immunol.* **1**, 465 (1983).
9. C. G. Fathman and J. G. Frelinger, *Annu. Rev. Immunol.* **1**, 633 (1983).
10. C. DeLisi and J. A. Berzofsky, *Proc. Natl. Acad. Sci. U.S.A.*, in press.
11. J. A. Berzofsky *et al.*, *J. Biol. Chem.* **257**, 3189 (1982); I. J. East *et al.*, *ibid.*, p. 3199.
12. J. A. Berzofsky, in *Monoclonal and Antiidiotypic Antibodies: Probes for Receptor Structure and Function*, J. C. Venter, C. M. Fraser, J. Lindstrom, Eds. (Liss, New York, 1984), pp. 1–19.
13. M. Sela, *Science* **166**, 1365 (1969).

14. D. H. Sachs *et al.*, *Proc. Natl. Acad. Sci. U.S.A.* **69**, 3790 (1972); B. Furie *et al.*, *J. Mol. Biol.* **92**, 497 (1975); J. G. R. Hurrell, J. A. Smith, S. J. Leach, *Biochemistry* **16**, 175 (1977); M. J. Darsley and A. R. Rees, *EMBO J.* **4**, 383 (1985); A. Hirayama, Y. Takagaki, F. Karush, *J. Immunol.* **134**, 3241 (1985); I. A. Wilson, personal communication.
15. J. A. Berzofsky and A. N. Schechter, *Mol. Immunol.* **18**, 751 (1981).
16. M. J. Crumpton and J. M. Wilkinson, *Biochem. J.* **94**, 545 (1965).
17. G. Köhler and C. Milstein, *Nature (London)* **256**, 495 (1975); F. Melchers, M. Potter, N. L. Warner, *Lymphocyte Hybridomas* (Springer-Verlag, Berlin, 1978); R. H. Kennett, T. J. McKearn, K. B. Bechtol, *Monoclonal Antibodies. Hybridomas: A New Dimension in Biological Analysis* (Plenum, New York, 1980).
18. G. Lando, J. A. Berzofsky, M. Reichlin, *J. Immunol.* **129**, 206 (1982).
19. J. A. Berzofsky and I. J. Berkower, in *Fundamental Immunology*, W. E. Paul, Ed. (Raven, New York, 1984), pp. 595–644.
20. S. J. Smith-Gill *et al.*, *J. Immunol.* **128**, 314 (1982); S. J. Smith-Gill, T. B. Lavoie, C. R. Mainhart, *ibid.* **133**, 384 (1984); D. W. Metzger *et al.*, *Eur. J. Immunol.* **14**, 87 (1984); A. G. Amit, R. A. Mariuzza, S. E. V. Phillips, R. J. Poljak, *Nature (London)* **313**, 156 (1985); M. Z. Atassi, *Immunochemistry* **15**, 909 (1978); ______ and C.-L. Lee, *Biochem. J.* **171**, 429 (1978).
21. The validity of defining sites by comparing evolutionary sequence variants was recently confirmed by P. V. Hornbeck and A. C. Wilson [*Biochemistry* **23**, 998 (1984)], who studied the affinity of lysozymes for the substrate analog chitotriose. Only amino acid substitutions within the binding site affected affinity, whereas even multiple substitutions outside the site had no significant effect.
22. D. C. Wiley, I. A. Wilson, J. J. Skehel, *Nature (London)* **289**, 373 (1981); P. A. Underwood, *J. Gen. Virol.* **62**, 153 (1982); L. M. Staudt and W. Gerhard, *J. Exp. Med.* **157**, 687 (1983); P. M. Colman, J. N. Varghese, W. G. Laver, *Nature (London)* **303**, 41 (1983).
23. Z. Al Moudallal, J. P. Briand, M. H. V. Van Regenmortel, *EMBO J.* **1**, 1005 (1982); D. Altschuh *et al.*, *Mol. Immunol.* **22**, 329 (1985).
24. S. J. Leach, *Biopolymers* **22**, 425 (1983); P. E. E. Todd, I. J. East, S. J. Leach, *Trends Biochem. Sci.* **7**, 212 (1982); P.-T. Shi *et al.*, *Mol. Immunol.* **21**, 489 (1984).
25. Z. Al Moudallal, J. P. Briand, M. H. V. Van Regenmortel, *EMBO J.*, in press.
26. T. J. White, I. M. Ibrahimi, A. C. Wilson, *Nature (London)* **274**, 92 (1978).
27. D. C. Benjamin and J. M. Teale, *J. Biol. Chem.* **253**, 8087 (1978); D. C. Benjamin, L. A. Daigle, R. L. Riley, in *Protein Conformation as an Immunological Signal*, F. Celada *et al.*, Eds. (Plenum, New York, 1983), pp. 261–280; T. Peters, Jr., R. C. Feldhoff, R. G. Reed, *J. Biol. Chem.* **252**, 8464 (1977); T. Kamiyama, *Immunochemistry* **14**, 91 (1977); N. Doyen, A. J. Pesce, C. Lapresle, *J. Biol. Chem.* **257**, 2770 (1982).
28. Accessibility of atoms of a protein to solvent or solute molecules of various sizes can be defined quantitatively [B. Lee and F. M. Richards, *J. Mol. Biol.* **55**, 379 (1971); M. L. Connolly, *Science* **221**, 709 (1983)].
29. T. P. Hopp and K. R. Woods, *Proc. Natl. Acad. Sci. U.S.A.* **78**, 3824 (1981); *Mol. Immunol.* **20**, 483 (1983).
30. S. Fraga, *Can. J. Chem.* **60**, 2606 (1982).
31. M. Sela and E. Mozes, *Proc. Natl. Acad. Sci. U.S.A.* **55**, 445 (1966).
32. E. Benjamini *et al.*, *Biochemistry* **7**, 1261 (1968).
33. C. Chothia and J. Janin, *Nature (London)* **256**, 705 (1975).
34. E. Westhof *et al.*, *ibid.* **311**, 123 (1984).
35. J. A. Tainer *et al.*, *ibid.* **312**, 127 (1984); J. A. Tainer *et al.*, *Annu. Rev. Immunol.* **3**, 501 (1985).
36. G. R. Moore and R. J. P. Williams, *Eur. J. Biochem.* **103**, 543 (1980).
37. H. L. Niman *et al.*, *Proc. Natl. Acad. Sci. U.S.A.* **80**, 4949 (1983).
38. D. E. Koshland, Jr., G. Nemethy, D. Filmer, *Biochemistry* **5**, 365 (1966).
39. F. R. N. Gurd and R. M. Rothgeb, *Adv. Protein Chem.* **33**, 73 (1979); M. Levitt, *J. Mol. Biol.* **168**, 621 (1983).
40. J. Monod, J. Wyman, J.-P. Changeux, *J. Mol. Biol.* **12**, 88 (1965).
41. R. M. Lewis, B. C. Furie, B. Furie, *Biochemistry* **22**, 948 (1983); J. Owens *et al.*, *J. Biol. Chem.* **259**, 13800 (1984).
42. M. J. Crumpton, in *The Antigens*, M. Sela, Ed. (Academic Press, New York, 1974), vol. 2, pp. 1–78.
43. J. Dean and A. N. Schechter, *J. Biol. Chem.* **254**, 9185 (1979).
44. F. M. Burnet, *The Clonal Selection Theory of Immunity* (Vanderbilt Univ. Press, Nashville, 1959).
45. A. B. Edmundson, K. R. Ely, J. N. Herron, *Mol. Immunol.* **21**, 561 (1984).
46. J. A. Berzofsky *et al.*, in *Protein Conformation as Immunological Signal*, F. Celada *et al.*, Eds. (Plenum, New York, 1983), pp. 165–180.
47. G. J. V. Nossal, *Annu. Rev. Immunol.* **1**, 33 (1983).
48. G. J. Urbanski and E. Margoliash, *J. Immunol.* **118**, 1170 (1977); R. Jemmerson and E. Margoliash, *J. Biol. Chem.* **254**, 12706 (1979).
49. H. M. Cooper *et al.*, *Mol. Immunol.* **21**, 479 (1984).
50. J. A. Berzofsky *et al.*, *J. Exp. Med.* **145**, 111 (1977); J. A. Berzofsky *et al.*, *ibid.*, p. 123; J. A. Berzofsky *et al.*, *Adv. Exp. Med. Biol.* **98**, 241 (1978); J. A. Berzofsky, unpublished observations.
51. J. A. Berzofsky, *J. Immunol.* **120**, 360 (1978); ______, L. K. Richman, D. J. Killion, *Proc. Natl. Acad. Sci. U.S.A.* **76**, 4046 (1979); Y. Kohno and J. A. Berzofsky, *J. Immunol.* **128**, 2458 (1982).
52. H. G. Bluestein *et al.*, *J. Exp. Med.* **135**, 98 (1972).
53. D. R. Milich *et al.*, *ibid.* **159**, 41 (1984).
54. R. B. Fritz *et al.*, *J. Immunol.* **134**, 2328 (1985).
55. J. A. Berzofsky, *Surv. Immunol. Res.* **2**, 223 (1983).
56. J. M. Cecka *et al.*, *Eur. J. Immunol.* **6**, 639 (1976); E. Sercarz *et al.*, *Ann. Immunol. (Paris)* **128C**, 599 (1977); A. Campos-Neto, H. Levine, S. F. Schlossman, *Cell. Immunol.* **69**, 128 (1982); F. Celada *et al.*, in *Regulation of the*

Immune System, H. Cantor, L. Chess, E. Sercarz, Eds. (Liss, New York, 1984), pp. 637–646; T. A. Ferguson *et al.*, *Cell. Immunol.* **78**, 1 (1983).
57. J. A. Kapp *et al.*, *J. Exp. Med.* **140**, 648 (1974).
58. U. Krzych, A. V. Fowler, E. E. Sercarz, *ibid.*, **162**, 311 (1985).
59. E. R. Unanue, *Annu. Rev. Immunol.* **2**, 395 (1984).
60. J. A. Berzofsky, in *The Year in Immunology 1984–85*, J. M. Cruse and R. E. Lewis, Jr., Eds. (Karger, Basel, 1985), pp. 18–24; H. Z. Streicher *et al.*, *Proc. Natl. Acad. Sci. U.S.A.* **81**, 6831 (1984).
61. N. Shastri, A. Miller, E. E. Sercarz, in *Regulation of the Immune System*, H. Cantor, L. Chess, E. E. Sercarz, Eds. (Liss, New York, 1984), pp. 153–161.
62. K. L. Rock, B. Benacerraf, A. K. Abbas, *J. Exp. Med.* **160**, 1102 (1984); B. A. Malynn and H. H. Wortis, *J. Immunol.* **132**, 2253 (1984).
63. E. Benjamini *et al.*, in *High Technology Route to Virus Vaccines*, G. R. Dreesman, J. G. Bronson, R. Kennedy, Eds. (American Society for Microbiology, Washington, D.C., in press).
64. H. G. Kunkel, M. Mannik, R. C. Williams, *Science* **140**, 1218 (1963); J. Oudin and M. Michel, *C. R. Acad. Sci.* **257**, 805 (1963); K. Rajewsky and T. Takemori, *Annu. Rev. Immunol.* **1**, 569 (1983).
65. L. J. Rubinstein, M. Yeh, C. A. Bona, *J. Exp. Med.* **156**, 506 (1982).
66. H. Kawamura *et al.*, *ibid.* **160**, 659 (1984); U. M. Babu and P. H. Maurer, *ibid.* **154**, 649 (1981); M. C. Bekoff, H. Levine, S. F. Schlossman, *J. Immunol.* **129**, 1173 (1982).
67. S. Tonegawa, *Nature (London)* **302**, 575 (1983).
68. D. Y. Loh *et al.*, *Cell* **33**, 85 (1983).
69. M. F. Gurish and A. Nisonoff, in *Idiotypy in Biology and Medicine*, H. Köhler, J. Urbain, P.-A. Cazenave, Eds. (Academic Press, New York, 1984), pp. 63–88.
70. T. Manser, S.-Y. Huang, M. L. Gefter, *Science* **226**, 1283 (1984); T. Manser *et al.*, *Immunol. Today* **6**, 94 (1985).
71. F. L. Owen and A. Nisonoff, *J. Exp. Med.* **148**, 182 (1978).
72. O. Mäkelä and K. Karjalainen, *Immunol. Rev.* **34**, 119 (1977).
73. R. E. Dickerson, in *The Proteins*, H. Neurath, Ed. (Academic Press, New York, ed. 2, 1964), vol. 2, pp. 603–778.
74. R. J. Feldmann *et al.*, *Proc. Natl. Acad. Sci. U.S.A.* **75**, 5409 (1978).
75. T. Takano, *J. Mol. Biol.* **110**, 537 (1977).
76. I am grateful to Drs. David Davies, Charles DeLisi, Arnold Feinstein, Elizabeth Getzoff, Richard A. Lerner, Alan N. Schechter, Eli Sercarz, Sandra Smith-Gill, John Tainer, and Marc Van Regenmortel for critical reading of the manuscript and for very helpful discussions and to R. Feldmann for computer graphics used in Fig. 1.

Part III

Developmental Biology and Cancer

9

Spatially Regulated Expression of Homeotic Genes in Drosophila

Katherine Harding, Cathy Wedeen, William McGinnis, and Michael Levine

A fundamental problem of development is how embryonic cells acquire their particular developmental fates as a result of their location within a developing embryo. A model system for analyzing the elaboration of this positional information during *Drosophila* development involves the morphogenesis of body segments. The adult fruit fly is composed of eight abdominal, three thoracic, and four to six head segments (*1*). Several of the constituent tissues of a given segment have morphological properties specific for that segment. For example, the epidermis elaborates cuticular structures, such as legs and antennae, that are distinct for a particular segment. In addition, the morphology of some of the mesodermal (*2*) and neural tissues (*3, 4*) may be specific for a given segment.

Homeotic genes are those that establish the diverse pathways by which each embryonic segment primordium develops a distinct adult phenotype (*5, 6*). Mutations of homeotic loci result in partial or complete transformations of the epidermal tissues of one segment into those of another. Homeotic transformations may include the neural and mesodermal tissues of the affected segment as well (*2, 3, 7*). For example, embryos that lack the *Antennapedia* (*Antp*) gene function display a transformation of the meso- and metathorax (T2 + T3) into homologous tissues of the prothorax (T1) (*8*).

Many homeotic genes appear within one of two clusters in the *Drosophila* genome, the bithorax complex (BX-C) (*5, 9*) or the Antennapedia complex (ANT-C) (*10, 11*). Genes of the BX-C are required for the specification of segments in the posterior regions of the fly (*5, 12, 13*). Lewis has identified a number

Science 229, 1236–1242 (20 September 1985)

of homeotic loci within the BX-C on the basis of embryonic and adult mutant phenotypes (*5*). Recently, a minimum of three essential domains of homeotic function within the BX-C have been identified by means of lethal complementation analyses: *Ultrabithorax (Ubx), Abdominal-A (abd-A), and Abdominal-B(Abd-B)* (*9*). The ANT-C is required for the specification of anterior body segments (*8, 14*). Several homeotic lethal complementation groups have been identified for the ANT-C (*8, 11, 14, 15*). These include the *Antp, Sex combs reduced (Scr), and Deformed* (*Dfd*) loci. Each ANT-C and BX-C homeotic lethal complementation group controls the development of a different subset of the embryonic segment primordia (Fig. 1a).

A central problem in elucidating the genetic control of segment morphogenesis is how the different ANT-C and BX-C loci come to function in primarily nonoverlapping domains along the body axis of the fly. The molecular cloning of ANT-C and BX-C loci has permitted a direct assessment of the spatial and temporal limits of homeotic gene expression. The previous demonstration that *Ubx* and *Antp* share direct nucleotide sequence homology (*16-19*) facilitated the isolation of ANT-C and BX-C loci. This homology occurs within a conserved protein coding region designated the homeo box. A total of seven genomic DNA fragments cross-hybridizes strongly with the *Antp* and *Ubx* homeo boxes (*20*). These seven regions correspond to the Antennapedia class of the homeo box gene family, all of which are located within either the ANT-C or the BX-C (*20*). It appears that each of the six lethal complementation groups of the ANT-C and BX-C (Fig. 1) contains an Antennapedia class homeo box. However, there are additional homeotic loci within the BX-C that do not contain the homeo box (Fig. 1a) (*21*).

We show that each of the ANT-C and BX-C homeotic loci that contains a homeo box specifies transcripts that accumulate in discrete regions of the embryonic central nervous system (CNS). To a close approximation, the regions of the CNS that contain transcripts encoded by each of these loci correspond to the embryonic segments that are disrupted in mutants for these genes. We propose that spatially restricted expression of each ANT-C and BX-C locus involves hierarchical, cross-regulatory interactions that are mediated by the homeo box protein domains encoded by these genes. Support for this model is based on analysis of the distribution patterns of *Antp* transcripts in mutant embryos that lack BX-C loci.

Isolation of a new ANT-C homeo box locus. Molecular clones for the *Dfd, Antp, Ubx, iab-2*, and *iab-7* loci have been previously isolated (*16, 20, 22–25*). In order to determine the spatial limits of expression for each homeotic lethal complementation group within the ANT-C and BX-C by in situ hybridization, it was necessary to obtain a molecular probe for the *Scr* locus. A genomic DNA fragment that appears to derive from *Scr* was isolated on the basis of homeo box sequence homology as described below.

A total of 6×10^4 recombinants from a *Drosophila*–Charon 4 DNA library (approximately six genome equivalents) were screened with the homeo box sequence as described (*16*). Approximately 50 cross-hybridizing recombinant phage were isolated. By in situ hybridization to polytene chromosomes, clone A40 was found to reside within the 84A/B cytogenetic region of chromosome 3. Previous cytogenetic analyses have demonstrated that this corresponds to the location of the ANT-C (*11*). As judged by high stringency cross-hybridization tests, A40 does not correspond to the previously isolated *Dfd*, *fushi tarazu* (*ftz*), and *Antp* loci.

The *Scr* locus has been mapped to a region of the ANT-C that lies between the *Dfd* and *Antp* loci (*14*, *15*, *24*). Mutant embryos that lack *Scr* function show homeotic transformations of the posterior-most head regions (the labium) and the prothorax (*8*, *14*). Since *Dfd* and *Antp* transcripts principally accumulate in the regions of the CNS corresponding to the embryonic segments that are most disrupted in Dfd^- and $Antp^-$ mutants (*16*, *26*, *27*, see below), we anticipated that *Scr* transcripts would be detected in the embryonic subesophageal ganglion.

To determine the time during embryonic development when A40 is transcribed, we hybridized Northern blots containing polyadenylated [poly$(A)^+$] RNA from successive embryonic stages with the homeo box region of the A40 clone. Multiple RNA species homologous to pA40 are found during embryonic development, especially in 6- to 12-hour embryos (Fig. 2). The largest transcript is 4.0 kb in length. In situ hybridization analyses (see below) confirm that mid-stage embryos contain the highest levels of A40 transcript. In addition, A40 transcripts are primarily detected within tissues of the labial head segment. Since A40 maps within the 84A/B cytogenetic interval, contains homeo box cross-homology, and is expressed in the embryonic segment most disrupted in Scr^- mutants (*8*, *14*), it is likely that A40 derives from the *Scr* locus of the ANT-C (*28*).

ANT-C and BX-C gene expression in wild-type embryos. The regions of developing embryos that accumulate transcripts specified by *Antp* and *Ubx* have been identified by in situ hybridization to tissue sections (*26*, *27*, *29*). The principal sites of $Antp^+$ and Ubx^+ expression appear to correspond to the embryonic segments that are most severely disrupted in $Antp^-$ and Ubx^- mutants. For the most part, *Antp* transcripts accumulate in the T1/T2 portion of the embryonic CNS (*26*, *27*, *30*). *Ubx* transcripts are detected principally in T3/A1 (*29*).

The identities of tissues that contain *Dfd*, *Scr*, *iab-2*, and *iab-7* RNA's were determined by an in situ hybridization technique (*26*). Serial tissue sections of wild-type embryos at various stages of development were separately hybridized to tritium-labeled genomic DNA fragments that contain a homeo box (Fig. 1b). A primary site of hybridization for each of these probes corresponds to neural tissues.

The embryonic CNS is composed of the dorsally located brain, which is connected to a ventral nerve cord by a subesophageal ganglion. The ventral cord is a composite of 11 repeating ganglia each of which contains approximately 300 nerve cells (*31*). The epidermal portions of each of the three thoracic and eight abdominal segments are innervated by a corresponding ganglion of the ventral cord.

At mid-embryonic stages, *Dfd* transcripts are detected in a central portion of the subesophageal ganglion, S1/S2 (Fig. 3, a and b). Transcripts homologous to the putative *Scr* probe, pA40, are detected in the S2/S3 region of the subesophageal ganglion of mid-stage embryos. A second focus of embryonic hybridization is observed over epidermal and mesodermal tissues of a posterior head

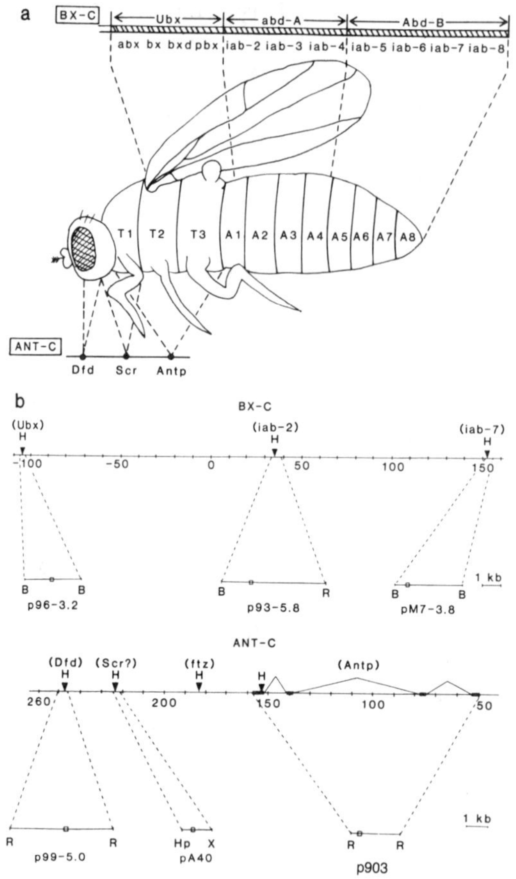

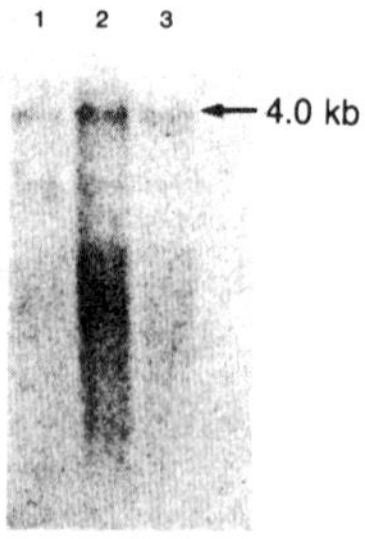

Fig. 1 (left). Genetic and molecular maps of the ANT-C and BX-C. (a) Primary domains of ANT-C and BX-C homeotic function. The cross-hatched horizontal bar represents a genetic map of the BX-C. Lewis has assigned a series of homeotic loci to the BX-C (*5*); these are shown along the bottom of the horizontal bar. The three lethal complementation groups identified within the BX-C by Sanchez-Herrero *et al.* (*9*) are shown above the horizontal bar. The domains of homeotic function were determined by mosaic analyses (*9*) and are indicated on a schematic representation of an adult fly. *Ubx* is critically required for the development of the posterior compartment of the mesothorax (T2p) (*39*) through the anterior compartment of the first abdominal segment (A1a). The *abd-A* function is required for morphogenesis of A1p through A4, and *Abd-B* is required for A5, through A8. The relationship between the Lewis (*5*) and Sanchez-Herrero *et al.* (*9*) genetic maps is uncertain. An approximate correlation is indicated by the horizontal arrows associated with each lethal complementation group. At least three essential homeotic functions have been assigned to the ANT-C: *Dfd*, *Scr*, and *Antp*. These are shown on the genetic map below the schematic of the fly. The primary domains of ANT-C function are indicated. *Antp* function is required for proper segment morphogenesis of the thorax (*8, 33, 44*). Analyses of Scr^- and Dfd^- mutant embryos suggest that these genes are required for the differentiation of the prothorax and posterior head regions (*8, 14*). (b) Molecular maps of the BX-C and the ANT-C. The horizontal lines correspond to overlapping cloned genomic DNA's that span the two complexes [for details see (*20, 22–24, 38*); units are kilobase pairs]. Arrowheads indicate the positions of homeo box copies within the two complexes. Homeo box–containing cloned genomic DNA's used as probes for in situ hybridizations are indicated below the horizontal lines; the positions within the chromosome walk from which these cloned DNA's are derived are indicated by the connecting dashed lines. The open boxes within each cloned DNA used as a probe correspond to the position of the homeo box within each fragment. The *ftz* homeo box region was not used in this analysis. The probe p99-5.0 is a 5-kb Eco RI genomic DNA fragment that derives from the *Dfd* locus; pA40 is a 1.4-kb Hpa I–Xho I genomic fragment that appears to derive from the homeo box region of the *Scr* locus (the exact position of the homeo box within the fragment has not been determined); p903 is a 2.3-kb *Antp* cDNA; p96-3.2 is a 3.2-kb Bam HI genomic fragment with homology to *Ubx* transcripts; p93-5.8 is a 5.8-kb Bam HI–Eco RI genomic fragment with homology to *iab-2* transcripts; pM7-3.8 is a 3.8-kb Bam HI genomic fragment with homology to *iab-7* transcripts (*20, 38*). Fig. 2 (right). Transcripts homologous to the putative *Scr* probe. Poly(A)$^+$ RNA was extracted from wild-type embryos, and 10-μg aliquots were fractionated on an agarose-formaldehyde gel and transferred to nitrocellulose (*20*). The lanes in each panel correspond to embryonic stages 1 (0 to 6 hours), 2 (6 to 12 hours), and 3 (12 to 24 hours). The blot was hybridized with the pA40 DNA fragment (Fig. 1b) after labeling with ^{32}P by nick-translation; nonspecifically bound probe was removed by stringent washing. The strongest hybridization signal (arrow) corresponds to a poly(A)$^+$ RNA species 4.0 kb in length. Autoradiographic exposure was for 10 days.

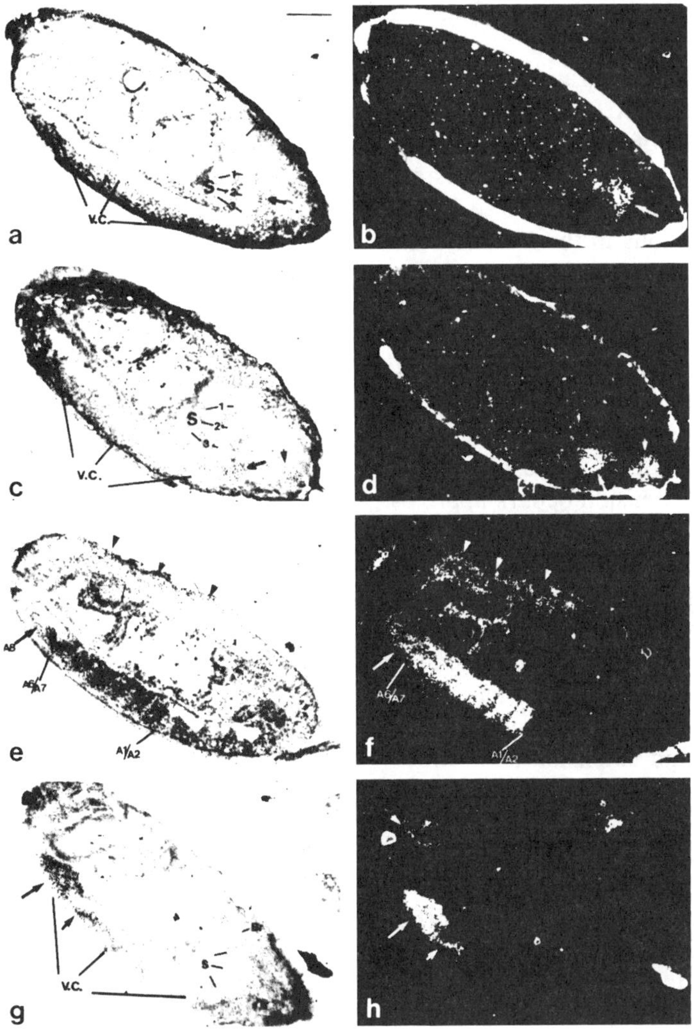

Fig. 3. Localization of *Dfd, Scr, iab-2*, and *iab-7* transcripts in 12- to 14-hour embryos. All sections are sagittal and are oriented such that anterior is to the right and dorsal is up. Because of this orientation, the left-to-right order of the body segments shown in each photomicrograph is opposite to the order in which the labels are read. For example, in a region labeled T1/T2, T1 is actually to the right (more anterior) of T2 in the tissue autoradiogram. The luminescence of the chorion in darkfield photomicrographs is due to refraction of light and does not indicate nonspecific or specific hybridization. The bar in (a) represents 0.1 mm. (a) Localization of transcripts homologous to the *Dfd* probe (p99-5.0). Hybridization signals are seen over the cells of the S1/S2 region of the subesophageal ganglion (arrow). No signals over background are detected in regions either anterior or posterior to S1/S2. (b) Dark-field photomicrograph of the same section as in (a). (c) Localization of transcripts homologous to the putative *Scr* probe, pA40. Hybridization signals are seen over the S2/S3 region of the subesophageal ganglion (arrow); hybridization is also seen to labial tissues (arrowhead). No other sites of hybridization are detected. (d) Dark-field photomicrograph of the same section as in (c). (e) Localization of transcripts homologous to the *iab-2* probe. The strongest hybridization signals occur over A1/A2 through A6/A7 ventral ganglia. Weaker signals are observed in the A8 ganglion (arrow) as well as in the hypodermal and mesodermal regions of segments A2 through A6/A7 (arrowheads). (f) Dark-field photomicrograph of the same section as in (e). (g) Localization of transcripts homologous to the *iab-7* probe. The strongest hybridization signals are seen in the A6/A7 and A8/A9 ventral ganglia (large arrow); weaker hybridization is observed in A5/A6 (small arrow) and in hindgut tissues (arrowhead). (h) Dark-field photomicrograph of the same section as in (g). Abbreviations: A, abdominal region of the ventral cord; Br, brain; S, subesophageal ganglion; T, thoracic region of the ventral cord; v.c., ventral cord.

segment, probably the labium (Fig. 3, c and d).

Strong *iab-2* hybridization signals are detected in the A1/A2 through the A6/A7 ventral ganglia of mid-stage embryos. Less labeling is observed in the posterior-most ventral ganglia (Fig. 3, e and f). RNA's homologous to the *iab-7* hybridization probe are detected primarily in the posterior-most ventral ganglia, including at least a posterior portion of the A7 ganglion and all of the composite A8/A9 ganglion. Less intense hybridization signals are observed over the A5 and A6 ventral ganglia (Fig. 3, g and h).

Germ band extension occurs approximately 5 hours after fertilization (*31*). During this time, the ectodermal and mesodermal cell layers that lie along the ventral aspect of the embryo expand posteriorly. About 7 to 8 hours after fertilization the germ band begins to retract to its original site of formation. It is during germ band extension and retraction when the ectodermally derived neuroblasts differentiate and subsequently elaborate ganglion cells by undergoing unequal mitotic divisions (*31*).

Embryos undergoing elongation of the germ band were hybridized with the cloned *Dfd*, A40, and *Antp* homeo box regions of the ANT-C. The *Dfd* probe principally hybridizes to the mesodermal, neural, and epidermal progenitors of a posterior head segment—possibly the maxilla (Fig. 4, a and b, arrows). The neural tissues of this segment primordium appear to give rise to the S1/S2 region of the subesophageal ganglion. Weaker labeling is also observed at what might be the epidermal portions of a more anterior region of the head (Fig. 4, a and b, arrowheads).

A40 transcripts appear to be distributed over a broader portion of the presumptive CNS during germ band extension as compared with more advanced stages of neurogenesis. Transcripts are present in two distinct segment primordia (Fig. 4, c and d); however, at the conclusion of germ band retraction, only the S2/S3 region contains detectable levels of A40 transcripts. As has been shown, *Antp* transcripts are detected at the highest steady-state levels in the thoracic regions of the germ band and accumulate at high levels only transiently in the presumptive abdominal ganglia during germ band extension and retraction (*27*) (Fig. 4, e and f).

During germ band extension and retraction, transcripts homologous to the *Ubx* homeo box region accumulate in a broad portion of the germ band which includes the progenitors of T2 and at least the first seven abdominal segments (Fig. 5, a and b). At similar embryonic stages, *iab-2* transcripts are detected in the A1/A2 through A7/A8 regions of the germ band (Fig. 5, c and d). During germ band extension, *iab-7* transcripts are detected primarily in the A6/A7, A7/A8, and A8/A9 segments. In addition, weak hybridization is observed to more anterior segments, including A4/A5. Weak hybridization is also observed to a region posterior to the A8/A9 segments, possibly including presumptive hindgut structures (Fig. 5, e and f).

Altered distribution of *Antp*$^{+}$ transcripts in BX-C^{-} embryos. BX-C^{-} embryos [Df(3R)P9 homozygotes] die during the terminal stages of embryogenesis (*5*). Lewis has shown that the epidermal tissues of embryonic segments T3 through A7 acquire features characteristic of the mesothorax (T2). More recently, it has been suggested that each of the transformed segments might acquire a composite pro- and mesothoracic (T1/

Fig. 4 (upper right). Localization of *Dfd*, A40, and *Antp* transcripts during germ band extension. The orientation of the sections and other features of presentation are as described in the legend of Fig. 3. The bar in (a) corresponds to 0.1 mm. (a) Localization of transcripts homologous to the *Dfd* probe. The strongest hybridization signals are seen in the region that gives rise to the S1/S2 region of the subesophageal ganglion and portions of the maxilla (arrow); signals are also detected in cells which appear to correspond to a more anterior region of the presumptive head (arrowheads). (b) Dark-field photomicrograph of the same section as in (a). (c) Localization of transcripts homologous to the putative *Scr* probe, pA40. Hybridization signals are detected in two regions of the presumptive ventral cord, one of which gives rise to S2/S3 (arrows). (d) Dark-field photomicrograph of the same section as in (c). (e) Localization of transcripts homologous to the *Antp* probe (p903). Strong hybridization signals are seen over the regions giving rise to S3/T1, T1/T2, and T2/T3; weaker hybridization signals are seen more posteriorly through A8. (f) Dark-field photomicrograph of the same section as in (e). Abbreviations: AMG, anterior midgut invagination; gb, germ band; PMG, posterior midgut invagination; St, stomadeum. Fig. 5 (lower right). Localization of *Ubx*, *iab-2*, and *iab-7* transcripts during germ band extension and retraction. The orientation of the sections and other features are as described in the legend to Fig. 3. The bar in (a) represents 0.1 mm. (a) Localization of transcripts homologous to the *Ubx* probe in an embryo undergoing germ band extension. Hybridization signals are detected over regions corresponding to the presumptive T2/T3 → A7/A8 segments. (b) Dark-field photomicrograph of the same tissue section as in (a). Arrows demarcate T2/T3 → A7/A8. (c) Localization of transcripts homologous to the *iab-2* probe in an embryo undergoing germ band retraction. Intense hybridization signals are seen in regions corresponding to A1/A2 → A6/A7 with slightly weaker signals seen in A7/A8. (d) Dark-field photomicrograph of the same section as in (c); arrows bracket A1/A2 → A7/A8. (e) Localization of transcripts homologous to the *iab-7* probe in an embryo with a fully extended germ band. Strong hybridization signals are detected in the regions corresponding to A7 and A8. Weaker hybridization signals are detected in regions anterior to A7 in a graded fashion with A7/A8 showing the strongest intensity of hybridization and A5/A6 the weakest. Weak hybridization is also detected in a region posterior to A8; this region might correspond to future hindgut structures. (f) Dark-field photomicrograph of the same section as in (e). Thick arrow indicates A8; thin arrow indicates A5/A6; arrowhead indicates possible hindgut labeling. Abbreviations: AMG, anterior midgut invagination; gb, germ band; PMG, posterior midgut invagination; Pr, proctodeum; St, stomadeum.

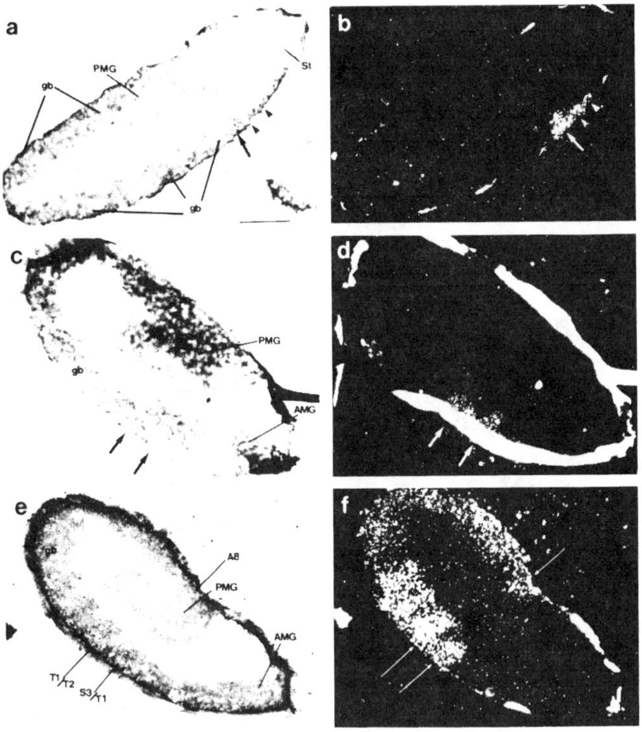

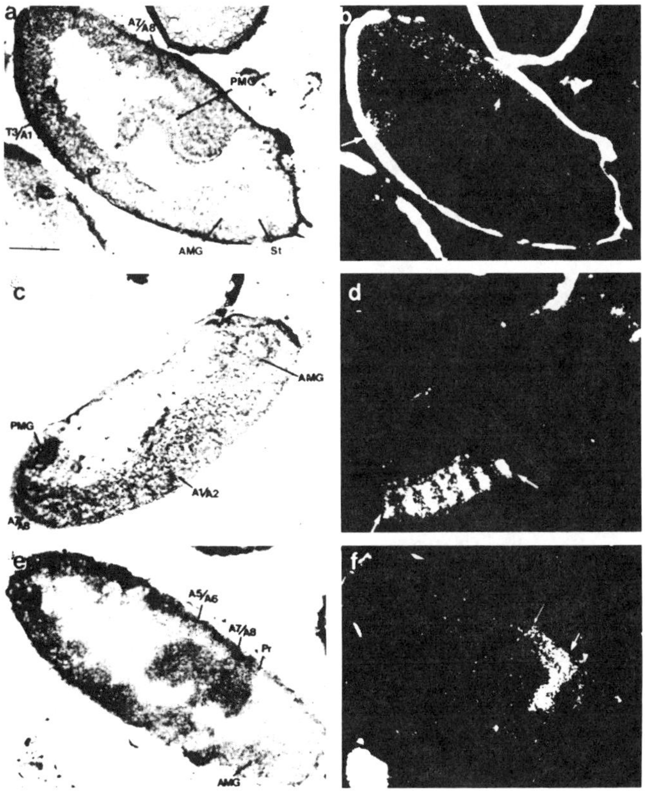

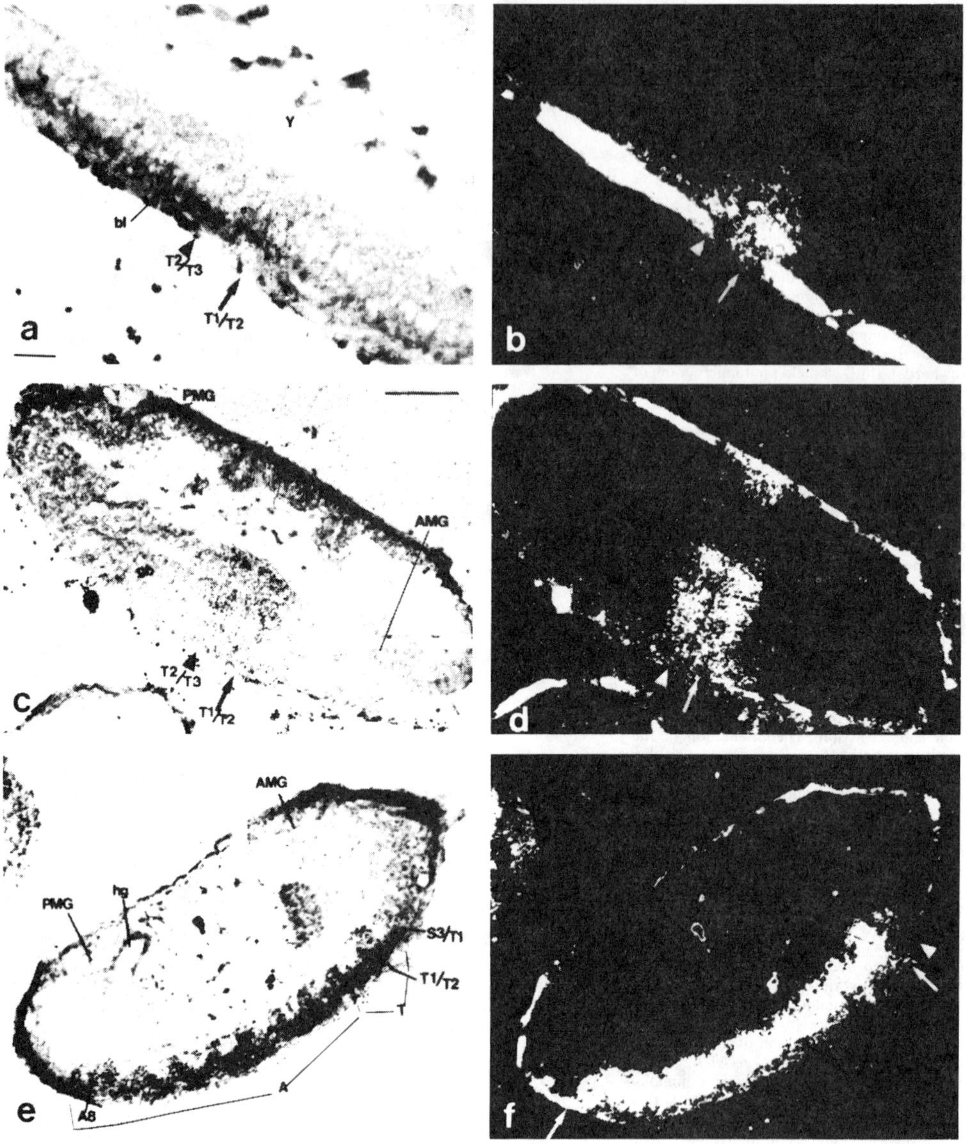

Fig. 6. Localization of *Antp* transcripts in BX-C$^-$ embryos. The orientation of the sections and other features are described in the legend to Fig. 3. Parallel sections through the same embryos did not show hybridization to either the *Ubx* or *iab-2* probes. (a) High magnification of a section through a BX-C$^-$ embryo at the cellular blastoderm stage; the bar represents 0.025 mm. Hybridization signals are seen over cells in the regions of the embryo which give rise to T1/T2 (arrow) and T2/T3 (arrowhead). (b) Dark-field photomicrograph of the same section as in (a). (c) A section through a gastrulating BX-C$^-$ embryo; bar indicates 0.1 mm. Hybridization signals are detected in regions giving rise to T1/T2 (arrow) and T2/T3 (arrowhead). (d) Dark-field photomicrograph of the same section as in (c). (e) A section through a BX-C$^-$ embryo undergoing germ band retraction. Uniformly intense signals are detected in T1/T2 through A8. Weaker signals are detected in cells of S3/T1. (f) Dark-field photomicrograph of the same section as in (e). Arrows demarcate T1/T2 to A8; the arrowhead indicates S3/T1 labeling. Abbreviations: A, abdominal region of the germ band; AMG, anterior midgut invagination; bl, blastoderm; hg, hindgut; PMG, posterior midgut invagination; T, thoracic region of the germ band.

T2) identity (*32, 33*). In advanced stage BX-C$^-$ embryos, *Antp*$^+$ transcripts persist at high steady-state levels in the region of the ventral cord that encompasses the T1/T2, T3, and first seven abdominal ganglia (*34*).

Antp transcripts are detected in the T1/T2 and T2/T3 segment primordia of BX-C$^-$ embryos at the cellular blastoderm stage (Fig. 6, a and b). Similar discrete T1/T2 and T2/T3 hybridization signals are detected in gastrulating P9/P9 embryos (Fig. 6, c and d).

In wild-type embryos undergoing retraction of the germ band, *Antp* transcripts are broadly distributed over the progenitors of the ventral cord. However, labeling of the presumptive abdominal ganglia is less intense as compared with future thoracic ganglia (Fig. 4, e and f). In contrast, P9/P9 embryos of this developmental stage show uniformly intense *Antp* hybridization signals from the future T1/T2 primordium posteriorly through A8 (Fig. 6, e and f).

Patterns of ANT-C and BX-C gene expression in the CNS. By mid-embryonic periods of development, each of the homeo box–containing homeotic genes of the ANT-C and BX-C comes to specify transcripts that appear to accumulate in largely discrete, nonoverlapping regions of the mature embryonic CNS (Fig. 7). The foci of transcript accumulation within the CNS seem to correspond to the embryonic segments that are most severely disrupted in mutants for the corresponding genes. During gastrulation and germ band extension, strong expression is also detected in epidermal and mesodermal tissues (Figs. 4, 5, and 6) (*27, 29*). By advanced stages of embryonic development, however, much of the labeling over the hypodermal and mesodermal tissues is no longer detectable. Because expression is seen to persist in the CNS over the course of embryonic and larval development, it serves as a convenient model for analyzing the spatial regulation of homeotic gene expression.

A striking feature of ANT-C and BX-C homeo box expression is the colinearity between the physical order of the genes within the chromosome and the embryonic segments where they are expressed along the body axis. Lewis first demonstrated such colinearity for the BX-C loci and their corresponding domains of function (*5*). The proximal-most (closest to the centromere) locus of the ANT-C that we have analyzed is *Dfd* (*15*). By mid-embryogenesis *Dfd* transcripts are principally detected in the anterior-most portion of the CNS that has been thus far shown to accumulate homeo box–containing transcripts (see Fig. 3). The *Scr* locus is distal to *Dfd* and transcripts homologous to the A40 clone accumulate in a region of the mature embryonic CNS that is just posterior to the *Dfd* domain. *Antp* transcripts are mostly confined to the T1/T2 region of the ventral cord. Similarly, the proximal-most BX-C homeo box region (*Ubx*) specifies the most anteriorly localized BX-C products within the CNS (T3/A1) (*29*). The more distal *iab-2* and *iab-7* loci specify transcripts that are detected primarily in ventral ganglia A1/A2 through A6/A7 and A6/A7 through A8/A9, respectively.

It has been suggested that diverse pathways of segment morphogenesis are established by different combinations of homeotic gene products (*5, 35–37*). In particular, Lewis has proposed that each embryonic segment primordium in the posterior half of the embryo contains a different combination of BX-C gene products (*5, 38*). According to the Lewis

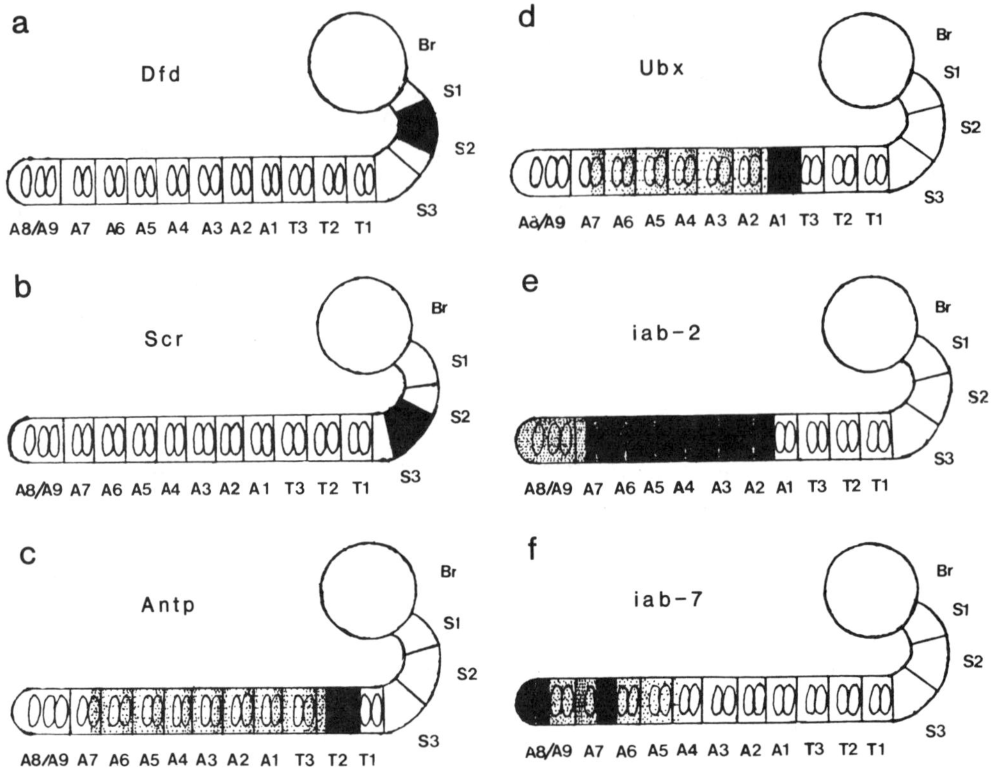

Fig. 7. Sites of ANT-C and BX-C expression in the embryonic CNS. The diagrams represent the CNS of a 14-hour embryo. The sites of *Dfd, Scr, Antp, Ubx, iab-2*, and *iab-7* transcript accumulation are indicated. Solid shading represents the strongest sites of hybridization for each probe. Stippled areas correspond to weaker levels of hybridization; the intensity of stippling represents qualitative but not quantitative differences in the extent of hybridization. Anterior is to the right, and dorsal is up. (a) *Dfd* transcripts appear to be localized to S1/S2. This assignment could be in error by half a segment; that is, the transcripts may be localized to S2. (b) Transcripts homologous to pA40 (*Scr*) are localized to S2/S3. Again, this assignment could be in error by half a segment; the transcripts could be localized to S3. (c) *Antp* transcripts accumulate most strongly in the posterior compartment of T1 (T1p) and the anterior compartment of T2 (T2a). Lower levels are detected, as indicated, through the anterior commissure of A7. (d) *Ubx* transcripts accumulate most strongly in T3p/Ala (*29*). Lower levels are seen posteriorly through the anterior commissure of A7. More transcripts are detected in ganglion cells associated with the anterior commissure of each of the A2 through A7 ganglia as compared with the posterior commissures (*29, 42, 43*). (e) *iab-2* transcripts accumulate strongly in A1p through A7a. There are probably more transcripts in the posterior compartment of each of the A2 through A6 ganglia than in the anterior compartment. Fewer transcripts are seen in A7p through A8/A9. (f) Transcripts homologous to *iab-7* accumulate strongly in A6p through A8/A9. However, labeling in this region is not uniform in that the A6p/A7a and A8p/A9 show higher levels of transcript than A7p/A8a. Fewer transcripts are seen anteriorly through at least A5a and possibly A4p. Abbreviations: Br, brain; S1, S2, S3, anterior, central, and posterior regions of the subesophageal ganglion, respectively; T1, prothoracic ganglion; T2, mesothoracic ganglion; T3, metathoracic ganglion; A1-A8, the first through eighth abdominal ganglia.

model there is a sequential activation of BX-C gene expression along the anterior-posterior axis of the embryo. Thus, the anterior-most segment anlage that is acted upon by the BX-C (T2p)(*39*) would contain products encoded by only one BX-C locus (*Ubx*), whereas the posterior-most segment primordium (A8) is expected to contain all BX-C gene products (*Ubx, iab-2*, and *iab-7*) (*21*).

During extension and retraction of the germ band, the transcript distribution patterns for the *Ubx*, *iab-2*, and *iab-7* loci are consistent with the Lewis model (see Fig. 5). Progenitors of the metathorax and first abdominal segment contain *Ubx* transcripts, but do not display detectable levels of either *iab-2* or *iab-7* transcripts. In contrast, progenitors of the seventh (and possibly eighth) abdominal segment contain levels encoded by each of the homeo box–containing BX-C loci.

Over the course of embryonic development, there is a successive restriction in the spatial limits of transcript accumulation for at least some of the ANT-C and BX-C homeo box–containing loci. As compared with earlier periods of embryogenesis, advanced stage embryos show relatively discrete transcript accumulation patterns for each ANT-C and BX-C gene that contains a homeo box (compare Figs. 4 and 5 with Fig. 3).

While not specifically predicted by the Lewis model, this observation is consistent with the demonstration that each ANT-C and BX-C homeotic locus has a discrete primary domain of function (see Fig. 1a). For example, during germ band extension, *Antp* transcripts are detected at almost uniform levels in the neural progenitors of segment primordia S3/T1 through A7/A8. In contrast, by late stages of embryogenesis, the T1/T2 ganglion cells display substantially stronger hybridization as compared with the ganglion cells of other segments. Thus, as embryogenesis proceeds, *Antp* becomes localized to the segments which are most disrupted by lethal mutations of the gene. It is therefore possible that the primary domain of function for a given homeotic gene is determined by the embryonic regions where its expression is sustained.

Cross-regulatory interactions among homeo box genes. It has been previously suggested that hierarchical interactions among homeotic genes might influence their spatially restricted domains of expression (*35, 40*). More recent evidence for such homeotic cross-regulatory interactions has been obtained by analysis of homeo box transcript distribution patterns in homeotic mutant embryos. *Antp* transcripts have been shown to stably accumulate in the posterior ventral ganglia of advanced stage embryos that lack all known genes of the BX-C. On this basis it was suggested that one or more BX-C gene products either directly or indirectly inhibit *Antp* expression (*34*). Here we present further support for this proposal.

The altered distribution of *Antp* transcripts in advanced stage BX-C$^-$ embryos could be either a direct or indirect consequence of deleting the BX-C. An example of an indirect effect is that the absence of the BX-C allows for the activation of *Antp* expression by a common trans-regulatory component that in wild-type embryos preferentially initiates BX-C gene expression. The data presented in Fig. 6 are not obviously consistent with this explanation. BX-C$^-$ embryos display a wild-type *Antp*$^+$ transcript distribution pattern during early developmental stages. Only after the sixth hour of embryogenesis do BX-C$^-$ individuals show a striking deviation from wild type in the *Antp*$^+$ transcript pattern. It is therefore unlikely that initiation of *Antp* expression is altered in BX-C$^-$ embryos. Thus, the altered *Antp*$^+$ pattern observed in Fig. 6, e and f, might result from the absence of products encoded by the BX-C rather than from the physical removal of the complex per se.

The molecular basis for possible cross-regulatory interactions among *Drosophila* homeotic loci is not known. Such interactions might be mediated by the different homeo box protein domains encoded by the various ANT-C and BX-C loci. There appears to be weak structural homology between the homeo domain and known (or suggested) DNA binding proteins (*19, 41*). It is possible that each homeo box–containing locus autoregulates its own expression, thereby permitting relatively high steady-state levels of product to accumulate within its primary domain of function. In contrast, homeo box–containing protein products might repress the expression of other homeo box–containing transcription units in a hierarchical manner. For example, perhaps *Antp* protein products act as positive regulators of *Antp* expression. This results in the maintenance of high levels of *Antp* product within the primary domain of *Antp* function (that is, T1/T2 ventral ganglion cells). In embryonic ventral ganglia posterior to T1/T2, *Antp* product accumulation might be hindered by high affinity binding of BX-C homeo box–containing proteins to the *Antp* promoter. As a result, *Antp* products are observed to only transiently accumulate at high levels in the posterior ventral ganglia of wild-type embryos.

Consistent with this proposal is the demonstration that *Antp* RNA's stably accumulate in posterior portions of the developing ventral cord in embryos that lack all genes of the BX-C (*34*). Moreover, we have shown that the altered distribution of *Antp* transcript accumulation in BX-C$^-$ embryos probably does not result from initiation of *Antp* expression within inappropriate segment primordia (see Fig. 6). Rather, the absence of products encoded by one or more BX-C loci appears to be either directly or indirectly responsible for the stable accumulation of *Antp* transcripts in posterior ventral ganglia.

At present, there appears to be no direct evidence to support or reject this model of "homeo box competition" between different homeotic genes of the Antennapedia class. The availability of cloned homeotic DNA sequences and the isolation of proteins encoded by homeotic loci (*42, 43*) should furnish direct information regarding the spatial localization of homeotic gene products to discrete regions of developing embryos.

References and Notes

1. G. F. Ferris, in *Biology of Drosophila*, M. Demerec, Ed. (Wiley, New York, 1950), pp. 368–419.
2. P. A. Lawrence and P. Johnson, *Cell* **36**, 775 (1984).
3. F. Jiminez and J. A. Campos-Ortega, *Arch. Entwicklungsmech. Org.* **190**, 370 (1981).
4. D. R. Kankel, A. Ferrus, S. H. Garen, P. J. Harte, P. E. Lewis, in *The Genetics and Biology of Drosophila*, M. Ashburnerc and T. S. Wright, Eds. (Academic Press, New York, 1980), vol. 2, pp. 295–363.
5. E. B. Lewis, *Nature (London)* **276**, 565 (1978).
6. W. H. Ouweneel, *Adv. Genet.* **18**, 179 (1976).
7. E. Teugels and A. Ghysen, *Nature (London)* **304**, 440 (1983).
8. B. T. Wakimoto and T. C. Kaufman, *Dev. Biol.* **81**, 51 (1981).
9. E. Sanchez-Herrero, I. Vernos, R. Marco, G. Morata, *Nature (London)* **313**, 108 (1985).
10. R. E. Denell, *Genetics* **75**, 279 (1973).
11. T. C. Kaufman, R. Lewis, B. Wakimoto, *ibid.* **94**, 115 (1980).
12. E. B. Lewis, *Am. Zool.* **3**, 33 (1963).

13. ———, in *Embryonic Development, Part A: Genetic Aspects* (Liss, New York, 1982), pp. 269–288.
14. B. T. Wakimoto, F. R. Turner, T. C. Kaufman, *Dev. Biol.* **102**, 147 (1984).
15. T. Hazelrigg and T. C. Kaufman, *Genetics* **105**, 581 (1983).
16. W. McGinnis, M. S. Levine, E. Hafen, A. Kuroiwa, W. J. Gehring, *Nature (London)* **308**, 428 (1984).
17. W. McGinnis, R. L. Garber, J. Wirz, A. Kuroiwa, W. J. Gehring, *Cell* **37**, 403 (1984).
18. M. P. Scott and A. J. Weiner, *Proc. Natl. Acad. Sci. U.S.A.* **81**, 4115 (1984).
19. A. Laughon and M. P. Scott, *Nature (London)* **310**, 25 (1984).
20. M. Regulski, K. Harding, R. Kostriken, F. Karch, M. Levine, W. M. McGinnis, *Cell*, in press.
21. In this report, we consider only the homeotic loci of the BX-C that contain a homeo box. There are additional homeotic loci within this complex (Fig. 1a), each of which has a well-defined domain of function (*5*). For example, a fly that lacks *bx* function shows a transformation of T3a into T2a. The Lewis model (see text) predicts that only *abx* and *bx* would be normally expressed in T3a but that all the loci of the BX-C would be expressed in A8.
22. W. Bender *et al.*, *Science* **221**, 23 (1983).
23. R. L. Garber, A. Kuroiwa, W. J. Gehring, *EMBO J.* **2**, 2027 (1983).
24. M. P. Scott *et al.*, *Cell* **35**, 763 (1983).
25. The molecular map positions of several *Dfd* mutant alleles indicate that the p99-5.0 probe (Fig. 1b) derives from the *Dfd* locus (M. Regulski and W. McGinnis, unpublished results).
26. E. Hafen, M. Levine, R. L. Garber, W. J. Gehring, *EMBO J.* **2**, 617 (1983).
27. M. Levine, E. Hafen, R. L. Garber, W. J. Gehring, *ibid.*, p. 2037.
28. A. Kuroiwa and W. J. Gehring have localized the *Scr* locus region within overlapping cloned genomic DNA's that span most of the ANT-C (personal communication). On the basis of high stringency cross-hybridization tests with the cloned *Scr* DNA's it has been determined that A40 derives from the homeo box region of the *Scr* locus (W. McGinnis, A. Kuroiwa, W. J. Gehring, unpublished results).
29. M. Akam, *EMBO J.* **2**, 2075 (1983).
30. Our in situ localizations of transcripts could be in error by half a segment. The term "T1/T2" indicates that the region of hybridization could be in tissues of T1 or T2 (or both). Similarly, A1/A2, for example, indicates labeling in all or portions of the first or second abdominal ganglia (or both).
31. D. F. Poulson, in *Biology of Drosophila*, M. Demerec, Ed. (Wiley, New York, 1950), pp. 168–274.
32. I. Duncan, *Genetics* **100**, 520 (1981).
33. G. Struhl, *J. Embryol. Exp. Morphol.* **76**, 297 (1983).
34. E. Hafen, M. Levine, W. J. Gehring, *Nature (London)* **307**, 287 (1984).
35. G. Struhl, *Proc. Natl. Acad. Sci. U.S.A.* **79**, 7380 (1982).
36. A. Garcia-Bellido, *Am. Zool.* **17**, 613 (1977).
37. ———, in *Cell Patterning*, S. Brenner, Ed. (Elsevier/North-Holland, Amsterdam, 1975), pp. 161–182.
38. F. Karch *et al.*, in preparation.
39. Each segment has two developmentally distinct subdivisions: an anterior and a posterior compartment (*36, 37*). The founder cells for each compartment are determined during early embryonic stages. The cells that give rise to one compartment do not intermix with those that give rise to the other. However, there are no overt morphological markers which delineate the compartments of a segment.
40. G. Morata and S. Kerridge, *Nature (London)* **300**, 191 (1982).
41. J. C. W. Shepherd, W. McGinnis, A. E. Carrasco, M. DeRobertis, W. J. Gehring, *Nature (London)* **310**, 70 (1985).
42. R. White and M. Wilcox, *Cell* **39**, 163 (1984).
43. P. A. Beachy, S. L. Helfand, D. S. Hogness, *Nature (London)* **313**, 545 (1985).
44. G. Struhl, *ibid.* **292**, 635 (1980).
45. We thank Ernst Hafen, Eric Frei, and Markus Noll for stimulating and provocative discussions; Mark Dworkin, Eva Dworkin-Rastl, Ernst Hafen, Helen Doyle, Tim Hoey, and Lily Mirels for reviewing the manuscript; and Diane Robins for encouragement. This work was supported by a grant from the National Institutes of Health.

21 May 1985; accepted 2 August 1985

10

Plasticity of the Differentiated State

Helen M. Blau, Grace K. Pavlath, Edna C. Hardeman, Choy-Pik Chiu, Laura Silberstein, Steven G. Webster, Steven C. Miller, and Cecelia Webster

Tissue-specific phenotypes result from a sequence of developmental stages. Totipotent cells in the early embryo give rise to stem cells specific to three distinct layers—endoderm, ectoderm, and mesoderm. Although the lineage, or progression from stem cell to tissue-specific phenotype is not always fixed (*1*), once a cell is determined, it is generally destined for specialization along a specific pathway, such as erythropoiesis or myogenesis. The option to generate other phenotypes no longer exists for the determined vertebrate cell, and its progeny stably inherit its limited potential. The determined cell will give rise to other phenotypes only under unusual experimental conditions, such as at sites of regeneration in amphibian limbs where transdifferentiation has occurred or after treatment with a drug such as 5-azacytidine; even then, only derivatives of the same embryonic lineage are obtained (*2*). At some point in development, the determined cell expresses its phenotype, and the genes necessary for its role in the function of a particular tissue are transcribed.

To obtain tissue-specific phenotypes, a sequence of regulatory mechanisms must exist that determine when in a cell's history specific genes are transcribed. The genetic composition of eukaryotic cells is generally stable and heritable. Chromosomes are not lost in the course of cell specialization. This is evident since entire frogs can be generated from the transplantation of nuclei of specialized intestine cells into enucleated oocytes and since a diversity of normal tissue-specific cell types can be generated from malignant tumor cells introduced into early mouse embryos (*3*). The

Science 230, 758–766 (15 November 1985)

current model for the differential expression of genes characteristic of tissues at different points in development requires regulation by DNA sequences on the same chromosome (*cis*-acting) and on different chromosomes (*trans*-acting). *Cis*-acting DNA sequences that impart tissue-specific regulation have been identified from the study of the expression of cloned genes after transfection into cultured cells (*4*). The diffusible products of *trans*-acting genes are assumed to be negative or positive regulators of the *cis*-acting gene sequences. Although some general mediators of gene transcription and gene-specific binding proteins have been characterized in eukaryotes (*5*), with the exception of the factors that bind the *Drosophila* alcohol dehydrogenase gene (*6*), no tissue-specific *trans*-acting regulators have yet been isolated. An understanding of how the expression of tissue-specific genes is activated is not only of fundamental biological interest but also of practical importance in implementing genetic engineering and possibly gene therapy.

Muscle provides a model system for studies of the mechanisms controlling the appearance of tissue-specific functions. For a number of species, developmentally distinct stages are readily recognized by their morphological and biochemical properties, and conversion from one stage to another can be mimicked under the controlled conditions of tissue culture (*7*) (Fig. 1). First, a mesodermal stem cell gives rise to a myoblast, destined for myogenesis. The determined myoblast is capable of recognizing and spontaneously fusing with other myoblasts leading to the production of a differentiated myotube. The multinucleated myotube no longer divides or synthesizes DNA but produces muscle proteins in large quantity. These include constituents of the contractile apparatus and specialized cell-surface components essential to neuromuscular transmission. Eventually the differentiated muscle cell exhibits characteristic striations and rhythmic contractions. A further step in this pathway is maturation: the contractile apparatus in muscle at different stages of development contains distinct isoforms of muscle proteins such as myosin and actin, encoded by different members of multigene families (*8, 9*).

Heterokaryons: A Model System for Studying Cell Specialization

To study the mechanisms regulating cell specialization, we developed a system in which nonmuscle cells can be induced to express muscle genes predictably and stably. We combine entire muscle cells with cells of other phenotypes and from other species through the use of polyethylene glycol. In these fused cells, or heterokaryons, expression of previously silent muscle genes is activated in response to tissue-specific factors present in muscle (*10–12*). The nuclear composition of the heterokaryons and the activation of genes can be determined by taking advantage of species-specific differences: the nonmuscle cells used are always human and the muscle cells are always mouse. Furthermore, the fusion product is stable; in contrast with typical interspecific hybrids (synkaryons), cell division does not occur and the parental cell nuclei remain intact and retain a full complement of chromosomes. Finally, changes in gene expression can be assayed immediately after fusion and monitored for relatively long periods thereafter.

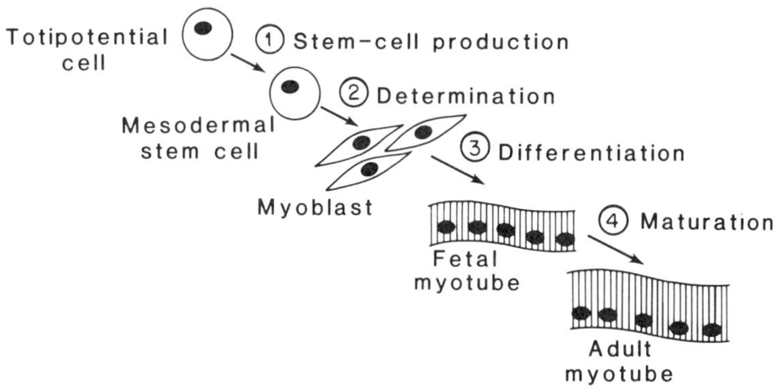

Fig. 1. Points of regulation in muscle cell specialization.

We have shown that muscle genes are activated in nonmuscle nuclei upon fusion with muscle. This activation must be mediated by diffusible, *trans*-acting molecules that are transported to nuclei through the cytoplasm, since the nuclei of the two cell types remain separate and distinct. These results show that gene expression by nuclei of highly specialized cells is remarkably plastic. Thus, the muscle genes in cells with very different roles (skin, cartilage, lung, and liver) are receptive to activation by muscle regulatory factors. The differences in the requirements for gene activation in these cell types provide insight into the molecular mechanisms that lead to the generation and maintenance of phenotypes.

Gene Activation in Heterokaryons

The mouse muscle cell line used to produce heterokaryons is a diploid subclone, C_2C_{12} (*13*). To test the influence of histogenetic state and of in vivo aging on muscle gene expression, heterokaryons were produced with mouse muscle and eight different human nonmuscle cell types, including keratinocytes (ectoderm), chondrocytes and fibroblasts (mesoderm), and hepatocytes (endoderm) (*14*). In addition, the four strains of fibroblasts used were either from different tissues (amniotic fluid, skin, and lung) or from different developmental stages (fetal skin and adult skin). One malignant cell type, HeLa, was also tested. To assess the ability of these nonmuscle cell types to be reprogrammed to express muscle functions in heterokaryons, it was important to verify their identity and determine that they continued to function as specialized cells in tissue culture. Accordingly, at the time of fusion with muscle cells, tissue-specific products were identified in cultures of chondrocytes, keratinocytes, and hepatocytes by immunofluorescence with antibodies to type II collagen, keratin, and albumin, respectively (*15*). Although phenotypic markers are not well characterized for fibroblasts, we determined that the two strains of skin fibroblasts used in the studies of developmental stage, from a fetus (14 weeks) and from an adult (66 years), were distinct cell types: they differed in their proliferative properties, cell size, and binding of the mitogens, insulin, and insulin-like growth factor 1 (*16, 17*).

Table 1. Cell types in which human muscle genes were activated in heterokaryons.

Biological function	Muscle gene product	Assay	Cell lineage tested	Phenotype tested
Enzyme	Creatine kinase Human MM Human MB Mouse-human hybrid MM	Electrophoresis and enzyme activity	Mesoderm	Amniotic fibroblast Fetal skin fibroblast Adult skin fibroblast Lung fibroblast HeLa (malignant) Chondrocyte
			Ectoderm	Keratinocyte
			Endoderm	Hepatocyte
Contractile apparatus	Myosin light chains Fetal 1s 2s 2f	Electrophoresis and monoclonal antibodies	Mesoderm	Amniotic fibroblast
	Actin mRNA's α-cardiac	cDNA probes	Mesoderm	Fetal skin fibroblast Adult skin fibroblast
			Ectoderm	Keratinocyte
	α-skeletal		Mesoderm	Fetal skin fibroblast Adult skin fibroblast
			Ectoderm	Keratinocyte
Membrane components	Cell surface antigens 24.1D5	Monoclonal antibodies	Mesoderm	Fetal skin fibroblast Lung fibroblast
	5.1H11		Mesoderm	Fetal skin fibroblast Adult skin fibroblast Lung fibroblast HeLa (malignant)
			Ectoderm	Keratinocyte
			Endoderm	Hepatocyte

We took advantage of species differences to assay the induction of expression of muscle genes in nonmuscle cells. A heterokaryon is readily identified by its mixed nuclear composition: mouse nuclei appear punctate and human nuclei are uniformly stained with the fluorescent dye Hoechst 33258 (Plate I). The novel activation of a muscle gene is evident by detection of a product specific to human muscle on the heterokaryon cell surface, a product which neither cell type alone produces. This gene product, 5.1H11, is an antigen present in small amounts on human myoblasts and in large amounts on myotubes (*18*). Unfused nonmuscle cells grown on the same dish with heterokaryons did not express muscle functions. Thus, assays of gene products could be performed at the level of a single cell in heterokaryons of defined nuclear composition. Gene activation was also analyzed in mass cultures at biochemical and molecular levels.

In each case tested, human muscle genes were activated (Table 1). Human myosin light chains—1s, 2s, 2f, and fetal—and human isozymes of creatine kinase (CK)—MB, MM, and a functional mouse-human hybrid MM—were distinguished from their mouse muscle counterparts by their mobility upon gel electrophoresis; their identities were confirmed either by reaction with monoclonal antibodies on Western blots or by in situ assays of enzyme activity (*10, 19*). The human muscle-specific transcripts of two sarcomeric actin genes, α-cardiac and α-skeletal actin, were detected by Northern blot and S1 nuclease analysis with species- and isotype-specific complementary DNA (cDNA) and genomic DNA probes (*20*). Expression of the human muscle-specific cell surface antigens, 24.1D5 and 5.1H11, was induced and could be monitored on single heterokaryons with monoclonal antibodies (*11, 17, 21, 22*). Thus, the genes that were activated encoded diverse products: enzymes critical to energy production, structural components of the contractile apparatus, and cell surface components. The relative amounts and sequence of expression of these different muscle gene products in heterokaryons paralleled myogenesis in pure cultures of human muscle cells.

The Kinetics of Gene Expression Differ

We examined the time course and frequency of muscle gene expression in different cell types contained in heterokaryons. The proportion of individual heterokaryons expressing the cell surface antigen 5.1H11 was determined between 1 and 15 days after fusion in 19 independent experiments in which more than 7000 individual heterokaryons were analyzed (Fig. 2). The three distinct patterns observed differed in the time course of gene expression, primarily because of differences in the lag period before 5.1H11 could be detected. In addition, the ultimate frequency of gene expression differed among cell types and approximated 95, 60, and 25 percent for fibroblasts, keratinocytes, and hepatocytes, respectively. Thus, tissue derivation and possibly embryonic origin have marked effects: fibroblasts, which are from the same embryonic lineage as muscle (mesoderm), exhibit faster kinetics and a higher ultimate frequency of 5.1H11 expression than keratinocytes (ectoderm) and hepatocytes (endoderm). In contrast, the kinetic curves for skin fibroblasts from two developmental stages were indistinguishable (*16, 17*).

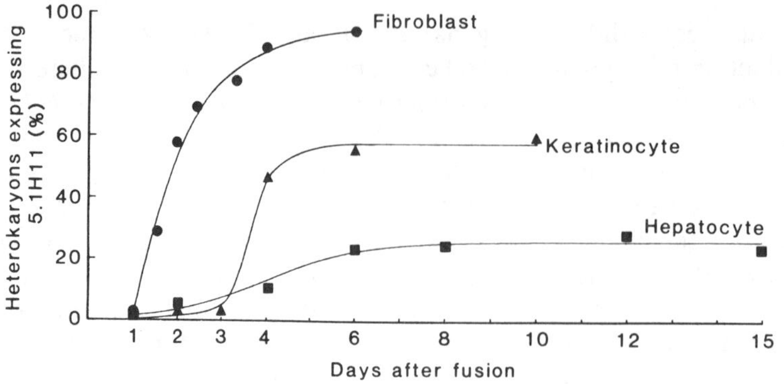

Fig. 2. Kinetics of 5.1H11 expression in heterokaryons containing different cell types. Individual heterokaryons containing nuclei from muscle and from either lung fibroblasts (●), keratinocytes (▲), or hepatocytes (■) were analyzed in replicate cultures for nuclear composition and the expression of 5.1H11 between 1 and 15 days after polyethylene glycol fusion. Curves are computer-derived best fit lines of the data. The size of the symbols includes ±1 standard error of the proportion calculated from the standard binomial equation.

To test whether the differences in the kinetics of 5.1H11 accumulation were due to phenotypic differences in the translation and subsequent processing of a cell surface protein, the expression of another muscle gene was examined at the transcriptional level. The relative levels of the messenger RNA's (mRNA's) for the muscle-specific α-cardiac and α-skeletal actins were studied in mass cultures of heterokaryons. We used species- and isotype-specific cDNA probes that recognize the transcripts of the human α-cardiac and α-skeletal actin genes, respectively, but not those of mouse or of other actin genes (*23*). Both α-cardiac and α-skeletal actin expression was evident in heterokaryons, and the time course and relative levels of the two transcripts differed (Fig. 3). These differ-

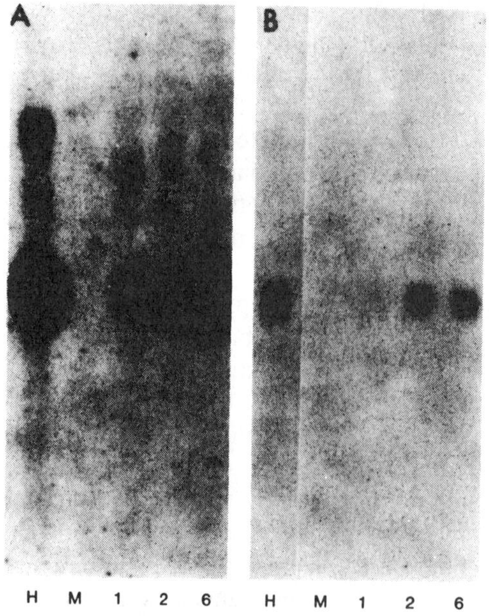

Fig. 3. Activation of human muscle-specific actin mRNA's in heterokaryons. Total RNA's were isolated from differentiated human muscle cells (H), differentiated mouse muscle cells (M), and mouse muscle–human lung fibroblast heterokaryon cultures on days 1, 2, and 6 after polyethylene glycol–mediated fusion. RNA's (5 μg from human muscle cells and 10 μg from each of the other samples) were electrophoresed on agarose gels, transferred to nitrocellulose, and hybridized with the human-specific cDNA probes to either α-cardiac actin (A) or α-skeletal actin (B). Autoradiograms were exposed to XAR-5 film at −80°C for 6 days.

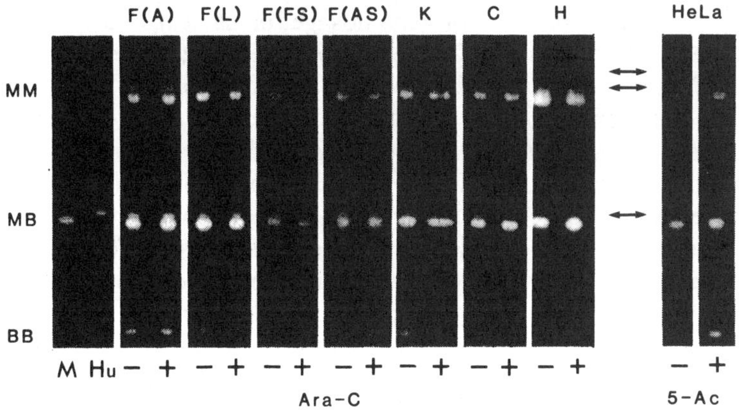

Fig. 4. Activation of human creatine kinase in different nonmuscle cells. Whole-cell extracts were electrophoresed on 5 percent nondenaturing polyacrylamide gels, and the CK isozymes were detected with ultraviolet illumination and a coupled enzyme reaction, yielding the reduced form of $NADP^+$ as its end-product. CK isozymes are shown for mouse (M) muscle cells and human (Hu) muscle cells. Heterokaryons formed between mouse muscle cells and fibroblasts from amniotic fluid F(A), lung F(L), fetal skin F(FS), and adult skin F(AS), keratinocytes (K), chondrocytes (C), and hepatocytes (H) are shown. Cultures marked + were exposed to cytosine arabinoside (Ara-C) for 24 hours before and 24 hours after fusion to ensure inhibition of DNA synthesis. Cultures marked − were not exposed to Ara-C during this period. Heterokaryons formed between mouse muscle cells and HeLa cells were either untreated (−) or exposed to 5 μM of 5-azacytidine (5-Ac) for 3 days followed by 1 day of drug removal (+). Equivalent enzyme activities were loaded in each lane. Arrows indicate CK isozymes containing human M subunits. Abbreviations: BB is the nonmuscle isozyme, M-subunit synthesis is initiated early in differentiation, and the dimers MB and MM are characteristic of differentiated muscle.

ences in sarcomeric actin expression were similar to those observed in pure cultures of human muscle cells (*20*). With this assay we compared the kinetics of accumulation of actin transcripts in heterokaryons produced with keratinocytes and with fibroblasts. The results paralleled those obtained for 5.1H11 expression. α-Cardiac actin transcripts were not detectable in keratinocyte heterokaryons on days 1 and 2 after fusion, when the levels in fibroblast heterokaryons were marked. By day 5, transcripts were detectable in keratinocyte heterokaryons as well (*16*). Since the differences in kinetics observed at the protein level were also observed at the mRNA level, they probably reflect differences among cell types in steps necessary for the activation of gene expression.

Mechanisms for Gene Activation

To examine potential mechanisms required for gene activation, we determined whether DNA replication was necessary to reprogram cells from different stages of development and of different phenotypes to express muscle genes (*11, 16, 17*). We exposed nonmuscle cells to the DNA synthesis inhibitor cytosine arabinoside (Ara-C) before and after fusion. The expression of human muscle CK was assayed in extracts of whole cells, and the expression of 5.1H11 was monitored on individual heterokaryons. The human M-CK gene was activated in all of the seven nonmalignant cell types tested, and three novel CK isozymes containing this subunit were detected: human MB, human MM, and mouse-human hybrid MM (Fig. 4). In all cases, similar amounts of the human isozymes were present regardless of whether DNA synthesis was inhibited. Furthermore, inhibition of DNA synthesis also had no effect on the frequency of 5.1H11 expression at the single-cell level. These results suggest that alterations in chromatin configuration requiring DNA replication are not necessary for the muscle genes in these cell types to be accessible to and respond to regulatory factors present in differentiated muscle cells.

In contrast to these results, CK and 5.1H11 muscle gene expression were never observed when we formed heterokaryons with HeLa cells, a malignant, aneuploid cell type (*12*). We examined whether prior treatment with 5-azacytidine (5Ac), a drug thought to reduce gene methylation (*24*), could alter the responsiveness of HeLa cells to muscle gene regulators in heterokaryons. Indeed, the expression of both 5.1H11 and CK isozymes containing human subunits was detected in heterokaryons formed with 5Ac-treated HeLa cells, but not in control heterokaryons containing untreated HeLa cells (Fig. 4) or in 5Ac-treated HeLa cells fused to themselves (*12*). Therefore gene activation in HeLa cells appears to require a mechanistic step not required by the other cell types tested: first a change induced by 5Ac is necessary and then interaction with muscle gene regulators.

Gene Dosage Influences Gene Expression

Further differences among cell types in the activation of muscle genes were apparent when we examined the effects of gene dosage, or nuclear ratio, on the frequency of muscle gene expression (*25*). Nuclear composition and the expression of the human muscle cell surface antigen 5.1H11 were monitored in the same heterokaryons with fluorescence microscopy at two different wavelengths (Plate I). The results for heterokaryons of different nuclear composition were pooled into five groups according to nuclear ratio, or the relative number of muscle to nonmuscle nuclei they contained (Fig. 5). The proportion in each group that expressed the antigen on a given day was determined. Examples for the 6-day time point are shown, beyond which the pattern of gene expression did not change.

The effect of gene dosage on 5.1H11 expression differed markedly for the three cell types. For fibroblasts, a high proportion of heterokaryons (95 percent) expressed 5.1H11 at all nuclear ratios, except in the increased fibroblast group (1: > 2), in which it was ~70 percent. In keratinocyte heterokaryons, gene expression in the 1:1 nuclear ratio group never exceeded 30 percent, whereas increased proportions of either muscle or keratinocyte nuclei resulted in the maximum expression of 70 percent. For hepatocyte heterokaryons, the frequency of gene expression was greatest (50 percent) when the relative number of muscle nuclei was increased and lowest (5 percent) when the proportion of hepatocyte nuclei was increased.

Although the interaction between two disparate cell types combined by fusion is likely to be complex, some noteworthy relationships are apparent from the studies of gene dosage (*16, 25*). (i) Even when the nonmuscle nuclei of each cell type outnumbered the muscle nuclei, muscle

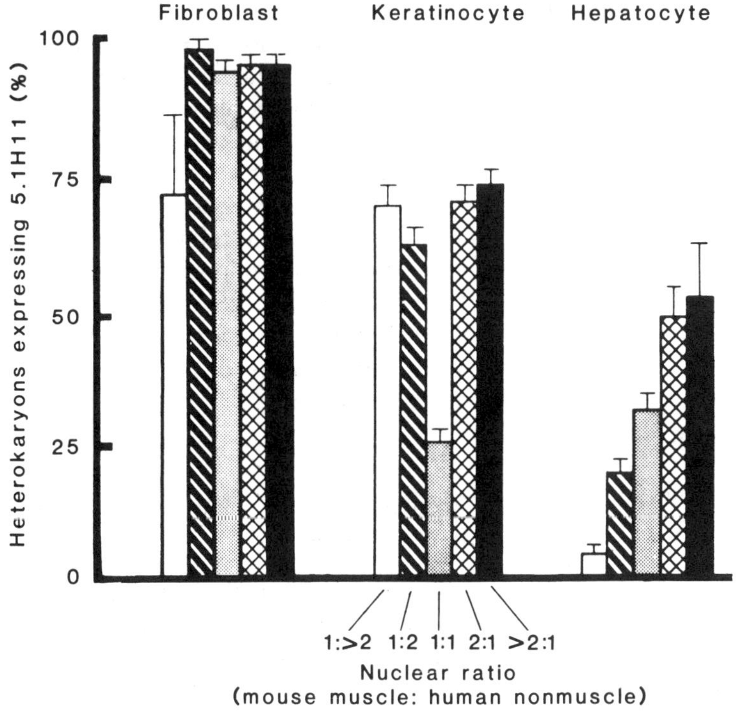

Fig. 5. Effect of nuclear ratio on 5.1H11 expression in heterokaryons containing different cell types. Individual heterokaryons containing nuclei from muscle and from either lung fibroblasts, keratinocytes, or hepatocytes were analyzed for nuclear composition and the expression of 5.1H11 6 days after fusion. Data were grouped into five ratios of muscle : nonmuscle nuclei. Error bars indicate the standard error of the proportion calculated from the standard binomial equation.

gene expression was induced, albeit with different frequencies. (ii) The kinetics of gene expression in fibroblast heterokaryons with increased proportions of nonmuscle nuclei were slower than when the nuclear ratio was 1:1. These results could be due to a requirement for gene activation of a threshold concentration of positive regulators: increased time would be necessary for the progressive accumulation of factors that must be shared among fibroblast nuclei. The rate with a 1:1 nuclear ratio exceeded that with increased muscle nuclei for fibroblast heterokaryons, which suggests that increased input from muscle was not optimal; instead a balance of factors contributed by each cell type might be required.

In hepatocyte heterokaryons, the frequency of activation of muscle gene expression was lowest when the proportion of hepatocyte nuclei was increased and highest when the proportion of muscle nuclei was increased. That these frequencies did not change with time suggests that gene expression was determined by the continued balance between components contributed by each cell type. Keratinocytes were more frequently activated to express muscle genes when either the proportion of keratinocyte or of muscle nuclei was increased in heterokaryons. The increased frequency of gene expression in heterokaryons with excess keratinocyte nuclei could be due to the existence of subpopulations of cells in keratinocyte cultures that differed in their ability to respond to muscle regulatory factors. Accordingly, the chance of obtaining a keratinocyte in which activation could occur would increase with the number of keratinocyte nuclei. Stratification within colonies of keratinocytes was evident in our cultures, and although the upper, more mature cells reportedly do not replate well (*26*), possibly some were incorporated into heterokaryons.

Expression of the Nonmuscle Phenotype

To test whether the phenotype of the nonmuscle cell persists in heterokaryons, we monitored the expression of albumin in individual hepatocyte heterokaryons. Six days after fusion, when the muscle product 5.1H11 was expressed at maximum frequency, albumin was still present in a large proportion of hepatocyte heterokaryons. By 15 days after fusion, 5.1H11 was still expressed at maximal frequency, but the proportion of albumin-containing heterokaryons was reduced. The persistence of albumin at 6 days was not due to the stability of previously synthesized proteins, since albumin is secreted within a few hours of its production. To detect albumin in heterokaryons, we had to inhibit secretion with monensin 2.5 hours before fixation (*16, 17*). Furthermore, the decline in albumin-containing heterokaryons with time was not due merely to extinction of the liver phenotype in the course of long-term culture, since parallel cultures of hepatocytes fused to themselves with polyethylene glycol (PEG) contained albumin on day 15. From these experiments we cannot determine whether the expression of the genes of one phenotype precludes the expression of the genes of the other. This determination would require assays of active transcription for genes characteristic of both nonmuscle and muscle phenotypes in single nuclei. Despite this limitation, our results indicate that the combination of specialized cells in a common cytoplasm does not result in the immediate degradation of differentiation-specific proteins. Instead, we conclude that the nonmuscle phenotype persists for a time in hepatocyte heterokaryons, but the muscle phenotype dominates.

Regulatory Circuits Between Nuclei in Heterokaryons

To examine whether the nonmuscle cell influences gene expression in the muscle cell, we examined the accumulation of mouse and human mRNA's for α-cardiac actin, a major component of the muscle contractile apparatus. We took advantage of the fact that primary human muscle cells and cultured mouse C_2C_{12} cells exhibited distinct patterns of α-cardiac actin transcript accumulation with time after myotube formation (*20, 27–29*). The amount of α-cardiac actin mRNA in each sample was normalized to the amount present in a human heart standard analyzed at the same time. In the mouse muscle C_2C_{12} cultures, α-cardiac actin transcripts accumulated rapidly, transiently peaked at a level comparable to 50 percent of the α-cardiac actin mRNA expressed in human heart within 24 hours, and declined (Table 2). In contrast, α-cardiac actin transcripts accumulated steadily in human myotubes over a 6-day period and exceeded the peak reached in the C_2C_{12} cell line 16-fold, a level comparable to 800 percent that observed in human

Table 2. Relative α-cardiac actin mRNA in pure muscle and in heterokaryon cultures. PEG treatment occurred after the mouse muscle myotubes had been in fusion medium for 3 days. Thus, to compare the mouse time scale in fusion medium (left) with that after PEG treatment (right), add 3 days.

Days in fusion medium	α-Cardiac actin mRNA in human heart (%)		Days after PEG treatment	α-Cardiac actin mRNA in human heart (%)	
	Mouse muscle C_2C_{12}	Human primary muscle		Mouse in heterokaryons	Human in heterokaryons
0	20	1	0	—	0
1	50	100	1	10	50
2	25	200	2	30	200
6	20	800	6	40	700

heart. In heterokaryon cultures, the time course of human transcript accumulation and the amount of transcript per non-muscle nucleus were similar to those observed in pure human, not mouse, muscle cultures. Of greatest surprise was the result obtained for mouse α-cardiac actin mRNA in heterokaryons: instead of continuing to decline, as in pure mouse muscle cultures, the amount of mouse transcripts increased after heterokaryon formation.

These findings suggest that the human α-cardiac actin gene responds to mouse muscle cytoplasmic factors by producing transcripts with the time course and at levels typical of human, not mouse, muscle cultures. These differences in transcript accumulation could arise in part from differences in the *cis*-acting regulatory regions of the mouse and human α-cardiac actin genes. In addition, the results are compatible with the hypothesis that *trans*-acting mouse muscle factors activate human muscle regulatory genes encoding factors that, in turn, act on mouse muscle genes or stabilize mouse α-cardiac actin transcripts. Finally, the "activated" phenotype seems to dominate, since the usual decline in the amount of mouse α-cardiac actin transcripts is overridden by the presence of the human nuclei.

Muscle Gene Regulators at Distinct Stages of Development

Like the experiments described above, the majority of studies of gene expression in specialized cells have utilized the differentiated stage of development (Fig. 1). This is true primarily because for all cell types, this stage is generally the best characterized and most readily obtained in tissue culture. The experiments described below show that other stages of muscle development—determination and maturation—are now amenable to analysis.

To test whether muscle cells at the determined stage of development elaborate distinct muscle gene regulators, we produced heterokaryons with fibroblasts and muscle cells at the myoblast (determined) and myotube (differentiated) stages (*22*). Convenient markers for this study were provided by two cell-surface antigens: 5.1H11 and 24.1D5. The 5.1H11 antigen is expressed on cultured myoblasts to some extent, but markedly increases in amount on myotubes. The 24.1D5 antigen is detected only on myoblasts and is no longer present once myoblasts fuse to form multinucleated differentiated myotubes. The human forms of both muscle cell–surface components were recognized with species-specific monoclonal antibodies (*17, 20*). The 24.1D5 antigen was detected only in myoblast heterokaryons (22 percent) (Table 3).

In contrast, 5.1H11 was expressed in both myoblast heterokaryons (19 percent) and myotube heterokaryons (92 percent). Thus, the expression of these two antigens in heterokaryons paralleled their expression in human myogenesis. By day 5, a maximum frequency of expression was reached in both cases and further increases were not observed with time. The lower frequency of gene expression in myoblast than in myotube heterokaryons may be due to myoblast nuclei that were in different phases of the cell cycle, whereas most of the myotube nuclei were presumably withdrawn from the cell cycle. These results suggest that muscle cells at myoblast and myotube stages of development differ in the concentration or type of at least two regulatory factors.

The study of the maturation stage of muscle development is facilitated by the myosin heavy chains. These proteins are major components of the contractile apparatus that are encoded by different genes in fetal, neonatal, and adult muscle tissues (*8*). Studies of maturation have been hindered because these transitions are not readily obtained in vitro; primarily fetal forms of muscle proteins are expressed in cultured myotubes. We have determined that C_2C_{12} cells express

Table 3. Evidence for two temporally distinct muscle gene regulators. Fetal skin fibroblasts were used except in myotube experiment 3, in which lung fibroblasts were used; S.E.M., standard error of the mean.

Experiment	Heterokaryons expressing 24.1D5: Percentage exhibiting detectable antigen	24.1D5: Number scored	5.1H11: Percentage exhibiting detectable antigen	5.1H11: Number scored
	Fusion partner: myoblast			
1	34	228	8	229
2	13	131	19	59
3	20	113	30	93
Mean ± S.E.M.	22 ± 6		19 ± 6	
Total		472		381
	Fusion partner: myotube			
1	1	556	90	407
2	0	800	98	166
3	0	600	88	376
Mean ± S.E.M.	0		92 ± 3	
Total		1956		949

developmentally distinct myosins. These myosin isoforms were detected with monoclonal antibodies generated to myosin purified from human muscle at different developmental stages (*30*). C_2C_{12} cells are stained with two distinct antibodies that recognize myosin isozymes present at fetal and neonatal stages of development (Plate II). That the intensity of staining differed among myotubes suggests that the relative concentrations of fetal and neonatal myosins differ. A progressive increase in the relative amount of neonatal myosin was observed with time in culture. The results were not duc to thc usc of the C_2C_{12} cell line, since they have been corroborated with human primary muscle cultures (*31*). Consequently, neural contact and complex substrates do not seem to be required for certain steps in muscle maturation. Since some of our antibodies to myosin are human-specific, it should now be possible to examine the activation of myosin isozymes in heterokaryons and the regulatory mechanisms underlying the maturation stages of muscle development.

Discussion

Several theories have been proposed to account for the development of cell diversity: changes in the genome that accompany quantal mitoses (*32*), sequential repression of genes (*33*) or modulation of gene expression by activators and repressors (*34*) with additional complex regulatory effects resulting from the relative concentration and affinity of these factors for specific sequences within genes (*4–6*). Some major questions are raised by these theories: How fixed and irreversible is the differentiated state or tissue-specific phenotype of a cell? What is required to maintain the differentiated state? What underlying molecular mechanisms regulate the transition from the determined to the differentiated to the mature cell phenotype? Do these mechanisms differ for different specialized cell types?

Our approach to these questions was inspired by reports by Weiss and others of experiments with somatic cell hybrids, in which the expression of genes atypical of a cell's normal phenotype could be induced (*35, 36*). A limitation in the comparative analysis of gene activation in different cell types is that highly aneuploid transformed cells were usually used. In addition, hybrid cells were frequently obtained after proliferation in mitogen-rich media, which led to variable retention and rearrangement of genetic material (*35, 37*), both of which are likely to have influenced gene expression (*38*). As a result, certain phenotypes were thought to be incapable of being activated, because of either their specialized state or their developmental stage (*37, 39*).

The activation of muscle gene expression in all of the eight nonmuscle cell types we tested is likely to be due to several factors. To facilitate the study of normal regulatory mechanisms, most nonmuscle cells were diploid primary cells. To overcome the problems of gene loss and rearrangement, we used a heterokaryon system that remained stable for the 2-week life-span of the cells and permitted a systematic analysis of gene expression over time. We cannot rule out the possibility that muscle is particularly well suited for internuclear communication and existence as a heterokaryon, since it is naturally a multinucleated cell in its differentiated state. However, the use of culture conditions which promoted muscle differentiation, not cell division, was probably critical to the success of our experiments. Proliferation and differentiation are antagonistic, mutually exclusive states in muscle cells (*40*). Furthermore, specific culture conditions that differ markedly among cell types are required for the optimal expression of phenotypes in vitro (*41*). Many investigators using somatic cell genetics to analyze the expression of differentiated functions used a nonspecific mitogen-rich culture medium. We propose that with culture conditions that favor the differentiation of the phenotype of interest, gene activation in stable heterokaryons could be achieved with cell types other than muscle.

The ability to activate muscle genes in heterokaryons is not a peculiarity of the C_2C_{12} cell line. We have recently been able to induce human muscle gene expression with another mouse muscle line, the myogenic derivative of a fibroblastic 10T1/2 cell treated with 5-azacytidine (*13*). However, the frequency of gene activation in fibroblast heterokaryons formed with 10T1/2 is half that observed with C_2C_{12} cells. We selected the C_2C_{12} clone for its potential to give rise to extensive differentiated contractile myotubes within a short period, and in this respect it exceeds all other myogenic lines with which we have had experience. Thus, C_2C_{12} may be an optimal source of muscle gene regulatory factors. Another myogenic cell line that may be enriched for regulators has been obtained by Wright, who selected for muscle cells resistant to an inhibitor of differentiation, bromodeoxyuridine (*42*). The selection of cells that undergo rapid differentiation should also prove useful for characterizing tissue-specific regulators in other phenotypes.

The plasticity in the function of the nucleus of specialized cells is remarkable. In each nonmalignant cell type tested, the differentiated state could be altered in the absence of DNA replication or cell division and the cells induced to express human muscle functions after exposure to muscle cytoplasm. Representatives of different embryonic lineages, phenotypes, and developmental stages were tested, including four strains of fibroblasts and chondrocytes (mesoderm), keratinocytes (ectoderm), hepatocytes (endoderm), and malignant HeLa cells. Differences among cell types were observed in the frequency, kinetics, and gene dosage requirements for the expression of human muscle genes. Cells that were more closely related to muscle—mesodermal derivatives—consistently expressed muscle genes sooner and at a greater frequency than the cells of ectodermal or endodermal origin.

Our experiments suggest that the modulation of gene expression which accompanies eukaryotic cell differentiation is mediated by a number of positive and negative regulatory factors in a concentration-dependent manner. When cells are mixed by fusion, the end result is likely to be due to complex interactions; the following simple interpretation of our experiments may nonetheless provide a useful conceptual framework. The delayed but high frequency of expression of muscle genes in heterokaryons with one muscle and multiple nonmuscle nuclei suggests that muscle regulators must reach a threshold concentration for gene activation to occur. The observation that muscle gene expression in fibroblast heterokaryons is fastest with a 1:1 ratio suggests that a balance of factors contributed by both cell types is optimal for that cell type. On the other hand, the frequency of gene expression in hepatocytes differs in heterokaryons containing different proportions of muscle and nonmuscle nuclei, and the differences persist with time, suggesting that hepatocytes produce negative regulators that must be titrated by the positive regulators in muscle cells. Muscle cells at different stages of development seem to synthesize different positive regulators. Finally, the reprogramming of the nonmuscle nucleus in a heterokaryon results in the synthesis by that nucleus of muscle regulatory factors. Thus, from an analysis of the combined effects on gene expression of two cell types fused to form heterokaryons, inferences regarding the regulators that generate and maintain these phenotypes can be drawn.

Gene expression in cells specialized for very different functions is not fixed and irreversible. Although under normal circumstances the pattern of gene expression of specialized cells is stable and heritable, it can be altered if the regulatory circuits between nucleus and cytoplasm are disrupted. Thus, changes in gene expression during development depend not only on the nucleus, but also on the cytoplasm, which plays an essential role as signal transducer. Our results extend the observations that nuclear gene expression is subject to modulation by the cytoplasm (*43*) to a system that may be particularly amenable to molecular analysis.

Future Prospects

Heterokaryons have utility as experimental models for investigating the regulation of gene expression in differentiated cells. The cells that are reprogrammed in heterokaryons should provide useful test systems for studying the mechanisms of action of the relevant molecules. Differences among cell types in the lag period before gene expression, the time course of accumulation of gene products, and the frequencies of gene expression at different nuclear ratios suggest different mechanistic steps. These could result from differences in gene conformation due to DNA modifications or binding proteins in addition to differences in the requirement for the type, concentration, or stoichiometry of *trans*-acting factors. We and others have begun to characterize the *cis*-acting regulatory sequences of muscle-specific genes (*44*). The isolation of the genes for *trans*-acting muscle-regulatory molecules at determined, differentiated, and mature stages of muscle development should now also be possible. Approaches that may prove useful for isolating mRNA's and genes for uncharacterized factors include microinjection of cytoplasmic components, use of subtractive hybridization to enrich for mRNA's for stage-specific regulators, and transfection of either cDNA's made to the enriched mRNA's or of genomic DNA into cells in conjunction with selective assays for gene expression (*45*). Elucidation of the molecular mechanisms underlying the reprogramming of nonmuscle cell types should lead to a better understanding of the generation and maintenance of these phenotypes and of different myogenic stages.

References and Notes

1. J. R. Whittaker, *Dev. Biol.* **93**, 463 (1982); J. E. Sulston, E. Schierenberg, J. G. White, J. N. Thomson, *ibid.* **100**, 64 (1983).
2. T. Yamada and D. S. McDevitt, *ibid.* **38**, 104 (1974); D. A. Dunis and M. Namenwirth, *ibid.* **56**, 97 (1977); S. M. Taylor and P. A. Jones, *Cell* **17**, 771 (1979); A. B. Chapman, D. M. Knight, B. S. Dieckmann, G. M. Ringold, *J. Biol. Chem.* **259**, 15548 (1984); S. F. Konieczny and C. P. Emerson, Jr., *Cell* **38**, 791 (1984).
3. J. B. Gurdon, *J. Embryol. Exp. Morphol.* **10**, 622 (1962); ——— and V. Uehlinger, *Nature (London)* **210**, 1240 (1966); M. J. Dewey, D. W.

Martin, Jr., G. R. Martin, B. Mintz, *Proc. Natl. Acad. Sci. U.S.A.* **74**, 5564 (1977).
4. S. L. McKnight and R. Kingsbury, *Science* **217**, 316 (1982); D. A. Goldberg, J. W. Posakony and T. Maniatis, *Cell* **34**, 59 (1983); Y. Gluzman and T. Shenk, Eds., *Enhancers and Eukaryotic Gene Expression* (Cold Spring Harbor Laboratory, Cold Spring Harbor, N.Y., 1983); R. Treisman, M. R. Green, T. Maniatis, *Proc. Natl. Acad. Sci. U.S.A.* **80**, 7428 (1983); D. D. Brown, *Cell* **37**, 359 (1984); M. Mercola, J. Goverman, C. Mirell, K. Calame, *Science* **227**, 266 (1985); A. Ephrussi, G. M. Church, S. Tonegawa, W. Gilbert, *ibid.*, p. 134; R. Grosschedl and D. Baltimore, *Cell* **41**, 885 (1985); C. M. Gorman, P. W. J. Rigby, D. P. Lane, *ibid.* **42**, 519 (1985); K. R. Yamamoto, *Annu. Rev. Genet.*, in press.
5. H. R. Pelham and D. D. Brown, *Proc. Natl. Acad. Sci. U.S.A.* **77**, 4170 (1980); D. R. Engelke, S.-Y. Ng, B. S. Shastry, R. G. Roeder, *Cell* **19**, 717 (1980); W. S. Dynan and R. Tjian, *Nature (London)* **316**, 774 (1985); A. B. Lassar, P. L. Martin, R. G. Roeder, *Science* **222**, 740 (1983); C. S. Parker and J. Topol, *Cell* **37**, 273 (1984); C. Wu, *Nature (London)* **311**, 81 (1984).
6. U. Heberlein, B. England and R. Tjian, *Cell* **41**, 965 (1985).
7. D. Yaffe, *Proc. Natl. Acad. Sci. U.S.A.* **61**, 477 (1968); R. Bischoff and H. Holtzer, *J. Cell Biol.* **41**, 188 (1969); I. R. Konigsberg, *Dev. Biol.* **26**, 133 (1971); C. P. Emerson and S. K. Beckner, *J. Mol. Biol.* **93**, 431 (1975); J. P. Merlie and F. Gros, *Exp. Cell Res.* **97**, 406 (1976); D. Yaffe and O. Saxel, *Nature (London)* **270**, 725 (1977); B. Nadal-Ginard, *Cell* **15**, 855 (1978); R. V. Storti *et al.*, *Cell* **13**, 589 (1978).
8. G. F. Gauthier and S. Lowey, *J. Cell Biol.* **74**, 760 (1977); R. G. Whalen *et al.*, *Nature (London)* **292**, 805 (1981); V. Mahdavi, M. Periasamy, B. Nadal-Ginard, *ibid.* **297**, 659 (1982); D. Bader, T. Masaki, D. A. Fischman, *J. Cell Biol.* **95**, 763 (1982); R. Matsuda, D. H. Spector, R. C. Strohman, *Dev. Biol.* **100**, 478 (1983); R. M. Wydro, H. T. Nguyen, R. M. Gubits, B. Nadal-Ginard, *J. Biol. Chem.* **258**, 670 (1983).
9. A. J. Minty, S. Alonso, M. Caravatti, M. E. Buckingham, *Cell* **30**, 185 (1982); B. M. Paterson and J. D. Eldridge, *Science* **224**, 1436 (1984); W. Bains, P. Ponte, H. Blau, L. Kedes, *Mol. Cell Biol.* **4**, 1449 (1984); M. Shani *et al.*, *Nuc. Acids Res.* **9**, 579 (1981); E. A. Fyrberg, J. W. Mahaffey, B. J. Bond, N. Davidson, *Cell* **33**, 115 (1983); P. Gunning, P. Ponte, H. Blau, L. Kedes, *Mol. Cell. Biol.* **3**, 1783 (1983).
10. H. M. Blau, C.-P. Chiu, C. Webster, *Cell* **32**, 1171 (1983).
11. C.-P. Chiu and H. M. Blau, *ibid.* **37**, 879 (1984).
12. ———, *ibid.* **40**, 417 (1985).
13. The mouse muscle cell line used in most experiments was originally obtained by Yaffe and Saxel [in (*7*)] and is a diploid subclone, C_2C_{12}, isolated and karyotyped in our laboratory and selected for its ability to differentiate rapidly and produce extensive contracting myotubes expressing characteristic muscle proteins. Where indicated, a myogenic derivative of a 10T1/2 cell provided by A. Lassar and H. Weintraub was used. Heterokaryons were produced with polyethylene glycol (PEG) (*10*) and maintained in a medium containing the selective agents cytosine arabinoside (Ara-C) and ouabain to eliminate unfused cells.
14. Tissue-specific cell types included chondrocytes isolated in our laboratory from neonatal cartilage under the tutelage of L. Smith. Keratinocytes were a strain (N1) obtained from H. Green, isolated from the outer layer of newborn foreskin. Hepatocytes (Hep G2) were a cell line isolated from a hepatoma, which is nontumorigenic and retains the ability to stably express 17 differentiated functions characteristic of liver [D. P. Aden, A. Fogel, S. Plotkin, I. Damjanov, B. S. Knowles, *Nature (London)* **282**, 615 (1979); B. B. Knowles, C. C. Howe, D. P. Aden, *Science* **209**, 497 (1980)]. Fibroblasts were obtained from the skin of a 14-week fetus and from a 66-year-old adult by P. Byers and M. Karasek, respectively. Fibroblasts from fetal lung were a cell strain (MRC-5) [J. P. Jacobs, C. M. Jones, J. P. Baille, *Nature (London)* **227**, 168 (1970)].
15. Reagents for the detection of tissue-specific products were cDNA probes to α-cardiac and α-skeletal actin genes (P. Gunning, A. Minty, and L. Kedes) and monoclonal antibodies to cell surface antigens 5.1H11 and 24.1D5 (F. Walsh), to myosin light chains (F. Stockdale), to collagen type II (R. Burgeson), and to keratin (T.-T. Sun). The human-specific antibody to albumin was generated by us from nonspecific commercial antisera obtained from Cappel Labs. Binding of insulin and insulin-like growth factor 1 were assayed with the aid of R. Roth.
16. H. M. Blau, C.-P. Chiu, G. K. Pavlath, E. C. Hardeman, *J. Cell. Biochem. Suppl.* 9B:35 (1985).
17. G. K. Pavlath, C.-P. Chiu, H. M. Blau, in preparation.
18. F. S. Walsh and M. A. Ritter, *Nature (London)* **289**, 60 (1981); O. Hurko and F. S. Walsh, *Neurology* **33**, 737 (1983).
19. H. M. Blau, C.-P. Chiu, G. K. Pavlath, C. Webster, *Adv. Exp. Med. Biol.* **182**, 231 (1985).
20. E. Hardeman, C.-P. Chiu, H. M. Blau, *J. Cell Biol. Abstr.* **101**, 207a (1985).
21. F. S. Walsh, S. E. Moore, R. Nayak, *Invest. Exploit. Antibody Comb. Sites*, in press.
22. G. K. Pavlath and H. M. Blau, *J. Cell Biol. Abstr.* **101**, 207a (1985).
23. P. Gunning *et al.*, *J. Mol. Evol.* **20**, 202 (1984).
24. D. V. Santi, C. E. Garrett, P. J. Barr, *Cell* **33**, 9 (1983).
25. G. K. Pavlath and H. M. Blau, *J. Cell Biol.*, in press.
26. Y. Barrandon and H. Green, *ibid.* **82**, 5390 (1985).
27. Human primary muscle cells were isolated and cultured [H. M. Blau and C. Webster, *Proc. Natl. Acad. Sci. U.S.A.* **78**, 5623 (1981); ———, G. K. Pavlath, *ibid.* **80**, 4856 (1983); H. M. Blau *et al.*, *Exp. Cell Res.* **144**, 495 (1983)].
28. W. Bains, P. Ponte, H. Blau, L. Kedes, *Mol. Cell. Biol.* **4**, 1449 (1984).
29. To assay mouse and human transcripts, three distinct DNA probes to the α-cardiac actin gene were obtained from P. Gunning, A. Minty, and L. Kedes. To compare actin transcript accumulation in mouse and human muscle cultures, a

cDNA probe capable of detecting both mouse and human cardiac actin transcripts was used. To assay actin transcripts, cDNA probes were labeled with ^{32}P by nick translation and hybridized on nitrocellulose filters to serial dilutions of total RNA isolated from a time course of muscle cultures for each species. To facilitate comparison among experiments, a sample of RNA from human heart was always included on the filters. The amount of α-cardiac actin RNA in the samples was quantitated by densitometric scanning of autoradiograms of the filters, and the results for each were normalized to a heart RNA standard on that autoradiogram. To examine the accumulation of human α-cardiac actin transcripts in heterokaryons, a second cDNA probe was used, which was species- and isotype-specific: at the hybridization stringency used, only human α-cardiac actin transcripts were detected (Fig. 3). To examine the expression of the mouse cardiac actin gene in the same heterokaryons, the RNA from the heterokaryons was subjected to S1 nuclease analysis. We used a third probe, which shared homology with a longer sequence of human than mouse α-cardiac actin mRNA. Thus, the sequences for mouse and human that hybridized well and were protected from treatment with S1, an enzyme that digests single-stranded nucleic acids, differed in size and could be distinguished by gel electrophoresis. The accumulation of human cardiac actin transcripts determined by this method paralleled that observed in the slot blots, providing confidence in this method, which requires substantially more manipulations of the RNA.

30. L. Silberstein, S. G. Webster, H. M. Blau, annual meeting of the American Association for the Advancement of Science, Los Angeles, May 1985; L. Silberstein and H. M. Blau, in *Molecular Biology of Muscle Development*, E. Emerson, D. A. Fischman, B. Nadal-Ginard, M. A. Q. Siddiqui, Eds. (Liss, New York, in press). Antibodies were generated to myosin purified from human muscle at fetal, neonatal, and adult developmental stages. Antibodies were determined to be specific for myosin heavy chain by enzyme-linked solid-phase immunoassay, by their reaction in immunoblots with proteins of 200,000 molecular weight, and by their distinctive staining pattern of striations within cultured myotubes and within fibers of cross-sectioned muscle tissues. The myosins recognized by these antibodies had distinct time courses of appearance in muscle tissues from mouse and human at different stages of development. The antibodies did not react with myosin in nonmuscle cells.
31. L. Silberstein, S. G. Webster, D. Arvanitis, H. M. Blau, *J. Cell. Biol. Abstr.* **101**, 45a (1985).
32. H. Holtzer, in *Stem Cells and Tissue Homeostasis*, B. I. Lord and C. S. Potter, Eds. (Cambridge Univ. Press. Cambridge, 1978), p. 1.
33. A. I. Caplan and C. P. Ordahl, *Science* **201**, 120 (1978).
34. M. Ptashne *et al.*, *Cell* **19**, 1 (1980).
35. M. C. Weiss, *Results Prob. Cell Differ.* **11**, 87 (1980); M. C. Weiss, *Somatic Cell Genetics*, C. T. Caskey and D. C. Robins, Eds. (Plenum, New York, 1982), p. 169.
36. N. R. Ringertz and R. E. Savage, Eds., *Cell Hybrids* (Academic Press, New York, 1976); C. J. Epstein, *The Consequences of Chromosome Imbalance: Principles, Mechanisms, and Models* (Cambridge Univ Press, Cambridge, in press), chap. 6.
37. G. J. Darlington, J. K. Rankin, G. Schlanger, *Som. Cell Genet.* **8**, 403 (1982).
38. B. McClintock, *Science* **226**, 792 (1984).
39. S.-A. Carlsson, N. R. Ringertz, R. E. Savage, *Exp. Cell Res.* **84**, 255 (1974); M. Mevel-Ninio and M. C. Weiss, *J. Cell. Biol.* **90**, 339 (1981); S. Junker, *J. Cell Sci.* **47**, 207 (1981); S. Linder, S. H. Zuckerman, N. R. Ringertz, *Proc. Natl. Acad. Sci. U.S.A.* **78**, 6286 (1981); J. B. Lawrence and J. R. Coleman, *Dev. Biol.* **101**, 463 (1984); W. Wright, *J. Cell Biol.* **98**, 427 (1984); W. Wright, *Exp. Cell Res.* **151**, 55 (1984).
40. T. A. Linkhart, C. H. Clegg, S. D. Hauschka, *Dev. Biol.* **86**, 19 (1981); B. H. Devlin and I. R. Konigsberg, *ibid.* **95**, 175 (1983); H. T. Nguyen, R. M. Medford, B. Nadal-Ginard, *Cell* **34**, 281 (1983).
41. B. M. Gilfix and H. Green, *J. Cell Physiol.* **119**, 1972 (1980); L. M. Reid *et al.*, *Ann. N.Y. Acad. Sci.* **349**, 70 (1980); M. S. Wicha *et al.*, *Proc. Natl. Acad. Sci. U.S.A.* **79**, 3213 (1982); E.Y.-H. Lee, G. Parry, M. T. Bissell, *J. Cell Biol.* **98**, 146 (1984).
42. W. E. Wright, *J. Cell Biol.* **100**, 311 (1985).
43. J. W. McAvoy, K. E. Dixon, J. A. Marshall, *Dev. Biol.* **45**, 330 (1975); E. M. De Robertis and J. B. Gurdon, *Proc. Natl. Acad. Sci. U.S.A.* **74**, 2470 (1977); J. R. Whittaker, *J. Embryol. Exp. Morphol.* **55**, 343 (1980); J. C. Gerhart, in *Biological Regulation and Development*, vol. 2, *Molecular Organization and Cell Function*, R. F. Goldberger, Ed. (Plenum, New York, 1980), p. 133; M. A. DiBerardino and N. J. Hoffner, *Science* **219**, 862 (1983); J. W. Shay, *Mol. Cell. Biochem.* **57**, 17 (1983); W. B. Wood, E. Schierenberg, S. Strome, *Molecular Biology of Development*, E. H. Davidson and R. Firtel, Eds. (Liss, New York, 1984), p. 37; J. B. Gurdon, T. J. Mohun, S. Fairman, S. Brennan, *Proc. Natl. Acad. Sci. U.S.A.* **82**, 139 (1985); J. B. Gurdon, S. Fairman, T. J. Mohun, S. Brennan, *Cell* **41**, 913 (1985); R. L. Gimlich and J. C. Gerhart, *Dev. Biol.* **104**, 117 (1984).
44. D. Melloul *et al.*, *EMBO J.* **3**, 93 (1984); U. Nudel *et al.*, *Proc. Natl. Acad. Sci. U.S.A.* **82**, 3106 (1985); A. Seiler-Tuyns, J. D. Eldridge, B. M. Paterson, *ibid.* **81**, 2980 (1985); S. F. Konieczny and C. P. Emerson, Jr., *Mol. Cell Biol.* **5**, 2423 (1985); A. J. Minty, H. M. Blau, L. Kedes, in preparation.
45. P. Kavathas and L. A. Herzenberg, *Proc. Natl. Acad. Sci. U.S.A.* **80**, 524 (1983); K. V. Anderson and C. Nusslein-Volhard, *Nature (London)* **311**, 223 (1984); A. Fainsod, M. Marcus, P.-F. Lin, F. H. Ruddle, *Proc. Natl. Acad. Sci. U.S.A.* **81**, 2393 (1984); T. Yokota *et al.*, *ibid.*, p. 1070; P. Kavathas, V. P. Sukhatme, L. A. Herzenberg, J. R. Parnes, *ibid.*, p. 7688; V. Episkopou, A. J. M. Murphy, A. Efstratiadis, *ibid.*, p. 4657; M. M. Davis *et al.*, *ibid.*, p. 2194; L. C. Kuhn, A. McClelland, F. H. Ruddle, *Cell* **37**, 95 (1984); J. Deschatrette, C. Fougere-Deschatrette, L. Corcos, R. T.

Schimke, *Proc. Natl. Acad. Sci. U.S.A.* **82**, 765 (1985).

46. We thank C. Goodman, R. Schimke, D. Spiegel, and K. Yamamoto for critical reading of the manuscript, B. Efron and H. Kraemer for aid with statistical analyses, A. Pavlath for expert assistance with the computer-derived kinetic analyses, J. Schuster and R. Spudich for technical assistance, and N. K. Williams and R. Joseph for helping to prepare the manuscript. We gratefully acknowledge the gifts of antibodies, cells, and hormones from P. Byers, R. Burgeson, H. Green, M. Karasek, B. Knowles, A. Lassar, R. Roth, T.-T. Sun, F. Stockdale, F. Walsh, H. Weintraub, and D. Yaffe. We extend special thanks to L. Kedes, A. Minty, and P. Gunning for the different isotype and species-specific actin DNA probes and expert assistance in their use. Supported by grants to H.M.B. from the National Institutes of Health (GM26717 and HD18179), the Muscular Dystrophy Association of America, the March of Dimes Foundation, and the National Science Foundation (DCB-8417089), NIH predoctoral training grants GM07149 to G.K.P. and S.G.W. and CA 09302 to C.W., NIH postdoctoral fellowships to E.C.H. and S.C.M., and a Muscular Dystrophy Association postdoctoral fellowship to L.S. H.M.B. is the recipient of an NIH research career development award (HD00580).

11

The Action of Oncogenes in the Cytoplasm and Nucleus

Robert A. Weinberg

Oncogene research has changed substantially over the past several years. Initial emphasis concerned the identification of oncogenes present in tumor virus genomes and in the genomes of a number of different types of tumor cells. Together, various experimental routes have led to the characterization of at least 30 different oncogenes originating from the cellular genome and 10 or more found in the genomes of DNA tumor viruses (*1*). Having catalogued these genes and their structures, workers are now moving into a new phase in which mechanistic problems are confronted: how do oncogenes and their encoded proteins convert normal cellular metabolism to that of a tumor cell? What regulatory pathways are perturbed by oncogenes, and how can their various modes of action be interrelated?

This review attempts to synthesize much of the currently available data on these issues. It is written with the belief that much of the information about oncogenes will eventually be understandable in terms of a small number of mechanisms and that the outlines of some of these are gradually becoming apparent. Much of the present discussion concerns the modes of activation of cellular oncogenes and the effects that these genes have on cellular physiology. The implications of all this for tumorigenesis involves less discussion, because we still understand relatively little about the connections between oncogene action and the outgrowth of tumors in vivo.

Nuclear and Cytoplasmic Oncogenes

The existence of the 40 and more oncogenes provokes an obvious question: do they represent as many as 40

Science 230, 770–776 (15 November 1985)

distinct mechanisms of transformation, or can they be grouped into a small number of functional classes on the basis of shared functional properties? One such classification that has arisen in recent years groups oncogenes on the basis of the nuclear or cytoplasmic localization of their gene products (Table 1). Perhaps surprisingly, this crude classification scheme correlates with similarities of function within each group. The terms used to label these as "nuclear" or "cytoplasmic" oncogenes are a bit misleading, in that they imply the location of the genes rather than the site of action of their gene products. I will use the terms here nevertheless.

Among the nuclear oncogenes of special interest are several cellular genes or variants thereof (*myc*, N-*myc*, *myb*, and Ela) that exhibit some structural homology with one another (*2*, *3*). These and several other nuclear oncogenes (*p53*, polyoma large T, and SV40 large T) exhibit similarities of function, although no one of them behaves precisely the same as any other in all respects (*4–6*). Among the most readily measured of these traits is that of immortalizing ability—the power to convert a tissue culture cell of limited replicative potential in vitro into one that can be passaged without limit in culture.

Associated with this immortalizing ability are often other functions that may affect the altered growth properties of tumor cells in vivo. For example, a recent characterization of the *myc* oncogene (*6*) revealed that this oncogene also allows embryo fibroblasts to grow at lower serum concentrations and at lower density in monolayer culture, echoing similar results obtained with some other oncogenes of this group. These nuclear oncogenes tend to be weak in their ability to induce anchorage independence of fibroblasts, in contrast to cytoplasmic oncogenes described below. Many of the encoded proteins are bound to nuclear structures (*7*).

Table 1. Classification of oncogenes mentioned here on the basis of the nuclear or cytoplasmic localization of their gene products. (The *sis* oncogene, of cellular origin, specifies an extracellular protein.)

Nuclear	Cytoplasmic
Viral oncogenes	
SV40 large T	Polyoma middle T
Polyoma large T	
Adenovirus Ela	
Cellular oncogenes	
myc, *myb*, N-*myc*	*ras*, *src*, *erb*B, *neu*,
p53, *ski*, *fos*	*ros*, *fms*, *fes*/*fps*, *yes*,
	mil/*raf*, *mos*, *abl*

A much larger group of oncogenes specifies proteins found in the cytoplasm. However, only a few of these genes have been subjected to a detailed scrutiny in which effects on cellular phenotype are studied. Among these are the three *ras* oncogenes, the *src* oncogene, and the middle T oncogene of polyoma. The cytoplasmic oncogenes are generally weak in their ability to immortalize cells (*4–6*, *8*) and strong in their ability to promote anchorage independence of fibroblasts. At least eight cytoplasmic oncogenes have been reported to induce secretion of growth factors in one or another cell type (*9*, *10*), but this trait has never been associated with the effects of a nuclear oncogene. It remains unclear whether yet other members of the large group of cytoplasmic oncogenes will follow these functional patterns or exhibit their own particular pattern of function.

Nuclear oncogenes have been found to

collaborate effectively with cytoplasmic oncogenes in malignant transformation of previously normal cells. Indeed, a nuclear or cytoplasmic oncogene, acting on its own, seems unable to induce full transformation under many conditions of culture (*4–6*). Cells that are effectively transformed by pairs of collaborating nuclear and cytoplasmic oncogenes range from rat embryo fibroblasts and chondroblasts to avian cells of the myelomonocytic and retinoblast lineage (*4, 5, 10*). The nuclear polyoma large T oncogene collaborates with the cytoplasmic middle T oncogene (*4*); the viral Ela, polyoma, and SV40 large T oncogenes and the cellular *myc*, N-*myc*, and *p53* oncogenes all collaborate with the cytoplasmic *ras* oncogene (*5, 11, 12*); and the nuclear *myc* and *myb* act synergistically with the cytoplasmic *src*, *erb*B, *fes*/*fps*, *yes*, *ros*, and *mil*/*raf* oncogenes (*10*). MH2 virus, a spontaneously arising, particularly potent retrovirus, has been found to have acquired both a nuclear (*myc*) and a cytoplasmic (*mil*/*raf*) oncogene from the cell genome (*13*). Use of recombinant DNA procedures has allowed construction of a hybrid avian/mammalian retrovirus also carrying these two oncogenes; the behavior of this virus also demonstrates a strong synergy between the two genes (*14*).

One apparent anomaly in this scheme is the SV40 large T oncogene, which, through the actions of a single protein, is able to induce "nuclear" functions such as immortalization and "cytoplasmic" functions such as anchorage independence (*15*). Of great interest is evidence indicating that the encoded large T antigen can be found both in the nucleus and at the plasma membrane. Mutations that inhibit the transport of T antigen into the nucleus appear to reduce its immortalizing ability while leaving intact its effects on anchorage independence and its ability to transform already immortalized cells (*16*). Consequently, this oncogene gains membership in both classes by sending its gene product to do work at two distinct cellular sites.

These various lines of evidence suggest that the activities of nuclear and cytoplasmic oncogenes are complementary rather than additive. One can speculate that certain cancer cell traits are more effectively induced by proteins acting in the nucleus while others are achieved more readily by gene products acting at cytoplasmic sites. There are, however, data that indicate exceptions to this. For example, certain pairs of nuclear oncogenes can collaborate with one another [for example, *fos* and polyoma large T (*17*)], and certain nuclear oncogenes can transform established, spontaneously immortalized cell types (*18*).

At greatest variance with the proposed requirement of multiple oncogenes for full transformation is the extensively documented ability of many retroviruses carrying single oncogenes to induce tumors in vivo (*19, 20*). Some recent work shows that a *ras* oncogene, acting alone, is able to convert an embryo fibroblast to the tumorigenic state (*20*). Such results show that, under the proper conditions, a single oncogene can induce all the phenotypes of transformation, not just some of them. Direct conflict exists between the two models of tumorigenesis: are two or more oncogenes required for making a cell tumorigenic, or will one suffice?

One resolution may come from consideration of another, overlooked factor in tumorigenesis—the environment of the cell that initially acquires an oncogene by mutation, infection, or transfection. Oncogene-bearing cells surrounded by normal neighbors do not grow into a large mass if they carry only a single oncogene (*4, 5, 11*). But if the normal neighbors are removed, either by killing them with a cytotoxic drug (*18, 20*) or by recruiting them into the tumor mass by viral spread in vivo (*21*), then a single oncogene often suffices. The environment of a cell may therefore strongly influence its responsiveness to an oncogene that it carries.

We know little about the identity of those cells within a tissue that are the targets of viral and of nonviral carcinogens. The nature of the cell-to-cell interactions that affect the clonal expansion of these target cells into tumors is even more obscure. Nonviral carcinogenesis appears to involve a number of distinct stages of tumor progression (*22*), and some of these stages may reflect an underlying requirement for the activation of multiple oncogenes. To date, only relatively few tumors of nonviral etiology have been found to carry multiple, independently activated oncogenes (*23*). Thus, the involvement of multiple, collaborating oncogenes in creating such tumors still requires extensive substantiation.

Mechanisms of Nuclear Oncogene Activation

Some of the nuclear oncogenes discussed here are found in the genomes of DNA tumor viruses, while others arise from alteration of normal cellular protooncogenes. Concentrating for a moment on those of cellular origin, one finds that they are all created by processes that lead to deregulation in the level of their encoded proteins.

The most well documented of these activating mechanisms concerns the deregulation of *myc* expression that results from chromosomal translocation [reviewed in (*24*)]. This genetic reshuffling often deprives the *myc* gene of its normal transcriptional promoter-enhancer regulators and replaces these with sequences from the immunoglobulin genes. Other consequences may also follow. Because the translocation often removes the initial (noncoding) exon of the gene, the resulting messenger RNA (mRNA) becomes restructured. Some have speculated that this may improve the utilization of this mRNA template (*25*), and such speculation is supported by recent work on the translation of several *myc* transcripts in vitro (*26*).

The *myc* gene may also become deregulated by the actions of a retrovirus that integrates its genome nearby in the chromosome, providing promoter-enhancer segments that override normal regulation (*27*). Other studies have demonstrated amplification of *myc*, N-*myc*, and *myb* genes in various tumor types (*3, 28*), leading in turn to increases in levels of encoded protein.

Most interesting are the alternative routes by which deregulation of the *p53* gene can be achieved. The *p53* protein is normally very labile metabolically, having a lifetime of less than 30 minutes. SV40 virus is able to increase greatly the steady-state levels of *p53* by causing its large T antigen to complex with *p53*, resulting in an increase in metabolic stability by about a factor of 50 (*29*).

The *p53* protooncogene itself can be activated experimentally by fusion with a strong promoter, resulting in greatly increased transcript and protein levels (*12*). This and analogous work on the *myc* and N-*myc* genes (*30*) provide the clearest demonstration of the importance of protein level and the lesser importance of altered protein structure. In all cases, clones of the normal cellular versions of these genes can be activated to full oncogenic potential by simple attachment to a strong, constitutive transcriptional promoter (*11, 12, 30*).

Taken together, these various lines of evidence persuade one that deregulation of the level of these nuclear oncogene proteins is sufficient to allow them to exert their effects on the cell. This deregulation does not always involve strong overexpression of the gene; instead, it may simply make the gene constitutively active and thus unresponsive to its normal regulators.

Activation of Cytoplasmic Oncogenes

Many of the well-studied cellular oncogenes of the cytoplasmic class appear to be effectively activated by mutations that affect the structure of their encoded proteins. Examples of this are now available for the *ras, src, erb*B, *abl, fes/fps*, and *neu* oncogenes. The exception to this is the *mos* oncogene, which acquires very strong oncogenic powers when linked to a constitutive promoter (*31*).

In the case of *ras*, a large number of spontaneously arising point mutations have been described in naturally occurring tumors. The point mutations affect amino-acid residues 12, 13, or 61 of the *ras*-encoded p21 protein, and they impart to this protein the ability to transform cells even when it is present in very low levels in the cytoplasm (*1, 32*).

An analogous situation pertains for the *src* gene. Structural alterations in the Rous sarcoma virus–encoded pp60 protein result in potent transforming activity. In contrast, strong overexpression of the normal gene product leaves one, at best, with a cell showing only partial transformation (*8, 33*). The *neu* gene presents a more extreme example, in which high levels of the protein encoded by the normal cell have no effect on cellular phenotype, while low levels of a structurally altered protein induce strong transformation (*34*). Yet other, analogous observations have been made on the *fes/fps* oncogene as well (*35*).

The *erb*B gene, to which *neu* is distantly related, has been the object of intense scrutiny. Here, once again, it appears that structural alteration, specifically truncation of the protein termini, is the favored route when attempting to activate the cloned normal gene (*36*). Indeed, when *erb*B becomes activated by retrovirus genome integration, the structure of the encoded protein is affected, in contrast to the events leading to viral activation of the nuclear protooncogene *myc* (*27, 36*), which have no effect on the structure of the *myc* protein.

Very recent studies of the *abl* gene, which belongs to the cytoplasmic group, provide an equally dramatic contrast with other, earlier work on *myc*. Both the *abl* and *myc* genes become altered in human malignancies as a consequence of chromosomal translocation. As mentioned above, translocations affecting *myc*, such as those occurring during the pathogenesis of Burkitt's lymphoma, deregulate the gene but leave its protein-encoding region intact (*24*). The results with the *abl* gene, as observed in cells of chronic myelogenous leukemia, are quite different. Here, the translocations cause the *abl* protein to be altered: its amino terminus is lost and is replaced by a protein sequence encoded by the foreign partner gene that participates in the translocation event (*37*). Overexpression of the unaltered, normal *abl*-encoded protein appears to have little or no effect on cellular phenotype (*38*).

Taken together, these various lines of evidence can be used to establish a useful generalization: deregulation in the nucleus, change of protein structure in the cytoplasm. Unfortunately, the picture is clouded a bit by evidence that several of the cytoplasmic genes can contribute to transformation by simple overexpression. A *ras* gene can be activated to an oncogene by linkage to a long terminal repeat (LTR) (*39*). Moreover, both *ras* and *erb*B oncogenes have been reported to be present in greatly amplified copy numbers in certain tumors (*36, 40, 41*). While overexpression represents a relatively ineffective way of activating these genes (*36, 42*), these results tell us that the distinctions between nuclear and cytoplasmic genes may not be as clearly drawn as we might like.

Significance of Activation Mechanisms

Protooncogenes play key roles in the growth control of normal cells, and the mechanisms that create cellular oncogenes surely provide us with important lessons on how they do so. Many of the normal genes and their gene products must function to pass on growth-stimulatory signals from upstream in a regulatory pathway to one or more targets downstream. The evidence provided by the activation mechanisms, as described above, provokes me to suggest that two quite different schemes for signal transduction are exhibited by the normal cytoplasmic and nuclear genes.

The cytoplasmic protooncogene products are expressed in relatively constant amounts in time. Their expression may vary somewhat depending on growth and differentiation state (*43*), but it would seem that physiological increases in levels of gene product do not per se produce a stream of growth-stimulatory signals. Instead, one can propose that the cytoplasmic protooncogene protein molecules are usually in a resting state, awaiting a direct stimulus from an upstream agonist. They then rise to an excited state, send out excitatory signals for a short period of time, and then lapse back to a state of relative inactivity.

Mechanisms must exist that limit the length of the excited state of these protein molecules to seconds or minutes. Such mechanisms would include the guanosine triphosphatase activity of *ras* proteins (see below), the internalization and kinase C–phosphorylation of receptor proteins such as that encoded by *erb*B (*44*), and other, still obscure, negative feedback mechanisms that strongly limit the length of the excited state of the proteins encoded by genes such as *src* and *abl*. By this logic, the lesions that create cytoplasmic oncogenes cause the constitutive excitation of their encoded proteins. This may happen by providing a constant, gratuitous, excitatory stimulus to the protein molecule or, perhaps more frequently, by trapping the molecule in its excited state.

The nuclear protooncogenes respond in a quite different way to growth stimulatory signals: they increase the steady-state concentration of their encoded proteins, often by great amounts (*45*). They achieve this by changes in transcription rate, by changes in posttranscriptional processing of RNA, and perhaps by the posttranslational stabilization of protein. The responses of these nuclear genes are by necessity much slower, and once achieved, are longer lasting, being measured in many minutes and even hours (*45, 46*).

As discussed above, genetic lesions that create nuclear oncogenes lead to an apparent constitutive expression of these genes by uncoupling them from their normal regulators. The protein molecules encoded by these nuclear genes may alter cell metabolism simply by their presence in the nucleus in enhanced concentration. At present, it is difficult to formulate a rationale for these two quite different response mechanisms in the nucleus and cytoplasm.

Reversible Activation of Normal Cytoplasmic Proteins: *ras* and *src*

The most intensively studied cytoplasmic genes are those of the *ras* family and *src*. Our understanding of these proteins in their normal and oncogenic configuration has changed substantially over the past 2 years. The pioneering work of Scolnick and his colleagues on the oncogenic *ras* proteins encoded by Harvey and Kirsten sarcoma viruses revealed, among other things, two facts of great importance: the proteins (termed p21) are membrane bound, apparently in large part to the plasma membrane, and they act to bind guanosine diphosphate (GDP) or triphosphate (GTP) (*47*). These facts strongly colored subsequent thinking in that they suggested analogy with the G proteins, which are known to transduce signals from various cell-surface receptors to adenyl cyclase (*48*). Moreover, they suggested a biochemical mechanism of action quite distinct from that of the much-studied tyrosine kinases.

The tyrosine kinases use their bound adenosine triphosphate (ATP) as a phosphate donor for modifying the tyrosine residues of target proteins. The GTP bound by the G proteins plays a quite different role: its presence indicates that the protein binding it has been raised to an excited state. The hydrolysis of this bound GTP by an activity intrinsic to the G protein signals a relaxation to a ground state and in this way can serve to limit the period of excitation.

Recent results allow us to rationalize the oncogenic activation of the *ras* genes in terms of such a scheme of alternating excited and relaxed states. Enzymology performed on the p21 proteins has revealed that the p21 protein encoded by the normal allele of the H-*ras* gene exhibits GTP binding ability and has the power to hydrolyze this GTP to GDP. The p21 protein encoded by an oncogenic allele and bearing an amino-acid substitution at residue 12 continues to bind GTP effectively, but its powers of hydrolysis are reduced by a factor of about 10 (*49*). This fact may explain the molecular mechanisms of *ras* oncogene activation (Fig. 1). The inability to hydrolyze GTP effectively would seem to trap the protein in its excited state by blocking the route normally responsible for relaxation.

An understanding of the regulation of the other well-studied cytoplasmic oncogene protein, pp60*src*, remains more elusive. This protein also seems to enjoy only transitory periods of activation, during which its tyrosine kinase activity is manifest (*50*), but in this instance no GDP-GTP cycle can be invoked to explain its control. The *src* oncogene has

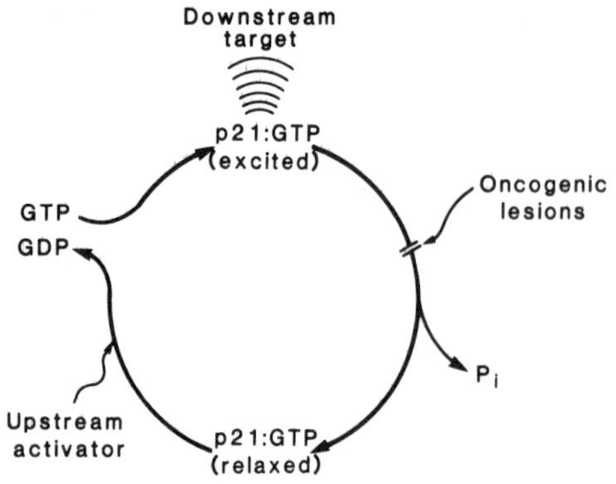

Fig. 1. A scheme of alternating excited and relaxed states of the *ras*-encoded p21 protein.

not been found to date in any spontaneously arising tumors, and investigators are consequently not provided with useful mutant alleles that might aid in understanding normal regulation.

One important clue may come from analysis of the normal cellular *src* protein as it functions in polyoma virus–transformed cells. It is now apparent that the middle T antigen encoded by polyoma virus binds tightly to the cellular *src* proteins in the cytoplasm of infected cells (*51*). The complexed pp50*src* exhibits a tyrosine kinase specific activity that may be as much as 30 to 50 times higher than that of the free protein (*52*). This viral middle T may be mimicking an analogously functioning, endogenous cellular regulatory subunit that is responsible for the reversible activation of the normal pp60*src* protein.

Effector Functions of Cytoplasmic Oncogenes

The results described here provide examples of how three intensively studied protooncogenes can become activated: the *myc* gene loses transcriptional control; the *ras* gene protein loses guanosine triphosphatase activity; and the *src* gene protein acquires a foreign, physiologically unresponsive regulatory subunit. However, none of these insights into regulation provides useful information on the effector functions of these proteins. How do the oncogene proteins elicit responses from the cell? Any answer to this question must consider the pleiotropic action of oncogenes, that is, their ability to evoke multiple changes and, by implication, to affect multiple molecular targets within the cell.

The first lessons on oncogene effector functions came from the discovery of Erikson and colleagues, who showed that pp60*src* carries tyrosine kinase activity (*53*). Mutations in the *src* gene that inactivate the kinase activity invariably eliminate the transforming potential of the protein (*54*), indicating that this activity is not an adventitious element in the transformation process.

Paradoxically, this work provided limited insight into mechanisms of cellular transformation. While a number of cellular proteins were found to be modified by the kinase activity, the modification of none of these proteins could be tied directly to malignant conversion (*55*). Indeed, some target proteins were found to be phosphorylated even in the absence of observable transformation (*56*).

One important clue may be provided by studies that connect the activity of the *src* protein with changes in the pathway that involves the phosphorylation and degradation of the membrane constituent phosphatidyl inositol (PI). Several experiments have shown that the *src* protein and the tyrosine kinase encoded by the related *ros* oncogene are associated with a lipid kinase activity that is able to convert PI to mono- and diphosphorylated forms (*57*). Work with polyoma-infected cells, in which the *src* protein has been activated, also indicates a shift in PI metabolism (*58*). A remaining question is whether these tyrosine kinase proteins function directly as lipid kinases. Alternatively, they may be copurified with or regulate other proteins having lipid kinase activity (*59*).

This work presents an attractive hypothesis that explains tyrosine kinase effector function. By activating lipid kinases, these oncogene proteins may be able to induce formation of inositol polyphosphates, which in turn yield the pharmacologically potent "second messengers" diacylglycerol and inositol trisphosphate (*60*). These in turn activate protein kinase C and mobilize intracellular calcium, thereby effecting many of the pleiotropic changes that are associated with oncogene action.

This theme of oncogene-mediated regulation of second messenger molecules has been echoed by recent studies on the *ras* protein. The finding of *ras* homologs in the yeast genome (*61*) catalyzed a major and highly productive effort to analyze yeast *ras* function (*62*). This work has exploited the elegant genetic manipulations that are possible when working with yeast and are not available to those working with mammalian cells. Results to date have shown that the analogies between the *ras* proteins and the adenyl cyclase G proteins are more than superficial, at least in yeast. The yeast *ras* protein functions as a strong positive regulator of adenyl cyclase; it may regulate other yeast functions as well (*63*). Its inactivation leads to defects in spore germination that can be cured by introduction of the homologous mammalian *ras* gene (*62*).

These are striking findings that demonstrate the power and elegance that results from the convergence of biochemistry and the molecular genetics of yeast. The immediate transferability of the adenyl cyclase results to mammalian systems remains in question, however, because there are often discordances between cellular cyclic adenosine monophosphate levels and malignant growth. Oncogenic transformation of mammalian cells would not seem to be explainable simply by derangement in adenyl cyclase. Nevertheless, the strong homologies between the yeast and mammalian *ras* systems, as evidenced by the interchangeability of some components, together with the powers of the yeast system, may soon reveal the elements that are indeed critical to *ras*-mediated malignant transformation in mammalian cells.

Effector Functions of Nuclear Oncogenes

We may also have a clue concerning the effector functions of the *myc* and analogously acting gene products (*64*). In this instance, one deals with a different locus of action, the nucleus, and with an entirely different mode of action. These insights were stimulated by studies of the Ela oncogene of human adenovirus, which showed an ability of its encoded gene product to be a *trans*-acting regulator of the transcription of other viral genes (*65*). The initial findings have been extended in several directions. First, the Ela oncogene has been found to stimulate expression of cellular genes (*65*). Second, observation of cells transfected with *myc* oncogenes indicates a greatly increased ability of the cell to promote expression of resident cellular genes as well as introduced genes, such as a heat-shock protein gene (*66*).

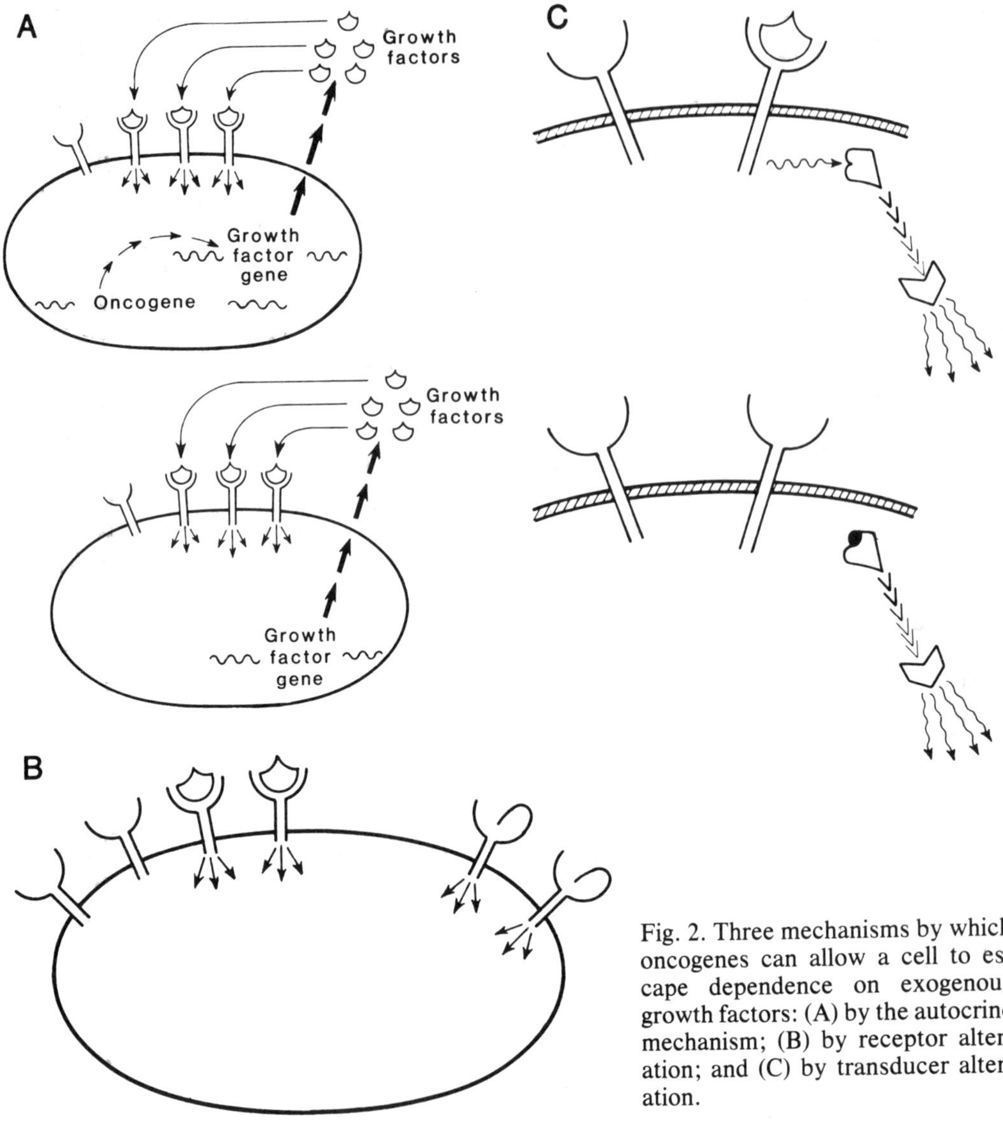

Fig. 2. Three mechanisms by which oncogenes can allow a cell to escape dependence on exogenous growth factors: (A) by the autocrine mechanism; (B) by receptor alteration; and (C) by transducer alteration.

These results appear to provide important clues to the functioning of the nuclear oncogenes of both viral and cellular origin. They suggest that the *myc* protein perturbs the activity or specificity of the cellular transcription apparatus and in this way mobilizes the expression of a bank of cellular genes whose products are critical to growth and differentiation. The normal cell genome carries multiple protooncogenes of this type (*myc*, N-*myc*, *myb*, *fos*, *p53*, and *ski*), and the abilities of most of these to affect transcription remain to be demonstrated. One might speculate that each of the proteins encoded by these genes may address the activation of a slightly different constituency of cellular genes.

The *myc* and E1a genes are not the only examples of oncogenes encoding nuclear proteins that serve on the one hand as *trans*-activators of transcription and on the other as agents of cellular immortalization. The same can be said of the large T oncogenes of polyomavirus and SV40 virus (*64, 67*). This provokes the question of whether the transcriptional activators of human T-cell leukemia virus type I (HTLV-I), HTLV-III, and pseudorabies virus (*68*) can also express an immortalizing function and will stimulate some to pursue the apparent mechanistic connection between cellular immortalization and *trans*-activation of transcription.

Oncogenes and Growth-Factor Autonomy

Grouping oncogenes into nuclear and cytoplasmic classes represents only one way to conceptualize the interrelations between these various genes. Other results from recent work provide a quite different way to view the mechanisms by which protooncogenes and their oncogenic alleles regulate growth.

This view stems from observations of many workers that a fundamental trait of tumor cells is a decreased dependence on growth factors (GF's) for the promotion of their growth (*69*). Following this view, normal metazoan cells, with the possible exception of certain embryonic cell types, may never proliferate unless prompted to do so by one or more types of GF present in their surroundings. Tumor cells, in contrast, acquire partial or complete autonomy that permits proliferation, even in the absence of any encountered GF's.

Accepting this, one then confronts the question of how oncogenes can confer such GF autonomy on the cell (*70*). Four different mechanisms come to mind, three of which are depicted in Fig. 2 (*70*). The first mechanism, termed autocrine, was proposed some years ago by Sporn and Todaro (*71*) and depends on the ability of a tumor cell to manufacture GF's. These can be secreted into the medium and then adsorbed to appropriate GF receptors present on the same cell that has just released them, thereby creating an uncontrolled autostimulation (Fig. 2A).

Oncogene-stimulated GF secretion can occur in two different ways. Cells transformed by certain cytoplasmic oncogenes (*ras*, *src*, middle T, *mos*, *fes*, *abl*, *fps*, *erb*B, *yes*, and *mil/raf*) release growth-stimulatory factors into their culture medium (*10, 72*). It is now clear that these oncogenes do not themselves encode the GF's. Instead, the released GF's are encoded by distinct genes whose expression is indirectly stimulated by these cytoplasmic oncogenes. The mechanism of this stimulation and its purpose in normal cell and tissue physiology remain unexplained.

A more direct autocrine route was indicated by the finding that a gene encoding the structure of a GF can itself become deregulated and converted to the status of an active oncogene. Thus, the oncogene *sis* was found to be an altered, deregulated version of the normal cellular gene specifying platelet-derived growth factor (PDGF) (*73*).

A generalization of this finding states that any cellular gene encoding a GF may be considered a protooncogene. If appropriately activated, it can force the constitutive elaboration of a GF. Should the cell in which this occurs also happen to display the cognate GF receptor, then a closed, positive feedback loop becomes established, providing the cell with a steady stream of growth-stimulatory signals and freeing the cell from its previous dependence on GF imported from elsewhere in the tissue or organism. Curiously, although many GF's have been identified, only the gene for PDGF has been found to date in the form of an active oncogene.

Oncogenes can confer GF autonomy in a second way, this one involving the receptors that are displayed on the cell surface and used by cells to recognize the presence of GF's in the extracellular space (Fig. 2B). In this instance, the receptors themselves become changed in a fashion that allows them to bombard continually the cell with growth stimulatory signals, even in the absence of any encountered GF. The cell is thus deluded by its malfunctioning receptor, being informed of high GF concentration when little or none is in fact present. In this way the GF receptor can assume the role of oncogene protein, and the sequences encoding it can assume the role of oncogene.

Three examples of this have now been reported. The first and precedent-setting example came from the work on the epidermal growth factor (EGF) receptor (*74*). Sequencing of a portion of this receptor demonstrated near identity with the protein specified by the *erb*B oncogene, known from its presence in the avian erythroblastosis virus genome. This showed clearly that a gene encoding a GF receptor should be considered as a protooncogene, capable of participating in cellular transformation when appropriately altered.

A second oncogene, termed *neu*, has been discovered in rat neuroblastomas and glioblastomas. This gene, related distantly to *erb*B, encodes a protein that has all the structural characteristics of a GF receptor; a ligand has not yet been identified (*75*). Most recently, the oncogene *fms*, originally found in a feline sarcoma virus, has been shown by elegant detective work to be an altered version of the normal cellular gene specifying the mononuclear phagocyte (CSF-1) GF receptor (*76*). Other tyrosine kinase oncogene proteins may eventually be associated with known GF receptors (*77*).

The third means by which oncogenes can confer GF autonomy involves proteins that lie within the cell and transduce signals from the GF receptors to targets farther downstream in the signaling pathway (Fig. 2C). Here once again the analogy with the G protein of the adenyl cyclase system becomes useful. Altered forms of these transducing proteins may acquire an autonomy that enables them to send out signals, even without prior prompting by a GF receptor. As before, the presence of the GF is not required for growth.

The *ras* proteins stand as good candidates for a role in the transduction of signals from cell surface receptors to intracellular targets. There are at least four *ras* p21 proteins (one H-*ras*, two K-*ras*, and one N-*ras*), and their distinctive carboxyl-terminal tails may permit each to interact with its own set of GF receptors. The remaining portion of these proteins is almost identical among the four, and this may indicate a common downstream effector function. To date, only one piece of evidence supports such a role for *ras* proteins—a report that the EGF receptor stimulates the nucleotide binding of the H-*ras*–encoded p21 (*78*).

The fourth and final way to achieve GF autonomy is associated with the nuclear oncogenes. We do not understand all the mechanisms that govern their expression and the levels of their encoded proteins. Certainly, very important mediators of nuclear protooncogene regulation are the GF's, which can strongly stimulate *myc*, *fos*, and p53 protooncogene expression (*45*). The transcriptional deregulation of these genes, described earlier, frees them from GF dependence. This may in turn relieve much of the normal GF requirement that must be satisfied in order for the cell to undertake a growth program.

Effects of Protooncogenes on Functions Other Than Growth

Because cellular oncogenes mediate abnormal cellular growth regulation, it follows that the corresponding protooncogenes must regulate normal growth patterns. Indeed, this notion has permeated much of the discussion in this article. Biological reality may, however, prove to be much more interesting: protooncogenes may be involved in a variety of cellular functions that are quite unrelated to growth and its regulation.

This idea stems from observations over the past several years that the cellular *src* gene is expressed in much higher levels in cells of the nervous system than in other tissues. Such observations have been made in chick, rat, and *Drosophila* cells (*79*). In many cases, *src* expression is high in fully differentiated cells, such as neurons that have no prospect of undertaking a growth program.

Because of these observations, it seems that the *src* protooncogene may be involved in some aspect of neuronal function that is unrelated in any way to growth control. It is even possible that the *src* gene, in its normal form, is never involved in any aspect of cellular growth regulation; its association with growth deregulation may be a consequence of the rare accident that caused its activation by a transducing retrovirus.

Perhaps the control circuits that were initially developed to regulate protozoan growth have been exploited repeatedly during the evolution of metazoa to regulate a variety of differentiated functions, such as neuronal signaling and exocytosis. Perhaps oncogenes and protooncogenes will provide useful insights into many more problems than just that of cancer.

References and Notes

1. G. Cooper, *Science* **217**, 801 (1982); J. M. Bishop, *Annu. Rev. Biochem.* **52**, 301 (1983); H. Land, L. F. Parada, R. A. Weinberg, *Science* **222**, 771 (1984); H. E. Varmus, *Annu. Rev. Genet.* **18**, 553 (1984); J. M. Bishop, *Cell*, in press.
2. R. Ralston and J. M. Bishop, *Nature (London)* **306**, 803 (1984).
3. M. Schwab *et al., ibid.* **305**, 245 (1983); N. E. Kohl *et al., Cell* **35**, 359 (1983).
4. A. Houweling, P. J. van der Elsen, A. van der Eb, *Virology* **105**, 537 (1980); M. Rassoulzadegan *et al., Nature (London)* **300**, 713 (1982); S. Palmieri, P. Kahn, T. Graf, *EMBO J.* **2**, 2385 (1983); M. Rassoulzadegan *et al., Proc. Natl. Acad. Sci. U.S.A.* **80**, 4354 (1983); A. J. van der Eb and R. Bernards, *Curr. Top. Microbiol. Immunol.* **110**, 23 (1984); S. Alema, F. Tato, D. Boettiger, *Mol. Cell. Biol.* **5**, 538 (1985); E. Gionti, G. Pontarelli, R. Cancedda, *Proc. Natl. Acad. Sci. U.S.A.* **82**, 2756 (1985).
5. H. Land, L. F. Parada, R. A. Weinberg, *Nature (London)* **304**, 596 (1983); H. E. Ruley, *ibid.*, p. 602.
6. E. Mougneau *et al., Proc. Natl. Acad. Sci. U.S.A.* **81**, 5758 (1984).
7. N. Ito, R. Spurr, R. Dulbecco, *ibid.*, **74**, 1259 (1977); P. Donner, I. Greiser-Wilke, K. Moelling, *Nature (London)* **298**, 262 (1982); H. D. Abrams, L. R. Rohrschneider, R. N. Eisenman, *Cell* **29**, 427 (1982); L. Feldman and J. Nevins, *Mol. Cell Biol.* **3**, 829 (1983); K. H. Klempnauer *et. al., Cell* **37**, 537 (1984); W. J. Boyle *et al., Proc. Natl. Acad. Sci. U.S.A.* **81**, 4265 (1984); T. Curran *et al., Cell* **36**, 259 (1984); R. Eisenman *et al., Mol. Cell Biol.* **5**, 114 (1985).
8. D. G. Thomassen *et al., Cancer Res.* **45**, 726 (1985); J. H. Pierce and S. A. Aaronson, *Mol. Cell Biol.* **5**, 667 (1985); T. Gilmer *et al., ibid.*, p. 1707; P. J. Johnson *et al., ibid.*, p. 1073; H. Land *et al.*, in preparation.

9. G. J. Todaro *et al.*, *Cold Spring Harbor Conf. Cell Proliferation* **6**, 113 (1979); P. L. Caplan, M. Anderson, B. Ozanne, *Proc. Natl. Acad. Sci. U.S.A.* **79**, 495 (1982); A. Roberts *et al.*, *Fed. Proc. Fed. Am. Soc. Exp. Biol.* **42**, 2621 (1983); D. R. Twardzik *et al.*, *Science* **216**, 894 (1982).
10. B. Adkins, A. Leutz, T. Graf, *Cell* **39**, 439 (1984); C. Bechade *et al.*, *Nature (London)* **316**, 559 (1985).
11. M. Schwab, H. E. Varmus, J. M. Bishop, *Nature (London)* **316**, 160 (1985); G. D. Yancopoulos *et al.*, *Proc. Natl. Acad. Sci. U.S.A.* **82**, 5455 (1985).
12. D. Eliyahu *et al.*, *Nature (London)* **312**, 646 (1984); L. F. Parada *et al.*, *ibid.*, p. 649; J. R. Jenkins, K. Rudge, G. A. Currie, *ibid.*, p. 651.
13. H. W. Jansen *et al.*, *EMBO J.* **2**, 1969 (1983); G. H. Mark and U. R. Rapp, *Science* **224**, 285 (1984).
14. U. R. Rapp *et al.*, *J. Virol.* **55**, 23 (1985).
15. W. W. Colby and T. Shenk, *Proc. Natl. Acad. Sci. U.S.A.* **79**, 5189 (1982); C. A. Petit, M. Gardes, J. Feunteun, *Virology* **127**, 74 (1983); S. Sugano and N.Yamaguchi, *J. Virol.* **52**, 884 (1984); M. Kriegler *et al.*, *Cell* **38**, 483 (1984).
16. H. R. Soule and J. S. Butel, *J. Virol.* **30**, 523 (1979); W. Deppert, *Virology* **104**, 497 (1980); R. E. Lanford and J. S. Butel, *ibid.* **119**, 169 (1982); D. Kalderon *et al.*, *Nature (London)* **311**, 33 (1984); R. E. Lanford, C. Wong, J. S. Butel, *Mol. Cell Biol.* **5**, 1043 (1985); L. Fischer-Fantuzzi and C. Vesco, *Proc. Natl. Acad. Sci. U.S.A.* **82**, 1891 (1985).
17. T. Jenuwein *et al.*, *Cell* **41**, 629 (1985).
18. E. J. Keath, P. G. Caimi, M. D. Cole, *ibid.* **39**, 339 (1984); D. Eliyahu, D. Michalovitz, M. Oren, *Nature (London)* **316**, 158 (1985).
19. H. Hanafusa, *Compr. Virol.* **10**, 401 (1977); *Molecular Biology of Tumor Viruses*, R. Weiss *et al.*, Eds. (Cold Spring Harbor Press, Cold Spring Harbor, N.Y., 1982), part 3.
20. D. A. Spandidos and N. M. Wilkie, *Nature (London)* **310**, 469 (1984).
21. J. Ponten, *J. Natl. Can. Inst.* **17**, 131 (1964).
22. L. Foulds, *Neoplastic Development* (Academic Press, London, 1969), vol. 1; J. Cairns, *Cancer: Science and Society*, (Freeman, San Francisco, 1978).
23. M. Murray *et al.*, *Cell* **33**, 749 (1983); Y. Taya *et al.*, *EMBO J.* **3**, 2943 (1984).
24. R. Perry, *Cell* **33**, 647 (1983); P. Leder *et al.*, *Science* **222**, 765 (1984).
25. H. Saito *et al.*, *Proc. Natl. Acad. Sci. U.S.A.* **80**, 7476 (1983).
26. A. Darveau, J. Pelletier, N. Sonenberg, *ibid.* **82**, 2315 (1985).
27. W. S. Hayward, B. G. Neel, S. M. Astrin *Nature (London)* **290**, 475 (1981); B. Neel and W. Hayward, *Cell* **23**, 323 (1981); H. E. Varmus, in *Mobile Genetic Elements*, J. A. Shapiro, Ed. (Cold Spring Harbor Laboratory, Cold Spring Harbor, N.Y., 1983), p. 411; D. Steffen, *Proc. Natl. Acad. Sci. U.S.A* **81**, 2097 (1984); Y. Li *et al.*, *ibid.*, p. 6808.
28. S. Collins and M. Groudine, *Nature (London)* **298**, 679 (1982); R. Dalla Favera, F. Wong-Staal, R. Gallo, *ibid.* **299**, 61 (1982); K. Alitalo *et al.*, *Proc. Natl. Acad. Sci. U.S.A* **80**, 1707 (1983); P. Pelicci *et al.*, *Science* **224**, 1117 (1984); M. Shibuya *et al.*, *Mol. Cell Biol.* **5**, 414 (1985).
29. N. C. Reich, M. Oren, A. J. Levine, *Mol. Cell Biol.* **3**, 2143 (1983).
30. W. Lee *et al.*, in preparation; L. Parada *et al.*, in preparation.
31. D. G. Blair *et al.*, *Science* **212**, 941 (1981).
32. R. Newbold, *Nature (London)* **310**, 628 (1984); P. H. Seeburg *et al.*, *ibid.* **312**, 71 (1984).
33. D. Shalloway, P. M. Coussens, P. Yaciuk, *Proc. Natl. Acad. Sci. U.S.A* **81**, 7071 (1984); H. Iba *et al.*, *ibid.*, p. 4424; R. C. Parker, H. E. Varmus, J. M. Bishop, *Cell* **37**, 131 (1984).
34. M. C. Hung and R. A. Weinberg, in preparation.
35. H. Hanafusa, personal communication; D. Foster *et al.*, *Cell* **42**, 105 (1985).
36. A. Ullrich *et al.*, *Nature (London)* **309**, 418 (1984); A. Ullrich, personal communication; T. W. Nilsen *et al.*, *Cell* **41**, 719 (1985).
37. C. R. Bartram *et al.*, *Nature (London)* **306**, 277 (1983); J. B. Konopka, S.M. Watanabe, O. N. Witte, *Cell* **37**, 1035 (1984); E. Shtivelman *et al.*, *Nature (London)* **315**, 550 (1985); N. Heisterkamp *et al.*, *ibid.*, p. 758.
38. Y. Ben Neriah and D. Baltimore, personal communication.
39. E. J. Chang *et al.*, *Nature (London)* **297**, 479 (1982).
40. M. Schwab *et al.*, *ibid.* **303**, 497 (1983); O. Fasano *et al.*, *Mol. Cell Biol.* **4**, 1695 (1984).
41. G. T. Merlino *et al.*, *Science* **224**, 417 (1984); C. R. Lin *et al.*, *ibid.*, p. 843.
42. C. J. Tabin and R. A. Weinberg, *J. Virol.* **53**, 260 (1984).
43. R. Mueller *et al.*, *Mol. Cell Biol.* **3**, 1062 (1983); M. Goyette *et al.*, *Science* **219**, 510 (1983); M. Goyette *et al.*, *Mol. Cell Biol.* **4**, 1493 (1984); J. Campisi, *Cell* **361**, 241 (1984); P. Yaswen *et al.*, *Mol. Cell Biol.* **5**, 780 (1985); B. Mozer *et al.*, *ibid.*, p. 885.
44. L. Beguinot *et al.*, *Proc. Natl. Acad. Sci. U.S.A.* **81**, 2384 (1984); C. Cochet *et al.*, *J. Biol. Chem.* **259**, 2553 (1984); J. C. Fearn and A. C. King, *Cell* **40**, 991 (1985).
45. K. Kelly *et al.*, *Cell* **35**, 603 (1983); B. H. Cochran *et al.*, *Science* **226**, 1080 (1984); M. E. Greenberg and E. B. Ziff, *Nature (London)* **311**, 433 (1984); W. Kruijer *et al.*, *ibid.* **312**, 711 (1984); R. Mueller *et al.*, *ibid.*, p. 716.
46. N. C. Reich and A. J. Levine, *Nature (London)* **308**, 199 (1984).
47. R. W. Ellis, D. R. Lowy, E.M. Scolnik, in *Advances in Viral Oncology*, G. Klein, Ed. (Raven, New York, 1982), p. 107.
48. E. M. Ross and A. G. Gilman, *Annu. Rev. Biochem.* **49**, 533 (1980); J. B. Hurley *et al.*, *Science* **226**, 860 (1984); A. G. Gilman, *Cell* **36**, 577 (1984).
49. T. Finkel, C. Der, G. M. Cooper, *Cell* **37**, 151 (1984); J. B. Gibbs *et al.*, *Proc. Natl. Acad. Sci. U.S.A.* **81**, 5704 (1984); V. Manne *et al.*, *ibid.* **82**, 376 (1985); J. P. McGrath *et al.*, *Nature (London)* **310**, 644 (1984); R. W. Sweet *et al.*, *ibid.* **311**, 273 (1984).
50. H. Iba *et al.*, *Mol. Cell Biol.* **5**, 1058 (1985).
51. S. A. Courtneidge and A. E. Smith, *Nature (London)* **303**, 435 (1983); *EMBO J.* **3**, 585 (1984).
52. J. B. Bolen *et al.*, *Cell* **38**, 767 (1984); J. B. Bolen and M. A. Israel, *J. Virol.* **53**, 114 (1985).
53. M. S. Collett and R. L. Erikson, *Proc. Natl. Acad. Sci. U.S.A.* **75**, 2021 (1978).

54. A. W. Stoker *et al.*, *Mol. Cell Biol.* **4**, 1508 (1984); M. A. Snyder *et al.*, *ibid.* **5**, 1772 (1985).
55. T. Hunter, *Sci. Am.* **251**, 70 (1984); ______ and J. A. Cooper, *Annu. Rev. Biochem.*, in press.
56. M. P. Kamps, J. E. Buss, B. M. Sefton, *Proc. Natl. Acad. Sci. U.S.A.* **82**, 4625 (1985).
57. Y. Sugimoto *et al.*, *ibid.* **81**, 2117 (1984); I. G. Macara, G. V. Marinetti, P.C. Balduzzi, *ibid*, p. 2728.
58. M. Whitman *et al.*, *Nature (London)* **315**, 239 (1985).
59. M. L. McDonald *et al.*, *Proc. Natl. Acad. Sci. U.S.A.* **82**, 3993 (1985); W. Koch and G. Walter, in preparation.
60. M. J. Berridge and R. F. Irvine, *Nature (London)* **312**, 315 (1984); Y. Nishizuka, *ibid.* **308**, 693 (1984); M. J. Berridge, *Biochem. J.* **220**, 345 (1984); P. W. Majerus, E. J. Neufeld, D. B. Wilson, *Cell* **37**, 701 (1984).
61. D. Defeo-Jones *et al.*, *Nature (London)* **306**, 707 (1983); S. Powers *et al.*, *Cell* **36**, 607 (1984); R. Dhar *et al.*, *Nucleic Acids Res.* **12**, 3611 (1984); A. Papageorge *et al.*, *Mol. Cell Biol.* **4**, 23 (1985).
62. T. Kataoka *et al.*, *Cell* **37**, 437 (1984); K. Tatchell *et al.*, *Nature (London)* **309**, 523 (1984); T. Toda *et al.*, *Cell* **40**, 27 (1985); T. Kataoka *et al.*, *ibid.*, p. 19; G. L. Temeles *et al.*, *Nature (London)* **313**, 700 (1985); D. Defeo-Jones *et al.*, *Science* **228**, 179 (1985); D. Broek *et al.*, *Cell* **41**, 763 (1985).
63. M. Wigler, personal communication.
64. R. E. Kingston, A. S. Baldwin, P. A. Sharp, *Cell* **41**, 3 (1985).
65. N. Jones and T. Shenk, *Proc. Natl. Acad. Sci. U.S.A.* **76**, 3665 (1979); A. J. Berk *et al.*, *Cell* **17**, 935 (1979); J. R. Nevins, *ibid.* **26**, 213 (1981); *ibid.* **29**, 913 (1982); R. B. Gaynor, D. Hillman, A. J. Berk, *Proc. Natl. Acad. Sci. U.S.A.* **81**, 1193 (1984); C. Svenson and G. Akusjarvi, *EMBO J.* **3**, 789 (1984).
66. R. E. Kingston *et al.*, *Nature (London)* **312**, 280 (1984).
67. J. Keller and J. Alwine, *Cell* **36**, 381 (1984); J. Brady *et al.*, *Proc. Natl. Acad. Sci. U.S.A* **81**, 2040 (1984); J. Brady and G. Khoury, *Mol. Cell Biol.* **5**, 1391 (1985).
68. L. Feldman *et al.*, *Proc. Natl. Acad. Sci. U.S.A.* **79**, 4952 (1982); J. G. Sodroski, C. A. Rosen, W. A. Haseltine, *Science* **225**, 381 (1984); J. G. Sodroski *et al.*, *ibid.* **227**, 171 (1985); S. K. Arya *et al.*, *ibid.* **229**, 69 (1985).
69. H. M. Temin, *J. Cell Physiol.* **75**, 107 (1970); R. Risser and R. Pollack, *Virology* **59**, 477 (1974); C. D. Scher *et al.*, *ibid.* **97**, 371 (1978); S. Powers, P. B. Fisher, R. Pollack, *Mol. Cell Biol.* **4**, 1572 (1984); R. Pollack *et al.*, in *Advances in Viral Oncology*, G. Klein, Ed. (Raven, New York, 1984), p. 3.
70. K. H. Heldin and B. Westermark, *Cell* **37**, 9 (1984).
71. M. B. Sporn and G. J. Todaro, *N. Engl. J. Med.* **303**, 878 (1980).
72. J. E. DeLarco and G. J. Todaro, *Proc. Natl. Acad. Sci. U.S.A.* **75**, 4001 (1978); A. B. Roberts *et al.*, *Fed. Proc. Fed. Am. Soc. Exp. Biol.* **42**, 2621 (1983).
73. M. D. Waterfield *et al.*, *Nature (London)* **304**, 35 (1983); R. F. Doolittle *et al.*, *Science* **221**, 275 (1983); I. Chiu *et al.*, *Cell* **37**, 123 (1984); J. S. Huang, S. S. Huang, T. F. Deuel, *ibid.* **39**, 79 (1984); A. Gazit *et al.*, *ibid.*, p. 89.
74. J. Downward *et al.*, *Nature (London)* **307**, 521 (1984).
75. A. Schechter *et al.*, *ibid.* **312**, 513 (1984).
76. S. J. Anderson *et al.*, *J. Virol.* **51**, 730 (1984); C. W. Rettenmier *et al.*, *Cell* **40**, 971 (1985); C. J. Sherr *et al.*, *ibid.* **41**, 665 (1985).
77. Y. Ebina *et al.*, *Cell* **40**, 747 (1985); A. Ullrich *et al.*, *Nature (London)* **313**, 756 (1985); W. S. Neckameyer and L. H. Wang, *J. Virol.* **53**, 879 (1985).
78. T. Kamata and J. R. Feramisco, *Nature (London)* **310**, 147 (1984).
79. P. Cotton and J. S. Brugge, *Mol. Cell Biol.* **3**, 1157 (1983); L. K. Sorge and P. S. Maness, *J. Cell Biol.* **99**, 150 (1983); L. K. Sorge, B. T. Levy, P. S. Maness, *Cell* **36**, 249 (1984); A. Barnekow and M. Schartl, *Mol. Cell Biol.* **4**, 1179 (1984); J. S. Brugge *et al.*, *Nature (London)* **316**, 554 (1985).
80. The author thanks many colleagues whose helpful discussions made possible the writing of this review, and B. Cahill for technical assistance. The author is an American Cancer Society research professor and is supported by the American Business Cancer Research Foundation and by National Cancer Institute grant CA39826.

12

The Granulocyte-Macrophage Colony-Stimulating Factors

Donald Metcalf

Science 229, 16–22 (5 July 1985)

Most of the red and white cells in the circulating blood are short-lived and need to be replaced constantly throughout life. This process of blood cell formation, termed hematopoiesis, is not only enormous in scale (there are 100 times more cells in the bone marrow of an adult than there are people in the whole world) but is also complex, since cells of nine distinct hematopoietic cell lineages, each with multiple maturation stages, are admixed apparently at random in the tightly packed bone marrow. Furthermore, hematopoiesis must be capable of rapid but controlled fluctuations to meet a wide variety of emergency situations ranging from blood loss to infections. A novel element in the system is that all blood cells originate from a small common population of multipotential stem cells that is formed during one short interval in early embryonic life and thereafter maintains hematopoiesis by an extensive capacity for self-generation. Derangements of this complex process of blood cell formation do occur and result in a range of medically important diseases from anemia to leukemia. However, the hematopoietic system usually functions with remarkable fidelity as a consequence of regulation by an overlapping system of control mechanisms.

Control of Hematopoiesis

Some of the mechanisms controlling hematopoiesis, particularly the behavior of stem cell populations, appear to involve cell contact regulation by microenvironmental cells located in the sites of blood cell formation. However, it is also apparent that much of the control of

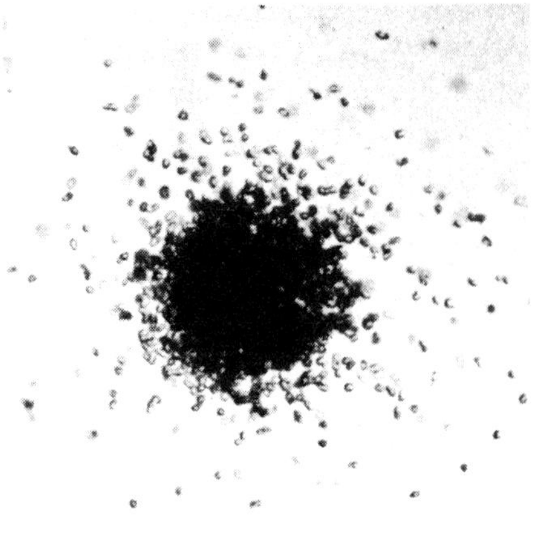

Fig. 1. A granulocyte-macrophage colony generated from a single progenitor cell stimulated to proliferate by purified GM-CSF.

blood cell formation is mediated by a group of interacting specific glycoproteins, multiple subsets of which control each of the major hematopoietic families.

Recognition and characterization of these hematopoietic regulators were made possible by the development, beginning in the mid-1960's, of a series of techniques for the clonal culture of hematopoietic cells in semisolid culture medium. In such cultures, individual progenitor cells of a particular hematopoietic lineage are able to proliferate and generate a clone of maturing progeny cells that remain in physical proximity to each other and are identifiable as a colony (Fig. 1). In the original technique, colonies of two types of white cells—granulocytes or macrophages, or both—were grown from ancestral progenitor cells (*1*, *2*) but modifications of this technique now permit stem and progenitor cells of all hematopoietic lineages from humans or other animals to proliferate clonally in vitro and generate mature progeny in a manner essentially identical to the comparable process in vivo (*3*).

Analysis of the events occurring in these cultures led to the recognition that hematopoietic cells are intrinsically incapable of unstimulated cell division. All cell division is dependent on continuous stimulation by appropriate specific regulatory molecules and, since colony formation was the method used to detect and characterize these molecules, they have been named the colony-stimulating factors (CSF's). The general subject of the CSF's has recently been reviewed (*3*) and the present article will be restricted to the granulocyte-macrophage colony-stimulating factors (GM-CSF's).

Granulocyte-Macrophage Colony-Stimulating Factors

In the mouse, four major GM-CSF's have been characterized as interacting to control the formation and function of granulocytes and macrophages (*4–7*). The granulocyte-macrophage population is an interesting model system since these two quite different cell types, each with an abundance of differentiation markers, originate from common bipotential progenitor cells that, in turn, arise by differentiation commitment from multipotential stem cells. The four GM-CSF's are listed in Table 1, but the list may be incomplete because additional variant CSF's have been noted in the mouse but not yet fully characterized. Information is less complete for the corresponding regulators in humans, but at least two distinct forms, CSF-α and CSF-β, have been identified (*8*).

Each of the four murine CSF's has been purified to homogeneity in small amounts from medium conditioned by various mouse cells with the use of multistep separative procedures coupled with bioassays of all fractions in agar cultures of bone marrow cells. Success in this work depended heavily on the use of high-performance liquid chromatography since, for example, with multi-CSF purified from lymphocyte-conditioned medium it was necessary to achieve a millionfold purification before a homogeneous product was obtained yielding a single amino acid sequence. All four CSF's are glycoproteins: GM-CSF, G-CSF, and multi-CSF are monomers of molecular weight 23,000 to 28,000, approximately 40 percent of which is carbohydrate; M-CSF (molecular weight 70,000) is a dimer of two polypeptide subunits, each with a molecular weight of approximately 14,000, the remainder of the molecule being carbohydrate. Deglycosylation experiments and analysis of CSF's synthesized by cells grown in the presence of tunicamycin have suggested that the carbohydrate portion of these CSF's is not necessary for action in vitro. Digestion by peptidases destroys the biological activity of the CSF's and no digestion fragments of GM-CSF have been observed that can stimulate GM proliferation. Treatment with mercaptoethanol reduces M-CSF to subunits that have no biological activity, and treatment of the three monomeric CSF's also destroys biological activity, suggesting the need in these monomers for disulfide-based tertiary structure for biological activity.

All normal tissues contain and synthesize one or more CSF's, but studies on the CSF's have been restricted by the minute amounts produced by even the richest tissue sources. For example, medium conditioned by lung tissue from mice injected with endotoxin is the richest source of both GM-CSF and G-CSF, but only 5 to 12 micrograms (*4*) and 2 to 4 micrograms (*5*), respectively, of these CSF's can be obtained from 1000 mouse lungs.

Cloning of CSF Genes

Complementary DNA (cDNA) and genomic DNA clones have been isolated for both GM-CSF (*9, 10*) and multi-CSF, the latter under the alternative names for multi-CSF of mast cell growth factor (*11*) and interleukin-3 (*12*). Transfection of these cDNA or genomic clones to monkey COS cells leads to the synthesis of GM-CSF or multi-CSF with similar biological properties to the purified molecules (*10, 13, 14*). The deduced amino

Table 1. The major murine GM-CSF's.

Name	Alternative acronyms*	Major progeny resulting from CSF stimulation	Molecular weight	Source of conditioned medium used in CSF purification	Amino acid sequence data obtained
Granulocyte-macrophage colony-stimulating factor, or GM-CSF	MGI-IGM (*57*)	Granulocytes and macrophages	23,000	Mouse lung (*4*)	Full (*9, 56*)
Granulocyte colony-stimulating factor, or G-CSF†	MGI-IG (*57*)	Granulocytes	25,000	Mouse lung (*5*)	NH_2-terminal
Multipotential colony-stimulating factor, or multi-CSF	IL-3 (*6*), BPA (*58*), HCGF (*59*), MCGF (*60*), PSF (*61*)	Granulocytes, macrophages, erythroid cells, eosinophils, megakaryocytes, mast cells, stem cells	23,000 to 28,000	WEHI-3B leukemia cell (*6*)	Full (*11, 12*)
Macrophage colony-stimulating factor, or M-CSF†	CSF-1 (*20*)	Macrophages	70,000	L cell (*7*)	NH_2-terminal

*MGI, macrophage-granulocyte inducer; IL-3, interleukin 3; BPA, burst-promoting activity; HCGF, hematopoietic cell growth factor; MCGF, mast cell growth factor; PSF, persisting cell-stimulating factor. †For NH_2-terminal sequence data on G-CSF and M-CSF (*62*).

acid sequences for the polypeptides of GM-CSF (124 amino acids, molecular weight 13,138) and multi-CSF (134 amino acids, molecular weight 15,142) share no significant homology and the molecules differ markedly in hydrophobicity profiles and predicted secondary structure. This is surprising since, in the restricted context of granulocyte-macrophage populations, GM-CSF and multi-CSF function in an apparently identical manner and can stimulate the proliferation of the same granulocyte-macrophage clones. From the NH_2-terminal sequence data on G-CSF and M-CSF, it is likely that these two CSF's also differ both from each other and from GM-CSF and multi-CSF. Thus four molecules superficially resembling one another in being highly active glycoproteins with considerable overlap in their functional control of granulocyte-macrophage populations appear to have had quite diverse evolutionary origins. No sequence homology is evident between the CSF's and interleukin-2 (IL-2, or T-cell growth factor, TCGF), the analogous glycoprotein regulator for T-lymphocytes, or between the CSF's and other growth regulators, for example, epidermal growth factor (EGF), platelet-derived growth factor (PDGF), or growth hormone.

The genes for both GM-CSF (*9*) and multi-CSF (*11, 12*) exist in single-copy form in the mouse genome. The GM-CSF's synthesized by the different adult mouse tissues appeared initially to be quite different molecules, although biochemical analysis suggested that the differences were likely to be based merely on tissue differences in glycosylation (*15*). The presence of only a single gene for GM-CSF supports this interpretation. Although GM-CSF and multi-CSF differ structurally, mitogen-stimulated T-lymphocyte clones can exhibit a closely coordinated synthesis of these two molecules (*16*). It is of interest in this context that both genes have been identified provisionally on chromosome 11, which raises the possibility that the two genes may be located adjacent to one another or that a common regulatory element is able to activate the transcription of both genes.

Responsiveness to Different CSF's

While most granulocyte-macrophage progenitors are bipotential and can respond to stimulation by more than one CSF, they exhibit considerable heterogeneity with respect to the number of progeny each generates (from 50 to 10,000 cells) and to the concentrations of CSF required to stimulate cell division (*3*). This latter variability is the basis for the familiar sigmoid dose-response curve between CSF concentration and the number of granulocyte-macrophage colonies developing in a culture dish, some clonogenic cells requiring 10- to 50-fold higher CSF concentrations than others to be stimulated to proliferate. This may be related to the tenfold variation in CSF receptor numbers observed by autoradiography on individual normal progenitor cells (Fig. 2).

There are also differences in the responsiveness of subsets of granulocyte-macrophage progenitor cells to stimulation by the different CSF's. This heterogeneity in target cell populations could provide some explanation for the existence of multiple types of CSF but, despite this, reciprocal clone transfer studies indicate that most progenitors can be stimulated by more than one CSF. This highly redundant control system might seem an unnecessarily complex manner in which to regulate cell proliferation but, as shall be discussed shortly, the CSF's have other important actions on granulocyte-macrophage cells allowing subtle competitive and potentiating interactions between the CSF's and thus a fine control of the production and activation of selected subsets of mature progeny. Furthermore, multi-CSF is not normally detectable in the circulation and may be produced and act primarily on local target cells adjacent to sites of multi-CSF production, whereas other CSF's, for example, G-CSF, circulate and may act simultaneously in multiple locations.

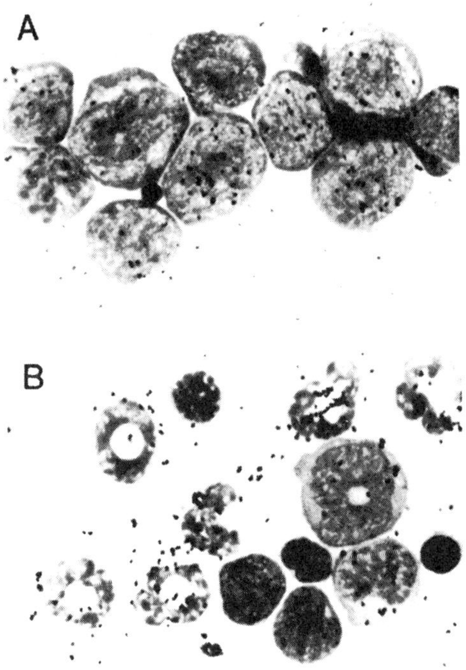

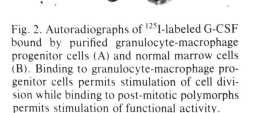

Fig. 2. Autoradiographs of ^{125}I-labeled G-CSF bound by purified granulocyte-macrophage progenitor cells (A) and normal marrow cells (B). Binding to granulocyte-macrophage progenitor cells permits stimulation of cell division while binding to post-mitotic polymorphs permits stimulation of functional activity.

Receptors for CSF's

The membranes of responding granulocytes and macrophages exhibit specific high-affinity receptors for the CSF's. Each CSF receptor appears to bind only a single species of CSF and since most granulocytes and macrophages are able to respond to more than one CSF, these cells simultaneously exhibit more than one type of CSF receptor. Until more detailed autoradiographic data are available, it remains uncertain how many hematopoietic cells exhibit only a single species of CSF receptor. Stimulation of granulocyte and macrophages by a combination of two different CSF's enhances the resulting proliferation (*3*), but these interactions are likely to be complex since exposure of cells to one CSF can lead to down-regulation of other CSF receptors (*17*). Receptors for the different CSF's differ in molecular weight: M-CSF, 165,000 (*18*); G-CSF, 150,000; multi-CSF, 50,000 to 70,000; and GM-CSF, 50,000 (*17*). There are also differences in the mean number of receptors on normal responsive cells, receptors for GM-CSF, G-CSF, and multi-CSF being low in number (100 to 500 per cell) (*17*) and those for M-CSF being somewhat higher (3,000 to 16,000 per cell) (*19*). Despite these low receptor numbers, half-maximal proliferative effects are achieved by G-CSF and GM-CSF with a receptor occupancy of only 5 to 10 percent. Degradation of M-CSF-receptor complexes appears to be very rapid (*20*), but G-CSF–receptor complexes are degraded much more slowly with a half-life of at least 6 hours (*21*). A slow turnover of CSF-receptor complexes of this latter

```
Mouse  AlaProThrArgSerProIleThrValThrArgProTrpLysHisValGluAlaIleLys
        *  *     *  *  *           *     *  *     *  *     *  *
Human  AlaProAlaArgSerProSerProSerThrGlnProTrpGluHisValAsnAlaIleGln

Mouse  GluAla---------LeuAsnLeuLeuAspAspMetProValThrLeuAsnGluGluVal
        *  *           *  *  *        *                 *  *     *
Human  GluAlaArgArgLeuLeuAsnLeuSerArgAspThrAlaAlaGluMetAsnGluThrVal

Mouse  GluValValSerAsnGluPheSerPheLysLysLeuThrCysValGlnThrArgLeuLys
        *  *     *        *                 *  *     *  *  *  *
Human  GluValIleSerGluMetPheAspLeuGlnGluProThrCysLeuGlnThrArgLeuGlu

Mouse  IlePheGluGlnGlyLeuArgGlyAsnPheThrLysLeuLysGlyAlaLeuAsnMetThr
                 *  *  *  *  *        *  *  *  *  *     *     *
Human  LeuTyrLysGlnGlyLeuArgGlySerLeuThrLysLeuLysGlyProLeuThrMetMet

Mouse  AlaSerTyrTyrGlnThrTyrCysProProThrProGluThrAspCysGluThrGlnVal
        *  *     *           *  *  *  *  *  *  *     *     *  *
Human  AlaSerHisTyrLysGlnHisCysProProThrProGluThrSerCysAlaThrGlnThr

Mouse  ThrThrTyrAlaAspPheIleAspSerLeuLysThrPheLeuThrAspIleProPheGlu
           *           *           *  *     *  *        *  *  *
Human  IleThrPheGluSerPheLysGluAsnLeuLysAspPheLeuLeuValIleProPheAsp

Mouse  CysLysLysProValGlnLys
        *        *  *  *
Human  CysTrpGluProValGlnGlu
```

Fig. 3. Comparison of amino acid sequence of mouse GM-CSF (*56*) with human GM-CSF (*22*) deduced from sequencing of cDNA clones. Asterisks indicate homologous amino acids. Fifty-four percent homology is evident as is conservation of the four cysteine residues.

type may permit sustained intracytoplasmic activation of mediator molecules to compensate for the low absolute number of bound receptors.

Considerable species-specificity is exhibited by the CSF's, but there are exceptions. In general, murine CSF's do not stimulate the proliferation of human granulocytes and macrophages. Sequence data from a cloned human GM-CSF cDNA (*22*) indicate 54 percent homology with murine GM-CSF in the protein coding region and conservation of the four cysteines present in the mouse polypeptide that appear to allow the disulfide bridging necessary for biological activity (Fig. 3). Despite these structural homologies, murine GM-CSF neither binds to nor stimulates the proliferation of human granulocytes and macrophages (*23*).

However, murine G-CSF and human CSF-β are exceptional in showing considerable cross-species reactivity. G-CSF is able to bind to human cells and stimulate the proliferation of one subset of human granulocyte-macrophage cells. Conversely, human CSF-β stimulates murine granulocytes and macrophages to form colonies resembling those stimulated by G-CSF and competes for G-CSF binding to both murine and human cells (*23*).

Multiple Functions of the CSF's

The CSF's were first detected because of their mandatory and unique role in stimulating hematopoietic cell proliferation. However, each CSF also exhibits three other important actions on responding cells: (i) promotion of cell survival, (ii) differentiation commitment, and (iii) stimulation of end-cell functional activity.

Most hematopoietic cells that are in active cell cycle when placed in cultures lacking an appropriate CSF usually fail to complete the cycle in progress. A minority of some cell types, for example, immature granulocytes, can slowly progress through one and occasionally two divisions in the absence of added CSF (*24, 25*), but this may reflect the slow turnover of CSF-receptor complexes existing prior to the removal of the cells for culture. Withdrawal of CSF from hematopoietic cells in vitro is usually followed by rapid cessation of DNA and protein synthesis (*20*), and cell death occurs exponentially with half-life periods varying from 6 to 24 hours according to cell type (*24, 25*). This may be due to failure to maintain cytoplasmic concentrations of adenosine triphosphate and a general breakdown in the membrane glucose transport system (*26*).

Colony-stimulating factor is required continuously to stimulate sustained hematopoietic cell proliferation in vitro. The proliferative effects of the CSF's are not the passive consequence of CSF-promoted cell survival. Cell survival maintained by other means does not result in hematopoietic cell proliferation (*26*), and there is a precise dose-response relation between CSF concentration and the magnitude of proliferation shown by responding granulocytes and macrophages, half-maximal proliferative effects for all four murine CSF's being observed at molar concentrations of approximately 10^{-12} (*3*).

There is only fragmentary information on the biochemical events occurring when CSF's stimulate cells to pass through a cell cycle and divide. Autophosphorylation of M-CSF receptors has been noted following binding of M-CSF (*27*), and the CSF's have been shown to induce changes in the rate of synthesis of a number of cytoplasmic and nuclear proteins (*28*). GM-CSF has also been observed to increase the phosphorylation of a number of proteins, prominent among which are the p21 *ras* gene product and a protein of 53,000 molecular weight (*28*). However, the complexity of these changes has thus far prevented identification of the events crucial for cell division.

Studies with micromanipulated paired daughter cells of granulocyte-macrophage progenitor cells have shown that the concentration of CSF determines the mean cell cycle time and the total number of progeny produced by a particular cell (*5, 29*). Culture of granulocyte-macrophage progenitor cells purified by fluorescence-activated cell sorting or of manipulated single progenitor cells has shown that the proliferative action of the CSF's is a direct one on responding cells (*3*), a conclusion supported by the presence of CSF receptors on these cells (Fig. 2).

Although CSF-stimulated proliferation of granulocytes and macrophages is accompanied by maturation of the cells in the resulting clones to mature granulocytes and macrophages, it remains uncertain whether the CSF's have a direct capacity to influence these maturation events. This is a difficult problem to resolve since hematopoietic cell survival in vitro depends on CSF, and the behavior of healthy cells cannot be analyzed in the absence of CSF. The question is of relevance because of the ability of certain of the CSF's to stimulate the proliferation of cells in other hematopoietic lineages. For example, GM-CSF can stimulate the formation of eosinophil colonies from fetal progenitor cells (*30*). GM-CSF can also stimulate the initial cell divisions of erythroid progenitor cells without altering their ultimate ability to form mature red cells (*31*). It may be that the genetic and biochemical programs controlling cell divisions are separate from those directing highly specialized functions in maturing cells such as the synthesis of eosinophil granules or hemoglobin.

The four granulocyte-macrophage CSF's differ widely in their ability to stimulate the proliferation of cells in other hematopoietic lineages (*3*). At one extreme, multi-CSF can also stimulate the proliferation of multipotential, erythroid, eosinophil, megakaryocyte, and mast cells (*13, 14*), while at the other extreme the proliferative effects of M-CSF are largely restricted to macrophage precursors (*7*). GM-CSF and G-CSF occupy an intermediate position in being able to initiate, but not sustain, proliferation in some multipotential, erythroid, eosinophilic, and megakaryocytic precursors (*30, 31*). Autoradiographic studies with ^{125}I-labeled CSF's indicate that these differences are based on differing distributions of membrane receptors for the various CSF's on different hematopoietic subpopulations.

However, the CSF's are not simply regulators of cell proliferation. Where granulocyte-macrophage progenitor cells are bipotential and able to form both granulocytic and macrophage progeny, paired daughter cell and reciprocal clone transfer studies have shown that the CSF's are able to induce irreversible commitment to one or other restricted pathway of differentiation. Thus, high GM-CSF concentrations force many cells to enter the granulocytic pathway whereas low concentrations permit the formation only of macrophage progeny (*29*). Similarly, stimulation by M-CSF of a bipotential cell to pass through two to three cell divisions irreversibly commits the cells to form macrophage progeny regardless of the CSF used subsequently to maintain proliferative stimulation (*32*). Exposure of a bipotential granulocyte-macrophage progenitor cell simultaneously to two CSF's, for example, M-CSF and G-CSF, allows competitive commitment to occur (*33*). The commitment process in these situations seems to require cell division, and commonly occurs asymmetrically with one daughter cell becoming committed while initially the other daughter remains bipotential (*33*). A similar phenomenon has been documented in an analysis of the manner in which G-CSF suppresses self-generation by myeloid leukemic stem cells and forces the production of differentiating progeny. Again, the process requires the presence of G-CSF for one or two cell cycles, is asymmetrical, and, having occurred, commitment to the production of differentiating progeny is irreversible (*34*). These observations on the irreversible effects of the CSF's suggest that, to mediate such effects, CSF's or, more likely, amplified cellular mediators of CSF action, may move to the nucleus and bind to regions of S-phase extended chromosomes adjacent to genes involved in these processes.

The CSF's can also stimulate a variety of functional activities of mature granulocytes and macrophages, for example, phagocytosis of bacteria or yeast by granulocytes and macrophages, antibody-dependent cytotoxic killing of tumor cells by granulocytes, or synthesis of prostaglandin E, plasminogen-activator, and other regulators by macrophages (*3, 35*). These effects occur rapidly, are associated with obvious membrane changes, for example, in adherence and self-agglutination, and could in many cases be mediated by changes occurring in regions close to the membrane.

Two important events in the life history of white cells are release from the marrow and, where necessary, entry of the cells into the tissues. There is no evidence that the CSF's influence release of cells from the marrow, but a purified human CSF can inhibit neutrophil migration (*36*). Since CSF is produced locally at inflammatory sites, this local CSF could functionally activate mature cells that have migrated to inflammatory foci and, by inhibiting further migration, ensure their retention in the region of inflammation.

Location and Control of CSF Production

All mouse tissues contain extractable CSF in concentrations higher than are present in the serum and, where various organs have been tested, all have been found to synthesize one or other form of CSF in vitro (*37*). Until in situ hybridization studies are performed with the use of CSF cDNA's as probes to identify messenger RNA, information is restricted on the specific cell types able to synthesize CSF. Only a few cell types can be purified and tested in the cell concentrations (10^5 to 10^6 per milliliter) required to detect low levels of CSF synthesis. Of cell types so tested, macrophages, T lymphoyctes, endothelial cells, fibroblasts, and skin epithelial cells have all been shown to synthesize one or more of the four major CSF's. A variety of tumors can sometimes synthesize CSF, and in animals or patients bearing such tumors, granulocyte levels can be grossly elevated (*3*).

Because macrophages produce CSF and both polymorphs and macrophages produce inhibitors of granulocyte-macrophage production, such as lactoferrin and prostaglandin E, it has been proposed that the granulocyte-macrophage system could be internally self-regulating with levels of mature cells controlling new cell production (*38*). There may be some situations in which self-regulation is prominent, but most of the evidence suggests that many, and possibly all, cell types have the ability to synthesize CSF, and that the level of CSF production is dictated by signals extrinsic to the granulocyte-macrophage system. For example, CSF production by macrophages and endothelial cells is strongly stimulated by exposure to endotoxin and other bacterial products, CSF production by T lymphocytes is stimulated by lectins or alloantigens, and CSF production by skin epithelial cells is increased after contact with promoting agents [for a review, see (*3*)].

The serum half-life of injected CSF is short, with the level falling in a biphasic mode with an initial half-life of 5 to 15 minutes followed by a slower phase with a half-life of 1 to 7 hours (*3, 17*). From studies with radioactively labeled GM-CSF and G-CSF, initial tissue localization is mainly in the liver, but this is followed by progressive relocation in and degradation by the kidney.

Two types of tissue control of granulocyte-macrophage populations by the CSF's have become evident. In the first, basal levels of cells are likely to be maintained by CSF production by stromal cells within the marrow cavity (*39*), a system apparently able to respond in a compensatory manner to depopulation of hematopoietic cells (*40*).

The second type of control system involves tissues throughout the body and is activated most commonly by exposure to microorganisms and their products, such as endotoxin. After the injection of endotoxin, a 1000-fold increase occurs in circulating CSF (*41*) and all tissues contain and synthesize elevated levels of CSF (*42*). These changes occur within minutes, and peak levels of serum and tissue CSF are attained by 3 to 6 hours. Where infections are minor and localized, it seems probable that locally involved tissues are the major source of the additional CSF. When infections are resolved, CSF levels promptly return to normal (*3*). This response pattern to clinical or subclinical infections appears to be a major factor in determining the "normal" circulating levels of CSF in humans and other animals. The increased CSF can be regarded as a defense response that is self-limiting when the stimulated granulocytes and macrophages have eliminated the microorganisms initiating the response.

Radioactively labeled CSF, when injected intravenously, can be shown to be bound by marrow cell populations (*21*) so the high circulating CSF levels elicited by microbial products can influence granulocyte and macrophage proliferation in the marrow and spleen. However, since the CSF's also have important stimulating effects on mature granulocytes and macrophages, it may be that, in the initial phases of an acute infection, the primary purpose of the elevated CSF levels in the circulation and tissues is to deliver an ultrarapid stimulus for increased functional activity by preexisting mature granulocytes and macrophages.

The CSF's and Myeloid Leukemia

Myeloid leukemias are clonal neoplasms of granulocyte-macrophage precursor cells. A currently popular view of the cancerous state is that the uncontrolled proliferation exhibited by a cancer cell population is ascribable to the action of viral or cellular oncogenes whose products are either specific growth factors or receptors for such factors. Cancer cell proliferation is viewed, not as an autonomous process, but as the response of cells to oncogene-induced autosynthesis of specific growth factors or receptors. It is postulated that the polypeptides involved could in some cases be structurally normal with a cellular oncogene becoming dysregulated by translocation or activation by viral enhancers, while in other situations the oncogene may be structurally abnormal because of mutational events.

Since the CSF's are the only known proliferative stimuli for granulocytes and macrophages, it is of interest to determine whether the myeloid leukemias are ascribable to oncogene-deranged autosynthesis of CSF or CSF receptors. There is no sequence homology between the CSF's and the known oncogenes, but no sequence data are yet available for CSF receptors.

Is there any evidence that myeloid leukemia cells are autostimulating because of an acquired capacity to synthesize their own CSF's? Data on this question are extensive and quite unambiguous for leukemia cells from patients with either acute or chronic myeloid leukemia. In no instance has it been documented that myeloid leukemic cells are capable of sustained autonomous proliferation in vitro. The proliferation of leukemic cells in vitro is, like normal cells, absolutely dependent on the addition of exogenous CSF (*43*), and the concentrations of CSF required are similar to those required to stimulate the proliferation of normal cells (*44*). These observations should not be misinterpreted as indicating that leukemia cells are unable to produce CSF. Studies on monocytes of leukemic origin have clearly documented that they have the capacity to produce CSF (*45*) and, if sufficient numbers of such cells are crowded together in a culture, the CSF's they produce can lead to apparently spontaneous proliferation by clonogenic leukemic or normal cells in the culture. However, the levels of CSF produced under these circumstances do not differ from those produced by corresponding normal monocytes. Furthermore, from the dependency of autostimulation on cell crowding it can be concluded that, to be active, CSF must be secreted and must then bind to appropriate membrane receptors. The absolute levels of CSF that can be produced by an emerging leukemic clone of a few cells would be insignificant compared with the CSF normally produced by adjacent stromal cells or with the amount reaching the cells via the circulation, and it is improbable that autoproduction of CSF by leukemia cells would represent a significant event in the emergence of a leukemic clone.

It must be emphasized, however, that since all primary myeloid leukemias are CSF-dependent, the emergence of a myeloid leukemic clone is absolutely dependent on CSF stimulation—whatever the source. This makes the CSF's essential cofactors in leukemia induction or the reemergence of leukemia in a relapse following unsuccessful therapy.

Little information is available on CSF receptor numbers on human myeloid leukemia cells, but studies with the cross-reactive murine G-CSF have indicated no significant difference in receptor numbers from those on comparable normal cells (*46*), a situation quite unlike that documented for IL-2 receptors on human T-leukemic cells transformed by human T-cell leukemia virus (*47*). With the murine myelomonocytic leukemia WEHI-3B, a cell line known to synthesize M-CSF and multi-CSF, receptor numbers for ^{125}I-labeled G-CSF are within the normal range (*48*); the numbers for M-CSF and multi-CSF are low, however (*49*), although it is possible that they are down-regulated. With monocytic leukemia cell lines not capable of constitutive CSF synthesis (*50*), CSF receptors are either undetectable or within the normal range (*49*).

Taken together, the data on the lack of autonomous growth ability and the unremarkable receptor numbers on leukemic cells argue strongly against a simple autocrine growth model for myeloid leukemia cells. A more likely possibility, regardless of whether or not leukemic cells themselves can synthesize CSF, is that an intrinsic abnormality exists in leukemic stem cells as a consequence of which CSF-stimulated proliferation results in an abnormally high ratio of self-generative divisions versus divisions leading to the production of differentiating progeny. In this model, CSF would be a necessary cofactor in the emergence of a leukemic clone since it is mandatory for cell proliferation, but an intrinsic defect, possibly caused by an oncogene product, would be responsible for the aberrant pattern of resulting cell divisions.

Evidence for the importance of abnormally high self-generation by clonogenic leukemic cells has come (i) from recloning studies that have amply documented the high capacity of leukemic stem (clonogenic) cells to self-generate and (ii) from the observations by Ichikawa (*51*) that certain cell-conditioned media are able to induce cells of the myeloid leukemia cell line M1 to differentiate to mature granulocytes and macrophages and simultaneously to lose their leukemogenicity when injected into mice. In subsequent studies, the ability of a wide variety of chemical and biological agents to induce terminal differentiation in murine myeloid leukemia cells has been extensively analyzed (*52*) as has the comparable process in human myeloid leukemia cell lines (*53*).

While the CSF's are not the only biological agents able to influence the differentiation of myeloid leukemic cells, the ability of various conditioned media containing CSF to induce differentiation and the dependency of leukemic cells on CSF raised the likelihood that one or other of the CSF's might be able to induce differentiation in myeloid leukemia cells. Analysis of the four purified murine CSF's using the WEHI-3B myelomonocytic leukemia showed that M-CSF and multi-CSF had no differentiation-inducing activity, GM-CSF had weak activity, and G-CSF had a remarkable capacity to induce differentiation.

In line with the proliferative effects of CSF's, G-CSF initially enhances the proliferation of WEHI-3B cells, but rapidly superimposed is a suppressive effect from the differentiation-inducing action of G-CSF. G-CSF has the capacity to

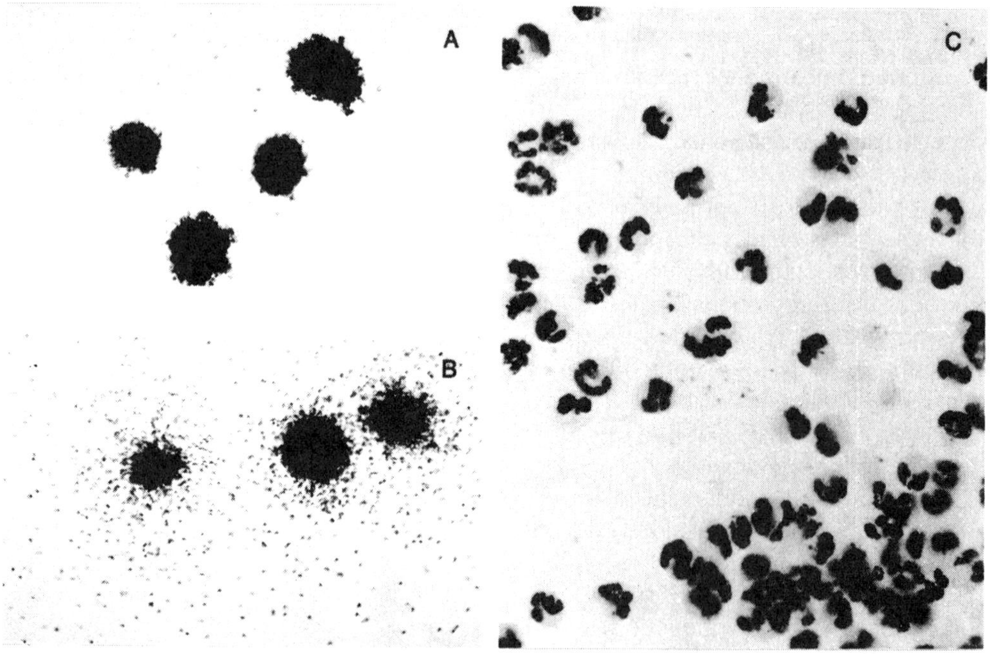

Fig. 4. Induction by purified G-CSF of differentiation in colonies of WEHI-3B murine myelomonocytic leukemic cells. Leukemic cells usually form compact colonies of undifferentiated cells (A), but in the presence of G-CSF colonies are surrounded by a corona of dispersing cells (B) because of differentiation of leukemic cells to maturing granulocytes and macrophages (C).

irreversibly shift the ratio of self-generative versus differentiative divisions in WEHI-3B leukemic stem cells with progressive suppression of self-generative divisions and the production of nondividing differentiated progeny (*5, 34*) (Fig. 4). As a consequence of this action, the leukemic population can be completely suppressed in vitro by quite normal concentrations of G-CSF (*54*). At present, the available amounts of purified G-CSF are too small to permit corresponding experiments in vivo, but the injection of crude preparations containing G-CSF has been shown to suppress other established transplanted myeloid leukemias in mice (*55*).

Conclusions

With the purification and cloning of several of the GM-CSF's, these glycoproteins are now firmly established as specific growth factors. The CSF's are the only known agents able to stimulate the proliferation of granulocyte-macrophage precursors, and they also have the capacity to induce differentiation commitment in these precursors and to stimulate the functional activity of mature granulocytes and macrophages. The hematopoietic culture systems that permitted the detection and characterization of the CSF's are also now well established as robust but elegant techniques permitting studies in defined medium on single cells producing differentiating progeny in response to stimulation by low concentrations of a single purified regulator. For the general cell biologist, clonal cultures of granulocytes or macrophages represent a model system with a potential possibly greater than any other for providing answers to some of the most fundamental questions in cell biology relating to the control of cell division and differentiation.

Given the advantages of the granulocyte-macrophage system in having many well-defined differentiation markers, it should be possible to make useful progress in a molecular analysis of the complex events occurring in responding cells following binding of the CSF's to their receptors and for this the availability of four similar, but competing, CSF's may be of particular value. It should also prove feasible to analyze the cellular events regulating CSF synthesis since, unlike some growth factors, a spectacular increase in CSF synthesis can be induced by agents such as endotoxin, lectins, and antigens (*9, 16, 41, 42*).

The availability of mass-produced recombinant CSF will now permit detailed studies in vivo on the actions of the CSF's. Two goals of high priority are to determine whether the CSF's will prove useful in enhancing hematopoietic regeneration following marrow damage or transplantation and in increasing host resistance to established infections. Of equal importance, in view of the demonstrated capacity of G-CSF to suppress myeloid leukemia populations, will be studies to establish whether the clinical use of CSF's will provide an effective adjuvant therapy in the management of myeloid leukemia. Here it will be necessary to keep in mind that the proliferation of myeloid leukemia cells is CSF-dependent and only CSF's with pronounced differentiation-inducing activity, such as murine G-CSF and human CSF-β, should be candidates for trial.

References and Notes

1. T. R. Bradley and D. Metcalf, *Aust. J. Exp. Biol. Med. Sci.* **44**, 287 (1966).
2. Y. Ichikawa, D. H. Pluznik, L. Sachs, *Proc. Natl. Acad. Sci. U.S.A.* **56**, 488 (1966).

3. D. Metcalf, *The Hemopoietic Colony Stimulating Factors* (Elsevier, Amsterdam, 1984).
4. A. W. Burgess, J. Camakaris, D. Metcalf, *J. Biol. Chem.* **252**, 1998 (1977); L. G. Sparrow *et al.*, *Proc. Natl. Acad. Sci. U.S.A.* **82**, 292 (1985).
5. N. A. Nicola *et al.*, *J. Biol. Chem.* **258**, 9017 (1983).
6. J. N. Ihle *et al.*, *J. Immunol.* **129**, 2431 (1982).
7. E. R. Stanley and P. M. Heard, *J. Biol. Chem.* **252**, 4305 (1977).
8. N. A. Nicola *et al.*, *Blood* **54**, 614 (1979); M-C Wu, A. M. Miller, A. A. Yunis, *J. Clin. Invest.* **67**, 1558 (1981).
9. N. M. Gough *et al.*, *Nature (London)* **309**, 763 (1984).
10. A. R. Dunn *et al.*, in *Growth Factors and Transformation* (Cold Spring Harbor Laboratory, Cold Spring Harbor, N.Y., 1985), pp. 227–234.
11. T. Yokota *et al.*, *Proc. Natl. Acad. Sci. U.S.A.* **81**, 1070 (1984); S. Miyatake *et al.*, *ibid.* **82**, 316 (1985).
12. M. C. Fung *et al.*, *Nature (London)* **307**, 233 (1984).
13. J. S. Greenberger *et al.*, *Exp. Hematol. (Copenhagen)* **13**, 7 (1985).
14. A. J. Hapel *et al.*, *Blood* **65**, 1453 (1985).
15. N. A. Nicola, A. W. Burgess, D. Metcalf, *J. Biol. Chem.* **254**, 5290 (1979).
16. A. Kelso and D. Metcalf, *Exp. Hematol. (Copenhagen)* **13**, 7 (1985).
17. N. A. Nicola, A. W. Burgess, F. Walker, unpublished data.
18. C. J. Morgan and E. R. Stanley, *Biochem. Biophys. Res. Commun.* **119**, 35 (1984).
19. P. V. Byrne, L. J. Guilbert, E. R. Stanley, *J. Cell Biol.* **91**, 848 (1981).
20. R. J. Tushinski *et al.*, *Cell* **28**, 71 (1982).
21. N. A. Nicola and D. Metcalf, *J. Cell. Physiol.* **124**, 313 (1985).
22. G. G. Wong *et al.*, *Science* **228**, 810 (1985).
23. D. Metcalf, C. G. Begley, N. A. Nicola, *Leuk. Res.* **9**, 521 (1985). C. G. Begley *et al.*, unpublished data.
24. D. Metcalf and S. Merchav, *J. Cell. Physiol.* **112**, 411 (1982).
25. D. Metcalf, *Blood* **65**, 357 (1985).
26. A. D. Whetton and T. M. Dexter, *Nature (London)* **303**, 629 (1983).
27. E. R. Stanley, personal communication.
28. B. Hoffman-Liebermann and L. Sachs, *Cell* **14**, 825 (1978); B. Hoffman-Liebermann, D. Liebermann, L. Sachs, *Dev. Biol.* **81**, 255 (1981); I. J. Stanley and A. W. Burgess, *J. Cell. Biochem.* **23**, 241 (1983); I. J. Stanley and N. A. Nicola, unpublished data.
29. D. Metcalf, *Proc. Natl. Acad. Sci. U.S.A.* **77**, 5327 (1980).
30. ______ and N. A. Nicola, *J. Cell. Physiol.* **116**, 198 (1983).
31. D. Metcalf, G. R. Johnson, A. W. Burgess, *Blood* **55**, 138 (1980).
32. D. Metcalf and A. W. Burgess, *J. Cell. Physiol.* **111**, 275 (1982).
33. D. Metcalf, S. Merchav, G. Wagemaker, in *Experimental Hematology Today 1982*, S. J. Baum, G. D. Ledney, S. Thierfelder, Eds. (Karger, Basel, 1982), pp. 3–9.
34. D. Metcalf, *Int. J. Cancer* **30**, 203 (1982).
35. E. Handman and A. W. Burgess, *J. Immunol.* **112**, 1134 (1979); A. F. Lopez *et al.*, *ibid.* **131**, 2983 (1983); J. I. Kurland *et al.*, *Proc. Natl. Acad. Sci. U.S.A.* **76**, 2336 (1979); H-S Lin and S. Gordon, *J. Exp. Med.* **150**, 231 (1979); J. A. Hamilton *et al.*, *J. Cell. Physiol.* **103**, 435 (1980).
36. J. C. Gasson *et al.*, *Science* **226**, 1339 (1984).
37. J. W. Sheridan and E. R. Stanley, *J. Cell. Physiol.* **78**, 451 (1971); J. W. Sheridan and D. Metcalf, *ibid.* **80**, 129 (1972); N. A. Nicola, A. W. Burgess, D. Metcalf, *J. Biol. Chem.* **254**, 5290 (1979); N. A. Nicola and D. Metcalf, *J. Cell. Physiol.* **109**, 253 (1981).
38. J. I. Kurland, in *Experimental Hematology Today 1978*, S. J. Baum and G. D. Ledney, Eds. (Springer-Verlag, New York, 1978), pp. 47–60; H. E. Broxmeyer *et al.*, *J. Exp. Med.* **148**, 1052 (1978).
39. S. H. Chan and D. Metcalf, *Blood* **40**, 646 (1972); T. M. Dexter, T. D. Allen, L. G. Lajtha, *J. Cell. Physiol.* **91**, 335 (1977); M. Lanotte, D. Metcalf, T. M. Dexter, *ibid.* **112**, 123 (1982).
40. S. H. Chan and D. Metcalf, *Cell Tissue Kinet.* **6**, 185 (1973); M. Onoda *et al.*, *J. Cell. Physiol.* **104**, 11 (1980).
41. D. Metcalf, *Immunology* **21**, 427 (1971); P. Quesenberry *et al.*, *N. Engl. J. Med.* **286**, 227 (1972).
42. J. W. Sheridan and D. Metcalf, *J. Cell. Physiol.* **80**, 129 (1972).
43. M. A. S. Moore, N. Williams, D. Metcalf, *J. Natl. Cancer Inst.* **50**, 603, (1973); D. Metcalf *et al.*, *Advances in Comparative Leukemia Research 1977*, P. Bentvelson, Ed. (Elsevier, Amsterdam, 1977), pp. 307–310; J. D. Griffin, R. P. Beveridge, S. F. Schlossman, *Blood* **60**, 30 (1982).
44. D. Metcalf *et al.*, *Blood* **43**, 847 (1974); G. E. Francis *et al.*, *Leuk. Res.* **4**, 531 (1980).
45. M. A. S. Moore, N. Williams, D. Metcalf, *J. Natl. Cancer Inst.* **50**, 591 (1973); D. W. Golde, B. Rothman, M. J. Cline, *Blood* **43**, 749 (1974); J. M. Goldman *et al.*, *ibid.* **47**, 381 (1976); G. L. Bianchi Scarra *et al.*, *Exp. Hematol.* **9**, 917 (1981).
46. D. Metcalf, N. A. Nicola, G. Begley, unpublished data.
47. R. C. Gallo *et al.*, *Blood* **62**, 510 (1983).
48. N. A. Nicola and D. Metcalf, *Proc. Natl. Acad. Sci. U.S.A.* **81**, 3765 (1984).
49. L. J. Guilbert and E. R. Stanley, *J. Cell Biol.* **85**, 153 (1980); E. W. Palaszynski and J. N. Ihle, *J. Immunol.* **132**, 1872 (1984); N. A. Nicola, F. W. Walker, D. Metcalf, unpublished data.
50. D. Metcalf and N. A. Nicola, in *Molecular Biology of Tumor Cells*, B. Wahren *et al.*, Eds. (Raven, New York, 1985), pp. 215–232.
51. Y. Ichikawa, *J. Cell. Physiol.* **74**, 223 (1969); *ibid.* **76**, 175 (1970).
52. See reviews, L. Sachs, *Nature (London)* **274**, 535 (1975); *J. Cell. Physiol. Suppl.* **1**, 151 (1982); M. Hozumi, *Adv. Cancer Res.* **38**, 121 (1983).
53. J. Lotem and L. Sachs, *Int. J. Cancer* **25**, 561 (1980); Y. Honma *et al.*, *Eur. J. Cancer Clin. Oncol.* **19**, 251 (1983); J. Abrahm and G. Rovera, in *Tissue Growth Factors*, R. Baserga, Ed. (Springer-Verlag, New York, 1981), pp. 405–425.
54. D. Metcalf, *Int. J. Cancer* **25**, 225 (1980).

55. J. Lotem and L. Sachs, *ibid.* **28**, 375 (1981).
56. N. M. Gough *et al.*, *EMBO J.* **4**, 645 (1985).
57. J. Lotem and L. Sachs, *Int. J. Cancer* **35**, 93 (1985).
58. N. N. Iscove *et al.*, *J. Cell. Physiol. Suppl.* **1**, 65 (1982).
59. G. W. Bazill *et al.*, *Biochem. J.* **210**, 747 (1983).
60. Y-P. Yung *et al.*, *J. Immunol.* **127**, 794 (1981).
61. I. Clark-Lewis and J. W. Schrader, *ibid.*, p. 1941.
62. A. W. Burgess, N. A. Nicola, E. Nice, R. Simpson, D. Metcalf, unpublished data.
63. Supported in part by the Anti-Cancer Council of Victoria, National Health and Medical Research Council, Canberra, and the National Institute of Health, Bethesda (grant CA-22556). I thank N. A. Nicola for reviewing the manuscript and N. M. Gough for assistance in preparing Fig. 3.

13

X-ray Structure of the Major Adduct of the Anticancer Drug Cisplatin with DNA: cis-$[Pt(NH_3)_2\{d(pGpG)\}]$

Suzanne E. Sherman, Dan Gibson, Andrew H.-J. Wang, and Stephen J. Lippard

cis-Diamminedichloroplatinum(II), *cis*-DDP or cisplatin, is a clinically important anticancer drug, being especially effective for the management of testicular, ovarian, and head and neck cancers (*1, 2*).

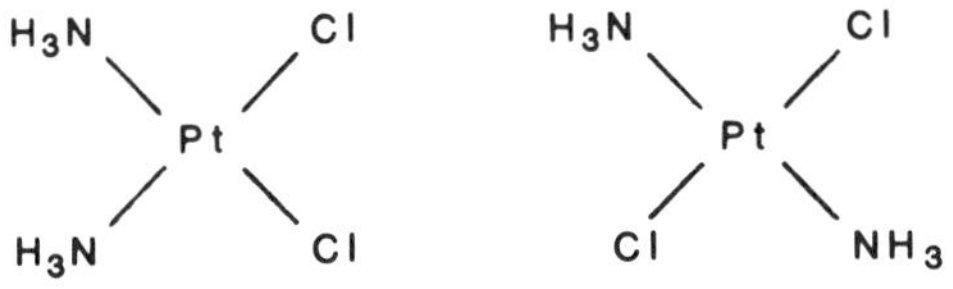

cis-DDP (cisplatin) *trans*-DDP

It is one of the most widely used antitumor drugs at the present time. The *trans* isomer, *trans*-DDP, is inactive. Considerable evidence points to DNA as being the main target of cisplatin in the tumor cell (*3*). Attention has therefore focused on the nature of, and the differences between, adducts formed by *cis*- and *trans*-DDP with DNA. By using a variety of enzymatic mapping techniques, we and others have shown (*4*) that the most common binding mode of *cis*-DDP with DNA involves loss of two chloride ions and formation of two Pt-N bonds to the N(7) atoms of two adjacent guanosine nucleosides on the same strand. For stereochemical reasons, this intrastrand d(GpG) cross-link cannot be formed by *trans*-DDP. Much structural information about the adduct of *cis*-DDP with synthetic oligodeoxynucleotides has been garnered through nuclear magnetic resonance (NMR) spectroscopic investigations (*5, 6*). In addition, recent molecular mechanics calculations on *cis*-$[Pt(NH_3)_2\{d(GpG)\}]$ adducts in two oligonucleotide duplexes and two single-stranded oligomers have provided theoretical insight about the structure (*7*).

Science 230, 412–417 (25 October 1985)

Conspicuously lacking thus far has been structurally definitive single crystal x-ray diffraction information about *cis*-DDP bound to DNA. In attempts to model the binding of the *cis*-$\{Pt(NH_3)_2\}^{2+}$ fragment to d(GpG), more than a dozen x-ray structural studies have been made on amine complexes of platinum bound to two 6-oxopurine bases, nucleosides, or nucleotides (*8, 9*); however, no oligodeoxynucleotide adduct has yet been crystallographically characterized. Three studies have been directed toward establishing the nature of cisplatin binding to nucleic acids by diffusing the drug into crystals of a B-DNA dodecamer (*10*) or into phenylalanine transfer RNA ($tRNA^{Phe}$) (*11, 12*). In all cases, high resolution information was precluded either by low or multiple occupancy (or both) of platinum binding sites in the crystal lattice or by the failure of the crystals to scatter beyond 4 to 6 Å resolution. Moreover, although several structures of antitumor drugs complexed with oligonucleotides have been described (*13–16*), no x-ray crystallographic information is yet available about a covalently linked DNA-drug adduct.

We now describe the synthesis and x-ray structural characterization to atomic resolution of *cis*-$[Pt(NH_3)_2\{d(pGpG)\}]$, a study that reveals the molecular geometry of the anticancer drug bound to its putative biological target on DNA. Although the structural analysis was inauspiciously complicated by the occurrence of four such complexes in the asymmetric unit of the cell, once solved we were fortunate in having four crystallographically independent views of the desired structure.

Synthesis and details of the x-ray crystal structure determination. The platinum dinucleotide complex was prepared by allowing the ammonium salt of d(pGpG) (Collaborative Research) to react at ~*p*H 4.5 with an equimolar amount of a 2 m*M* aqueous solution of *cis*-$[Pt(NH_3)_2(NO_3)_2]$ at 37°C for 80 minutes. The product, *cis*-$[Pt(NH_3)_2\{d(pGpG)\}]$, was purified by anion exchange and reversed-phase high-performance liquid chromatography (HPLC) and characterized by its proton nuclear magnetic resonance (NMR) spectrum (*17*). An identical NMR spectrum was obtained upon redissolving crystals (see below) of the complex.

Single crystals were grown from solutions that contained 9 m*M* *cis*-$[Pt(NH_3)_2\{d(pGpG)\}]$, 33 m*M* (*p*H 3.8) glycine · HCl buffer, 33 m*M* NaCl, and 9.2 percent 2-methyl-2,4-pentanediol (2-MPD), with equilibration to 70 percent 2-MPD. At this *p*H, the platinum dinucleotide complex is expected to be neutral with the terminal phosphate being monoprotonated. Upon equilibration, two visually distinct crystal forms were obtained. Initially, long prismatic crystals (crystal type 2) were formed. At longer equilibration times and with higher 2-MPD concentrations and the addition of 17 m*M* $MgCl_2$, rectangular parallelepipeds (crystal type 1) also deposited.

Both forms were examined by x-ray crystallography (Table 1). A crystal of type 1, having approximate dimensions of 0.6 by 0.5 by 0.4 mm, was sealed in a glass capillary containing a drop of mother liquor. Intensity data were collected at 16°C from one crystal on a Nicolet P3 diffractometer in the ω-scan mode with CuKα radiation. Several data sets were also taken on crystals of type 2, but at this stage we have only a partial solution of the structure, which we do not discuss further. The two crystal lattices are clearly related to one another, as can be

Table 1. X-ray crystallographic information about *cis*-[Pt(NH_3)$_2${d(pGpG)}].

Property	Crystal type 1	Crystal type 2
Unit cell	a = 31.326 Å b = 35.679 Å c = 19.504 Å Volume = 21,799 Å^3 Z = 16	a = 30.55 Å b = 33.90 Å c = 41.25 Å Volume = 42,709 Å^3 Z = 32
Space group	$P2_12_12$	$C222_1$ or $C222$
Reflections (No.)*	14,950 (10,236)	4,677
Resolution	0.94 Å	1.37 Å
Current R factor†	8.4%	

*Number collected and, in parentheses, number used in refinement for which $F_0 > 4\sigma(F_0)$. †$R = \Sigma \| F_0| - |F_c \| / \Sigma | F_0 |$.

seen from their unit cell dimensions.

The atomic positions of the four independent platinum atoms were obtained by direct methods with MULTAN (*18*). The remaining nonhydrogen atoms were located from a series of subsequent difference Fourier maps. Unconstrained full-matrix least-squares refinement was carried out in blocks with SHELX-76 (*19*). Only the platinum, phosphorus, and phosphate oxygen atoms were refined anisotropically, whereas all other atoms were assigned isotropic thermal parameters. After all 192 nonhydrogen atoms of the four independent *cis*-[Pt(NH_3)$_2${d(pGpG)}] molecules were located, one glycine molecule and 26 water molecules were found in the lattice and refined isotropically. At the present stage of refinement, the deoxyribose rings of the 3′-guanosine residues have large thermal parameters, and not all of the solvent molecules have been located. Although work on the structure is continuing, atomic coordinates have been deposited with the Cambridge Crystallographic Data Centre.

Unit cell contents and packing. The four crystallographically independent *cis*-[Pt(NH_3)$_2${d(pGpG)}] molecules, designated 1, 2, 3, and 4 in Fig. 1, form a tightly packed aggregate held together by an extensive network of hydrogen bonding and intermolecular base-base stacking interactions. The platinum-platinum distances range from 5.56 to 7.64 Å. The ammine ligands are oriented toward the center of the aggregate while the sugar-phosphate backbones are directed outward toward solvent channels and neighboring aggregates. A noncrystallographic C_2 axis along a relates molecule 1 with 2 and molecule 3 with 4 (Fig. 1). Around the periphery of the aggregate there are hydrogen bonds [donor-acceptor distances, 2.62(5) − 3.08(5) Å] between terminal 5′-phosphate groups and the N(1)-imino and exocyclic N(2)-amino hydrogen atoms of neighboring molecules. Specifically, the 5′-phosphate groups of molecules 3 and 4 interact with the respective 3′-guanosine hydrogen atoms of molecules 1 and 2, and the 5′-phosphate oxygen atoms of molecules 1 and 2 form hydrogen bonds to the 5′-guanosine residues of molecules 2 and 1, respectively. At the core of the aggregate there is partial stacking of the guanine rings of molecules 1 with 2 and 3 with 4 (Fig. 1). In addition, 20 hydrogen bonds

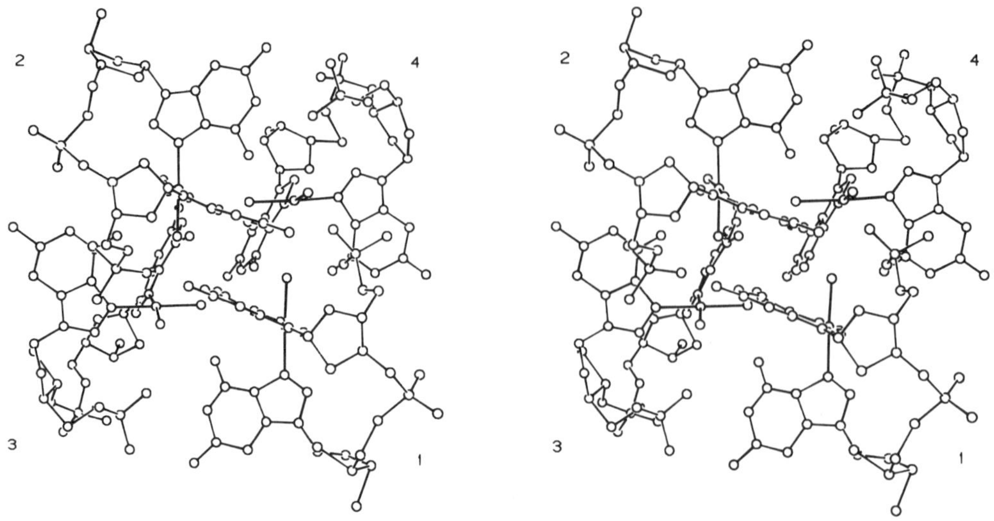

Fig. 1. Stereo view of the four crystallographically independent *cis*-[Pt(NH_3)$_2${d(pGpG)}] molecules in the unit cell. The view is down the *a* axis of the unit cell, revealing the pseudo-twofold symmetry of the aggregate. Water and glycine molecules in the lattice are not shown.

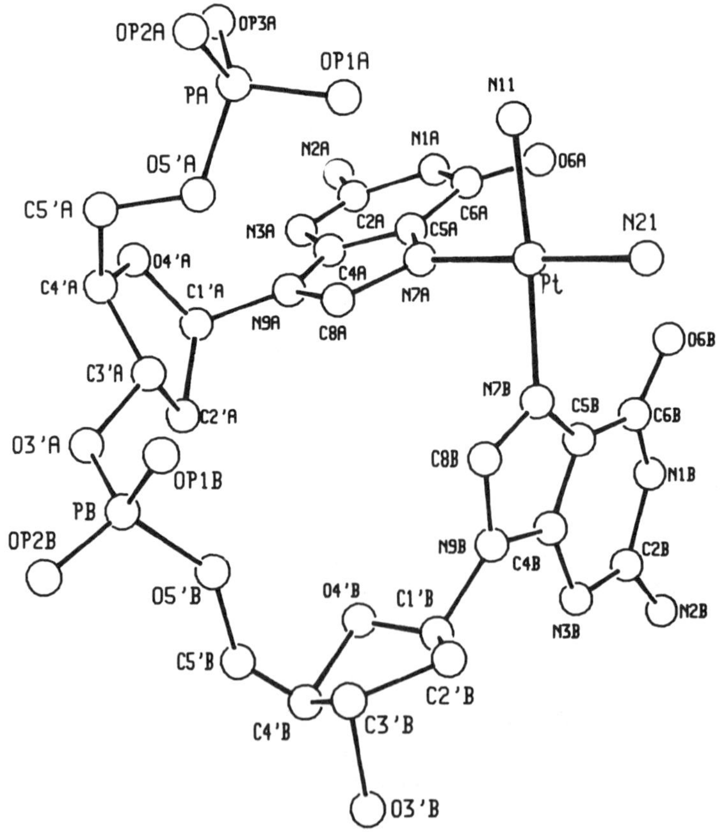

Fig. 2. Molecular structure of one of the four *cis*-[Pt(NH_3)$_2${d(pGpG)}] molecules portraying the atom labeling scheme.

[2.70(4) – 3.34(4) Å] occur between the ammine ligands and the guanine O(6) oxygen atoms of neighboring molecules. Most of these hydrogen bonds are formed between the 5′- and 3′-O(6) atoms of molecules 1, 2, 3, and 4 and the ammine ligands of molecules 3, 4, 2, and 1, respectively.

Between tetrameric aggregates, some hydrogen bonding occurs, but the aggregates are largely separated from one another by solvent channels in the crystal lattice. More specifically, hydrogen bonding occurs between the terminal O(3′)-hydroxyl protons and terminal phosphate oxygen atoms, and between guanine N(2) amino protons and N(3) nitrogen atoms, of neighboring aggregates. Water and glycine molecules occupy the solvent channels and are hydrogen bonded to the phosphate oxygen atoms, coordinated ammine protons, terminal O(3′) hydroxyl groups, and guanine ring heteroatoms. One or two water molecules also form bridges between the two phosphate groups along the sugar-phosphate backbone within each of the four individual platinum complexes.

The molecular geometry of *cis*-[Pt(NH$_3$)$_2${d(pGpG)}]. The structure of one of the four independent molecules is shown in Fig. 2, and selected geometric information is given in Table 2. In all four molecules the guanosine nucleosides are in the *anti* conformation, and the two O(6) oxygen atoms are on the same side (head-to-head orientation) of the platinum coordination plane. The geometry about each platinum atom is square-planar, with two ammine ligands and two guanosine N(7) atoms comprising the coordination sphere. The average Pt-NH$_3$ and Pt-N(7) bond lengths are 2.04 and 2.00 Å, respectively. No atom deviates by more than 0.06 Å from the best plane through platinum and its four bonded nitrogen atoms. The dihedral angles between coordinated guanine bases; defined according to conventions (*20*), range from 76° to 87°. Thus, coordination of *cis*-{Pt(NH$_3$)$_2$}$^{2+}$ to {d(pGpG)}$^{2-}$ completely disrupts the base stacking of the two adjacent nucleotides (Plate III b). Dihedral angles between planes through the PtN$_4$ and guanine rings atoms occur in pairs for the four molecules, reflecting the pseudo-twofold symmetry of the crystal lattice. For pairs of molecules 1 and 2 and 3 and 4, respectively, these angles are 111° and 111° and 80° and 77° for the 5′-guanosine and 86° and 95° and 58° and 60° for the 3′-guanosine nucleoside. With these values the ammine-O(6) contacts exceed 3.1 Å, precluding a significant amount of hydrogen bonding between the ammine ligands and the exocyclic O(6) oxygen atoms of the guanine bases. Intramolecular hydrogen bonding does occur to varying degrees between the coordinated ammines and the oxygen atoms of the 5′-phosphate groups, however, with N · · · O distances of 2.72(5) (Pt1 in Plate III b), 2.83(3) (Pt2), and 3.03(3) (Pt3), and 3.22(5) (Pt4) Å.

From the torsion angles within the five-membered deoxyribose sugar rings we may calculate the pseudorotation angles, *P*, and amplitudes, ψ (*21, 22*). In all four molecules the 5′-nucleotide is in the N, or C(3′)-endo, conformation. More precisely, there are two classes of sugar puckers. Molecules 1 and 2 have P-5′ = −12° and −8° and are characterized by relatively strong hydrogen bonding between the ammine ligands and 5′-phosphate groups. Weaker hydrogen bonding of this kind occurs for the other two molecules, for which the respective pseudorotation angles are P-5′ = 23° and 27°. The intramolecular phosphorus-

Table 2. Geometric information about *cis*-[Pt(NH_3)$_2${d(pGpG)}] from the crystal structure described in this article, *cis*-[Pt(NH_3)$_2${d(GpG)}] from NMR and model building (*5*), and two *cis*-DDP-oligonucleotide adducts from molecular mechanics calculations (*7*). Bond lengths are in angstroms and all angles are given in degrees.

Item	Coordination geometry and dihedral	
	< Pt-N > (range)	< N-Pt-N > (range)
Molecule 1	2.03 (2.01–2.05)	90 (88–92)
Molecule 2	2.01 (1.97–2.08)	90 (89–91)
Molecule 3	2.00 (1.92–2.09)	90 (87–92)
Molecule 4	2.04 (1.94–2.09)	90 (88–94)

Item	5′ Nucleotide			Dinucleotide			
	P	ψ	χ	β	γ	δ	ε
Molecule 1	−12	33	−94 (4)	141 (3)	56 (4)	94 (4)	−142 (3)
Molecule 2	−8	35	−89 (3)	209 (2)	65 (4)	104 (4)	−144 (3)
Molecule 3	23	46	−138 (4)	143 (3)	55 (4)	91 (5)	−126 (4)
Molecule 4	27	48	−142 (3)	166 (3)	36 (6)	101 (5)	−128 (4)
NMR + model building: d(GpG) − Pt(NH_3)$_2$*	1 (7)	37	−110	—	52	87	−157
Calculations: d(pGpG) − Pt(NH_3)$_2$	−3	40	−143	170	59	88	−115
Calculations: Pt-decamer†	−2	44	−152	137	55	84	−147

*Values computed from coordinates supplied by J. Reedijk.
†Values given are for *cis*-{Pt(NH_3)$_2$}$^{2+}$ bound to (GpG) in [d(TCTCGGTCTC)-d(GAGACCGAGA)], see (7).

phosphorus distances range from 5.58(4) to 6.18(2) Å. The 3′-nucleotide sugars display a wider range of pseudorotation angles, characteristic of S, or C(2′)-endo, conformations but centered about the C(1′)-exo conformation, 84° < *P* < 138°. The larger errors in torsion angles and the greater thermal motion of the 3′-nucleotide sugars in all four molecules suggest that they have greater conformational flexibility than the better behaved 5′-nucleotides in *cis*-[Pt(NH_3)$_2$-{d(pGpG)}]. Alternatively, there may be some unresolved disorder in these sugars, as reflected by a few unusual bond distances and angles.

Table 2 also lists the sugar-phosphate backbone and glycosyl bond torsion angles, defined according to standard convention (*21*). Despite their different sugar conformations, the χ-values of all eight nucleotides fall within or close to the "high-*anti*" range (*22*) except for the 5′ residues of molecules 3 and 4. Coordination of the two N(7) nitrogen atoms to platinum, which closes a large, 17-membered chelate ring (Plate III a), undoubtedly contributes to these glycosyl bond torsion angles.

Comparisons. With the crystal structure of a d(GpG)-containing oligonucleotide complex of *cis*-diammineplatinum-

Dihedral angles between planes are defined according to the conventions in (*20*), and P and ψ according to (*21, 22*). The symbol PtN_4 refers to platinum and its four bonded nitrogen atoms. Torsion angles are defined as P^{α}-$O5'^{\beta}$-$C5'^{\gamma}$-$C4'^{\delta}$-$C3'^{\epsilon}$-$O3'$-$^{\zeta}P$ and χ for the glycosyl O1′-C1′-N9-C4. Except when defined, numbers in parentheses are standard deviations.

angles between planes

3′-Gua/5′-Gua	5′-Gua/PtN_4	3′-Gua/PtN_4
76.2 (5)	111.0 (5)	85.6 (6)
81.4 (5)	111.1 (6)	95.4 (7)
86.7 (6)	80.2 (6)	57.8 (6)
80.5 (6)	76.8 (6)	59.7 (6)

geometry

	3′ Nucleotide						
ζ	P	ψ	χ	α	β	γ	δ
−65 (3)	84	38	−93 (5)	−85 (8)	−143 (6)	30 (11)	147 (7)
−64 (3)	138	28	−110 (4)	−77 (6)	−141 (4)	84 (16)	108 (15)
−71 (5)	130	49	−117 (4)	−57 (7)	−168 (6)	47 (8)	150 (7)
−69 (4)	136	43	−127 (4)	−49 (5)	−161 (4)	42 (7)	137 (6)
−53	149	34	−115	−68	−168	58	137
−70	148	39	−132	−65	−176	57	137
−53	147	29	−126	−63	−161	56	128

(II) in hand, it is interesting to compare the geometrical results with previously made predictions. The most accurate insights were provided by NMR spectroscopic investigations and molecular mechanics calculations. Proton NMR studies of *cis*-[$Pt(NH_3)_2$\{d(pGpG)\}] correctly forecast that the chelated complex would have the two guanine bases in an *anti-anti* configuration with corresponding head-to-head disposition of the O(6) oxygen atoms (*17*). A more detailed analysis of NMR data on the related complex *cis*-[$Pt(NH_3)_2$\{d(GpG)\}] revealed that the 5′-nucleoside has the N (C3′-endo) conformation but that the 3′-nucleoside is conformationally more flexible, with ~70 percent S (C2′-endo) character (*5*). Indeed, the picture constructed from NMR and related model building studies is in good accord with our crystallographic details (Table 2). This agreement suggests that the solid state structures of the four cisplatin-DNA adducts determined here most likely reflect the salient features of their geometry in solution.

One aspect of the *cis*-[$Pt(NH_3)_2$-\{d(pGpG)\}] structure not yet identified by NMR spectroscopy, but predicted from molecular mechanics calculations on platinated duplexes as well as single-stranded oligonucleotides (*7*), is the occurrence of a hydrogen bond between an ammine ligand and the 5′-phosphate group of the coordinated dinucleotide. This intramolecular hydrogen bond is clearly present in at least three of the four molecules that we studied. Hydrogen bonding between an ammine ligand and the 5′-phosphate group of coordinated mononucleotides had been postulated to explain the greater stability of unsubstituted compared to alkylated amines in a series of related palladium(II) complexes (*23*). Since substitution of protons by alkyl groups diminishes the antitumor activity of platinum coordination compounds (*24*), the intramolecular $NH_3 \cdots$ phosphate hydrogen bond may function in stabilizing the DNA adduct of cisplatin required for its biological mechanism of action.

Molecular mechanics calculations have been carried out on *cis*-diammineplatinum(II) coordinated to the N(7) atoms of adjacent guanosine nucleosides in d(pGpG), d(ApGpGpCpCpT), and two double helical oligonucleotides, an octameric and a decameric duplex (*7*). The predicted conformational angles (Table 2) are in relatively good agreement with the values observed from our x-ray study.

This result suggests that the geometry of the platinated d(GpG) fragment elucidated in this work could be retained in double helical DNA, with attendant local disruptions of Watson-Crick base pairing or duplex kinking as described (*7*). Base pair disruptions have been detected in studies with antibodies to nucleosides (*25*).

Finally, we inquire which predictions that were based on x-ray crystallographic studies of *cis*-DDP soaked into crystals of DNA or RNA are upheld by our results. The postulated (*10*) intramolecular hydrogen bond between a platinum-ammine ligand and the O(6) oxygen atom of the guanine base, possibly with an intervening water molecule as a bridge, is not observed. Because of the tight, hydrogen-bonded cluster formed among the four molecules in our crystal structure, however, it is not possible to rule out such a hydrogen-bonding network on platinated duplex DNA. Apart from the conclusion that platinum binds to N(7) of purines, there is little information in the two studies of *cis*-DDP soaked into $tRNA^{Phe}$ (*11, 12*) that can be assessed by our results.

Relevance to mechanism of action of *cis*-DDP. Our x-ray structural results, taken alone, offer little insight into the biological mechanism of action. Within the context of several recent findings, however, they do provide an important piece of the puzzle from which a picture of how *cis*-DDP might work has begun to emerge. Studies with an in vivo SV40 model system reveal that equal amounts of *cis*- and *trans*-DDP bound to the SV40 chromosome are equally effective at inhibiting replication, but that an order of magnitude more *trans*- than *cis*-DDP is required in the medium to produce equivalent inhibition of DNA synthesis (*26*). Furthermore, these results were shown to be the consequence of much more efficient repair of *trans*- versus *cis*-DDP from DNA in this system. What kind of DNA adduct would render *cis*-DDP difficult to repair yet capable of inhibiting replication? The intrastrand cross-linked *cis*-$[Pt(NH_3)_2\{d(pGpG)\}]$ is a very reasonable candidate. Both molecular mechanics calculations (*7*) and NMR studies (*27, 28*) reveal that such an adduct, the structural details of which have been elucidated here, can be accommodated in duplex DNA with only very localized disruption of the double helix. Such local melting of DNA at the site of platination by *cis*-DDP, while lethal to the cell because it inhibits replication, might elude the repair enzymes, especially if they were repressed in neoplastic cells (*29*). On the other hand, *trans*-DDP cannot, for stereochemical reasons, link two adjacent guanosine nucleosides in DNA. Studies in vitro reveal that it forms a greater variety of intrastrand cross-links between bases with one or more intervening nucleotides (*30*). Possibly these or other adducts made by *trans*-DDP in vivo disrupt the duplex more profoundly than the intrastrand cross-linked *cis*-$[Pt(NH_3)_2\{d(pGpG)\}]$. The *trans*-DDP adducts may therefore be more readily recognized and repaired by the cell, requiring the addition of very high doses to produce an equivalent effect on DNA replication. Further structural and biological studies of single-stranded and duplex DNA modified by *cis*- or *trans*-DDP will permit evaluation of this model.

References and Notes

1. P. J. Loehrer and L. H. Einhorn, *Ann. Intern. Med.* **100**, 704 (1984).
2. M. P. Hacker, E. B. Douple, I. H. Krakoff, Eds., *Platinum Coordination Compounds in Cancer Chemotherapy* (Nijhoff, Boston, 1984).
3. J. J. Roberts and M. F. Pera, Jr., in *Platinum, Gold and Other Metal Chemotherapeutic Agents*, S. J. Lippard, Ed. (ACS Symposium Series No. 209, American Chemical Society, Washington, D.C., 1983), pp. 3–25.
4. A. L. Pinto and S. J. Lippard, *Biochim. Biophys. Acta* **780**, 167 (1984).
5. J. H. J. den Hartog *et al.*, *Nucleic Acids Res.* **10**, 4715 (1982).
6. J. H. J. den Hartog, C. Altona, G. A. van der Marel, J. Reedijk, *Eur. J. Biochem.* **147**, 371 (1985) and references cited therein.
7. J. Kozelka, G. A. Petsko, S. J. Lippard, G. J. Quigley, *J. Am. Chem. Soc.* **107**, 4079 (1985), and unpublished results.
8. B. L. Heyl, K. Shinozuka, S. K. Miller, D. G. Van Derveer, L. G. Marzilli, *Inorg. Chem.* **24**, 661 (1985) and references cited therein.
9. B. Lippert, G. Randaschl, C. J. L. Lock, P. Pilon, *Inorg. Chim. Acta* **93**, 43 (1984).
10. R. M. Wing, P. Pjura, H. R. Drew, R. E. Dickerson, *EMBO J.* **3**, 1201 (1984).
11. J. R. Rubin, M. Sabat, M. Sundaralingam, *Nucleic Acids Res.* **11**, 6571 (1983).
12. J. C. Dewan, *J. Am. Chem. Soc.* **106**, 7239 (1984).
13. G. J. Quigley *et al.*, *Proc. Natl. Acad. Sci. U.S.A.* **77**, 7204 (1980).
14. F. M. Takusagawa, M. Dabrow, S. Neidle, H. M. Berman, *Nature (London)* **296**, 466 (1982).
15. A. H.-J. Wang *et al.*, *Science* **225**, 1115 (1984).
16. M. L. Kopka, C. Yoon, D. Goodsell, P. Pjura, R. E. Dickerson, *Proc. Natl. Acad. Sci. U.S.A.* **82**, 1376 (1985).
17. J. P. Girault, G. Chottard, J-Y. Lallemand, J-C. Chottard, *Biochemistry* **21**, 1352 (1982).
18. P. Main, L. Lessinger, M. M. Woolfson, G. Germain, J-P. Declercq, MULTAN77. *A System of Computer Programs for the Automatic Solution of Crystal Structures from X-ray Diffraction Data* (University of York, York, England;and University of Louvain, Louvain, Belgium, 1977).
19. G. M. Sheldrick, in *Computing in Crystallography*, H. Schenk *et al.*, Eds. (Delft University Press, Delft, The Netherlands, 1978), pp. 34–42. We thank G. M. Sheldrick for kindly providing a version of SHELX-76 extended by D. Rabinovich and K. Reich to handle 400 atoms and 500 variable parameters.
20. J. D. Orbell, L. G. Marzilli, T. J. Kistenmacher, *J. Am. Chem. Soc.* **103**, 5126 (1981).
21. IUPAC-IUB joint commission on biochemical nomenclature report, *Eur. J. Biochem.* **131**, 9 (1983).
22. W. Saenger, *Principles of Nucleic Acid Structure* (Springer-Verlag, New York, 1984), chap. 2 and 4.
23. S-H. Kim and R. B. Martin, *Inorg. Chim. Acta* **91**, 11 (1984).
24. M. J. Cleare and J. D. Hoeschele, *Bioinorg. Chem.* **2**, 187 (1973).
25. W. I. Sundquist, S. J. Lippard, B. D. Stollar, in preparation.
26. R. B. Ciccarelli, M. J. Solomon, A. Varshavsky, S. J. Lippard, *Biochemistry*, in press.
27. J. H. J. den Hartog *et al.*, *J. Am. Chem. Soc.* **106**, 1528 (1984).
28. B. Van Hemelryck *et al.*, *ibid.*, p. 3037.
29. R. S. Day, III, *et al.*, *Nature (London)* **288**, 724 (1980).
30. A. L. Pinto and S. J. Lippard, *Proc. Natl. Acad. Sci. U.S.A.* **82**, 4616 (1985).

31. This work was supported by U.S. Public Health Service grant CA 34992 (to S.J.L.) from the National Cancer Institute. We thank Professor G. Petsko for helpful discussions and assistance in preparing the computer graphics drawings, Dr. J. Kozelka for molecular mechanics calculations and advice, Dr. P. Shing Ho for experimental assistance, Dr. P. Mascharak for providing $[Pt(NH_3)_2(NO_3)_2]$, and the Engelhard Corporation for a loan of K_2PtCl_4 from which this complex was prepared. S.E.S. acknowledges support under National Cancer Institute Training CA-09112.

1 July 1985; accepted 28 September 1985

14

Immunoglobulin Heavy-Chain Enhancer Requires One or More Tissue-Specific Factors

Mark Mercola, Joan Goverman, Carol Mirell, and Kathryn Calame

The control of eukaryotic gene expression involves several classes of regulatory elements that act in *cis* to modulate transcriptional activity (*1*). An interesting class of positive regulatory elements, termed enhancers, elevate transcription from eukaryotic promoters (*2*). Enhancers were initially described as a necessary component of some viral promoters, notably those of SV40, polyoma, and retroviruses (*3*), and were subsequently found to enhance transcription of many heterologous viral and cellular promoters (*2, 4*). Cellular enhancers have been described in the immunoglobulin heavy (H)- and κ-chain genes, where they have been postulated to confer transcriptional competence on rearranged V genes during B-cell development (*5, 6*).

Variable (V) and constant (C) gene segments, which undergo rearrangement to form an active immunoglobulin gene, are widely separated on the same chromosome in unrearranged (germline) DNA (*7*). During B-cell maturation, transcription from a V_H or V_κ promoter occurs after the V gene is joined with D (diversity) or J (joining region) gene segments and is juxtaposed with an enhancer located upstream of C_μ or C_κ. In contrast, unjoined V genes are not transcribed at detectable levels even in terminally differentiated plasma cells (*8*). This observation has prompted the suggestion that joining brings the enhancer sufficiently close to a V-gene promoter to allow its activation.

However, there is evidence that activation of immunoglobulin gene transcription has requirements in addition to V-gene rearrangement. (i) Somatic cell hybrids between immunoglobulin-secreting lymphoid cells and nonlymphoid

Science 227, 266–270 (18 January 1985)

cells usually lose the capacity for antibody expression even though the chromosome containing the rearranged gene is retained (*9*); (ii) a joined κ gene introduced into transgenic mice is expressed only in lymphoid cells (*10*); and (iii) joined immunoglobulin genes are efficiently transcribed following transfection into lymphoid, but not fibroblast, cell lines (*11*). All or part of this tissue specificity may involve the enhancers since the heavy- and κ-chain enhancers function preferentially in plasmacytoma rather than in fibroblast cell lines (*5*). There is now evidence for soluble factors that bind to the enhancer for a ribosomal gene transcribed by polymerase I (*12*) and to the SV40 and murine sarcoma virus (MSV) viral enhancers (*13*). A central goal of our work was to determine whether cellular factors were required for activity of the immunoglobulin heavy-chain enhancer and, if they were, to ask if they were preferentially present in lymphoid cells. Such putative enhancer factors might comprise part of a stable transcription complex (*14*), cause alterations in chromatin structure, or position the gene at a site on the nuclear matrix appropriate for transcription.

By means of a sensitive, reliable procedure for the functional assay of enhancer activity in plasmacytomas and nonlymphoid cells we have shown that VDJ joining brings a V_H gene from a region lacking enhancers into functional proximity with the single enhancer in the region between J_H and C_μ. These data support the model that joined V-gene promoters are activated by nearby enhancers. We performed in vivo competition experiments with a modification of this procedure to demonstrate that molecules present in lymphoid, but not fibroblast, cell lines bind to the heavy-chain enhancer and are required for its activity. There is a requirement for the plasmacytoma-specific molecule (or molecules) by the immunoglobulin enhancer and by the SV40 enhancer in lymphoid cells.

An enhancer is 3′ of J_H, *but not 3′ of a germline V gene.* Several segments from the intervening sequence between J_H and C_μ and from the region 3′ to an unrearranged V_H gene segment had enhancer activity when assayed in the SV40-transformed monkey kidney cell line, COS (*6*). These results suggested that there might be multiple enhancers within the immunoglobulin heavy-chain locus. In order to assay these sequences for enhancing activity in a lymphoid cell environment, a modified protocol was developed for transient transfection and subsequent analysis of murine plasmacytoma P3X63-Ag8 cells with vectors containing the chloramphenicol acetyltransferase (CAT) gene and cloned regions from heavy-chain genes (legend to Fig. 1).

Portions of the joined V1-C_μ heavy-chain gene (Fig. 1A) from the hybridoma HPCM2, which produces antibodies to phosphorylcholine (*15*), were subcloned into the vector pA10CAT-2 (*16*). The location and sequence of the V1 gene promoter have been determined (*17*). The vector pA10CAT-2 directs transcription of the CAT gene with the SV40 early promoter but lacks a functional enhancer. Since CAT activity is normally absent in mammalian cells, the amount of CAT enzyme produced upon transfection with these constructs is a sensitive measure of the enhancing potential of the cloned sequence. The region to be assayed spanned a distance 2 kilobases (kb) upstream of the V1 promoter and extended 10 kb downstream

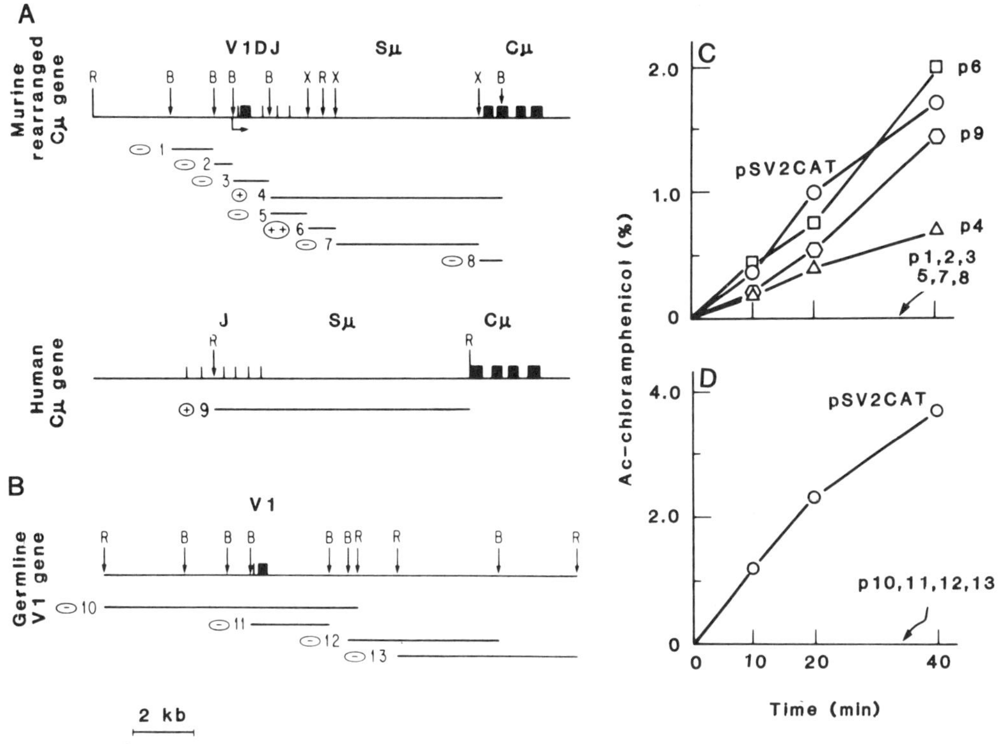

Fig. 1. Enhancer activity in the rearranged and germline heavy-chain genes. Regions of genomic clones subcloned into pA10CAT-2 were assayed for enhancing activity in plasmacytoma P3X63-Ag8 (γ1,κ). Exponentially growing plasmacytoma cells (6 × 10^6) were transfected with equimolar amounts of the various supercoiled plasmid constructs (corresponding to 80 μg of a 6.6-kb plasmid) by the calcium phosphate coprecipitation technique essentially as described (*29*). Briefly, a precipitate was formed by the dropwise addition of 1 ml of a solution of DNA and $CaCl_2$ (0.25*M*) to 1 ml of 2× HBS (2× HBS is 50 m*M* Hepes, 280 m*M* NaCl, 1.5 m*M* Na_2HPO_4, *p*H 7.05 ± 0.05). Cells, washed in serum-free medium, were gently resuspended in the DNA precipitate mixture and incubated for 20 minutes at 37°C; 8 ml of Dulbecco's modified Eagle medium supplemented with fetal calf serum (10 percent) was then added. Cells were fed 1 day later. Approximately 44 hours after transfection, the cells were harvested and a soluble extract was prepared and assayed for CAT activity as described (*16*) but with the following modifications. Extracts were heated for 7 minutes at 60°C to inactivate an endogenous deacetylase activity. After brief centrifugation at 4°C, 70 μl of the extract supernatants were assayed for CAT activity. Activity was corrected for the number of cells harvested by normalizing to 10^7 cells. These experiments have been repeated a minimum of three times with DNA from different plasmid preparations. (A) Regions of the murine and human μ genes assayed for enhancer activity. Lines under the maps of the genomic clones indicate the regions (1 to 9) subcloned in pA10CAT-2 in both orientations 3′ to the CAT gene. The symbols −, +, and ++ indicate levels of enhancer activity as measured in (B). (B) Results of the CAT enzyme assays for the constructs (p1 to p9) derived from regions shown in (A). Data are given for constructs containing inserts in the sense orientation relative to the CAT gene. (C) Regions of the germline V1 gene segment assayed for enhancer activity. (D) Results of the CAT enzyme assays for the constructs shown in (C). The ordinates in (B) and (D) are the percent of [^{14}C]chloramphenicol converted to the acetylated form.

to the second C_μ exon. Only those fragments containing the 1.0-kb Xba I fragment are competent as enhancers (Fig. 1B), confirming previous reports of an enhancer in this region (*5, 6*). The presence of additional sequences active in COS cells may be due to replication of the vector or interaction of the constitutively expressed SV40 T antigen with the promoter region. Thus, only one region of enhancer activity functional in lymphoid cells occurs near a rearranged μ gene. In our assay, the Xba I 1.0-kb enhancer region exhibits between 60 and 200 percent of the activity of the SV40 enhancer, which contains two tandem copies of the 72–base pair (bp) repeat. Like its viral counterparts, the heavy-chain element can enhance transcription in either orientation and from a position either 5′ or 3′ to the coding region (*18*).

Similar results were also obtained with the analogous region from the human μ gene. The intervening sequence from a human μ genomic clone was subcloned in both orientations 3′ to the CAT gene in pA10CAT-2, which produced a vector analogous to construct p4 derived from murine sequences (Fig. 1A). The orientation of fragments $\geq$ approximately 3 kb in pA10CAT-2 affects the level of enhancement, probably because enhancer-containing regions are located at significantly different distances from the promoter (*19*). Accordingly, we present data from constructs with inserts in the same position and in the sense orientation relative to the CAT gene transcription unit. Following transfection of these constructs into plasmacytoma P3X63-Ag8 cells, CAT assays indicated comparable activity in the μ-intron regions of both species (compare constructs p4 and p9 in Fig. 1B). Constructs containing inserts in the antisense orientation gave 20 percent (human) or 70 percent (mouse) less activity (*18*). These results confirm earlier reports identifying an enhancer in the human intron between J_H and the μ switch region (S_μ) (*20*).

The hypothesis that transcriptional competence is conferred upon the V_H gene promoter by its association with the J_H-C_μ enhancer following rearrangement predicts that no enhancer occurs within a functional distance 3′ of the promoter in its germline configuration. To test this prediction, a region of 17 kb surrounding the germline V1 gene segment was examined for the presence of enhancers functional in P3X63-Ag8 cells (Fig. 1C). As before, large fragments were cloned and tested in both orientations to control for the possibility that an enhancer would be overlooked if situated too far from the SV40 early promoter in the plasmid vector. No enhancing activity was detected in any of the subclones examined (Fig. 1D). These data indicate that, at least for the V1 gene segment, the promoter is moved from a region devoid of enhancer influence to a position adjacent to a strong enhancer as a consequence of VDJ joining.

Heavy-chain enhancer is not utilized in all cells. V-gene transcription could be a direct consequence of rearrangement with an enhancer or could require the contribution of the B-cell environment. Efficient transcription of rearranged light- and heavy-chain genes is obtained when they are transfected into plasmacytomas (*5, 11, 21*), but undetectable or very low transcription is seen upon transfection of rearranged λ, κ, and γ2b genes into fibroblasts (*11, 22, 23*). We obtained similar results when a rearranged μ heavy-chain gene from HPCM2 was transfected into mouse L cells. Although transformants contained one to ten copies of the rearranged gene, they contained less than 0.5 percent of the amount of μ transcripts present in a plasmacytoma control as shown by S1 nuclease-digestion analyses of RNA (*18*). These results suggest that, in addition to the requirement for proximity to an enhancer, transcription from V_H promoters requires factors not present in the fibroblasts.

A similar conclusion can be drawn from experiments assaying heavy-chain enhancer function in nonlymphoid cells. The murine heavy-chain enhancer functions preferentially in mouse lymphoid relative to murine Ltk^- (*5*), human HeLa (*5*), and simian CV1 (*6*) cell lines. Table 1 extends these results to include additional fibroblast, murine, and human lymphoid lines. The immunoglobulin enhancers were more active in murine plasmacytomas than in the fibroblast lines, although low-level enhancer activity was observed in the murine fibroblast lines. Neither the murine nor the human heavy-chain enhancer was active in the human B-cell lines tested, although the SV40 enhancer functioned at levels similar to those observed in the murine lymphoid lines. The reason for this result is not clear although it is known that the human lines produce significantly less antibody than the murine plasmacytomas. The ability of some cell lines to discriminate between the viral and heavy-chain elements is consistent with the hypothesis that discrete cellular factors might be required for efficient enhancer function.

Heavy-chain enhancer activity depends on factor binding. An in vivo competition experiment (Fig. 2) was designed to test the hypothesis that cellular molecules interact directly with enhancer elements. A constant amount of a test plasmid containing an enhancer and the assayable CAT gene was cotransfected into cells with increasing amounts of competitor enhancer sequence-containing vector that did not have the CAT gene. Plasmid vector lacking both the CAT gene and enhancer sequences was also added in order to introduce an equimolar amount of plasmid DNA into the cells for each transfection. If factors required for enhancer function were present at limiting concentrations in the cell, the competing enhancers would cause decreased CAT activity because the test enhancer would not be fully saturated.

Table 1. Heavy chain enhancer activity in various cell lines.

Cell line*	Enhancer†		
	Murine	Human	SV40
	Lymphoid		
Murine			
P3X63-Ag8	+	+	+
S107	+	N.D.	+
Human			
AF10	N.D.	−	+
8226	−	N.D.	+
HH1040	−	N.D.	+
IE 15.1	N.D.	−	+
JL 10.1	N.D.	−	+
	Fibroblast		
Murine			
Ltk^-	+/−	+/−	+
ML	+/−	+/−	+
Human			
HeLa	−	−	+
143	−	−	+
Simian			
CV1	−	−	+

*P3X63-Ag8, S107, AF10, and 8226 are myeloma lines; HH1040, IE 15.1, and JL 10.1 are lymphoblastoid lines. †For enhancing activity, + indicates activity equivalent to that of the SV40 enhancer in the particular cell type, +/− indicates levels 10 to 40 percent that of the SV40 enhancer, and − indicates no activity detected; N.D., not determined.

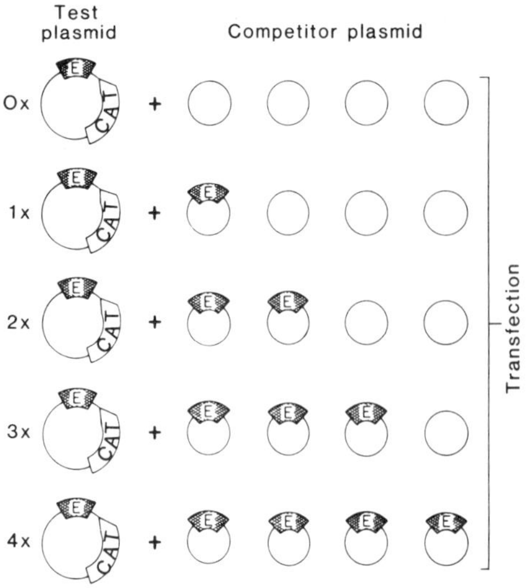

Fig. 2. Schematic outline of the in vivo titration experiments. A constant amount of test plasmid containing the CAT gene was co-transfected with variable amounts of competing enhancer-containing plasmids into P3X63-Ag8 cells. Plasmid DNA devoid of eukaryotic sequences was added so that equimolar amounts of plasmid molecules were introduced into the cells for each transfection. Transfected cells were processed for the CAT assay as described in the legend to Fig. 1.

Nonsaturating levels of CAT activity were attained when test plasmids were transfected under the conditions of the competition experiments, but in the absence of competing enhancers (*18*). Thus, the concentration of putative cellular factors was not initially limiting.

First we examined whether additional copies of the murine heavy-chain enhancer could decrease its own activity. The test construct p6, containing the Xba I 1.0-kb fragment of the HPCM2 heavy-chain gene (Fig. 1A), was transfected into P3X63-Ag8 cells with competing DNA. The enhancer-containing competitor plasmid, pX1.0, contained two copies of the Xba I 1.0-kb fragment. Increasing the amount of pX1.0 in the transfection mixture resulted in a decrease in CAT activity (Fig. 3A). This decrease was observed at all concentrations of pX1.0 tested; at a 12-fold excess of competing enhancer, the level of enhancer-dependent CAT activity declined to 20 to 25 percent of initial levels. To rule out artifacts, a plasmid containing the 1.2-kb Bam HI fragment from the murine heavy-chain gene, a region shown to lack enhancing activity (construct p3 from region 3 of Fig. 1A), was substituted for pX1.0 in the assay. This plasmid did not decrease CAT activity demonstrating that the transfection reaction was not affected by the presence of additional plasmids (*18*). Our interpretation of these results is that the binding of one or more cellular factors present in limiting concentration within the cell is essential for enhancer function.

The corresponding enhancer regions of the human and murine heavy-chain genes share extensive nucleotide sequence homology (*20*). The activity of the human immunoglobulin enhancer is comparable to that of the analogous mouse sequence in a murine cell line (Fig. 1). Thus, the two enhancers may utilize the same factor in plasmacytoma cells. To test this possibility, the 8.3-kb intron region of the human gene inserted into pA10CAT-2 3′ to the CAT gene sequences (construct p9 from region 9 of Fig. 1A) was used as test enhancer. A molar amount of plasmid equivalent to that used with the murine enhancer was transfected under identical conditions. In P3X63-Ag8 cells, a 12-fold excess of competing murine enhancer resulted in a decrease in CAT activity to 9 to 21 percent of that observed in the absence of competing enhancers (Fig. 3B). This result indicates that at least one factor is

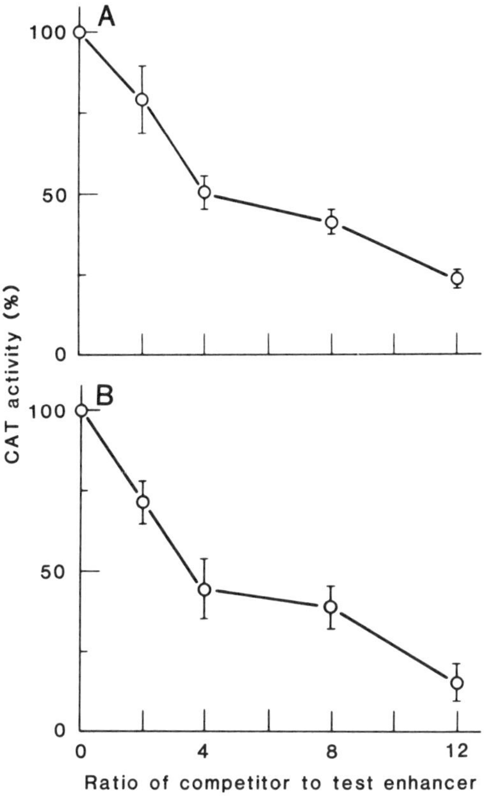

Fig. 3. Competition of the heavy-chain enhancers for a plasmacytoma factor or factors required for enhancer function. Transfected cells were assayed for CAT activity as described in the legend to Fig. 1. The slopes of the enzyme assay curves were used to determine the CAT activity for each transfection. The percentage of activity relative to uncompeted transfection was plotted against the molar ratio of competitor enhancer sequence to test enhancer sequence in the transfection mixture. Error bars represent the variation observed in three separate experiments with P3X63-Ag8 cells and DNA isolated from different plasmid preparations. CAT activity dependent on the murine heavy-chain enhancer (A) and the human enhancer (B) was determined with a pBR322-derived plasmid containing two copies of the Xba I 1.0-kb fragment from the murine μ gene as competitor (pX1.0). Ratios of competitor to test sequence greater than 12 were not achieved because transfection of larger quantities of DNA resulted in cell death. In (A) and (B), 30 μg of construct p6 and 63 μg of construct p9 were used as test plasmids, respectively.

required in both murine and human cells, suggesting that the mechanism for enhancer recognition and function is conserved across the two species.

Heavy-chain enhancer factor in lymphoid but not fibroblast cells. Activity of the immunoglobulin enhancer is restricted to certain lymphoid cell lines (Table 1). The viral SV40 enhancer, however, has a wider host range, permitting it to function in most fibroblast and lymphoid cell lines. Some quantitative host-cell tropism has been observed for the SV40 enhancer as compared to other viral enhancers and may be the result of the interaction of the enhancer with cellular factors (*13*). To determine whether the SV40 and heavy-chain enhancers are recognized by common or different factors in the lymphoid cell, we used the murine heavy-chain enhancer as the competitor of SV40 enhancer activity. A molar amount of pSV2CAT equivalent to that used for the test plasmids containing the heavy-chain enhancers was cotransfected with competing DNA into P3X63-Ag8 cells. Increasing amounts of the heavy-chain enhancer resulted in a decrease in the CAT signal dependent on the SV40 enhancer (Fig. 4A). The activity at a 12-fold excess of immunoglobulin enhancer was 4 to 18 percent of that seen without competing enhancer. The magnitude of reduction in test SV40 enhancer activity was comparable to that seen for the heavy-chain enhancers. The simplest interpretation is that, in the lymphoid cell, both SV40 and heavy-chain en-

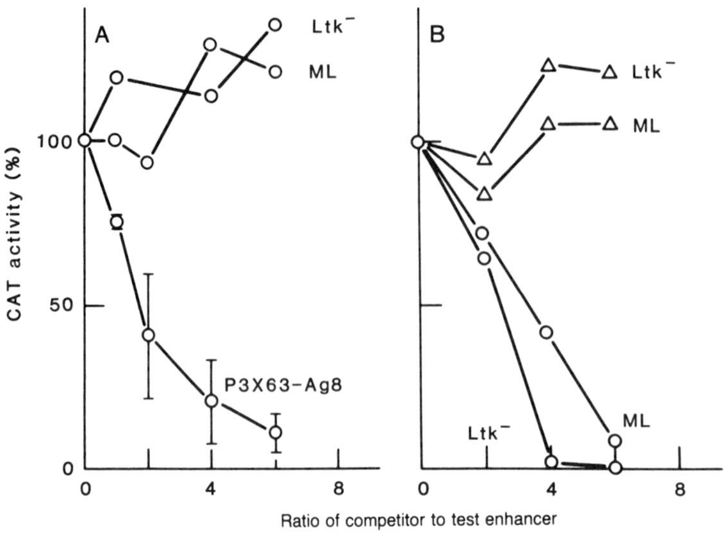

Fig. 4. The heavy-chain enhancer and the SV40 enhancer compete for plasmacytoma, but not fibroblast, factors. The competition experiment is similar to that of Fig. 3 except that a plasmid containing the two 72-bp repeat enhancers of SV40 (pSV2CAT) was used as the test construct. (A) Competition results using the Xba I 1.0-kb fragment from the murine μ gene as competitor enhancer. Since pSV2CAT contains the two 72-bp repeats, the ratio of competing to test enhancer sequences is the molar ratio of enhancer-containing to test plasmid molecules in the transfection mix. (B) Competition results using a plasmid containing the SV40 enhancer region, pSV2NEO (○), or a control plasmid, pA10NEO (△), as competitor. The pSV2NEO and pSV2CAT contain identical SV40 regulatory regions; pA10NEO is similar to pSV2NEO except that the fragment generated by cleavage with Pvu II and Sph I, containing most of the tandem 72-bp enhancer element, has been deleted. Either 23 or 6 μg of pSV2CAT was cotransfected into P3X63-Ag8 cells and the fibroblast lines Ltk⁻ and ML, respectively, with the competitor DNA. The ML line is a nonvirally transformed murine liver cell line which had been in culture for fewer than 200 population doublings.

hancer activities depend on a common factor.

We wished to determine whether the factor present in lymphoid cells is the same as that in fibroblast lines where the SV40, but not the immunoglobulin, enhancer is active. Two murine fibroblast lines, Ltk⁻ and a liver-cell line, ML, were chosen for titration experiments analogous to those in P3X63-Ag8 cells with the SV40 element as the test enhancer and the murine heavy-chain enhancer as competitor. In contrast to the results with the plasmacytoma line, increasing the amount of competing heavy-chain enhancer did not affect SV40 enhancer activity in these cells (Fig. 4A). However, use of another plasmid construction containing the SV40 enhancer as competitor (pSV2NEO) resulted in the complete inhibition of SV40 enhancer-dependent transcription (Fig. 4B). To control for the possibility that sequences present in pSV2NEO other than the SV40 72-bp repeat enhancer might be competing for transcription factors required by pSV2CAT, the vector pA10NEO, which contains the same sequences as pSV2NEO except that all but 21 bp of the tandem 72-bp repeats have been deleted, was used as competitor (Fig. 4B). Increasing amounts of pA10NEO did not affect SV40 enhancer-dependent CAT activity. Furthermore, total inhibition of pSV2CAT activity occurred with pSV2NEO as competitor and pA10NEO as the enhancer-lacking filler plasmid (*18*). These experiments suggest the presence of one or more fibroblast factors required for SV40 enhancer function. However, only in the plasmacytoma cell can the heavy chain enhancer compete for *trans*-acting factors required for SV40 enhancer function. We interpret these results to indicate that different enhancer-binding factors are present in the two types of cell lines tested.

Discussion. Transcription from immunoglobulin V_H promoters is thought to depend on an enhancer sequence (*5, 6*). Our results show that a region of 11 kb surrounding the rearranged μ heavy-chain gene from plasmacytoma HPCM2 contains only one enhancer active in plasmacytoma cells (Fig. 1). Joined V_H promoter regions are brought 2 to 3 kb upstream from the enhancer depending on the particular J_H gene segment used. The region 12 kb downstream from the germline V1 promoter did not contain detectable enhancing activity in lymphoid cells (Fig. 1D). Thus, the V1 promoter is only proximal to an active enhancer after VDJ joining, supporting the hypothesis that enhancers mediate V_H gene promoter activation.

The exact contribution of the enhancer to immunoglobulin expression during B-cell maturation is unclear. Recently, B-cell lines have been reported in which the known heavy-chain enhancer was deleted from the active μ gene yet normal levels of μ chains were made (*24*). This is inconsistent with our results and with the requirement of a transfected γ2b gene for an enhancer (*5*). It is difficult to reconcile these observations, although several explanations are possible. Sequence and functional analyses of the deleted μ genes may yield information regarding the nature of the deleted region, promoter-enhancer interaction, and the role of enhancers during development.

V-gene joining and proximity to an enhancer are not sufficient to activate V1 gene transcription. We and others (*23*) have found that transcription of a rearranged heavy-chain gene is not detected following transfection into Ltk^- cells, indicating that a B-cell environment is required for significant expression. The cellular preference of the heavy-chain enhancers (Table 1) suggests that at least one component of the B-cell environment is a factor required for enhancer function.

In vivo competition experiments revealed the presence of a limited concentration of molecules that bind to the heavy-chain enhancer and are required for its activity (Fig. 3). In the plasmacytoma cell, transcription dependent on the SV40 enhancer was also prevented with the heavy-chain enhancer as competitor (Fig. 4A) indicating that at least one common factor is utilized by the heavy-chain and SV40 enhancers. Furthermore, the similar extent of the decrease in the SV40 and heavy-chain enhancer activity (Figs. 3A and 4A) indicates that relatively few, if any, factors exist in the plasmacytoma cell which can substitute for the heavy-chain factor in mediating SV40 enhancer activity. Similar experiments with two murine fibroblast lines demonstrated the existence of a fibroblast enhancer factor functionally distinct from the lymphoid factor (Fig. 4A).

Eukaryotic transcription probably involves multiple factors which interact in complex ways (*14*). Our experiments demonstrate that tissue-specific enhancer recognition factors can be a critical part of this process. Since our assay is a functional one, we identify activities but cannot assign these activities to discrete molecules. We cannot tell if multiple factors are required for enhancer activity or if what we call a factor is a heterogeneous population of factors with overlapping affinities for enhancers. Furthermore, enhancing activity may reflect the presence of a soluble molecule or a component of the nuclear matrix. However, these experiments identify two activities: (i) an activity required for enhancer-dependent transcription present in limiting concentrations, which may or may not reside on the same molecule as (ii) a tissue-specific activity that binds to the enhancer.

On the basis of the differential ability of the heavy-chain enhancer to compete with SV40 enhancer activity in the lymphoid and fibroblast lines, we propose that the plasmacytoma and fibroblast cell lines contain different molecules which bind to enhancers and mediate their function. The SV40 enhancer could have evolved the capacity to bind and utilize the factors from both cell types. Thus, the SV40 element may be the prototype of a class of enhancers which can use the factor or factors present in a wide range of cell types. The heavy-chain enhancer, in contrast, would represent a class of enhancers specific for a particular cell lineage.

Modulation of immunoglobulin enhancer-specific factors could be involved in regulating the levels of immunoglobulin expression during B-cell development. In the murine pre–B-cell line 70Z/3, the κ light-chain gene is rearranged but not transcribed at detectable levels (*25*). Stimulation with lipopolysaccharide induces κ transcription and a deoxyribonuclease I hypersensitive site at the position of the κ enhancer, possibly reflecting the interaction of the enhancer with an induced factor (*25*). It is intriguing to postulate that levels of immunoglobulin factors that bind to the enhancer increase during B-cell ontogeny, contributing to the increased levels of immunoglobulin transcription observed in the terminally differentiated plasma cell. An increase in the cellular concentration of the immunoglobulin enhancer-specific factor or factors could be responsive to the effects of the T-cell growth and differentiation factors thought to be involved in B-cell maturation (*26*).

Recently, viral enhancers having much more stringent tissue specificities than the SV40 enhancer have been described and other cellular enhancers demonstrating tissue specificity have been reported (*27*). There is also increasing evidence for numerous enhancer elements in the genomes of higher organisms (*2*), and cellular factors have been shown to be required for activity of an RNA polymerase I enhancer (*12*) and the SV40 and MSV enhancers (*13*). Based on these data and our competition results, we propose that different types of cells may contain one or more specific enhancer factors which allow particular genes or sets of genes to be expressed in a tissue-specific or developmentally regulated manner (or both). Regulation of enhancer factors is suggested by the hormone dependence of the murine mammary tumor virus (MMTV) enhancer element (*28*).

References and Notes

1. T. Shenk, *Current Top. Microbiol. Immunol.* **931**, 25 (1981).
2. Y. Gluzman, T. Shenk, Eds., *Enhancers and Eukaryotic Gene Expression* (Cold Spring Harbor Laboratory, Cold Spring Harbor, N.Y., 1983).

3. P. Gruss, R. Dhar, G. Khoury, *Proc. Natl. Acad. Sci. U.S.A.* **78**, 943 (1981); C. Benoist and P. Chambon, *Nature (London)* **290**, 304 (1981); B. Levinson, G. Khoury, G. Vande Woude, P. Gruss, *ibid.* **295**, 568 (1982); E. H. Chang, R. W. Ellis, E. M. Scolnick, D. R. Lowy, *Science* **210**, 1249 (1980); D. G. Blair, M. Oskarsson, T. G. Wood, W. L. McClements, P. J. Fischinger, G. G. Vande Woude, *ibid.* **212**, 941 (1981); G. Payne, J. Bishop, H. Varmus, *Nature (London)* **295**, 209 (1982); J. de Villiers and W. Schaffner, *Nucleic Acids Res.* **9**, 6251 (1981).
4. J. Banerji, S. Rusconi, W. Schaffner, *Cell* **27**, 299 (1981).
5. J. Banerji, L. Olsen, W. Schaffner, *ibid.* **33**, 729 (1983); S. Gillies, S. Morrison, V. Oi, S. Tonegawa, *ibid.*, p. 717.
6. M. Mercola, X.-F. Wang, J. Olsen, K. Calame, *Science* **221**, 663 (1983).
7. T. Honjo, *Annu. Rev. Immunol.* **1**, 499 (1983).
8. E. L. Mather and R. P. Perry, *Nucleic Acids Res.* **9**, 6855 (1981).
9. S. Junker, *Exp. Cell Res.* **139**, 51 (1982).
10. U. Storb, R. O'Brien, M. McCullen, K. Gollahan, R. Brinster, *Nature (London)* **310**, 238 (1984); R. Brinster, K. Ritchie, R. Hammer, R. L. O'Brien, B. Arp, U. Storb, *ibid.* **306**, 332 (1983).
11. J. Stafford and C. Queen, *ibid.*, p. 77.
12. P. Labhart and R. Reeder, *Cell* **37**, 285 (1984).
13. H. R. Scholer and P. Gruss, *ibid.* **36**, 403 (1984).
14. D. D. Brown, *ibid.* **37**, 359 (1984); T. Matsui, J. Segall, P. A. Weil, R. G. Roeder, *J. Biol. Chem.* **255**, 11992 (1980).
15. P. J. Gearhart, N. D. Johnson, R. Douglas, L. Hood, *Nature (London)* **291**, 29 (1981).
16. L. A. Laimins, G. Khoury, C. Gorman, B. Howard, P. Gruss, *Proc. Natl. Acad. Sci. U.S.A.* **79**, 6453 (1982); C. Gorman, L. Moffat, B. Howard, *Mol. Cell. Biol.* **2**, 1044 (1982).
17. C. Clarke, J. Berenson, J. Goverman, P. D. Boyer, S. Crews, G. Siu, K. Calame, *Nucleic Acids Res.* **10**, 7731 (1982).
18. Data not shown.
19. B. Wasylk, C. Wasylk, P. Augereau, P. Chambon, *Cell* **32**, 503 (1983).
20. A. C. Hayday *et al.*, *Nature (London)* **307**, 334 (1984).
21. A. Ochi *et al.*, *Proc. Natl. Acad. Sci. U.S.A.* **80**, 6351 (1983).
22. D. Picard and W. Schaffner, *ibid.*, p. 417.
23. S. D. Gillies and S. Tonegawa, *Nucleic Acids Res.* **11**, 7981 (1983).
24. M. R. Wabl and P. D. Burrows, *Proc. Natl. Acad. Sci. U.S.A.* **81**, 2452 (1984).
25. K. J. Nelson, E. L. Mather, R. P. Perry, *Nucleic Acids Res.* **12**, 1911 (1984); T. G. Parslow and D. K. Granner, *Nature (London)* **299**, 449 (1982).
26. M. Howard and W. Paul, *Annu. Rev. Immunol.* **1**, 307 (1983); F. Melchers and J. Andersson, *Cell* **37**, 715 (1984).
27. S. D. Gillies, V. Folsom, S. Tonegawa, *Nature (London)* **310**, 594 (1984); M. D. Walker, T. Edlund, A. M. Boulet, W. R. Rutter, *ibid.* **306**, 557 (1983); J. Lenz, D. Celander, R. L. Crowther, R. Patarca, D. W. Perkins, A. Sheldon, W. A. Haseltine, *ibid.* **308**, 467 (1984).
28. K. S. Zaret and K. R. Yamamoto, *Cell* **38**, 29 (1984).
29. V. T. Oi, S. L. Morrison, L. A. Herzenberg, P. Berg, *Proc. Natl. Acad. Sci. U.S.A.* **80**, 825 (1983).
30. We are grateful to A. Saxon, N. Brown, and D. Haggerty for supplying cell lines; A. Berk and J. Gralla for critically reading the manuscript; E. Kakkis, C. Peterson, X.-F. Wang, J. Prehn, and S. Eaton for helpful discussions; and V. Windsor for preparation of the manuscript. Supported by NIGMS grant GM29361, an NIGMS cellular and molecular biology predoctoral traineeship (M.M.), NIGMS tumor immunology postdoctoral traineeships (J.G. and C.M.), and a Leukemia Society scholarship (K.C.).

14 August 1984; accepted 6 November 1984

Part IV

Hormones and Metabolism

15

Atrial Natriuretic Factor: A Hormone Produced by the Heart

Adolfo J. de Bold

The homeostatic control of body sodium and water and of blood pressure involves a complex interaction of hormonal and neural mechanisms. The major determinants include central and autonomic nervous function, cardiac output, blood vessel tonicity, renal function, the renin-angiotensin-aldosterone system, catecholamines, and antidiuretic hormone.

It was recently found that the heart atrial muscle produces a polypeptide hormone—atrial natriuretic factor (ANF)—that interacts with several of the above determinants allowing for both long- and short-term regulation of salt and water balance and of blood pressure. The properties of ANF include potent diuretic (natriuretic) and hypotensive actions as well as an inhibitory effect on renin and aldosterone secretion (*1–3*).

A substance with some of the properties of ANF has long been sought on theoretical grounds deriving from physiological and pathophysiological observations. The discovery of ANF, however, derived from functional morphological studies of the cardiac muscle cell. These studies are important to obtaining a full understanding of ANF.

Functional-Morphological Overview

Most of the muscle cells (cardiocytes) forming the atrial and ventricular muscle of the mammalian heart are differentiated for mechanical work. Morphologically this is evident from their large content of contractile elements. The rest of the cardiocytes do not share this morphological differentiation to the same degree. In

Science 230, 767–770 (15 November 1985)

1893, Kent (*4*) described a muscular connection between atria and ventricles, which were known to be otherwise separated from each other by connective tissue. Kent's observations led to studies by other investigators, who established that the microscopic appearance of cardiocytes forming the conducting system of the heart differs markedly from that of the cardiocytes making up the bulk of the myocardium. The morphology of the sinoatrial node was aptly described by Keith and Flack (*5*) in 1907 as "a remarkable remnant of primitive fibers." In 1910 Lewis *et al.* (*6*) showed that these fibers were the source of electrical events in the heart and thus defined the cardiac pacemaker.

The cardiocyte population of the mammalian heart is now regarded as a group of cells showing an essentially continuous spectrum of differentiation (*7*) from the primitive-looking cardiocyte found in the sinoatrial node to the cardiocyte fully differentiated as a special type of striated muscle cell. In more general terms, the cardiocyte population of the mammalian heart displays to different degrees the expression of three basic properties: excitability, conductivity, and contractility. These properties are used to explain the physiological and physiopathological phenomena associated with the heart as an organ that plays the central role in blood circulation. However, in 1956, B. Kisch (*8*) pointed out a morphological difference between atrial and ventricular cardiocytes in the heart of the guinea pig that could not be fitted into the functional framework then established. He observed, as others did subsequently, that atrial cardiocytes in mammals, unlike ventricular cardiocytes, have morphological features of secretory cells (*9, 10*). The most obvious expression of this differentiation is the presence of membrane-bound storage granules—the specific atrial granules—which, after conventional processing for electron microscopy, display an electron-dense core and measure 250 to 500 nanometers (Fig. 1). These granules are more concentrated in the central sarcoplasmic core of atrial cardiocytes. Often, the granules are found associated with a prominent Golgi complex from which they arise. Numerous profiles of rough endoplasmic reticulum are also normally found. These features of the central sarcoplasmic core of the mammalian atrial cardiocyte are more common in the general population of cardiocytes found in the auricles. Cardiocytes in the sinoatrial node and in specialized path of conduction are far less developed in this sense so that the expression of a secretory function is most often associated with cells that are also differentiated for contraction. Yet cells differentiated for contraction in ventricular muscle do not show elements of secretory cells. However, ventricular as well as atrial cardiocytes in nonmammalian vertebrates, including reptiles, amphibians, birds, teleost fishes, and elasmobranchs, do appear to have a dual contractile-secretory function even though, morphologically, this duality is most obvious in mammals. Even in mammals there are variations in the morphological expression of the secretory function. Small rodents, for example, have far more atrial granules—up to 4 percent of the cardiocyte volume—than large mammals such as cattle, which have very few granules per cardiocyte. Human cardiocytes have an intermediate content. In rats, granule content varies with age. It doubles between ages 6 and

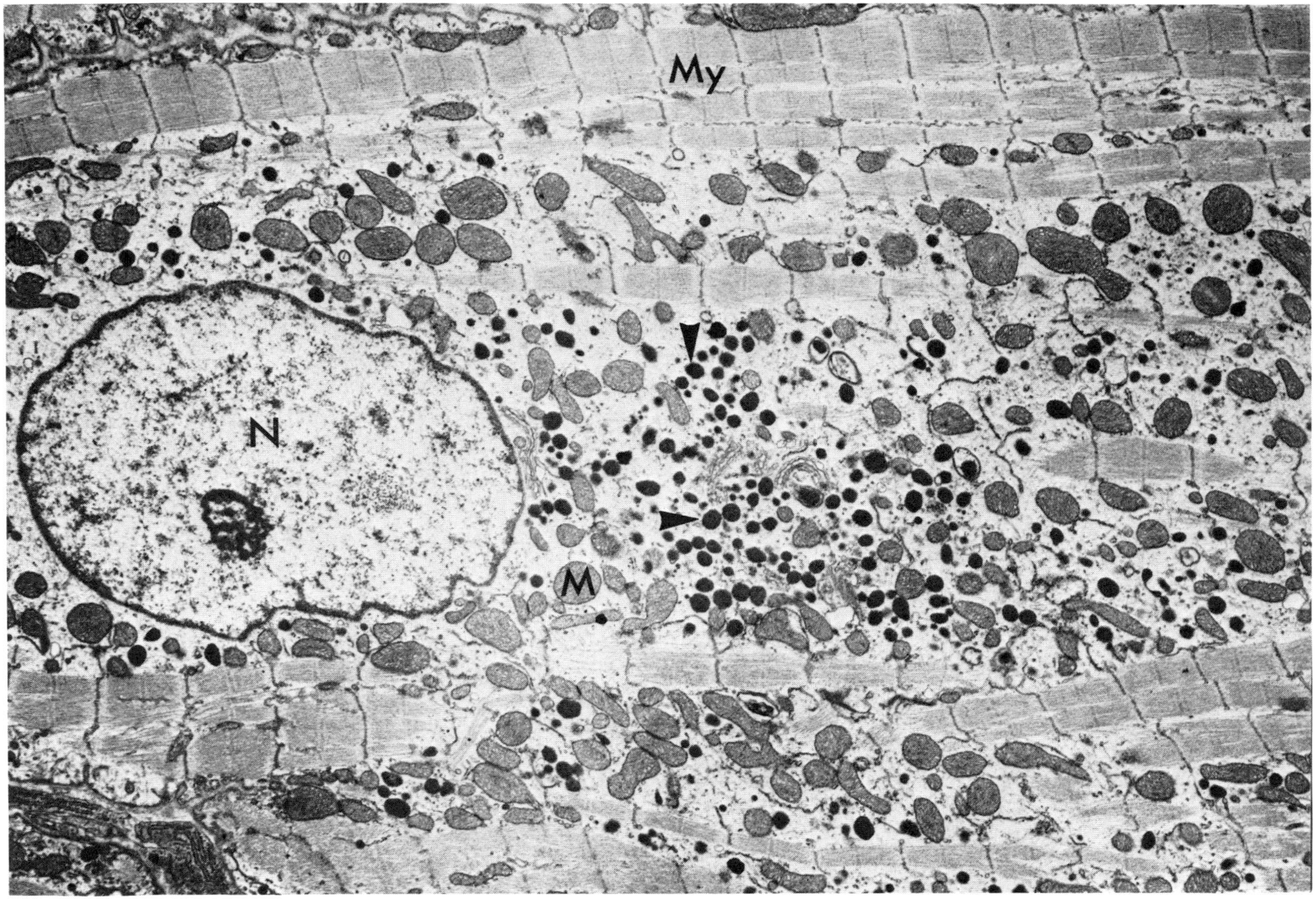

Fig. 1. Electron microscopic view of a rat atrial cardiocyte. *N*, nucleus; *M*, mitochondria; *My*, myofibrils. The central sarcoplasmic core displays morphological features associated with secretory cells that include a large number of storage granules (arrowheads), referred to as specific atrial granules (×1500).

10 weeks and, although the number plateaus at adulthood, morphometric procedures show significant variations in animals of the same age (*11*).

The dual secretory-contractile function of atrial cardiocytes in the rat is suggested from autoradiographic studies after these cells are exposed to [^{3}H]leucine (*12*). There is a striking difference in the stable protein-labeling pattern of mitochondria and myofibrils as compared with the rapid turnover of activity of the Golgi complex and specific atrial granules. The labeling of atrial granules shows kinetics similar to that found in cells producing polypeptide hormones. No such labeling pattern is found associated with any cell compartment of ventricular cardiocytes. Histochemically, atrial granules display properties in common with polypeptide hormone–containing granules. The selective staining of these granules with lead-hematoxylin-tartrazine and the relatively large sampling size afforded by the light microscope have been used to develop an unbiased morphometric procedure to measure the degree of granulation of atrial cardiocytes (*11*). These measurements were carried out because many experimental procedures were reported to affect the number of granules (*10, 13*). This approach demonstrated unequivocally that some experimental maneuvers leading to changes in water and electrolyte balance significantly altered the specific atrial granule population in the rat (*14*). These and other histochemical observations led to the hypothesis that the specific atrial granules stored a basic polypeptide containing tryptophan and sulfur amino acids and that this polypeptide was involved in the regulation of water and electrolyte balance. The experimental testing of this hypothesis led to the finding that atrial extracts contained a factor of proteinaceous nature that elicited natriuresis and a hypotensive effect when injected into rats (*1, 2*). This factor was named atrial natriuretic factor.

Specific Atrial Granules and ANF

Various methods have shown that ANF is stored within specific atrial granules. The net specific natriuretic activity of atrial extracts is highest for animals with the highest number of granules (*15*). In addition, both atrial and ventricular homogenates of nonmammalian vertebrate hearts have natriuretic activity. Tissue fractionation studies show that the highest natriuretic activity is associated with fractions containing purified granules (*16*) and immunocytochemical studies with antisera directed against ANF peptides clearly localize ANF within the atrial granules (*17–19*).

Immunocytochemical studies and radioimmunoassay techniques have suggested extracardiac localizations of ANF, notably in the central nervous system and kidney (*19*), but quantitatively the main store of ANF in mammals is the atria and the amount of ANF precursor messenger RNA (mRNA) is highest in this tissue (*20*). In other tissues, ANF mRNA is orders of magnitude lower, although this does not necessarily reflect its importance relative to function. In nonmammalian species ANF synthesis may also occur in ventricular muscle, and the extracardiac site of production may be quantitatively different from that found in mammals.

ANF peptides isolated from atrial muscle show molecular sizes of 2500 to

13,000 daltons. Purification and sequencing of peptides with 21 to 31 residues have been carried out in various laboratories (*21–25*). Evidence exists that some of these peptides are artifactual cleavage products generated during isolation (*21*). However, biological testing of these peptides has helped to establish that a disulfide-looped sequence of 17 amino acids plus various COOH- and NH_2-terminal extensions are necessary for biological activity.

Cloning and sequence analysis of complementary DNA encoding the ANF precursor have provided a more complete understanding of ANF peptides (*20, 21, 24, 26, 27*) (Fig. 2). From these studies it has been established that ANF is synthesized in a prepro form containing 152 amino acids in the rat and 151 amino acids in the human. A high degree of homology exists between the rat and human sequences. These precursors contain 24– and a 25–amino acid putative signal sequences, respectively. It is the COOH-terminal portion of this molecule that contains the biologically active sequences.

The main form of ANF in rat atrial homogenates and in isolated granules is a 126–amino acid peptide called cardionatrin IV (*17, 21*) or γ rat atrial natriuretic peptide (*27*). It is derived from prepro-ANF by removal of the signal peptide and residues 151 and 152, which are arginines. The absence of these residues in the peptides so far isolated indicates early removal after biosynthesis. The circulating form of ANF appears to be cardionatrin I (*22, 28*) or α rat atrial natriuretic peptide (*27*), a 28–amino acid peptide that comprises residues 123 to 150 of prepro-ANF. This peptide is also the major form detected in perfusates of isolated atria (*29*), so that it is likely that the 126-residue peptide is processed to the 28–amino acid peptide before release from cardiocytes. A similar process may take place in human atria from which an α human natriuretic peptide has been isolated (*24*). The latter shares 27 of 28 amino acids with rat cardionatrin I.

Plasma levels of ANF measured by radioimmunoassay vary widely from approximately 25 to 100 picograms per milliliter of plasma in humans and 100 to 1000 pg/ml in rats. Differences in species, measuring techniques, and sampling protocols account for some of this variability. It is not clear what stimulates ANF secretion from atria, but in vitro investigations indicate that adrenaline, arginine vasopressin, acetylcholine, and atrial distension all lead to increase in ANF release (*30*). The peptides have a half-life on the order of minutes in blood. Binding to specific receptors in known target organs, including kidney, blood vessels, and adrenal cortex has been demonstrated. Several additional binding sites have been described but are not as yet clearly defined as specific receptor sites (*31*).

The main systemic effects ascribed to ANF result from observations in vitro and in vivo. In the whole animal, these effects include a potent and rapid diuretic and natriuretic action of short duration accompanied by hypotension and, often, bradycardia as well as depressant effects on aldosterone and renin secretion. The cellular or molecular basis of these systemic effects are not clearly understood. Whether renal hemodynamic or tubular mechanisms account for the natriuretic action of ANF is unknown, and a possible role for both these mechanisms has been suggested (*32*). The effect of ANF on aldosterone and renin, as well as on smooth muscle, pertains to phenomeno-

Met Gly Ser Phe Ser Ile Thr Lys Gly Phe Phe Leu Phe Leu Ala Phe
1

Trp Leu Pro Gly His Ile Gly Ala **Asn Pro Val Tyr Ser Ala Val Ser**
25

Asn Thr Asp Leu Met Asp Phe Lys Asn Leu Leu Asp His Leu Glu Glu

Lys Met Pro Val Glu Asp Glu Val Met Pro Pro Gln Ala Leu Ser Glu

Gln Thr Asp Glu Ala Gly Ala Ala Leu Ser Ser Leu Ser Glu Val Pro

Pro Trp Thr Gly Glu Val Asn Pro Ser Gln Arg Asp Gly Gly Ala Leu

Gly Arg Gly Pro Trp Asp Pro Ser Asp Arg Ser Ala Leu Leu Lys Ser

Lys Leu Arg Ala Leu Leu Ala Gly Pro Arg Ser Leu Arg Arg Ser Ser
123

Cys Phe Gly Gly Arg Ile Asp Arg Ile Gly Ala Gln Ser Gly Leu Gly

Cys Asn Ser Phe Arg Tyr Arg Arg
150

Fig. 2. Rat prepro-ANF sequence. Cardionatrin IV begins at Asn^{25} and ends at Tyr^{150}. This carboxyl terminus is shared by cardionatrin I (underlined) which begins at Ser^{123}. These peptides appear to be the major storage and release products, respectively, of ANF. The disulfide bond indicated between Cys^{129} and Cys^{145} is essential for biological activity.

logical knowledge, although some of these effects are clearly related to activation of guanylate cyclase, elevation of guanosine 3′,5′-monophosphate, and inhibition of adenylate cyclase (*33*). The finding that dopamine receptor antagonists interfere with the natriuretic response to ANF suggests that the autonomic nervous system might mediate the actions of ANF (*34*).

Conclusion

Although much remains to be learned about the systemic actions of ANF and factors that affect the release of ANF from atrial cardiocytes, several mechanisms link the atria and, possibly ANF, to plasma volume regulation. This link may be demonstrated, for example, by maneuvers leading to changes in atrial distension and intrathoracic blood volume, which, in turn, influence renal and cardiovascular function and the renin-angiotensin-aldosterone system in a way that mimics the effects of ANF. The polyuria associated with paroxysmal atrial tachycardia has been linked with increased plasma ANF (*35*). Both long-term and short-term variation in extracellular volume affect the atrial stores, synthesis, or circulating levels of ANF. Moreover, the inability of the kidney to bring about an escape from the sodium-retaining state accompanying chronic cardiac failure may be viewed as related to ANF. All of this evidence, although multifactorial, can be extrapolated to indicate a physiological role for ANF in the regulation of sodium and water balance and thus to allow the development of new therapies for clinical entities such as hypertension and heart failure.

The finding of secretory-like morphological characteristics in heart muscle cells in all species studied, together with the highly conserved nature of the known sequences of ANF peptides, hints at a fundamental evolutionary strategy used to maintain water and electrolyte balance. In sharks, for example, ANF stimulates chloride secretion from the rectal gland (*36*). It is reasonable to conclude that ANF plays regulatory roles in accordance with each species' strategy for maintaining water and salt balance as well as with the ionic environment to which the species is exposed.

References and Notes

1. A. J. de Bold, H. B. Borenstein, A. T. Veress, H. Sonnenberg, *Life Sci.* **28**, 89 (1981).
2. A. J. de Bold, *Proc. Soc. Exp. Biol. Med.* **170**, 133 (1982).
3. K. Atarashi, P. J. Mulrow, R. Franco-Saenz, R. Snadjar, J. Rapp, *Science* **224**, 992 (1984); J. C. Burnett, Jr., R. P. Granger, T. J. Opgenorth, *Am. J. Physiol.* **247**, F863 (1984); T. L. Goodfriend, M. E. Elliott, S. A. Atlas, *Life Sci.* **35**, 1675 (1984); L. Chartier, E. Schiffrin, G. Thibault, R. Garcia, *Endocrinology* **115**, 2026 (1984); M. Volpe *et al.*, *Hypertension* **7** (Suppl. I), I-43 (1985); U. Ackermann *et al.*, *Can. J. Physiol. Pharmacol.* **62**, 819 (1984).
4. A. F. S. Kent, *J. Physiol. (London)* **14**, 233 (1893).
5. A. Keith and M. W. Flack, *J. Anat. Physiol.* **41**, 172 (1907).
6. T. Lewis, B. S. Oppenheimer, A. Oppenheimer, *Heart* **2**, 147 (1910).
7. J. M. Berger and G. Rona, in *Methods and Achievements in Experimental Pathology*, E. Bajusz and G. Jasmin, Eds. (Karger, Basel, 1971), vol. 5, p. 540.
8. B. Kisch, *Exp. Med. Surg.* **14**, 99 (1956).
9. J. D. Jamieson and G. E. Palade, *J. Cell Biol.* **23**, 151 (1964); V. J. Ferrans, R. G. Hibbs, L. M. Buja, *Am. J. Anat.* **125**, 47 (1969); G. D. Bompiani, C. Rouiller, P. Y. Hatt, *Arch. Mal. Couer Vaiss.* **52**, 1257 (1959); M. Tomisawa, *Arch. Histol. Jpn. (Niigata, Jpn.)* **30**, 449 (1969).
10. S. A. Bencosme and J. M. Berger, in *Methods and Achievements in Experimental Pathology*, E. Bajusz and G. Jasmin, Eds. (Karger, Basel, 1971), vol. 5, p. 173.
11. A. J. de Bold, *J. Mol. Cell. Cardiol.* **10**, 717 (1978).
12. ——— and S. A. Bencosme, *Recent Advances in Studies on Cardiac Structure and Metabolism*, P. E. Roy and P. Harris, Eds. (University Park Press, Baltimore, 1975), vol. 8, p. 129.
13. J.-P. Marie, H. Guillemot, P.-Y. Hatt, *Pathol. Biol.* **24**, 549 (1976).
14. A. J. de Bold, *Proc. Soc. Exp. Biol. Med.* **161**, 508 (1979).
15. ——— and T. A. Salerno, *Can. J. Physiol. Pharmacol.* **61**, 127 (1983).
16. R. Garcia *et al.*, *Experientia* **38**, 1071 (1982); A. J. de Bold, *Can. J. Physiol. Pharmacol.* **60**, 324 (1982).
17. A. J. de Bold *et al.*, *Proceedings of the Naito International Symposium on Natural Products and Activities* (Univ. of Tokyo Press, Tokyo, in press).
18. M. L. de Bold and A. J. de Bold, in *Immunocytochemistry in Tumor Diagnosis*, J. Russo, Ed. (Nijhoff, The Hague, 1985), p. 202; C. Chapeau *et al.*, *J. Histochem. Cytochem.* **33**, 541 (1985).
19. M. Sakamoto *et al.*, *Biochem. Biophys. Res. Commun.* **128**, 1281 (1985); C. B. Saper, D. G. Standaert, M. G. Currie, D. Schwartz, D. M. Geller, P. Needleman, *Science* **227**, 1047 (1985); I. Tanaka, K. S. Misono, T. Inagami, *Biochem. Biophys. Res. Commun.* **124**, 663 (1984); D. M. Jacobowitz *et al.*, *Neuroendocrinology* **40**, 92 (1985).
20. K. Nakayama *et al.*, *Nature (London)* **310**, 699 (1984).
21. T. G. Flynn, P. L. Davies, B. P. Kennedy, M. L. De Bold, A. J. de Bold, *Science* **228**, 323 (1985).
22. T. G. Flynn, M. L. de Bold, A. J. de Bold, *Biochem. Biophys. Res. Commun.* **117**, 859 (1983).
23. M. G. Currie *et al.*, *Science* **223**, 67 (1984); D. M. Geller *et al.*, *Biochem. Biophys. Res. Commun.* **121**, 802 (1984); S. A. Atlas *et al.*, *Nature (London)* **309**, 717 (1984).
24. K. Kangawa, A. Fukuda, H. Matsuo, *Nature (London)* **313**, 397 (1985).

25. K. Kangawa, A. Fukuda, N. Minamino, H. Matsuo, *Biochem. Biophys. Res. Commun.* **119**, 933 (1984); K. S. Misono, R. T. Grammer, H. Fukumi, T. Inagami, *ibid.* **123**, 444 (1984); K. S. Misono, H. Fukumi, R. T. Grammer, T. Inagami, *ibid.* **119**, 524 (1984).
26. B. P. Kennedy *et al.*, *ibid.* **122**, 1076 (1984); M. Maki *et al.*, *Nature (London)* **309**, 722 (1984); C. E. Seidman, K. D. Bloch, K. A. Klein, J. A. Smith, J. G. Seidman, *Science* **226**, 1206 (1984); M. Yamanaka *et al.*, *Nature (London)* **309**, 719 (1984); R. A. Zivin *et al.*, *Proc. Natl. Acad. Sci. U.S.A.* **81**, 6325 (1984).
27. K. Kangawa *et al.*, *Nature (London)* **312**, 152 (1984).
28. D. Schwartz *et al.*, *Science* **229**, 397 (1985); P. L. Davies *et al.*, *Proc. Can. Fed. Biol. Soc.* **28**, 164 (1985).
29. A. J. de Bold *et al.*, unpublished observations.
30. A. Pettersson *et al.*, *Acta Physiol. Scand.* **124**, 309 (1985); J. R. Dietz, *Am. J. Physiol.* **247**, R1093 (1984); H. Sonnenberg, R. F. Krebs, A. T. Veress, *IRCS Med. Sci.* **12**, 783 (1984); H. Sonnenberg, A. T. Veress, *Biochem. Biophys. Res. Commun.* **124**, 443 (1984).
31. A. de Lean *et al.*, *Endocrinology* **115**, 1636 (1984); M. A. Napier *et al.*, *Proc. Natl. Acad. Sci. U.S.A.* **81**, 5946 (1984); R. Quirion *et al.*, *Peptides* **5**, 1167 (1984); K. M. M. Murphy, L. L. McLaughlin, M. L. Michener, P. Needleman, *Eur. J. Pharmacol.* **111**, 291 (1985).
32. T. Maack *et al.*, *Kidney Int.* **27**, 607 (1985).
33. P. Hamet *et al.*, *Biochem. Biophys. Res. Commun.* **123**, 515 (1984); R. J. Winquist *et al.*, *Proc. Natl. Acad. Sci. U.S.A.* **81**, 7661 (1984); S. A. Waldman, R. M. Rapoport, F. Murad, *J. Biol. Chem.* **259**, 14332 (1984); M. B. Anand-Srivastava, D. J. Franks, M. Cantin, J. Genest, *Biochem. Biophys. Res. Commun.* **121**, 855 (1984); A. Friedl, C. Harmening, B. Schuricht, B. Hamprecht, *Eur. J. Pharmacol.* **111**, 141 (1985); J. Tremblay *et al.*, *FEBS Lett.* **181**, 17 (1985); E. H. Ohlstein and B. A. Berkowitz, *Hypertension* **7**, 306 (1985); H. Matsuoka *et al.*, *Biochem. Biophys. Res. Commun.* **127**, 1052 (1985); Y. Hirata, M. Tomita, S. Takada, H. Yoshimi, *ibid.* **128**, 538 (1985).
34. M. Marin-Grez, J. P. Briggs, G. Schubert, J. Schnermann, *Life Sci.* **36**, 2171 (1985).
35. E. L. Schiffrin *et al.*, *N. Engl. J. Med.* **312**, 1196 (1985); T. Yamaji *et al.*, *Lancet* **1985-I**, 1211 (1985).
36. R. Solomon *et al.*, *Am. J. Physiol.*, **249**, R348 (1985).

16

The LDL Receptor Gene: A Mosaic of Exons Shared with Different Proteins

Thomas C. Südhof, Joseph L. Goldstein, Michael S. Brown, and David W. Russell

Cell surface receptors are multifunctional proteins with binding sites that face the external environment and effector sites that couple the binding to an intracellular event. Many receptors have an additional function: they transport bound ligands into cells (*1*). Such receptor-mediated endocytosis requires that the proteins have specific domains that allow them to cluster within clathrin-coated pits on the plasma membrane and in many cases to recycle to the cell surface after ligand delivery (*2*).

The multiple functions of coated pit receptors imply that they will have multiple domains, each with a single function. The structural features responsible for these functions are currently the subject of intense study. Recent insights have emerged from the complementary DNA (cDNA) cloning of the messenger RNA's (mRNA) for several receptors and the subsequent determination of their amino acid sequences. These studies have revealed surprising homologies between the primary structures of receptors and other proteins. For example, the receptor for plasma low-density lipoprotein (LDL), a cholesterol transport protein, contains one region that is homologous to the precursor of a peptide hormone, epidermal growth factor (EGF) (*3, 4*), and another region that is homologous to complement component C9, the terminal component of the complement cascade (*5*). The cell surface receptor for immunoglobulin A/immunoglobulin M is homologous to the immunoglobulins themselves (*6*). Finally, the receptor for EGF is homologous to a viral and cellular gene, *erb-B*, that produces a protein with tyrosine kinase activity (*7*).

These findings suggest that coated pit receptors share functional domains with

Science 228, 815–822 (17 May 1985)

other proteins. One likely mechanism for such sharing is through the duplication and migration of exons during evolution (*8*). Although the cDNA's for five coated pit receptors have been isolated and sequenced (*4, 6, 9*), the organizations of the genes encoding these proteins are not yet known. The elucidation of the gene structures of coated pit receptors should reveal the relationships between exons and protein domains and provide insight into the evolution of this important class of cell surface molecules.

Here, we report the exon organization of the gene for the human LDL receptor, a classic example of a cell-surface protein that mediates endocytosis through coated pits. A close correlation between functional domains in the LDL receptor protein and the exon-intron organization of the gene is revealed. In an accompanying report (*10*), we show that genomic sequences that are shared between the human LDL receptor and the human EGF precursor have a similar exon-intron organization, suggesting that coated pit receptor genes may have been assembled during evolution from itinerant exons encoding discrete protein domains.

Protein domains of LDL receptor. We recently isolated a full-length 5.3-kilobase (kb) cDNA for the human LDL receptor (*4*). The amino acid sequence as deduced from the nucleotide sequence revealed that the receptor is synthesized as a precursor of 860 amino acids. The first 21 amino acids constitute a typical hydrophobic signal sequence that is cleaved from the protein prior to its appearance on the cell surface, leaving an 839-amino-acid mature protein with five recognizable domains.

The first domain of the mature receptor contains the binding site for apoproteins B and E of LDL and of related lipoproteins. This domain consists of ~300 amino acid residues (*4*), which is assembled from multiple repeats of 40 residues each. Each repeat has six cysteine residues, all of which are involved in disulfide bonds (*4*). This repeated 40-amino-acid unit bears a strong homology to a single 40-amino-acid sequence that occurs within the cysteine-rich region of human complement component C9, a plasma protein of 537 amino acids (*5*).

The second domain of the human LDL receptor is a sequence of about 400 amino acids that was found to be homologous to the precursor for mouse EGF (*3, 4*). The EGF precursor is a protein of 1217 amino acids that may be synthesized as a membrane protein with a short cytoplasmic tail at the COOH-terminus (*11–14*). The EGF sequence of 53 amino acids lies immediately external to the hydrophobic membrane segment, from which it is presumably released by proteolysis. In earlier studies we found that the human LDL receptor was homologous to a large part of the external domain of the mouse EGF precursor, but not to EGF itself. Within a stretch of 400 amino acids, 33 percent of the residues are identical between the two proteins (*3, 4*).

The third domain of the LDL receptor is a sequence of 48 amino acids that contains 18 serines and threonines, many of which appear to serve as attachment sites for *O*-linked carbohydrate chains (*3, 4*). The fourth domain is a 22-amino-acid hydrophobic membrane-spanning region, and the fifth domain is a 50-amino-acid COOH-terminal cytoplasmic tail (*3, 4*).

Genomic cloning of LDL receptor. A series of bacteriophage λ and cosmid clones that span most of the LDL receptor gene were isolated (Fig. 1). These

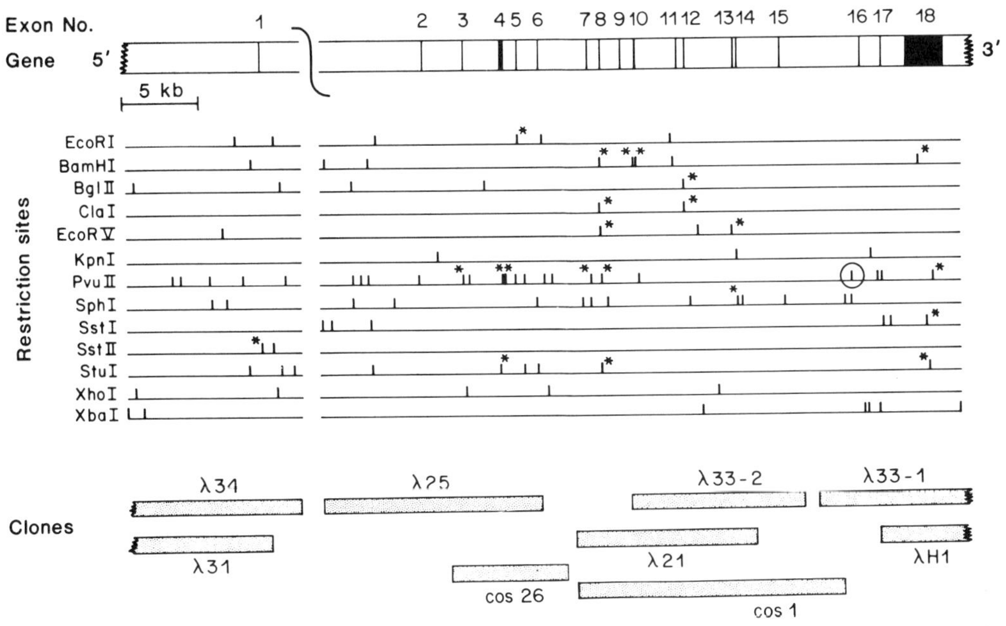

Fig. 1. Map of the human LDL receptor gene. The gene is shown in the 5′ to 3′ orientation at the top of the diagram and is drawn to scale. Exons are denoted by filled-in areas, and introns by open areas. The regions encompassed by genomic DNA inserts in the seven bacteriophage λ and two cosmid clones are indicated at the bottom. Cleavage sites for 13 selected restriction endonucleases are shown. Asterisks denote sites that are present in the cDNA. The encircled Pvu II site is polymorphic in human populations (*30*). The diagonal line between exons 1 and 2 represents a gap of unknown size not present in any of the genomic clones. Additional cleavage sites for the restriction enzymes shown may be present in this gap and in intron 6 (Table 1, legend). The λ clones were isolated from 1.2×10^7 plaques of a human genomic bacteriophage λ library (*31*). Cos1 was isolated from 6×10^6 colonies of a human cosmid library (*32*). Cos26 was isolated from 0.9×10^6 colonies of a human cosmid library (*33*). The libraries were screened with ^{32}P-labeled probes derived from the human LDL receptor cDNA, pLDLR-2 (*4*). Probes were isotopically labeled by nick translation (*34*) or hexanucleotide priming (*35*) and screening was carried out with standard procedures (*34*). Positive clones were plaque-purified or isolated as single colonies. Thirty fragments from the nine genomic clones were subcloned into pBR322 and characterized by restriction endonuclease digestion, Southern blotting (*34*), and DNA sequencing of exon-intron junctions (see Table 1). The restriction map was verified by comparing overlapping and independently isolated genomic clones and by Southern blotting analysis of genomic DNA isolated from normal individuals.

clones include the 18 exons that encode the LDL receptor protein and most of the 17 introns that separate them. Within two introns (located between exons 1 and 2 and between exons 6 and 7), there are gaps that are not covered by the genomic clones.

Exons within the cloned genomic DNA were identified in plasmid subclones by restriction mapping and by Southern blotting with cDNA probes (Fig. 1). The nucleotide sequences at the exon-intron boundaries (Table 1) were established by DNA sequence comparison of cDNA and genomic subclones. The 5′ donor and 3′ acceptor splice sites

Table 1. Exon-intron organization of the human LDL receptor gene. The nucleotide sequences of exon-intron junctions were determined by the enzymatic method (*36*) with the use of recombinant bacteriophage M13 clones containing genomic fragments as templates and the universal primer (*37*) together with a family of 25 LDL receptor–specific oligonucleotide primers. Exons 1 to 3 and 5 to 17 were sequenced in their entirety; intron 13 was sequenced in its entirety. Exon sequences are in capital letters; intron sequences are in lowercase letters. The number shown immediately below the DNA sequence denotes the nucleotide position at which the intron interrupts the LDL receptor mRNA (*4*). The numbers shown in parentheses in the extreme right-hand column denote the position of the indicated amino acid in the mature LDL receptor protein (*4*). The size for intron 1 is a minimal estimate based on concordance between data obtained from Southern blotting experiments with genomic DNA and data obtained from the restriction maps of the genomic clones. Approximately 2 kb of intron 6 was present in the genomic clones (Fig. 1); the amount of DNA not present in the genomic clones (0.7 kb) was estimated by Southern blotting of genomic DNA using probes derived from exons 6 and 7.

Exon number	Exon size (bp)	Sequence at exon-intron junction		Intron length (kb)	Amino acid interrupted
		5′ Splice donor	3′ Splice acceptor		
1	145–160	ACT GCA Ggtaagg. . . . 67	.tttcctctctctcagTG GGC GAC 68	>10	Val (2)
2	123	ACG TGC Tgtgagt. . . . 190	.ctgtctcttctgtagTG TCT GTC 191	2.5	Leu (43)
3	123	GGC TGT Cgtaagt. . . . 313	.catccatccctgcagCC CCC AAG 314	2.4	Pro (84)
4	381	AAC TGC Ggtatgg. . . . 694	.tgtcctgttttccagCT GTG GCC 695	0.95	Ala (211)
5	123	GTT AAT Ggtgagc. . . . 817	.ctctggctctcacagTG ACA CTC 818	0.85	Val (252)
6	123	GAG TGC Ggtgagt. . . . 940	.cctggccctgcgcagGG ACC AAC 941	2.7	Gly (293)
7	120	TGC GAA Ggtgatt. . . . 1060	.ttctctctcttccagAT ATC GAT 1061	0.45	Asp (333)
8	126	GCT GTG Ggtgagc. . . . 1186	.tccccggacccccagGC TCC ATC 1187	1.2	Gly (375)
9	172	ATC TGC AGgtgagc. . . . 1358	.ctcctcctgcctcagC ACC CAG 1359	0.9	Ser (432)
10	228	GTT CAT GGgtgcgt. . . . 1586	.ctgtcctcccaccagC TTC ATG 1587	2.6	Gly (508)
11	119	ACC CTA Ggtatgt. . . . 1705	.cacttgtgtgtctagAT CTC CTC 1706	0.6	Asp (548)
12	140	GTC TTT GAGgtgtgg. . . . 1845	.ttgctgcctgtttagGAC AAA GTA 1846	3.0	Glu (594)
13	142	CCA AGA Ggtaagg. . . . 1987	.cttcttctgccccagGA GTG AAC 1988	0.134	Gly (642)
14	153	CTC ACA Ggtgtgg. . . . 2140	.tatttattctttcagAG GCT GAG 2141	2.8	Glu (693)
15	171	CAC CAA Ggtaaag. . . . 2311	.gcttctctcctgcagCT CTG GGC 2312	5.5	Ala (750)
16	78	CCC ATC Ggtaagc. . . . 2389	.tgcctctccctacagTG CTC CTC 2390	1.4	Val (776)
17	158	TAC CCC TCGgtgagt. . . . 2547	.accatttgttggcagAGA CAG ATG 2548	1.7	Ser (828)
18	2535				

in each of the 17 introns conform to the GT . . . AG rule (G, guanine; T, thymine; A, adenine) and agree well with consensus sequences compiled for the exon-intron boundaries of other genes (*15*).

Characterization of 5′ end of LDL receptor gene. Figure 2 shows the nucleotide sequence of the 5′ end of the human LDL receptor gene, with the A of the initiator methionine codon designated as position +1 and the nucleotide positions to the 5′ side of this region designated by negative numbers. To determine the point at which transcription initiates, we performed an S1 nuclease

```
-687           GGATCCC ACAAAACAAAAAATATTTTT TTGGCTGTACTTTTGTGAAG ATTTTATTTAAATTCCTGAT TGATCAGTGTCTATTAGGTG  -601

-600  ATTTGGAATAACAATGTAAA AACAATATACAACGAAAGGA AGCTAAAAATCTATACACAA TTCCTAGAAAGGAAAAGGCA AATATAGAAAGTGGCGGAAG  -501

-500  TTCCCAACATTTTTAGTGTT TTCCTTTTGAGGCAGAGAGG ACAATGGCATTAGGCTATTG GAGGATCTTGAAAGGCTGTT GTTATCCTTCTGTGGACAAC  -401

-400  AACAGCAAAATGTTAACAGT TAAACATCGAGAAATTTCAG GAGGATCTTTCAGAAGATGC GTTTCCAATTTTGAGGGGGC GTCAGCTCTTCACCGGAGAC  -301

-300  CCAAATACAACAAATCAAGT CGCCTGCCCTGGCGACACTT TCGAAGGACTGGAGTGGGAA TCAGAGCTTCACGGGTTAAA AGCCGATGTCACATCGGCCG  -201

-200  TTCGAAACTCCTCCTCTTGC AGTGAGGTGAAGACATTTGA AAATCACCCCACTGCAAACT CCTCCCCCTGCTAGAAACCT CACATTGAAATGCTGTAAAT  -101

                                                   *
-100  GACGTGGGCCCCGAGTGCAA TCGCGGGAAGCCAGGGTTTC CAGCTAGGACACAGCAGGTC GTGATCCGGGTCGGGACACT GCCTGGCAGAGGCTGCGAGC  -1
                                                                               *

      Met Gly Pro Trp Gly Trp Lys Leu Arg Trp Thr Val Ala Leu Leu Leu Ala Ala Ala Gly Thr Ala     Intron 1
  +1  ATG GGG CCC TGG GGC TGG AAA TTG CGC TGG ACC GTC GCC TTG CTC CTC GCC GCG GCG GGG ACT GCA Ggtaaggcttgctcca
```

Fig. 2. Nucleotide sequence of the 5′ end of the human LDL receptor gene. Nucleotide position +1 is assigned to the A of the ATG codon specifying the initiator methionine; negative numbers refer to 5′ flanking sequences. Amino acids encoded by the first exon and the position of intron 1 are indicated on the bottom line. Vertical arrows above the sequence indicate sites of transcription initiation as determined by S1 nuclease mapping (Fig. 3). Vertical arrows below the sequence indicate sites of transcription initiation as determined by primer extension (Fig. 4). Asterisks denote an apparent S1 nuclease hypersensitive site (above sequence) or a strong stop point for reverse transcriptase (below sequence). Two AT-rich regions that are located 20 to 30 nucleotides upstream of the mRNA start points are boxed. Solid horizontal arrows denote three imperfect direct repeats of 16 nucleotides each. Dashed arrows denote two imperfect inverted repeats of 14 nucleotides each. The DNA sequence was determined by a combination of the chemical (*36*) and enzymatic (*37*) methods. Two M13 subclones derived from the bacteriophage genomic clone λ34 were used as templates together with the universal primer (*38*) and five LDL receptor-specific oligonucleotide primers to establish the sequence by the enzymatic method. Selected regions were then sequenced again by the chemical method; 90 percent of the sequence was determined on both strands of the DNA.

analysis with poly(A)$^+$ RNA (polyadenylated) isolated from SV40-transformed human fibroblasts or adult human adrenal glands (Fig. 3). The probe was a single-stranded DNA labeled with ^{32}P at the 5′ end and encompassing nucleotides −682 to +43 of the genomic sequence (Fig. 2). The sizes of the protected fragments were estimated by comparison with a dideoxynucleotide sequence established with an oligonucleotide primer and a recombinant bacteriophage M13 template (right-hand lanes in Fig. 3). We observed a cluster of three major S1 nuclease–protected fragments whose lengths corresponded to protection up to positions −79 to −89 of the genomic sequence. A small amount of an additional S1 nuclease-protected product was seen at a position corresponding to −57. Poly(A)$^+$ RNA from fibroblasts and adrenal tissue gave the same S1 nuclease-protected fragments. All of the protected fragments were drastically reduced in amount when the poly(A)$^+$ RNA was obtained from fibroblasts grown in the presence of sterols, conditions that reduce the amount of receptor mRNA (*4*).

To determine which S1 nuclease-protected fragments reflected transcription initiation sites, we performed a primer extension experiment (Fig. 4). We used poly(A)$^+$ RNA from cultured human A-431 carcinoma cells and SV40-transformed human fibroblasts grown in the absence or presence of sterols. To obtain a primer that would give the required sensitivity, we isolated a single-stranded, uniformly ^{32}P-labeled fragment of 93 nucleotides that was complementary to the 5′ end of the protein coding region of the mRNA. When this primer was extended on the poly(A)$^+$ mRNA template, three major products were observed whose length corresponded to transcription initiation sites between positions −84 to −93. In addition, we observed a shorter extension fragment corresponding to a 5′ end at position −29. When the A-431 cells or fibroblasts were grown in the presence of sterols, none of these primer-extended products was seen, confirming that they were derived from the receptor mRNA.

The three sites of transcription initiation as determined by the S1 nuclease technique and the three sites determined by the primer extension are indicated in Fig. 2. In general, these sites are well correlated with each other, except that the S1 nuclease technique systematically yields fragments that are 4 to 5 base pairs shorter than those obtained by primer extension. This may represent some "nibbling" by the S1 nuclease at the ends of the protected fragments. In contrast, the single shorter fragments seen in the S1 nuclease and primer extension experiments did not correspond to each other (see asterisks in Fig. 2). It is possible that the shorter S1 nuclease-generated fragment represents an S1 nuclease hypersensitive site and that the shorter primer extended product arises as a consequence of a strong stop sequence for reverse transcriptase in the mRNA template.

Considered together, the S1 nuclease and primer extension experiments indicate that there is no intron in the 5′ untranslated region of the LDL receptor gene and that transcription initiates at three closely spaced sites located between positions −79 and −93. This conclusion is supported by Northern blotting results showing that synthetic oligonucleotide or restriction fragment probes derived from DNA sequences upstream of position −102 do not hybridize to the LDL receptor mRNA, whereas probes

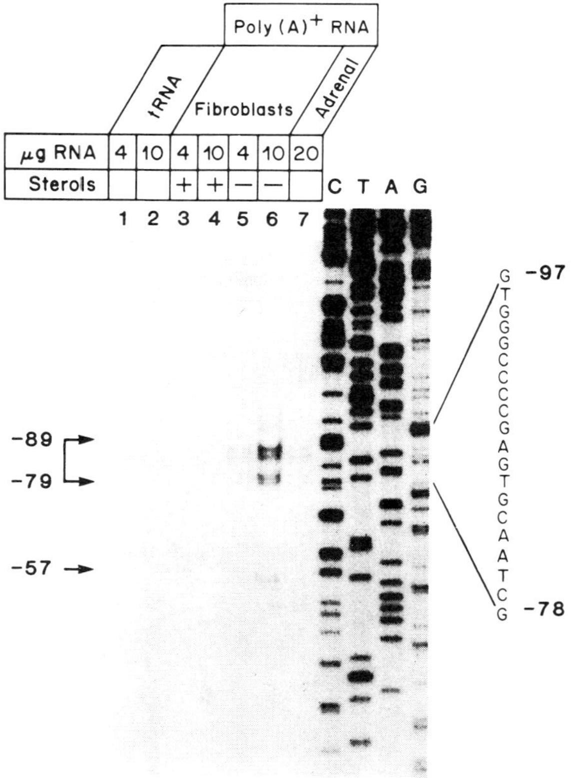

Fig. 3. Sites of transcription initiation in the human LDL receptor gene as determined by S1 nuclease analysis. The indicated amount of yeast tRNA (lanes 1 and 2), poly(A)$^+$ RNA from SV40-transformed human fibroblasts (lanes 3 to 6), or poly(A)$^+$ RNA from adult adrenal glands (lane 7) was annealed to a 5′ end-labeled with ^{32}P, single-stranded DNA probe corresponding to nucleotides −682 to +43 of Fig. 2. The fibroblasts were grown in the presence or absence of sterols as indicated. The RNA-DNA hybrids were digested with S1 nuclease, and the resistant products were subjected to electrophoresis through a 10 percent polyacrylamide–7*M* urea gel and detected by autoradiography for 72 hours at −20°C. SV40-transformed fibroblasts were set up in roller bottles (3×10^6 cells per bottle) and grown under standard conditions with fetal calf serum (10 percent) for 48 hours (*39*). On day 2, one-half of the roller bottles were switched to medium containing 10 percent calf lipoprotein-deficient serum and 10 μ*M* compactin in the absence of sterols. The other half of the roller bottles were switched to medium containing 10 percent newborn calf serum in the presence of 25-hydroxycholesterol (3 μg/ml) plus cholesterol (12 μg/ml). The cells were incubated for 24 hours and harvested for the preparation of poly(A)$^+$ RNA (*4*). Adult adrenal glands (obtained from human cadavers at the time of removal of kidneys for transplantation) were frozen at −70°C until preparation of poly(A)$^+$ RNA. The 5′ end-labeled, single-stranded ^{32}P probe of 725 nucleotides was prepared by priming an M13 clone containing a fragment of the LDL receptor gene corresponding to nucleotides −686 to +66 (Fig. 2) with a ^{32}P-labeled synthetic oligonucleotide complementary to nucleotides +19 to +43 (Fig. 2). The synthetic oligonucleotide was labeled at the 5′ end with [γ-^{32}P]ATP (7000 Ci/mmol) and polynucleotide kinase (*34*) to a specific radioactivity of $\sim 5 \times 10^6$ cpm/pmol. After primer extension with the Klenow fragment of DNA polymerase I in the presence of each of the four deoxynucleoside triphosphates (15 μ*M*), the resulting double-stranded DNA was cleaved with Bam HI, and the radioactive probe fragment was purified by electrophoresis on a denaturing polyacrylamide gel and subsequent electroelution (*34*). A portion of the probe (10^5 cpm) was coprecipitated with the indicated amount of tRNA or poly(A)$^+$ RNA in ethanol at −70°C. The precipitated material was resuspended in 20 μl of 80 percent formamide, 0.4*M* NaCl, 40 m*M* 1,4-piperazinediethane-sulfonic acid (*p*H 6.4), and 1 m*M* EDTA and hybridized at 65°C for 36 hours. The samples were diluted with 9 volumes of 0.25*M* NaCl, 30 m*M* potassium acetate (*p*H 4.5), 1 m*M* $ZnSO_4$, and 5 percent glycerol; treated with 200 units of S1 nuclease (Bethesda Research Laboratories) at room temperature for 60 minutes (*40*); precipitated with ethanol; and analyzed on a sequencing gel. The protected fragments were compared with the adjacent dideoxy nucleotide–derived sequencing ladder obtained with the same primer and M13 template used to generate the probe. Numbers on the left denote the estimated nucleotide position corresponding to the 5′ end of the protected fragments according to the numbering scheme of Fig. 2. The sequence in the −78 to −97 region is shown on the right.

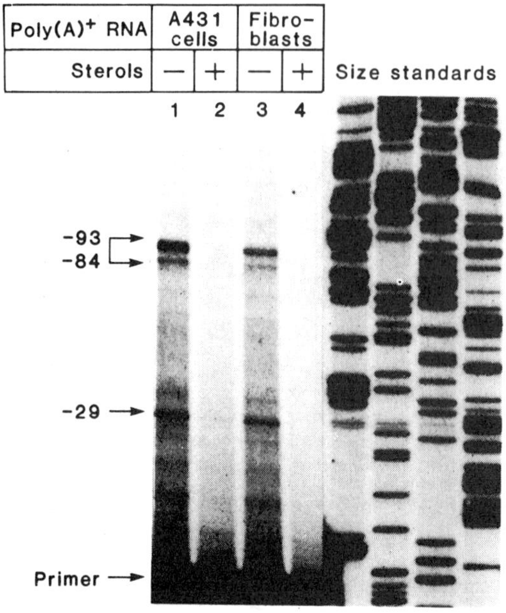

Fig. 4. Sites of transcription initiation in the human LDL receptor gene as determined by primer extension analysis. Poly(A)$^+$ RNA was obtained from human A-431 epidermoid carcinoma cells (lanes 1 and 2) or SV40-transformed human fibroblasts (lanes 3 and 4) that had been grown in the absence or presence of sterols. The RNA was annealed to a uniformly labeled, single-stranded ^{32}P probe [corresponding to nucleotides +9 to +101, Fig. 2 and (*4*)] that served as a primer for extension by reverse transcriptase. The primer-extended products were subjected to electrophoresis through a 10 percent polyacrylamide–7*M* urea gel and detected after autoradiography for 60 hours at −20°C. The poly(A)$^+$ RNA from A-431 cells and SV40-transformed fibroblasts cultured in the absence or presence of sterols was prepared as described in the legend of Fig. 3. A single-stranded, uniformly ^{32}P-labeled primer complementary to nucleotides +9 to +101 of the human LDL receptor mRNA was derived from an M13 cDNA clone (*38*) as described (*41*). A portion (approximately 2×10^6 cpm) of the 93-nucleotide ^{32}P-primer was precipitated with ethanol together with 10 μg of the indicated poly(A)$^+$ RNA; resuspended in 5 μl of a buffer containing 50 m*M* tris-chloride (*p*H 8.0), 50 m*M* KCl, 5 m*M* $MgCl_2$, and 20 m*M* dithiothreitol; and sealed in a glass capillary. The reaction mixture was denatured by boiling for 2 minutes, and primer-template complexes were allowed to form at 65°C for 4 hours. After this annealing period, the entire solution was transferred to a plastic microfuge tube containing 2.5 μl of a 1 m*M* solution of the four deoxynucleoside triphosphates and 6 units of avian myeloblastosis virus reverse transcriptase (Molecular Genetic Resources). Primer extension was allowed to occur for 50 minutes at 37°C and was stopped by the addition of 6 μl of a formamide-dye mix. The sample was boiled for 5 minutes, quickly chilled on ice, and subjected to electrophoresis on a sequencing gel. Size standards, shown on the right, were generated by dideoxy nucleotide sequencing (*37*) of a known M13 recombinant clone. Numbers on the left denote the calculated position corresponding to the limits of primer extension according to the numbering scheme of Fig. 2. The intense band at the bottom of lanes 1 to 4 represents the ^{32}P-labeled primer used in the experiment.

derived from sequences downstream of position −64 do hybridize to the mRNA (data not shown). Approximately 20 to 30 base pairs to the 5′ side of the mRNA initiation sites at −79 to −93 are two AT-rich sequences (TTGAAAT and TGTAAAT) that might serve as TATA boxes (*16*). The existence of two such closely spaced sites might explain the multiple closely spaced transcription initiation sites. To the 5′ side of these AT-rich sequences, we did not find the sequence CCAAT (C, cytosine), which is conserved in some but not all eukaryotic genes (*16*).

The findings derived from the experimental data shown in Figs. 3 and 4 confirm previous results that the amount of mRNA for the LDL receptor is reduced when cells are incubated with sterols (*4, 17*). If the control of receptor mRNA expression is similar to that of other genes, then the amount of mRNA is likely to be determined by the rate of transcription, and DNA sequences in the 5′ end of the gene are likely to be responsible for this regulation. In the 5′ end of the gene, there are three imperfect direct repeat sequences of 16 nucleotides each that are located just upstream of the clustered sites of transcription initiation (Fig. 2, solid arrows). This same region contains an imperfect inverted repeat sequence of 14 nucleotides (Fig. 2, dashed arrows). Repeat sequences of 12 nucleotides in the 5′ end of the mouse metallothionien gene have recently been shown to mediate heavy metal inducibility (*18*). Thus, the direct repeats noted above in the 5′ end of the LDL receptor gene may play a role as sterol regulatory elements. Linking the LDL receptor sequences to heterologous marker genes may provide insight into the regulation of mRNA expression by sterols.

Exon organization and protein domains. The arrows in Fig. 5 show the position of each intron in the LDL receptor gene in relation to the domains of the protein that were identified earlier on the basis of protein sequence and proteolysis studies (*3, 4*). The introns interrupt the protein coding sequence in such a way that many of the protein segments are revealed as products of individual exons. Exon 1 encodes the short 5′ untranslated region plus the signal sequence of the protein. The first intron interrupts the coding sequence two amino acids distal to the end of the signal sequence; thus, the signal sequence is contained in a discrete exon (Fig. 5).

An informative set of introns occurs in the cysteine-rich repeat region that contains the LDL binding domain. The initial analysis of the protein sequence derived from cDNA clones suggested a total of eight repeats in this region, of which the first seven were strongly homologous (*4*). The eighth repeat showed weaker homology but retained a cysteine-rich character. Analysis of the exon structure now reveals that the first seven of these sequences belong to one repeat sequence family (Figs. 5 and 6). Each repeat in this family is about 40 residues long, and each contains six cysteine residues spaced at similar intervals. In 19 of the 40 amino acid positions, a single amino acid occurs in more than 50 percent of the repeats, allowing a consensus sequence to be written (Fig. 6A). This

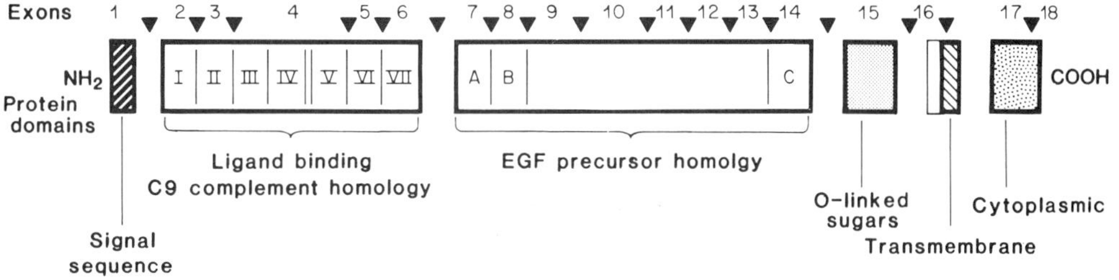

Fig. 5. Exon organization and protein domains in the human LDL receptor. The six domains of the protein are delimited by thick black lines and are labeled in the lower portion. The seven cysteine-rich, 40-amino acid repeats in the LDL binding domain (Fig. 6) are assigned the roman numerals I to VII. Repeats IV and V are separated by eight amino acids. The three cysteine-rich repeats in the EGF precursor homology domain (Fig. 8) are lettered A to C. The positions at which introns interrupt the coding region are indicated by arrowheads. Exon numbers are shown between the arrowheads.

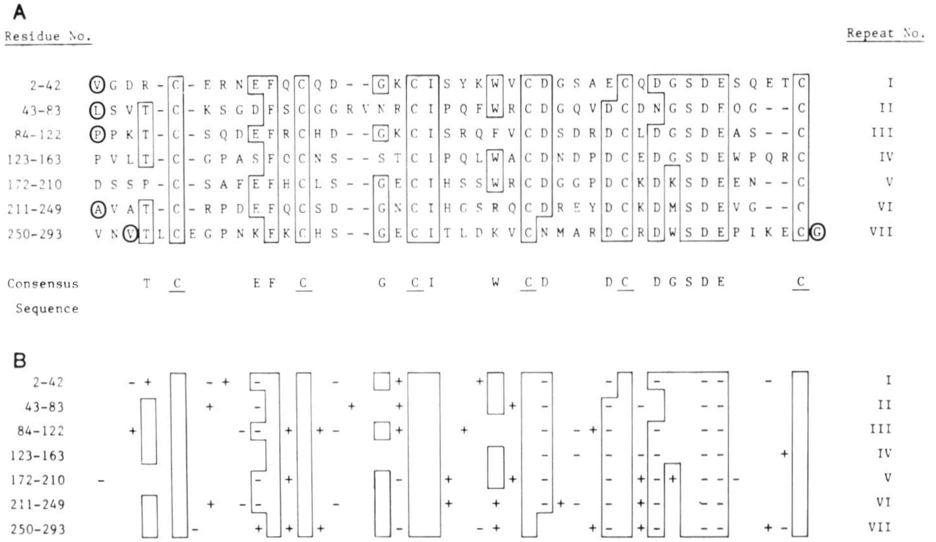

Fig. 6. Location of introns in the cysteine-rich repeat region of the binding domain of the LDL receptor. The amino acids constituting each of the seven repeat units are numbered in the left column according to the translated sequence of the receptor cDNA (*4*). (A) Optimal alignment was made by the computer programs ALIGN and RELATE (*4*) with modifications based on the location of introns. Amino acids that are present at a given position in more than 50 percent of the repeats are boxed and shown as a consensus on the bottom line. Cysteine residues (C) in the consensus sequence are underlined. The positions at which introns interrupt the coding sequence of the gene are denoted by the encircled amino acids. The single letter amino acid code translates to the three letter code as follows: A, Ala; C, Cys; D, Asp; E, Glu; F, Phe; G, Gly; H, His; I, Ile; K, Lys; L, Leu; M, Met; N, Asn; P, Pro; Q, Gln; R, Arg; S, Ser; T, Thr; V, Val; W, Trp; Y, Tyr. (B) The net charge of each of the amino acids in (A) is shown. All of the conserved amino acids that are charged bear a negative charge; none are positively charged.

consensus shows a striking preponderance of acidic residues; all of the charged residues that are conserved bear a negative charge (Fig. 6B). This clustering of negative charges is particularly strong in the sequence Asp-Cys-X-Asp-Gly-Ser-Asp-Glu, which occurs near the COOH-terminal end of each repeat. These clustered negative charges may account for the ability of the LDL receptor to bind to closely spaced positively charged residues on apoprotein E, a high affinity ligand for the LDL receptor (*4, 19*).

As originally noted by Stanley *et al.* (*5*), each of the seven 40–amino acid repeats in the LDL receptor is strongly homologous to a single 40-residue unit in complement factor C9. Of the 19 conserved amino acids in the LDL receptor repeats, 14 are found in the C9 sequence, including the highly conserved negatively charged cluster (Fig. 7).

Support for the seven-repeat model of the binding domain for the LDL receptor comes from analysis of intron placement (Fig. 6A, circled residues). Introns are located immediately before the first repeat (two residues after the signal sequence cleavage point) and precisely after repeats I, II, V, VI, and VII (Fig. 6). Thus, four of the repeats are each encoded by individual exons. The other three repeats (III, IV, and V) are all contained in a single exon. The introns between repeats III, IV, and V must have been lost after the primordial exon had undergone duplication to produce the seven repeats. Loss of introns during evolution has been postulated to occur (*20*). Interestingly, all of the introns in this region interrupt codons at the same location, that is, after the first base of the codon (Table 1). Thus, any or all of the exons in the binding domain could possi-

```
LDL Receptor Consensus
  x x T C x x x E F x C x x G x C I x x x W x C D x x x D C x D G S D E x x C
  E D D C - G N D F Q C S T G R C I K M R L R C N G D N D C G D F S D E D D C
Complement factor C9 (residues 77-113)
```

Fig. 7. Comparison of consensus sequence in the binding domain of the LDL receptor (Fig. 6A) with the homologous sequence from complement factor C9 (*5*).

bly be spliced in or out of the mRNA without disturbing the reading frame. Such alternative splicing could produce receptors with different affinities or specificities for apoproteins B or E, the two ligands for the LDL receptor. Differential in-frame splicing of exons in protein coding regions has been reported for several other genes (*21*).

In previous studies we found that the sequence of amino acid residues 290 to 690 of the human LDL receptor was homologous to the sequence of the mouse EGF precursor (*4*). This region contains three repetitive sequences of about 40 amino acids, that are designated A, B, and C in Fig. 5 and displayed in detail in Fig. 8. Each of these repeats contains six cysteine residues spaced at similar intervals. The A, B, and C sequences are homologous to repeat sequences designated 1 to 4 in the EGF precursor (*11*). A similar sequence occurs in three proteins of the blood clotting system: factor X, factor IX, and protein C (Fig. 8). Doolittle *et al.* (*13*) originally discovered that the cysteine-rich sequences shown in Fig. 8 were shared between the EGF precursor and the three proteins of the blood clotting system. A striking feature of repeats A, B, and C of the LDL receptor is that each of these structures is contained within a single exon (Fig. 5). A similar exon-intron organization for repeats 2, 3, and 4 occurs in the human EGF precursor gene (*10*). Moreover, recent studies by Anson *et al.* (*22*) indicate that the repeat found in human factor IX is also encoded by a discrete exon.

The repeat sequence shown in Fig. 8 is not the same as the recently described "growth factor–like" repeat found in EGF, α-transforming growth factor, tissue plasminogen activator, and the 19-kD vaccinia virus protein (*13, 23*). The proteins of the blood clotting system—factor IX, factor X, and protein C—appear to contain one copy of the growth factor–like sequence (*13, 23*) and one copy of the LDL receptor-like sequence shown in Fig. 8, whereas the EGF precursor contains four copies of the growth factor–like sequence and four copies of the LDL receptor-like sequence (*13*).

The LDL receptor undergoes several posttranslational carbohydrate processing events that cause a shift in its apparent molecular weight on sodium dodecyl sulfate–polyacrylamide gels from 120,000 to 160,000 during synthesis and transport to the cell surface (*24*). Most of this apparent molecular weight increase is due to the addition of complex carbohydrate chains in *O*-linkage to serine and threonine residues (*24*). The majority of these *O*-linked sugars have been localized by protease and lectin blotting studies to a region of the receptor that contains 18 clustered serine and threonine residues (*3*). This domain is located just above the transmembrane sequence of the protein. In the receptor gene, exon

Protein	Species	Residue	Amino Acid Sequence
LDL Receptor (A)	Human	297–331	C - - - L D N N G G C S H V C . (8) . C L C P D G F Q L V A Q - R R C
LDL Receptor (B)	Human	337–371	C - - - Q D P - D T C S Q L C . (8) . C Q C E E G F Q L D P H T K A C
LDL Receptor (C)	Human	646–690	C E R T T L S N G G C Q Y L C . (14) . C A C P D G M L L A R D M R S C
EGF-Precursor (1)	Mouse	366–401	C - - - A T Q N H G C T L G C . (8) . C T C P T G F V L L P D G K Q C
EGF-Precursor (2)	Mouse	407–442	C - - - P G N V S K C S H G C . (8) . C I C P A G S V L G R D G K T C
EGF-Precursor (3)	Mouse	444–482	C - - S S P D N G G C S Q I C . (9) . C D C F P G Y D L Q S D R K S C
EGF-Precursor (4)	Mouse	751–786	C - - - L Y R N G G C E H I C . (8) . C L C R E G F V K A W D G K M C
Factor X	Human	89–124	C - - - S L D N G D C D Q F C . (8) . C S C A R G Y T L A D N G K A C
Factor IX	Human	88–124	C - - - N I K N G R C E Q F C . (9) . C S C T E G Y R L A E N Q K S C
Protein C	Bovine	98–133	C - - - S A E N G G C A H Y C . (8) . C S C A P G Y R L E D D H Q L C
Consensus			C - - - x x x N G G C x x x C . (8) . C x C x x G Y/F x L x x D x K x C

Fig. 8. Amino acid alignment of segments A, B, and C from the LDL receptor with homologous regions from the EGF precursor and several proteins of the blood clotting system. The number of amino acids comprising the variable region in the middle of each sequence is shown in parentheses. The standard one-letter amino acid abbreviations are used (Fig. 6). Amino acids that are present at a given position in more than 50 percent of the sequences are boxed and shown as a consensus at the bottom line. Cysteine residues (C) are underlined in the consensus sequence. Sequence data for the LDL receptor was taken from (*4*); sequence data for the other proteins were taken from the original references cited in (*13*, *23*).

15 encodes 58 amino acids that encompass all 18 of the clustered serine and threonine residues (Fig. 5).

A hydrophobic sequence of 22 amino acids flanked by arginine and lysine residues forms the transmembrane domain of the human LDL receptor (*4*). Deletion of this region in a naturally occurring mutation results in the synthesis of a receptor that is secreted from the cell (*25*). The transmembrane domain is encoded by exons 16 and 17 (Fig. 5). The intron that separates these exons interrupts the codon specifying the ninth residue of this 22–amino acid hydrophobic domain.

The last protein domain in the LDL receptor consists of 50 amino acids located on the cytoplasmic side of the plasma membrane (*3, 4*). The amino acid sequence of this region is highly conserved between LDL receptors of different species (*26*). This domain serves to target the LDL receptor to coated pits on the cell surface (*27*). The cytoplasmic domain is encoded by exons 17 and 18. Exon 17 encodes 13 amino acids of the transmembrane domain and the first 39 amino acids of the cytoplasmic domain. Exon 18, the largest exon in the gene (Table 1), encodes the remaining 11 amino acids at the COOH-terminus of the protein and the 2.5 kb of DNA sequence that corresponds to the 3′ untranslated region of the mRNA, including three copies of the Alu family of middle repetitive DNA sequences (*4, 25*).

LDL receptor as a member of two supergene families. On the basis of studies of immunoglobulins and related proteins, Hood *et al.* (*28*) have defined the concept of a supergene family as "a set of . . . genes related by sequence (implying common ancestry), but not necessarily related in function." By this criterion the LDL receptor is a member of at least two supergene families. The LDL binding domain belongs to a supergene family whose only other member now known is complement component C9. The three repeated sequences in the domain of EGF precursor homology belong to a supergene family that includes the three proteins of the blood clotting system (factor IX, factor X, and protein C) as well as the EGF precursor. In the LDL receptor each of these regions is contained on one or more exons whose intron boundaries imply that they were free to move within the genome and to join other genes, thus creating a supergene family.

On the basis of these findings, it may become necessary to expand the concept of supergene families to include *regions* of proteins and to consider that a given protein may contain discrete regions derived from the exons of different supergene families. As originally proposed by Gilbert (*8*), the existence of introns permits functional domains encoded by discrete exons to shuffle between different proteins, thus allowing proteins to evolve as new combinations of preexisting functional units. The LDL receptor is a vivid example of such a mosaic protein.

Implications for genetics of familial hypercholesterolemia. The elucidation of the structure of the LDL receptor gene should be useful in further studies of mutations in this gene that underlie familial hypercholesterolemia (FH), a common cause of atherosclerosis and heart attacks (*29*). In the one mutation so far described at the molecular level, the defect involves a 5-kb deletion that removes several exons near the 3′ end of the gene (*25*). This deletion resulted from a recombination between two middle repetitive sequences of the Alu family. The finding of repeated exons in the LDL binding domain and in the EGF precursor homology region raises the possibility that some FH mutations may have arisen from deletions or duplications resulting from unequal crossing-over and recombination between these homologous segments. The availability of a detailed gene map should now permit the characterization of other mutations through Southern blotting and cloning of genomic DNA isolated from cells of FH patients.

References and Notes

1. J. L. Goldstein, R. G. W. Anderson, M. S. Brown, *Nature (London)* **279**, 679 (1979).
2. M. S. Brown, R. G. W. Anderson, J. L. Goldstein, *Cell* **32**, 663 (1983).
3. D. W. Russell *et al.*, *ibid.* **37**, 577 (1984).
4. T. Yamamoto *et al.*, *ibid.* **39**, 27 (1984).
5. K. K. Stanley *et al.*, *EMBO J.* **4**, 375 (1985); R. G. DiScipio *et al.*, *Proc. Natl. Acad. Sci. U.S.A.* **81**, 7298 (1984).
6. K. E. Mostov, M. Friedlander, G. Blobel, *Nature (London)* **308**, 37 (1984).
7. J. Downward *et al.*, *ibid.* **307**, 421 (1984).
8. W. Gilbert, *ibid.* **271**, 501 (1978).
9. A. Ullrich *et al.*, *ibid.* **309**, 418 (1984); E. C. Holland, J. O. Leung, K. Drickamer, *Proc. Natl. Acad. Sci. U.S.A.* **81**, 7338 (1984); A. McClelland, L. C. Kuhn, F. H. Ruddle, *Cell* **39**, 267 (1984); C. Schneider *et al.*, *Nature (London)* **311**, 675 (1984).
10. T. C. Südhof, D. W. Russell, J. L. Goldstein, M. S. Brown, R. Sanchez-Pescador, G. I. Bell, *Science* **228**, 893 (1985).
11. J. Scott *et al.*, *ibid.* **221**, 236 (1983).
12. A. Gray, T. J. Dull, A. Ullrich, *Nature (London)* **303**, 722 (1983).
13. R. F. Doolittle, D.-F. Feng, M. S. Johnson, *ibid.* **307**, 558 (1984); R. F. Doolittle, *Trends in Biochem. Sci.*, in press. Doolittle *et al.* described ten repeat units in the mouse EGF precursor that fall into two classes A and B. The four class A repeats, designated g–j, correspond to the "growth factor–like" repeats found in α-transforming growth factor, tissue plasminogen activator, and the 19-kD vaccinia virus (*23*). Four of the six class B repeats, which Doolittle, Feng, and Johnson designated c to f, correspond, respectively, to repeats 1 to 4 of Scott *et al.* (*11*). These four repeats show the greatest homology with repeats A to C in the LDL receptor (Fig. 8). The three proteins of the blood clotting system (factor IX, factor X, and protein C) contain one copy each of the class A and B repeats (see figure 2 in Doolittle *et al.*).
14. L. B. Rall *et al.*, *Nature (London)* **313**, 228 (1985).
15. S. M. Mount, *Nucleic Acids Res.* **10**, 459 (1982).
16. T. Shenk, *Curr. Topics Microbiol. Immunol.* **93**, 25 (1981).
17. D. W. Russell *et al.*, *Proc. Natl. Acad. Sci. U.S.A.* **80**, 7501 (1983).
18. G. W. Stuart *et al.*, *ibid.* **81**, 7318 (1984); A. D. Carter *et al.*, *ibid.*, p. 7392.
19. T. L. Innerarity *et al.*, *J. Biol. Chem.* **259**, 7261 (1984).
20. W. F. Doolittle, *Nature (London)* **272**, 581 (1978).
21. C. R. King and J. Piatigorsky, *Cell* **32**, 707 (1983); R. M. Medford *et al.*, *ibid.* **38**, 409 (1984); H. Nawa, H. Kotani, S. Nakanishi, *Nature (London)* **312**, 729 (1984).
22. D. S. Anson *et al.*, *EMBO J.* **3**, 1053 (1984).
23. M. C. Blomquist, L. T. Hunt, W. C. Barker, *Proc. Natl. Acad. Sci. U.S.A.* **81**, 7363 (1984).
24. H. Tolleshaug *et al.*, *Cell* **30**, 715 (1982); R. D. Cummings *et al.*, *J. Biol. Chem.* **258**, 15261 (1983).
25. M. A. Lehrman *et al.*, *Science* **227**, 140 (1985).
26. T. Yamamoto *et al.*, in preparation.
27. M. A. Lehrman *et al.*, *Cell*, in press.
28. L. Hood *et al.*, *ibid.* **40**, 225 (1985).
29. J. L. Goldstein and M. S. Brown, in *The Metabolic Basis of Inherited Disease*, J. B. Stanbury *et al.*, Eds. (McGraw-Hill, New York, ed. 5, 1983), pp. 672–712.
30. H. Hobbs *et al.*, in preparation.
31. R. M. Lawn *et al.*, *Cell* **15**, 1157 (1978).
32. Y.-F. Lau and Y. W. Kan, *Proc. Natl. Acad. Sci. U.S.A.* **80**, 5225 (1983).
33. F. G. Grosveld *et al.*, *Nucleic Acids Res.* **10**, 6715 (1982).
34. T. Maniatis, E. F. Fritsch, J. Sambrook, *Molecular Cloning: A Laboratory Manual* (Cold Spring Harbor Laboratory, Cold Spring Harbor, N.Y., 1982), pp. 1–545.
35. A. P. Feinberg and B. Vogelstein, *Anal. Biochem.* **132**, 6 (1983).
36. A. M. Maxam and W. Gilbert, *Methods Enzymol.* **65**, 499 (1980).
37. F. Sanger, S. Nicklen, A. R. Coulson, *Proc. Natl. Acad. Sci. U.S.A.* **74**, 5463 (1977).
38. J. Messing, *Methods Enzymol.* **101**, 20 (1983).
39. J. L. Goldstein, S. K. Basu, M. S. Brown, *ibid.* **98**, 241 (1983).
40. A. Berk and P. Sharp, *Cell* **12**, 721 (1977).
41. G. M. Church and W. Gilbert, *Proc. Natl. Acad. Sci. U.S.A.* **81**, 1991 (1984).

42. We thank Daphne Davis, James Cali, and Gloria Brunschede for technical assistance, M. Lehrman and H. Hobbs for helpful discussions, and T. Maniatis, F. Grosveld, and Y. W. Kan for providing the λ and cosmid libraries. Supported by NIH research grants HL 20948 and HL 31346, a fellowship from the Deutsche Forschungsgemeinschaft (T.C.S.), and a Research Career Development Award from the NIH (HL 01287) (D.W.R.).

12 March 1985; accepted 4 April 1985

17

Human von Willebrand Factor (vWF): Isolation of Complementary DNA (cDNA) Clones and Chromosomal Localization

David Ginsburg, Robert I. Handin, David T. Bonthron, Timothy A. Donlon, Gail A.P. Bruns, Samuel A. Latt, and Stuart H. Orkin

The hemostatic system has evolved to minimize blood loss following vascular injury. In higher vertebrates, including man, the system is quite complex and requires the interaction of circulating platelets, a series of plasma coagulation proteins, endothelial cells, and components of the vascular subendothelium. The initial and critical event in hemostasis is the adhesion of platelets to the subendothelium. It occurs within seconds of injury and provides a nidus for platelet plug assembly and fibrin clot formation.

The factor VIII molecular complex is composed of two distinct protein components, the antihemophilic factor (AHF or VIIIC) and von Willebrand factor (vWF), and plays a major role in both platelet adhesion and fibrin formation (*1*). Two of the most common inherited clinical bleeding disorders are the result of a deficiency in the activity of one or the other of these components.

The VIIIC molecule is an important regulatory protein in the coagulation cascade. After activation by trace quantities of thrombin, it accelerates the rate of factor X activation by factor IX, eventually leading to the formation of the fibrin clot. Classic hemophilia (VIIIC deficiency) is an X chromosome–linked disorder that affects one in 10,000 males, and has been recognized as a major source of hemorrhagic morbidity and mortality since biblical times. Treatment consists of supportive measures and usually requires frequent transfusion with blood products. The latter results in a high incidence of infectious complications in this population, including various forms of hepatitis and acquired immune deficiency syndrome.

Science 228, 1401–1406 (21 June 1985)

The vWF molecule is an adhesive glycoprotein synthesized by endothelial cells and megakaryocytes. It serves as a carrier in plasma for VIIIC and facilitates platelet-vessel wall interactions. By binding to subendothelial structures and to the platelet surface it promotes shear-dependent platelet adhesion to the vessel wall. Discrete domains of vWF which bind to platelet receptor sites on glycoprotein Ib and on the glycoprotein IIb–IIIa complex, as well as to binding sites on collagen, have been noted. A variety of abnormalities in vWF activity can result in von Willebrand's disease (vWD). This disorder is generally inherited in an autosomal dominant fashion and may affect as many as one in 2000 individuals. The mild forms may frequently go undiagnosed. Severely affected patients may require frequent blood product support with its associated risks.

Whereas the VIIIC molecule is a single-chain 220-kilodalton (kD) protein, vWF activity is expressed in a heterogeneous series of multimers with molecular sizes ranging from 450 to 20,000 kD. These multimers are assembled from a single glycoprotein subunit of approximately 220 kD. vWF accounts for 99 percent of the mass of the plasma factor VIII molecular complex (*1*). The VIIIC messenger RNA (mRNA) has been found in various tissues including liver, placenta, and a T-cell hybridoma line (*2*). The vWF protein has only been detected in endothelial cells, megakaryocytes, and in tumors derived from vascular tissue (*1, 3*). In cultured endothelial cells, vWF is synthesized as a large precursor (240 to 260 kD) which is processed to the mature 220-kD subunit and assembled into multimers which then enter plasma (*4*). Endothelial cells are the major site of plasma vWF synthesis. Blood platelet alpha granules contain approximately 15 percent of circulating vWF. There is little information regarding megakaryocyte biosynthesis of vWF or the role of this platelet vWF pool in hemostasis.

As noted above, vWD is a complex and heterogeneous group of hereditary bleeding disorders (*1*). In the most common variety (type I), patients have reduced vWF activity, but secrete the full range of multimers into plasma. Several additional forms of vWD have been defined from clinical studies. Although their molecular defects are not fully understood, they all exhibit a qualitative or quantitative abnormality of the vWF molecule. The most common variants have a selective loss of the high molecular weight multimers of vWF due to a failure of assembly (type IIa) or rapid clearance due to aberrant platelet binding (type IIb). Occasional patients with an autosomal recessive form of vWD have been described. Analyses of several large kindreds have failed to demonstrate definite linkage of any form of vWD to other genetic markers (*5*).

The study of vWF biochemistry, structure, and activity has been particularly difficult because of its large size, heterogeneous nature, and poorly defined function. Until recently, there was no information on the primary structure of either the VIIIC or vWF proteins, or on the molecular basis of their respective clinical defects. The isolation of genomic and complementary DNA (cDNA) clones for VIIIC has now been reported (*2*), and molecular techniques should soon be applicable to the prenatal diagnosis of hemophilia A. We have isolated cloned vWF cDNA sequences, which will now permit a molecular genetic approach to the study of vWD and to the analysis of vWF structure and function.

Expression cloning of vWF sequences. In order to isolate vWF cDNA we used the λgt11 bacteriophage expression vector with a specific antibody as the detection system. In this expression system, devised by Young and Davis (*6*), proteins are produced as fusion products at the COOH-terminus of bacterial β-galactosidase. A number of cDNA's have been cloned in this way (*6–7*), including those for several coagulation proteins (*8*).

We constructed two large λgt11 bacteriophage libraries of cDNA derived from cultured human umbilical vein endothelial cells (HUVEC), which are a recognized source of vWF. For the first library cDNA was synthesized with oligodeoxythymidylate [oligo(dT)] as primer [oligo(dT) library], and a pool of random hexanucleotides was used as primer for the second (random library). The random library was expected to contain clones randomly distributed along the length of the mRNA template, a feature that would avoid the bias toward the 3′ end inherent in libraries primed with oligo(dT).

A primary culture of HUVEC (*9*) was grown and passaged in Medium 199 with 20 percent fetal bovine serum in the presence of bovine endothelial cell growth factor (*10*) and fibronectin by the method of Maciag *et al.* (*11*). Growth was markedly enhanced by addition of heparin as described by Thornton *et al.* (*12*). Cultured cells were positive for vWF antigen by immunofluorescence (*11*), and the conditioned media from the cultures contained vWF antigen as determined by ELISA assay. After four additional passages, cells were harvested and total RNA prepared by guanidine-HCl extraction (*13*). Polyadenylated [poly(A)+] mRNA was isolated from total endothelial cell RNA by oligo(dT)-cellulose column chromatography (*14*). cDNA was synthesized by the method of Okayama and Berg (*15*) as modified by Gubler and Hoffman (*16*). As noted above, one library was prepared with oligo$(dT)_{12-18}$ as primer for the first strand synthesis [oligo (dT) library] while a second cDNA pool was synthesized with a random mixture of hexanucleotides (P-L Biochemicals) as primer for first strand synthesis (random library). T4 DNA polymerase was used to create blunt ends and the cDNA's were ligated to synthetic Eco RI linkers after protection of internal Eco RI sites. The linker-ligated cDNA's were digested with Eco RI and separated from free linkers by Sepharose CL4B chromatography. Complementary DNA's were then ligated to Eco RI–digested, alkaline phosphatase–treated λgt11 vector DNA (*6*), packaged in vitro (*14*), and plated on *Escherichia coli* host strain Y1088 (*6*). Each HUVEC cDNA library contained approximately 3×10^6 to 4×10^6 independent recombinant clones. Nonrecombinant background, as assessed by growth on isopropyl thio-β-D-galactopyranoside (IPTG)–X-Gal plates (*6*), was approximately 30 percent. Inserts of ten randomly chosen cDNA clones were 1 to 3 kb in length.

For detection of bacterial clones harboring recombinant phage encoding products of the fusion of vWF and β-galactosidase, we used an affinity purified heteroantiserum to human vWF (anti-vWF) (*17*). Approximately 3×10^6 recombinant clones from the above λgt11 endothelial cell cDNA libraries were screened as phage plaques on *E. coli* host strain Y1090 with anti-vWF at a 1:1000 dilution, by the method of Young and Davis (*6*) (Fig. 1A). One of nine

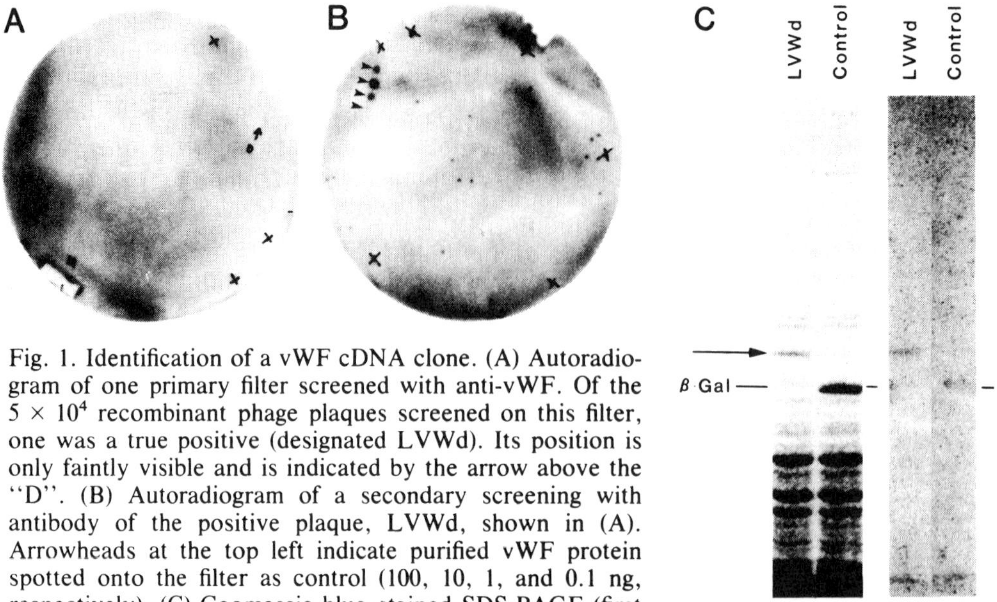

Fig. 1. Identification of a vWF cDNA clone. (A) Autoradiogram of one primary filter screened with anti-vWF. Of the 5×10^4 recombinant phage plaques screened on this filter, one was a true positive (designated LVWd). Its position is only faintly visible and is indicated by the arrow above the "D". (B) Autoradiogram of a secondary screening with antibody of the positive plaque, LVWd, shown in (A). Arrowheads at the top left indicate purified vWF protein spotted onto the filter as control (100, 10, 1, and 0.1 ng, respectively). (C) Coomassie blue–stained SDS-PAGE (first and second lanes) and Western blot with anti-vWF (third and fourth lanes) of lysate from a lysogenic bacterial strain carrying LVWd (lanes 1 and 3) or wild-type λgt11 (lanes 2 and 4); the position of a bacterial β-galactosidase standard is indicated. The arrow marks the position of the LVWd fusion protein product.

candidate positive plaques was strongly positive on a repeat antibody screening (Fig. 1B). Phage DNA of this clone (LVWd) was prepared by standard methods (*14*). Lysogens of LVWd and wild-type λgt11 were prepared in the *E. coli* bacterial host strain Y1089 (*6*). Lysogens were grown in LB medium at 32°C to $OD_{550} = 0.5$. Phage were induced by temperature shift to 45°C for 20 minutes, and β-galactosidase production was induced by the addition of IPTG (2 to 5 m*M* final concentration). After an additional 1 hour at 38°C, bacteria were harvested and resuspended in TBS with 0.2 m*M* phenylmethylsulfenyl fluoride. Lysates were prepared by two freeze-thaw cycles and sonication, and examined by gel electrophoresis and "Western blot" analysis (*17*). The β-galactosidase fusion protein material synthesized by the LVWd lysogen stained specifically with anti-vWF (Fig. 1C).

LVWd is authentic vWF cDNA. The nucleotide sequence was determined (Fig. 2) for the 549-base-pair cDNA insert of LVWd and an additional 81 bp just 3′ to it, which was obtained from the overlapping clone pVWE6 (Fig. 3). A single open reading frame encoding 193 amino acids was followed by a single termination codon. In LVWd this sequence was in the same orientation and reading frame as the β-galactosidase gene with which it was fused. This is consistent with expression of a fusion protein product. To provide evidence independent of immunodetection that this cDNA encodes vWF, the primary COOH-terminal amino acid sequence was determined by limited carboxypeptidase Y digestion of purified vWF protein

A

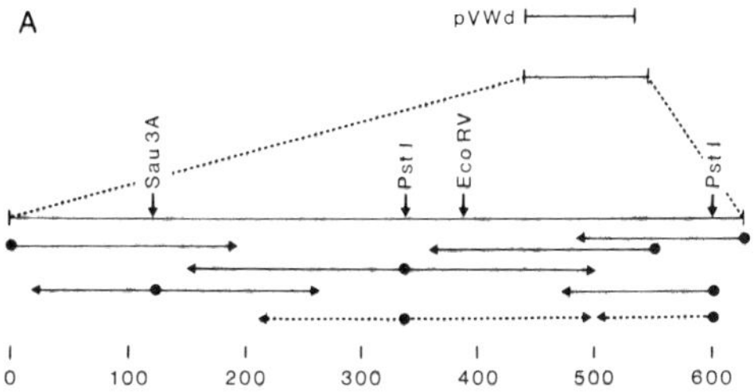

B

```
          10        20         30         40         50          60
           *         *          *          *          *           *
AA TTC CGG AAG ACC ACC TGC AAC CCC TGC CCC CTG GGT TAC AAG GAA GAA AAT AAC ACA GGT
           K   T   T   C   N   P   C   P   L   G   Y   K   E   E   N   N   T   G
                                                                   ▲
          70        80         90        100        110         120
           *         *          *          *          *           *
   GAA TGT TGT GGG AGA TGT TTG CCT ACG GCT TGC ACC ATT CAG CTA AGA GGA GGA CAG ATC
    E   C   C   G   R   C   L   P   T   A   C   T   I   Q   L   R   G   G   Q   I

         130       140        150        160        170         180
           *         *          *          *          *           *
   ATG ACA CTG AAG CGT GAT GAG ACG CTC CAG GAT GGC TGT GAT ACT CAC TTC TGC AAG GTC
    M   T   L   K   R   D   E   T   L   Q   D   G   C   D   T   H   F   C   K   V

         190       200        210        220        230         240
           *         *          *          *          *           *
   AAT GAG AGA GGA GAG TAC TTC TGG GAG AAG AGG GTC ACA GGC TGC CCA CCC TTT GAT GAA
    N   E   R   G   E   Y   F   W   E   K   R   V   T   G   C   P   P   F   D   E

         250       260        270        280        290         300
           *         *          *          *          *           *
   CAC AAG TGT CTG GCT GAG GGA GGT AAA ATT ATG AAA ATT CCA GGC ACC TGC TGT GAC ACA
    H   K   C   L   A   E   G   G   K   I   M   K   I   P   G   T   C   C   D   T

         310       320        330        340        350         360
           *         *          *          *          *           *
   TGT GAG GAG CCT GAG TGC AAC GAC ATC ACT GCC AGG CTG CAG TAT GTC AAG GTG GGA AGC
    C   E   E   P   E   C   N   D   I   T   A   R   L   Q   Y   V   K   V   G   S

         370       380        390        400        410         420
           *         *          *          *          *           *
   TGT AAG TCT GAA GTA GAG GTG GAT ATC CAC TAC TGC CAG GGC AAA TGT GCC AGC AAA GCC
    C   K   S   E   V   E   V   D   I   H   Y   C   Q   G   K   C   A   S   K   A

         430       440        450        460        470         480
           *         *          *          *          *           *
   ATG TAC TCC ATT GAC ATC AAC GAT GTG CAG GAC CAG TGC TCC TGC TGC TCT CCG ACA CGG
    M   Y   S   I   D   I   N   D   V   Q   D   Q   C   S   C   C   S   P   T   R

         490       500        510        520        530         540
           *         *          *          *          *           *
   ACG GAG CCC ATG CAG GTG GCC CTG CAC TGC ACC AAT GGC TCT GTT GTG TAC CAT GAG GTT
    T   E   P   M   Q   V   A   L   H   C   T   N   G   S   V   V   Y   H   E   V
                                                ▲
         550       560        570        580        590         600
           *         *          *          *          *           *
   CTC AAT GCC ATG GAG TGC AAA TGC TCC CCC AGG AAG TGC AGC AAG TGA GGC TGC TGC AGC
    L   N   A   M   E   C   K   C   S   P   R   K   C   S   K   ◆

         610       620        630
           *         *          *
   TGC ATG GGT GCC TGC TGC TGC CGG AAT T
```

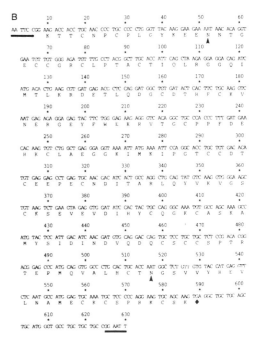

Fig. 2. (A) Restriction map of the insert of clone LVWd with the 3′ end of the overlapping clone pVWE6 (see Fig. 3), and the sequencing strategy employed. Solid lines and arrows indicate regions sequenced by the method of Maxam and Gilbert (*30*) with solid circles indicating the end-labeled restriction site. Dotted lines and arrows indicate regions sequenced by the method of Sanger *et al.* (*31*) with PstI fragments subcloned into M13 mp11 (*32*). (B) DNA sequence for 618 bp at the 3′ end of vWF cDNA. The predicted amino acid sequence for 193 amino acid residues at the COOH-terminus of vWF is shown in the single letter amino acid code (*18*). The six nucleotides at the beginning and end of the DNA sequence correspond to the synthetic Eco RI linker introduced by the cloning procedure. The single termination codon is marked by a diamond. Arrowheads indicate two potential *N*-glycosylation sites.

and subsequent high-performance liquid chromatography (HPLC) analysis of cleaved amino acids. The sequence obtained (K-C-S-K) (*18*) was identical to that immediately preceding the stop codon in our cDNA (Fig. 2B). In addition, the predicted COOH-terminal amino acid sequence E-C-K-C-S-P-R-K-C-S-K (Fig. 2B) exactly matches the 11-residue COOH-terminal primary amino acid sequence reported by Titani *et al.* (*19*). We conclude that LVWd encodes vWF and not another protein species inadvertently detected by antibody screening.

Overlapping cDNA clones spanning the vWF transcript. The insert of LVWd was used as probe to rescreen the HUVEC libraries. Positive phage were

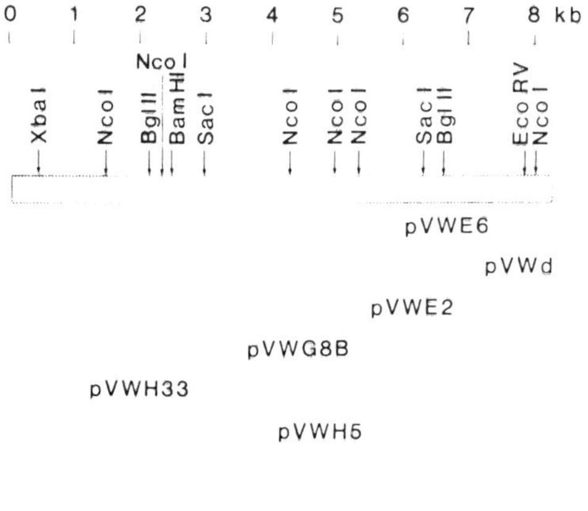

Fig. 3. Restriction map of vWF cDNA. Overlapping cDNA clones obtained by rescreening of the HUVEC library were used to construct this map. The locations of the original clone pVWd and selected overlapping clones are shown. Of the six vWF cDNA clones illustrated here, only pVWH33 was isolated from the random library. Others were obtained from the oligo(dT) library. Complementary DNA inserts of LVWd and other recombinant phage isolates were subcloned (for example, pVWd) into Eco RI–cut and bovine alkaline phosphatase–treated pUC-13 plasmid (P-L Biochemicals). The probes and strategy used are described in the text. Fragments were ^{32}P-labeled by an oligonucleotide primer labeling method (*33*).

detected with an abundance of about 0.3 percent. Seven of the vWF recombinant phage were plaque-purified and the four largest cDNA inserts were subcloned into the plasmid vector pUC-13. A 270-bp Eco RI–Pst I fragment from the 5′ end of pVWE2 (see Fig. 3) was used as probe to rescreen both HUVEC cDNA libraries, and a third series of overlapping vWF cDNA clones was identified. A 400-bp Stu I fragment from the 5′ end of one of these clones (pVWG8b) was then used as probe for a fourth round of screening of the primary libraries, and another set of cDNA clones further toward the 5′ end was obtained. These overlapping clones span 8.2 kb of vWF mRNA (Fig. 3).

The nucleotide sequence shown in Fig. 2B contains the translation termination signal TGA (*18*) at a position corresponding to 8.15 kb in the vWF cDNA map (Fig. 3). The remaining 5′ potential coding sequence is sufficient in length to encode a polypeptide of 300 kD, assuming a mean residue molecular weight of 111 based on the amino acid composition of vWF (*20*). The termination codon is followed by 34 bases of 3′ untranslated region (Fig. 2B). Neither the conserved endonucleolytic poly(A) addition signal AATAAA (*21*) nor a 3′ poly(A) tract are contained in this sequence. We surmise that additional 3′ untranslated sequences exist in vWF mRNA.

As is discussed above, the COOH-terminal sequence obtained by primary amino acid sequence analysis of purified vWF polypeptide is identical to the predicted amino acid sequence immediately preceding the stop codon in our cDNA (Fig. 2B). These data indicate that no posttranslational processing occurs at the COOH-terminal end of the vWF polypeptide.

vWF protein structure. Previous biochemical studies have demonstrated that vWF is rich in proline (6.8 percent) and cysteine (7.3 percent) (*20*) with each subunit participating in 69 intrachain and interchain disulfide bonds (*20, 22*). The predicted sequence of vWF shown in Fig. 2B contains 5.2 percent proline and 12.4 percent cysteine in 193 residues. This COOH-terminal fragment also contains two potential *N*-glycosylation sites (Fig. 2B).

Expression of vWF mRNA. When used as a hybridization probe in Northern blot analysis (Fig. 4), vWF cDNA detects a single large (approximately 8 to 10 kb) endothelial cell–specific RNA species. The observed signal intensity is in accord with the 0.3 percent frequency of positive phage detected on rescreening of the HUVEC cDNA library. No hybridization was observed with RNA's from human fibroblasts, HeLa cells, a human T-cell leukemia line (Fig. 4), or human kidney (*23*).

Gene structure and chromosomal localization. Portions of the vWF cDNA have been used as probes in Southern blot analyses to assess the complexity of the cellular vWF gene and its chromosomal localization. Initial use of LVWd as a probe revealed a simple pattern of hybridization, consistent with the presence of a single copy in the human genome (Fig. 5). Probes spanning almost the complete vWF mRNA detected a complex array of 17 or more bands (Fig. 5). This pattern indicates a large gene interrupted by at least 16 introns and spanning a minimum of 80 kb of human genome. No rearrangement or deletion of vWF sequences was detected in DNA obtained from a patient with a severe variant of vWD (Fig. 5). The patient had no detectable vWF antigen (VIIIR:Ag, 0

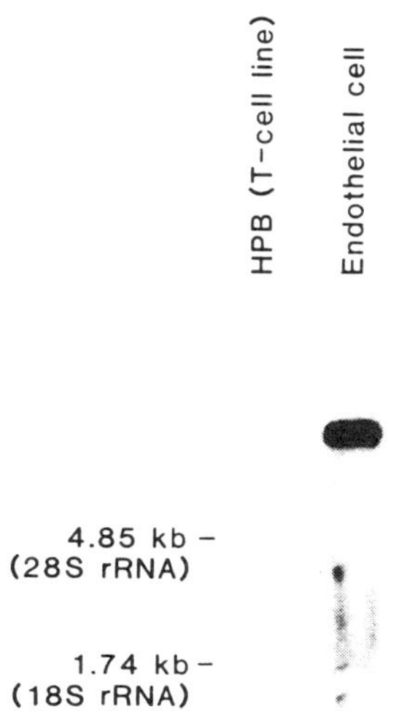

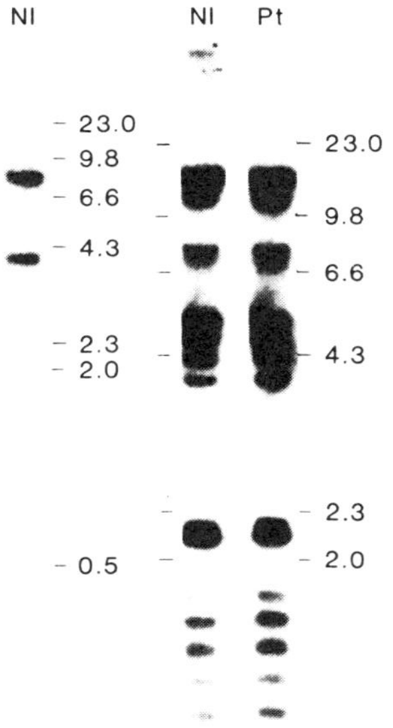

Fig. 4 (left). Northern blot analysis. Total cell RNA from HPB-ALL (a T-cell line) (lane 1), endothelial cells (HUVEC) (lane 2), fibroblasts (lane 3), and HeLa cells (lane 4) were hybridized with pVWd cDNA insert as probe. The location of ribosomal RNA size markers is indicated. A single hybridizing band of approximately 8 to 10 kb was evident in the endothelial cell lane. No signal was detectable in the other lanes. Total RNA was prepared from endothelial cells (HUVEC) and nonendothelial cells (HeLa cells, fibroblasts, and the T-cell line HPB-ALL), by lysis in a buffer containing 6*M* guanidine-HCl (*13*). The RNA's (20 μg of total RNA per lane) were denatured with formaldehyde, subjected to electrophoresis in agarose, and transferred to nitrocellulose filters (*14*). The filters were hybridized with ^{32}P-labeled LVWd insert. Hybridization was in 10 percent dextran, 50 percent formamide, 5× standard saline citrate (SSC) (0.75*M* NaCl, 0.075*M* sodium citrate) at 42°C; the final washing was done in 0.1× SSC at 68°C. Autoradiography with an intensifying screen at −80°C was for 24 hours (above), or 10 days (not shown). Fig. 5 (right). Southern blot analysis (*14*) of normal and vWD patient genomic DNA. Lane 1 shows normal human DNA digested with Bam HI with LVWd insert as probe. DNA's in lanes 2 and 3 have been digested with Eco RI and probed with nearly complete vWF cDNA (that is, pooled radiolabeled inserts of pVWH5 and pVWH33; see Fig. 3). The DNA in lane 2 was prepared from a normal individual, and the DNA in lane 3 from a patient with severe vWD. Size marker positions are indicated to the right of lane 1 and lanes 2 to 3. Hybridization and washing conditions were identical to those described in Fig. 4 above, except that dextran was omitted.

percent); this is measured with the use of an antibody to the human factor VIII-vWF complex, with pooled normal plasma as control. Ristocetin cofactor activity (VIIIR:RCo) was 10 percent of control. This is a functional assay of vWF based on the observation that the antibiotic ristocetin induces platelet aggregation in the presence of vWF (*1*). Analysis of DNA from a second patient with severe vWD (VIIIR:Ag, 0 percent, VIIIR:RCo, 0 percent), similarly yielded a pattern indistinguishable from normal (*23*).

The chromosomal assignment of the vWF gene was established by use of somatic hybrid mapping panels (*24*) and by in situ chromosomal hybridization (*25, 26*) (Fig. 6, A to C). The hybrid cell studies indicated that the vWF locus lies on 12p or proximal 12q (Fig. 6A). By the in situ hybridization analysis the vWF gene was localized to the region 12p12 → 12pter (Fig. 6, B and C).

Discussion. We report here the isolation of nearly full-length cDNA for human vWF and initial characterization of the vWF genetic locus. Several aspects of our cloning experiments deserve comment.

The initial vWF cDNA clone was isolated by antibody screening in the λgt11 expression vector. Only one positive clone was identified in 3×10^6 recombinant phage. This low frequency of positive plaques on initial screening of the HUVEC libraries with anti-vWF contrasts with the much higher abundance of vWF clones in the same libraries when assayed by molecular hybridization. This discrepancy is partially explained by the fact that only one of six vWF cDNA inserts is expected to be in the appropriate orientation and reading frame for expression as a fusion protein product. Additional factors are likely to contribute more substantially. Other features of the cDNA insert may influence the level of fusion protein in *E. coli*. Specifically, short peptide extensions may yield fusion products that are more stable or produced in larger amounts. In support of this hypothesis, a number of other workers using the λgt11 system have noted small inserts in their initial antibody screening isolate (*7*). In addition, epitopes expressed by cDNA clones corresponding to different regions of coding sequence may vary in reactivity with anti-vWF. Another important factor may be the characteristics of our particular anti-vWF preparation, that is, its relative avidity and titer for the multiple individual epitopes in vWF. It is possible that a different antibody preparation might detect additional clones in our cDNA library by this screening method.

Partial sequence of vWF cDNA in conjunction with primary COOH-terminal peptide sequence [our own and (*19*)], establishes that our cDNA's are derived from authentic vWF mRNA and indicates that the extensive posttranslational processing of the vWF precursor is restricted to the NH_2-terminus. DNA sequence analysis of 2.7 kb at the 5′ end of our cDNA clone pVWH33 (*23*) reveals one continuous open reading frame. In addition, a putative hydrophobic leader sequence is present immediately after the first ATG in clone pVWH33. These data suggest that the coding region extends close to the 5′ end of the cDNA. Given the location of the termination codon at position 8.15 kb (Figs. 2 and 3), these data imply that the primary vWF translation product is at least 300 kD in size. Thus, the intracellular vWF precursor observed in pulse-chase experiments

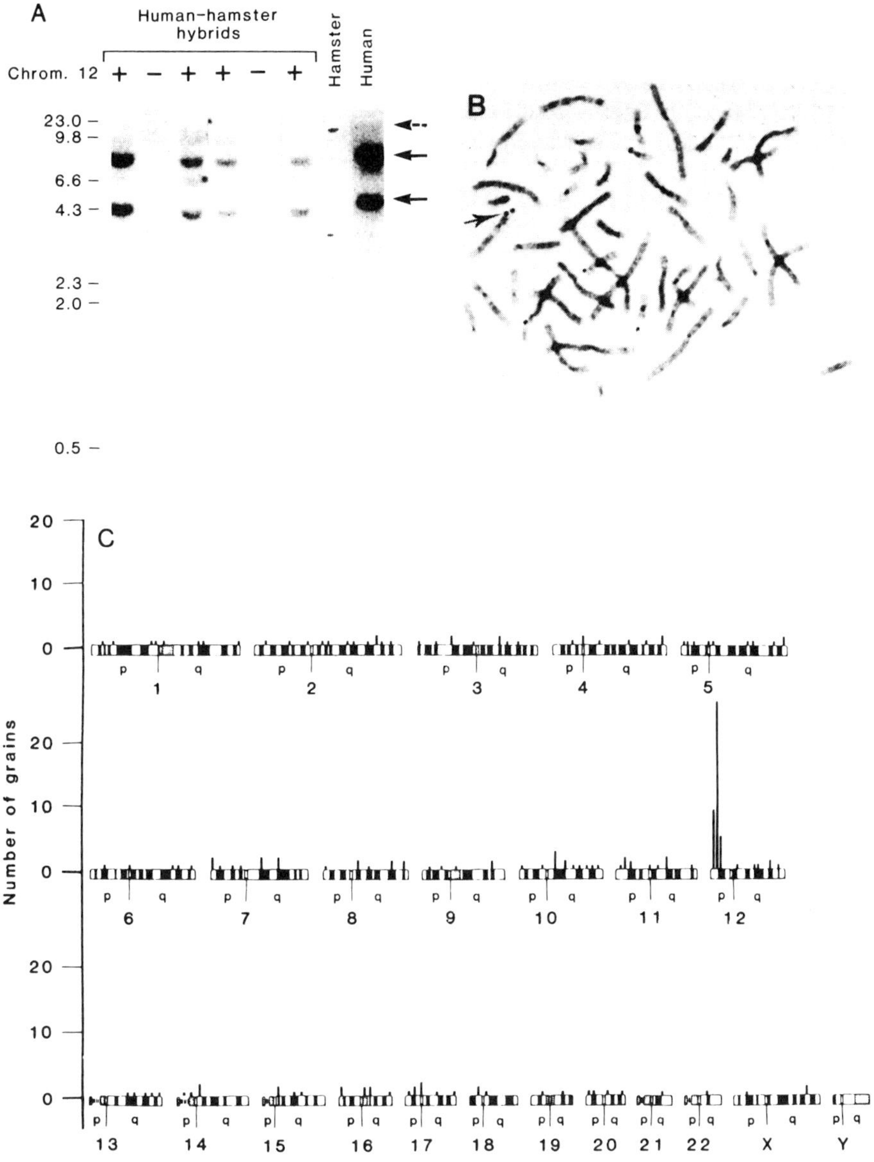

Fig. 6. Chromosomal mapping of the human vWF genetic locus. (A) Hybridization of the insert of pVWE6 with Bam HI restricted DNA's from human-hamster somatic cell hybrids (lanes 1 to 6), hamster cells (lane 7), and human cells (lane 8). The hybrid cells corresponding to the DNA's in lanes 1, 3, 4, and 6 contain human chromosome 12p or proximal 12q while those corresponding to lanes 2 and 5 lack this chromosome. The segregation of the two human genomic components recognized by this probe was completely concordant with that of chromosome 12p–proximal 12q in the 21 clones analyzed and discordant with the segregation of each of the other autosomes, the X and the Y (discordancy fractions 0.20 to 0.63). The solid arrows indicate the two human components and the dashed arrow the hamster component. Size marker positions are shown at the left. (B) In situ hybridization was performed by a modification of the method of Harper and Saunders (*25*) as previously described (*26*) with tritium-labeled pVWE6 as probe; positive grains over distal 12p are indicated by an arrow. (C) Histogram of positive grains from a total of 134 metaphases evaluated. Thirty-nine (29 percent) contained tritium over the region 12p12 → 12pter and, of a total of 192 grains, 40 (21 percent) were found over this region. No other major sites of hybridization were seen.

may be larger than the previous estimates of 240 to 260 kD as determined by gel electrophoresis (*1, 4*). Alternatively, an additional NH_2-terminal processing step may be required to generate the 240- to 260-kD precursor. Further studies are necessary to define these processing events.

Of the normal tissues examined, vWF mRNA was detected only in cultured endothelial cells, consistent with the cellular distribution of vWF as determined by antibody staining methods (*1, 3*). vWF mRNA of normal size was also detected in a biopsy specimen of Kaposi's sarcoma (*23*). Although controversial, the malignant cell in Kaposi's sarcoma has been postulated to be of endothelial cell origin (*3*). Our observations support this hypothesis.

The human vWF gene appears large and interrupted by multiple introns (Fig. 5) and resides on chromosome 12 (12p12 → 12pter) (Fig. 6). A report assessing vWF expression in hybrids of human endothelium and rodent fibroblasts suggested an assignment of vWF to chromosome 5 (*27*). Cosegregation of chromosomes 5 and 12 in human-rodent hybrid lines (*28*) may account for this finding. A number of other genetic loci have been mapped to 12p (*29*) but none of these have been previously linked to vWF. The proposed linkage between glutamate-pyruvate-transaminase [mapped to chromosome 8 or 16 (*29*)] and vWD (*5*) now appears to be unlikely. It is possible, however, that some subtypes of vWD are the result of genetic defects at loci involved in post-translational processing, modification, or transport of vWF. Such defects could result in a vWD phenotype that is unlinked to the vWF gene. The two patients studied here showed no evidence for a gross alteration at the vWF locus. They may have either a small deletion or insertion, or a single nucleotide substitution in the vWF gene or, alternatively, a defect at another locus as the molecular basis of their vWD.

Further studies of vWF biosynthesis and the molecular basis of vWD should benefit from the availability of vWF cDNA clones. The complete amino acid sequence of vWF, predicted from the cDNA, will help define the nature of vWF precursor processing, secretion, and multimer assembly. Application of DNA probes to the analysis of vWD offers the potential for sorting out the phenotypic heterogeneity of the disorder on a molecular genetic level, and for defining those regions of vWF that directly influence protein assembly and function.

References and Notes

1. T. S. Zimmerman, Z. M. Ruggeri, C. A. Fulcher, *Prog. Hematol.* **13**, 279 (1983).
2. J. Gitschier *et al., Nature (London)* **312**, 326 (1984); W.I. Wood *et al., ibid.*, p. 330; G. A. Vehar *et al. ibid.*, p. 337; J. J. Toole *et al., ibid.*, p. 342.
3. L. S. Guarda *et al., Am. J. Clin. Pathol.* **76**, 197 (1981); M. Miettinen *et al., ibid.* **79**, 32 (1983).
4. D. D. Wagner and V. J. Marder, *J. Biol. Chem.* **258**, 2065 (1983); D. C. Lynch *et al., ibid.*, p. 12757.
5. L. R. Goldin *et al., Am. J. Med. Genet.* **6**, 279 (1980); M. S. Verp *et al., Clin. Genet.* **24**, 434 (1983).
6. R. A. Young and R. W. Davis, *Proc. Natl. Acad. Sci. U.S.A.* **80**, 1194 (1983); in *Genetic Engineering*, J. Setlow and A. Hollaender, Eds. (Plenum, New York, in press), vol. 7; *Science* **222**, 778 (1983).
7. J. E. Schwarzbauer *et al., Cell* **35**, 421 (1983); R. C. Peterson *et al., Proc. Natl. Acad. Sci. U.S.A.* **81**, 4363 (1984).
8. S. P. Leytus *et al., Proc. Natl. Acad. Sci. U.S.A.* **81**, 3699 (1984); D. Foster and E. W. Davie, *ibid.*, p. 4766.
9. HUVEC's were originally obtained from M. Gimbrone [M. A. Gimbrone, Jr., R. S. Cotran, J. Folkman, *J. Cell Biol.* **60**, 673 (1974)] and were contributed by R. Weinstein.
10. T. Maciag, G. A. Hoover, R. Weinstein, *J. Biol. Chem.* **257**, 5333 (1982); bovine endothelial cell

growth factor and fibronectin were both provided by R. Weinstein.
11. T. Maciag *et al.*, *J. Cell Biol.* **91**, 420 (1981).
12. S. C. Thornton, S. N. Mueller, E. M. Levine, *Science* **222**, 623 (1983).
13. R. C. Strohman *et al.*, *Cell* **10**, 265 (1977).
14. T. Maniatis, E. F. Fritsch, J. Sambrook, *Molecular Cloning: A Laboratory Manual* (Cold Spring Harbor Laboratory, Cold Spring Harbor, N.Y., 1982).
15. H. Okayama and P. Berg, *Mol. Cell. Biol.* **2**, 161 (1982).
16. U. Gubler and B. J. Hoffman, *Gene* **25**, 263 (1983).
17. Rabbit heteroantiserum prepared against human factor VIII-vWF was passed over gelatin-Sepharose to remove contaminating fibronectin, and affinity-purified by absorption and elution from a column of vWF-Sepharose. vWF-Sepharose was prepared by incubation (overnight at 40°C) of 1 mg of purified vWF with 10 ml of CNBr-activated Sepharose 4B (Pharmacia) at *p*H 7.6 and subsequent washing with 1 liter of TBS (50 m*M* tris, *p*H 8.0, 150 m*M* NaCl). The antiserum (2 ml) was applied to 1 ml of packed gel and washed extensively, and the antibody was eluted at *p*H 11. Purified vWF protein was digested with staphylococcal V8 protease, reduced, fractionated by electrophoresis on 7 percent polyacrylamide gels (PAGE) in the presence of sodium dodecyl sulfate (SDS) and β-mercaptoethanol, transferred to nitrocellulose (Western blot), and stained with the affinity-purified vWF antibody preparation (anti-vWF); an immunoperoxidase indicator system was used. Multiple bands were detected, including all those seen on a duplicate gel stained with Coomassie blue, indicating that this antibody preparation recognized multiple epitopes on the vWF molecule. This anti-vWF was used to screen the λgt11 endothelial cell cDNA libraries as described in the text. This anti-vWF was also used in a "Western blot" analysis of the β-galactosidase fusion protein product of LVWd (Fig. 1C). The blot procedure was identical to that described above except that incubation with affinity purified, ^{125}I-labeled horse antibody to rabbit Ig followed by autoradiography was used as the indicator system.
18. The single letter abbreviations for the amino acid residues are: alanine, A; arginine, R; asparagine, N; aspartic acid, D; cysteine, C; glutamic acid, E; glutamine, Q; glycine, G; histidine, H; isoleucine, I; leucine, L; lysine, K; methionine, M; phenylalanine, F; proline, P; serine, S; threonine, T; tryptophan, W; tyrosine, Y; valine, V. The single letter abbreviations for the DNA bases are A, adenine; C, cytosine; G, guanine; and T, thymine.
19. K. Titani *et al.*, paper presented at the American Heart Association National Meeting, Miami, November 1984.
20. M. E. Legaz *et al.*, *J. Biol. Chem.* **248**, 3946 (1973).
21. N. J. Proudfoot, *Nature (London)* **307**, 412 (1984).
22. J. Loscalzo, M. Fisch, R. I. Handin, *Biochemistry*, in press.
23. D. Ginsburg, D. T. Bonthron, R. I Handin, S. H. Orkin, unpublished data.
24. N. Kanda *et al.*, *Proc. Natl. Acad. Sci. U.S.A.* **80**, 4069 (1983).
25. M. E. Harper and G. F. Saunders, *Chromosoma* **83**, 431 (1981).
26. T. A. Donlon *et al.*, *Am. J. Hum. Genet.* **35**, 1097 (1983).
27. N. A. Kefalides, *Fed. Proc. Fed. Am. Soc. Exp. Biol.* **40**, 327 (abstract) (1981).
28. G. P. Bruns, unpublished data.
29. P. S. Gerald and K. H. Grzeschik, *Cytogenet. Cell Genet.* **37**, 103 (1984); T. B. Shows, P. J. McAlpine, R. L. Miller, *ibid.*, p. 340.
30. A. M. Maxam and W. Gilbert, *Methods Enzymol.* **65**, 499 (1980).
31. F. Sanger, S. Nicklen, A. R. Coulson, *Proc. Natl. Acad. Sci. U.S.A.* **74**, 5463 (1977).
32. J. Messing, *Methods Enzymol.* **101**, 20 (1983).
33. A. P. Feinberg and B. Vogelstein, *Anal. Biochem.* **132**, 6 (1983).
34. We thank Dr. Robert Weinstein and Karen Wenc for assistance in growing HUVEC's, Dr. Alex Markham for help with the identification of clones isolated in the library rescreening, Dr. Paula Bockenstedt for assistance with the antibody preparation and fusion protein analysis, Dr. Alan Ezekowitz for help with the Western blots, Drs. Carl O'Hara and Jerome Groopman for providing the Kaposi's sarcoma specimen, and Dr. Joseph Loscalzo and Jean-Paul Bouissel for help with the carboxypeptidase amino acid sequence analysis. Supported by NIH grants to D.G., S.H.O., R.I.H., and S.A.L., and a Junior Faculty Clinical Fellowship Award (D.G.) from the American Cancer Society.

26 April 1985; accepted 23 May 1985

Part V

Biotechnology

18

Biotechnology in Food Production and Processing

Dietrich Knorr
and Anthony J. Sinskey

The use of biotechnology in the manufacture of food and beverages has been practiced for more than 8000 years with vinegar, alcoholic beverages, sourdough, and cheese production being the most prominent examples (*1*). Biotechnological processes are now being used to produce other fermented products, food and feed additives, and processing aids (Table 1). In fact, the food processing industry, which has annual sales of $300 billion in the United States and about £30 billion in Great Britain, is the oldest and largest user of biotechnological processes (*2*).

An important issue today is the impact that modern biology will have on the food industry. Recent advances in molecular biology, fermentation science, tissue culture systems, and bioengineering offer great potential for application to several areas of food production and processing. A key question related to this issue is, what are the constraints hindering applications of biotechnology to the food industry? In this article, we attempt to address these points and to review the current role of biotechnology in the production and processing of food.

Biotechnology in Food Production

Biotechnology can significantly influence the food supply, including the production and preservation of raw materials and the alteration of their nutritional and functional properties. In addition, development of production aids, processing aids, and direct additives such as enzymes, flavors, polysaccharides, pigments, and antioxidants can improve the overall utilization of raw materials.

Science 229, 1224–1229 (20 September 1985). This article was adapted from a position paper presented at the Institute of Food Technologists workshop on research needs, 11 to 14 November 1984, Arlington Heights, Illinois.

Table 1. World production, market value, and end use of selected products of the biotechnology-based food industry (*3*).

Products	Production (metric tons per year)		Market size (millions U.S. $)			Primary end use
	1974	1981	1974	1981	1990 (estimated)	
Amino acids		455×10^3*	290	1.9×10^3† 1.8×10^3*	2.2×10^3‡	Feed additive, food enrichment and flavoring agent, feed preservative
Citric acid	265×10^3	300×10^3†				Food additive, processing aid
Enzymes		65×10^3	132	310 to 400	1.5×10^3	Processing aid
Vitamins				668† 1.1×10^3		Feed and food additive, food enrichment agent
Baker's yeast	0.97×10^6	1.75×10^6†	380			Food additive, enrichment agent
Beer	75×10^6	87×10^6†		27×10^3§*	44×10^3§	Beverage
Cheese	11×10^6	12×10^6				Food
Fermented foods				3.5×10^3	6×10^3	Food
Miso‖		5.72×10^3				Food
Soy sauce‖		1.18×10^6*				Food

*Data for 1982. †Data for 1979. ‡Fermented chemicals. §Alcoholic beverages. ‖Japan only.

Raw materials. Plant products derived from fewer than 30 plant species provide more than 90 percent of the human diet. Eight cereal crops supply more than half the world's calories (*4*). Animal products contribute over 56 million tons of edible protein and over 1 billion megacalories of energy annually (*5*). In addition, the increasing importance of marine food products and single cell proteins (SCP) as raw materials has been stressed (*6*).

Currently, the role of biotechnology in raw material production is directed toward (i) increasing productivity through improved efficiency of nutrient use and conversion, (ii) increasing productivity through improved plant resistance, and (iii) identifying new food sources with desirable properties.

Feed efficiency and productivity of animals has been increased substantially (*7*). Furthermore, SCP, derived from the dried cells of microorganisms for use as protein sources in human food and animal feeds, are cultivated on a large scale by using both photosynthetic and nonphotosynthetic microorganisms (*8*). Extensive work is under way to increase the ability of plants to fix atmospheric nitrogen for their metabolic use (*9*), and cultured plants and plant cells are being considered for food production (*10*).

Additional efforts include the genetic improvement of animal breeds; improvements in the reproductive efficiency of livestock; the use of vaccines and monoclonal antibodies in the diagnosis, prevention, and control of animal diseases; and the improvement of crop species through the regulation of endogenous genes, the transfer of DNA from one species to another (for example, fusion of cells, transfer of subcellular organelles, vector-mediated DNA transfer), the improvement of plant resistance factors (for example, plant and microbial produced pesticides), and the improvement of photosynthetic efficiency (*7, 9, 11*). In addition, the use of the genetic diversity in plants, new plant and animal food sources, and improved food production technologies (such as aquaculture, hydroponics, continuous tissue culture, and solid-state fermentation) are being sought continuously (*9, 10, 12*).

Raw material modification and improvement can be applied to convert raw materials, to increase stress resistance, and to improve their functionality and nutritional quality.

In the processing of raw materials for food, polymeric carbohydrates may be removed, included in the product as dietary fiber, or converted to other products, such as sugars. The ability to convert these polysaccharides and to impart specific structural changes can result in improvements in the functionality of carbohydrates in foods and in increases in product yields (*12*). In vitro selection has been applied to improve the salt and cold tolerance and the herbicide and drought resistance of crop plants. Work on the improvement of such functional properties as color, flavor, and texture of raw materials is being conducted, as is work on increasing the essential nutrients and reducing the undesirable constituents in raw materials (*13*).

Raw material preservation by biological processes is critical to agriculture and the food processing industry. The production of silage, the fermentation of cocoa and coffee beans, the "fermentation" (oxidation) of tea leaves, and the conversion of raw material into SCP or feedstock chemicals (*1, 14*) suggest the diversity and magnitude of microbial fermentation processes applied for raw material preservation and quality improvement (*1, 14*).

Additives and production/processing aids. Additives and production aids used in raw material production include such materials as vaccines and growth regulators in animal production and microbial insecticides and herbicides in plant production, and they all are subject to intensive investigation (*11, 15*). Historically, such food additives as fatty acids and other organic acids, vanilla (for flavor), and vitamins B_2, B_{12}, C, and D have been produced through biotechnological processes. Intensive work is now being carried out on the production of plant metabolites by tissue culture, including flavors, pigments, vitamins, enzymes, antioxidants, antimicrobials, and lipids and on the microbial production of flavors, pigments, vitamins, amino acids, antioxidants, biosurfactants, and polysaccharides (*16*). For example, the plant metabolite shikonin, a bright red naphthaquinonine compound used as a dye and as an antibacterial and anti-inflammatory agent, is currently produced on an industrial scale from cultured *Lithospermum erythrorhizon* cells (*17*). Blue-green algae that produce tocopherols have been isolated, and a potential vitamin E precursor has been found in various genera of bacteria and yeasts. This suggests the potential for the development of a one-step fermentation process to fulfill the demand for vitamin E (*18*).

Aspartame (L-aspartyl-L-phenylalanine methyl ester) is a low-calorie dipeptide sweetener that has recently been approved as a food additive. Precursors such as aspartic acid and phenylalanine are produced by fermentation processes, and the microbial production of the dipeptide aspartyl-phenylalanine has been performed at the laboratory level by recombinant DNA processes. Worldwide demands for L-phenylalanine are projected to increase from 50 metric tons in 1981 to 7900 metric tons in 1990 (*19*).

Improved processes for the production of other amino acids, such as L-lysine and L-threonine, have been developed recently (*20*).

Polysaccharides commonly derived from algae or botanical sources and used as functional agents are now being produced commercially through microbiological processes. The search for new microbial polysaccharides is an area of active research. Recent advances toward understanding the specific steps involved in the biosynthesis of specific polysaccharides offer promise for the control and manipulation of the structure and form of the final polysaccharide product (*21*). There are a number of food-related applications of polysaccharides; they include the microencapsulation of flavors, immobilization of enzymes, entrapment of whole cells, and aiding of flocculation in food process waste management (*22*).

Yeasts have been used traditionally in the production of alcoholic beverages, and attention recently has been given to the genetic manipulation of *Saccharomyces* yeast cells to increase the efficiency of the brewing process and to prepare low-calorie, or so-called light, beers. Currently, glucoamylase enzymes from microbial sources (for example, *Aspergillus niger* and *Aspergillus awamori*) are used in the production of many light beers. These enzymes are fairly thermostable and are not destroyed by normal beer pasteurization at 60° to 62°C, so the beers become sweeter upon storage owing to the release of glucose units from dextrins by the glucoamylase. Thus, the production of a thermosensitive glucoamylase by a brewer's yeast could be of significant value (*23*).

Enzymes are used extensively in food production and processing. The ones most widely applied are amylases, glucose oxidases, proteases, pectic enzymes, and lipases. Excellent reviews on the production and utilization of food enzymes are available (*24*). Immobilized enzymes and immobilized whole cells have received significant attention as valuable biocatalysts for the food processing industry (*25*). The advantages of the application of immobilized systems include continuous operation, reuse of the biocatalyst, ease of process control, improved biocatalyst stability, and reduced waste disposal problems (*26*).

Immobilized biocatalyst technology also can be applied successfully to the production of secondary plant metabolites (*27*). In addition, the production of enzymes with enhanced stability to temperature and other processing conditions is receiving much attention (*28*).

The significant impact that biotechnology can have on the production of a food ingredient is exemplified by the development of high-fructose corn syrup (HFCS) technology. The production of HFCS involves the application of two amylases and glucose isomerase to effect the liquification and subsequent saccharification of cornstarch to yield an approximately equimolecular mixture of fructose and glucose. Because fructose is sweeter than glucose, HFCS is about as sweet as a sucrose syrup of the same solids content, and it has found wide use in processed foods. About 2.5 million metric tons of HFCS (dry basis) were produced in 1981, compared to about 72,000 metric tons in 1976. Over a 10-year period (1970 to 1980), HFCS increased its share of U.S. per capita consumption of nutritive sweeteners from almost nonexistence to 16.4 percent, while sucrose usage decreased from 84.1 percent to 68.0 percent (*2, 25*). The production of HFCS through the use of enzyme technology is one of the greatest commercial successes of immobilized biocatalyst technology (*29*).

Production methods. With the availability of high-performance bioreactors for large-scale fermentation processes (*30*), current emphasis is on computer process control, although there is still a need to improve bioreactor performance by overcoming the limitations of heat and mass transfer (*31*). This is especially important with new biocatalysts and the scale-up of such processes as animal and plant cell culture systems (*32*). The engineering problems are especially challenging when non-Newtonian systems are involved (*33*).

Extensive work on immobilization techniques, reactor design, and cell membrane permeabilization will help to overcome the current problems in continuous animal and plant cell cultures; these problems include shear sensitivity, slow growth rates, and the intracellular storage of metabolites (*34*).

Biotechnology in Food Processing

Biotechnology in food processing can significantly affect food product composition, quality, and functionality by providing tools for product modification, preservation, and stabilization, as well as for safety, characterization, and quality control. In addition, processing methods, especially separation and fermentation processes and waste treatment and utilization can contribute to the improvement of food products.

Product modification. Significant advances have been made in the modification of food components, such as proteins, polysaccharides, fats, and oils. Protein modifications, for example, include limited enzymatic hydrolysis to alter food functionality; the reverse process, the so-called plastein reaction, has been proposed as a method to create proteinlike materials to develop new food products. Modification of properties of proteins by combining information on crystal structure and protein chemistry with artificial gene synthesis is also being explored (*3, 35*).

Meat tenderization with papain is one example of the large-scale application of enzymatic hydrolysis to modify product functionality. Other potential processes are the enzymatic reduction of limonoid bitterness in citrus products to improve flavor (*36*) and the modification of the fatty acid composition of triglycerides by lipases. One example is the enzymatic modification of olive oil and stearic acid to a fat similar to cocoa butter, particularly the formation of 1-palmitoyl-2-oleyl-3-stearoly-*rac*-glycerol, the major triglyceride of cocoa butter, which has been obtained on reacting oleic anhydride with 1-palmitoyl-3-stearoyl-*rac*-glycerol in the presence of lipase. Furthermore, the development of a two-stage microbial process for producing glycerides having cocoa butter characteristics requires mention (*37*).

Product preservation. Historically, there has been extensive use of microbial metabolism for food preservation and stabilization, especially for dairy, meat, fish, fruit, and vegetable products (*38*). The efficiency of microorganisms used in the food fermentation industries potentially can be enhanced by genetic manipulation of starter cultures. However, additional fundamental knowledge of the genetics, biochemistry, and molecular biology of organisms used as starter cultures is required (*39*).

Product safety, characterization, and quality control. Besides the use of classical methods to ensure the quality and safety of food and to identify food components (*40*), three recent developments are relevant to product safety, characterization, and quality control: (i) the potential application of monoclonal antibodies to determine optimal crop harvesting and product freshness (*41*), (ii) the use of biosensors and DNA hybridization techniques for quality control (*42*), and (iii) the potential of tissue culture and genetic methods for nutrient and toxicity assessment (*43*). In addition, the regulatory and safety aspects of biotechnology and their impact on the nutritional quality of the resulting food products are being examined (*44*).

Processing methods. Mechanical unit operations used for product purification and recovery include sedimentation, centrifugation, and filtration, along with dialysis, flotation, and ultrafiltration (*45*). Biomass separation is commonly aided by bioflocculation or by the use of synthetic polyelectrolytes. Recently, natural polyelectrolytes such as chitin and chitosan have been investigated as substitutes for synthetic polyelectrolytes (*45, 46*). Application of aqueous two-phase (liquid-liquid) systems for the extractive purification of enzymes (*47*) and the supercritical extraction (*48*) of food ingredients are becoming increasingly important. In supercritical extraction, carbon dioxide is favored as the dense gas because it is nontoxic, nonexplosive, cheap, readily available, and easily removed from extracted products (*49*). Supercritical extraction is currently used on an industrial scale for decaffeinating coffee and tea. Scale-up of high-performance liquid chromatography (HPLC) separation processes is also being explored (*50*).

Nonlipolytic enzymes have been used to enhance the extractability of oil from seeds (*51*), and pectolytic enzymes are applied to increase yields in the processing of liquid fruit and vegetable products (*52*). In addition, cofermentation processes have been suggested to aid the separation and purification of secondary metabolites (*53*).

Treatment and utilization of process waste. Because of the large volumes involved in the production and processing of food, generated wastes create disposal and pollution problems. In addition, there is a substantial loss of essential nutrients. For example, 20 million metric tons of whey, the fluid that results from the separation of curd when converting milk into cheese, accumulate annually in the United States (*54*). Whey contains more than half of the nutrients of the milk used in cheese production, including 1 percent protein and 5 percent lactose. Approximately 50 percent of the total whey solids is disposed of in various industrial and municipal waste-treatment operations (*55*).

Biomass recovery, especially isolation of valuable protein by-products, has been carried out in the food processing industry for an extended period of time. The isolation of protein concentrates from potato processing wastes, for example, has been used on an industrial scale for several decades, and the product's potential for food application has been investigated extensively (*56*). During the past decade, ultrafiltration has become useful in food processes, especially for the recovery of whey protein from cheese, cottage cheese, or industrial casein processing wastes (*57*). By-product recovery has also been explored for application to the processing of meat, cereal, dairy, fruits and vegetables, and fish and shellfish, as well as fermentation operations (*58*).

The multifunctional potential of food processing wastes for by-product recovery and conversion can be illustrated by the case of chitin—poly-β (1,4) *N*-acetyl-D-glucosamine—which is a waste product of the shellfish industry and one of the most abundant polysaccharides in the world. It has been shown to have numerous potential food applications, such as being used as a possible dietary fiber, a functional ingredient, and an immobilizer of enzymes. Chitosan (partially deacetylated chitin) has been effective in aiding the separation of colloidal and dispersed particles from food process wastes and has the potential for being used for the microencapsulation of flavor and for the entrapment of whole cells (*59*). Chitin bioconversion to SCP has also been reported, and numerous additional applications of chitin and chitosan are being investigated (*60*).

Bioconversion of food processing wastes includes the use of substrates such as starch or whey. The so-called Symba process utilizes a symbiotic culture of two yeasts, *Endomycopsis fibuliger* and *Candida utilis*, to convert potato starch into SCP. A *Kluyveromyces fragilis* and *Candida intermedia* symbiotic culture, which is characterized by an exclusively oxidative lactose metabolism, is being used for production of protein-enriched whey (*61*). Other examples of process waste bioconversions are the application of molasses and corn steep liquor as substrates in many fermentation processes and the production of vinegar (from "waste" wine) (*1*). Furthermore, the anaerobic digestion of food wastes to provide methane for fuel use is now being used on a commercial scale (*62*).

Research Needs

As the above discussion indicates, there have been numerous achievements in the field of food biotechnology and there exist many more potential opportunities. One critical factor, however, is the formulation of achievable objectives to aid the rational improvement of food production and processing technologies and to reach desired goals for product quality. What is needed is a rational program for the food sector that has a structure based on identification of essential research needs. Also needed is increased investment in food research and development, which currently constitutes only about 0.3 percent of industry-based business in the United States, as well as long-term commitments to research projects in food biotechnology.

At a workshop sponsored by the Institute of Food Technologists, scientists from industry, government, and academia made the following statement concerning the research needs in this important area (*63*):

> Biotechnology directed toward the general area of food can bring significant economic benefits at both the macro and micro levels. The U.S. national (macroeconomic) interests can be served by more reliable supplies of critical food and food ingredients; by the development of methods of production which do not minimize the production capability of the growth environment; and by more efficient use of capital employed in food processing.
>
> In addition, faster innovation, particularly in agricultural raw material development, can occur, thereby maintaining a competitive international position. Also, lower energy consumption in food processing can be expected, as well as the provision of an added value usage for agricultural commodities currently in surplus.
>
> At the microeconomic level of individual food sectors benefits will come from more effective production, improved ability to meet the consumers' demands for natural foods and food ingredients, less waste, improved processing characteristics, consistent quality, and a greater nutritional value.

The following programs reflect research needs at the various steps in the path that leads from agricultural production to the consumer:

1) *Application of biotechnology to the structural-functional relationship of food material.* This program aims to improve the utilization of biomaterials by applying modern biotechnological principles to control the functional performance of foodstuffs. In addition, biotechnology will contribute analytical tools and processing procedures that will aid in the implementation of this new knowledge.

2) *Cell physiology and biochemistry of agricultural raw materials.* The potential exists to lower the cost of agricultural raw materials, both plant and animal, by application of biotechnological techniques. Potential targets for improvement are (i) solids content, sensory properties (color, flavor, texture), environmental adaptation, secondary metabolites (vitamins), and postharvest storability in crops and (ii) feed efficiency, palatability, fat/protein ratios, fertility, and maturation time of juveniles in animals.

To realize these benefits, a vast increase is necessary in our understanding (at the molecular level) of the cellular physiology, including biosynthetic and regulatory pathways, of the appropriate animal and plant species.

3) *Improvement of enzymatic processing.* Enzyme processes can reduce the high cost of traditional food processes and also permit development of totally novel foods and food ingredients. To expand the range of possible processes and to improve on the economics of current enzyme-based processes, increased basic knowledge is needed on enzyme isolation and characterization, the mechanisms of enzyme action, and enzyme incorporation into food processes. Specific needs are to understand the mechanisms of enzyme inactivation; to utilize enzymes for biosynthetic processes and redox reactions relevant to foods, including the low-cost production and recycling of cofactors; and to develop new process procedures using immobilized whole cells. Fundamental studies are needed on the control of mass transfer in food systems, maintenance of catalytic activity, and prevention of contamination. Also needed are computer modeling and understanding of the mechanisms of action of food processing enzymes in sufficient detail to permit systematic protein engineering to improve enzymes.

4) *Improvement of food-grade microorganisms.* Microorganisms—bacteria, yeasts, and fungi—are all used extensively in various aspects of food processing. To improve the economics (yield and productivity) and new product characteristics achievable with these organisms, major advances are needed in our understanding of their biochemistry and genetics. Specific research needs are (i) to establish recombinant DNA technologies and a fundamental understanding of microorganisms useful in food fermentation and preservation processes; (ii) to quantitatively describe the microbial ecology and biochemistry of mixed-culture and solid-state fermentations important in foodstuffs; (iii) to isolate, select, and genetically manipulate organisms capable of synthesizing food additives—such as biopolymers, colorants, natural flavorings, and preservatives—by fermentation and cell culture; and (iv) to develop economically viable bioprocesses as sources of raw materials for the food processing industry.

5) *Methods development.* To improve

the production costs, nutritional value, and cost in use of some of the major agricultural crops, particularly cereals, further fundamental advances in cell culture methods and recombinant DNA technologies are necessary. Specific research needs are (i) vector development and transformation procedures for cereal crops, (ii) improved regulation and expression of foreign genes, and (iii) techniques to regenerate and propagate crops that cannot now be so handled. To reduce the time and cost of developing new crop species, rapid screening methods are required to identify the desired genotype at the cell culture stage.

6) *Food safety*. There is an urgent need to improve and to accelerate techniques of food safety assessment. Biotechnology can contribute to food safety by increasing the sensitivity and specificity of such assays and by developing faster and more meaningful methodologies based on DNA hybridization, sequencing, and monoclonal antibody techniques.

The most critical research needs, in addition to basic studies on the structure-function relationship of food materials, are fundamental studies in the cell physiology and biochemistry of agricultural raw materials and improvement of food-grade microorganisms.

Conclusions

Biotechnology applied to food production and processing clearly encompasses a very large and diverse field. The utilization of the capabilities of biological systems is rapidly expanding into a variety of food applications, and consequently many new food sources, processes, and products are being developed. In addition, the identification of critical research needs will help to enhance food production and processing.

We have attempted to highlight biotechnology in food production and processing in broad terms, and consequently we recognize that we have only touched on many of the exciting involvements of biotechnology in providing, securing, and improving the world's food supply.

References and Notes

1. H. J. Rehm and P. Präve, in *Handbuch der Biotechnologie*, P. Präve, U. Faust, W. Sittig, D. A. Sukatsch, Eds. (Oldenburg Verlag, Munich, West Germany, ed. 2, 1984), p. 1; A. L. Demain and N. A. Solomon, *Sci. Am.* **245**, 67 (September 1981); H. J. Rehm and G. Reed, *Biotechnology*, vol. 5, *Food and Feed Production with Microorganisms* (Verlag Chemie, Weinheim, West Germany, 1983); A. H. Rose, *Industrial Microbiology* (Butterworths, Washington, D.C., 1961); H. J. Rehm, *Industrielle Mikrobiologie* (Springer Verlag, Berlin, 1967); D. Knorr, Ed., *Food Biotechnology* (Dekker, New York, in press).
2. P. Dunnill and M. Rudd, *Biotechnology & British Industry* (Science and Engineering Research Council, London, 1984), p. 23; B. J. Liska and W. W. Marion, *Food Technol.* **39(6)**, 3R (1985).
3. M. Castagne and F. Gautier, in *Biotechnology in Europe*, D. Behrens, K. Buchholz, H. J. Rehm, Eds. (Dechema, Frankfurt, 1983), pp. 107–117; H. A. C. Thijssen and J. A. Roels, paper presented at the Third International Congress on Engineering and Food, Dublin, September 1983; *1981 Statistical Yearbook* (United Nations, New York, 1983); H. Ruttloff, *Nutrition* **5**, 411 (1981); Office of Technology Assessment, *Genetic Technology: A New Frontier* (Westview, Boulder, Colo., 1982), pp. 107–114; *Commercial Biotechnology: An International Analysis* (Office of Technology Assessment, Washington, D.C., 1984), pp. 195–214; U. Faust and P. Präve, unpublished manuscript; D. Fukushima, *Food Rev. Int.* **1**, 149 (1985).
4. W. R. Coffman, in *Agriculture in the Twenty-First Century*, J. W. Rosenblum, Ed. (Wiley, New York, 1983), pp. 105–111.
5. National Research Council, *Priorities in Biotechnology Research for International Development* (National Academy Press, Washington, D.C., 1982), p. 87; N. Neushul, in *Agriculture in the Twenty-First Century*, J. W. Rosenblum, Ed. (Wiley, New York, 1983), pp. 149–156.
6. R. R. Colwell, *Science* **222**, 19 (1983); ——, E. R. Parisier, A. J. Sinskey, *Biotechnology in the Marine Sciences* (Wiley, New York, 1984); R. R. Colwell, E. R. Parisier, A. J. Sinskey, *Biotechnology of Marine Polysaccharides* (Hemisphere, Washington, D.C., 1985).

7. J. M. Elliot, in *Agriculture in the Twenty-First Century*, J. W. Rosenblum, Ed. (Wiley, New York, 1983), pp. 111–117.
8. J. H. Litchfield, *Science* **219**, 740 (1983); M. Castagne and F. Gautier, in (*3*); *Commercial Biotechnology: An International Analysis* (Office of Technology Assessment, Washington, D.C., 1984), pp. 202–205; S. Yanchinski, *Biotechnology* **2**, 933 (1984); W. J. Aston and A. P. F. Turner, in *Biotechnolology and Genetic Engineering Review*, G. E. Russell, Ed. (Intercept, Newcastle-upon-Tyne, 1984), vol. 1, pp. 65–88.
9. K. A. Barton and W. J. Brill, *Science* **219**, 671 (1983); S. H. Wittwer, in *Agriculture in the Twenty-First Century*, J. W. Rosenblum, Ed. (Wiley, New York, 1983), pp. 337–367.
10. *Commercial Biotechnology: An International Analysis* (Office of Technology Assessment, Washington, D.C., 1984), pp. 161–191; W. R. Sharp, D. A. Evans, and P. V. Ammirato, *Food Technol.* **38** (No. 2), 112 (1984); R. J. Mapletoft, *Bio/Technology* **2**, 149 (1984).
11. D. M. Yermanos, M. Neushul, R. D. Macelroy, in *Agriculture in the Twenty-First Century*, J. W. Rosenblum, Ed. (Wiley, New York, 1983), pp. 144–165; M. L. Shuler, J. W. Pyne, G. A. Hallby, *J. Am. Oil Chem. Soc.* **61**, 1724 (1984); A. F. Byrne and R. B. Koch, *Science* **135**, 215 (1962); D. Mulcahy, A. Wesenberg, J. Prybys, *Food Eng.* **56** (No. 6), 101 (1984).
12. S. P. Shoemaker, in *Biotech 84* (Online Publications, Pinner, United Kindgom, 1984), pp. 593–600; M. A. Innis *et al.*, *Science* **228**, 21 (1985).
13. R. S. Chaleff, *Science* **219**, 676 (1983); T. E. Teutorus and P. M. Townsley, *Bio/Technology* **2**, 696 (1984); D. A. Evans and W. R. Sharp, *Science* **221**, 949 (1983); R. A. Teutonico and D. Knorr, *Food Technol.* **38**, (No. 2), 120 (1984); D. vonWettstein, *Experientia* **39**, 687 (1983); F. A. Bliss, *HortScience* **19**, 43 (1984); R. A. Teutonico and D. Knorr, *Food Technol.* **39**(10), 127 (1985).
14. T. K. Ng, R. M. Busche, C. C. McDonald, R. W. F. Hardy, *Science* **219**, 733 (1983); J. H. Litchfield, *ibid.*, p. 740; C. A. Batt and A. J. Sinskey, *Food Technol.* **38**(2), 108 (1984); J. C. Jain and T. Takeo, *J. Food Biochem.* **8**, 243 (1984).
15. A. L. Demain, *Science* **219**, 709 (1983); L. K. Miller, A. J. Lingg, L. A. Bulla, Jr., *ibid.*, p. 715.
16. O. Sahai and M. Knuth, *Biotechnol. Prog.* **1**, 1 (1985); H. Ruttloff, *Die Nahrung* **26**, 575 (1982); F. Drawert and R. Berger, in *Flavor 81 3rd Weurman Symposium*, P. Schreiber, Ed. (de Gruyter, Berlin, 1981), pp. 508–527; J. Van Brunt, *Bio/Technology* **3**, 525 (1985); M. F. Balandrin, J. A. Klocke, E. S. Wurtele, W. H. Bollinger, *Science* **228**, 1154 (1985).
17. M. E. Curtin, *Bio/Technology* **1**, 649 (1983).
18. R. Powls and E. R. Redferan, *Biochem. J.* **104**, 24C (1967); B. A. Ruggeri, thesis, University of Delaware, Newark, (1984); E. J. Dasilva and A. Jensen, *Biochem. Biophys. Acta* **239**, 345 (1971); P. E. Hughes and S. B. Tove, *J. Bacteriol.* **151**, 1397 (1982).
19. J. R. Pellon and A. J. Sinskey, unpublished manuscript; M. J. Doel *et al.*, *Nucleic Acids Res.* **48**, 363 (1981); L. D. Stegink and L. J. Filer, Jr., *Aspartame Physiology and Biochemistry* (Dekker, New York, 1984); A. Klausner, *Bio/Technology* **3**, 301 (1985).
20. K. Shimazaku, Y. Nakamura, Y. Yamada, U.S. Patent 4,411,997 (25 October 1983); T. Tsuchida, K. Miwa, and S. Nakamori, U.S. Patent 4,452,890 (5 June 1984).
21. J. K. Baird, P. A. Sandford, I. W. Cottrell, *Bio/Technology* **1**, 778 (1983); D. P. Cheney, in *Biotechnology of Marine Polysaccharides*, R. R. Colwell, E. R. Parisier, A. J. Sinskey, Eds. (Wiley, New York, 1985), pp. 161–175; N. Basta, *High Technology* **5** (No. 2), 66 (1985); J. B. Tucker, *ibid.*, p. 34; G. W. Gooday, *Prog. Ind. Microbiol.* **18**, 85 (1983).
22. C. Rha, in *Biotechnology of Marine Polysaccharides*, R. R. Colwell, E. R. Parisier, A. J. Sinskey, Eds. (Hemisphere, Washington, D.C., 1985), pp. 283–311; D. Knorr, *ibid.*, pp. 313–332; *Impact of Biotechnology on the Production and Application of Biopolymers* (Bioinformation Associates, Boston, 1984).
23. C. J. Panchal, I. Russel, A. M. Sills, G. G. Stewart, *Food Technol.* **38** (No. 2), 99 (1984).
24. S. Schwimmer, *Source Book of Food Enzymology* (AVI, Westport, Conn., 1981); B. Volesky, J. H. T. Luong, A. Hutt, *CRC Crit. Rev. Biotechnol.* **2**, 119 (1984); R. L. Ory and A. J. St. Angelo, Eds., *Enzymes in Food and Beverage Processing* (American Chemical Society, Washington, D.C., 1977).
25. A. C. Olson and R. A. Korus, in *Enzymes in Food and Beverage Processing*, R. L. Ory and A. J. St. Angelo, Eds. (American Chemical Society, Washington, D.C., 1977), pp. 100–131; A. Kilara and K. M. Shahan, *CRC Crit. Rev. Food Sci. Nutr.* **10**, 161 (1979); H. O. Hultin, *Food Technol.* **37** (No. 10), 66 (1983).
26. M. L. Shuler, O. P. Sahai, G. A. Hallsby, in *Biochemical Engineering III*, K. Venkatsubramanian, A. Constantinides, W. R. Vieth, Eds. (New York Academy of Sciences, New York, 1983), pp. 373–382; S. M. Miazga and D. Knorr, paper presented at the 1984 International Congress of Pacific Basin Societies, Honolulu, December 1984.
27. J. E. Prenosil and H. Pedersen, *Enzyme Microbiol. Technol.* **5**, 323 (1983); P. Brodelius and K. Nilsson, *Eur. J. Appl. Microbiol. Biotechnol.* **17**, 275 (1983).
28. B. Wasserman, *Food Technol.* **38** (No. 2), 78 (1984).
29. A. M. Klibanov, *Science* **219**, 722 (1983); W. Carasik and J. O. Carroll, *Food Technol.* **37** (10), 85 (1983).
30. D. N. Bull, R. W. Thoma, and T. E. Stinnet, *Adv. Biotechnol. Process* **1**, 1 (1985); E. Bjurstrom, *Chem. Eng.* **92**, 126 (1984); B. C. Buckland, *Bio/Technology* **2**, 875 (1984).
31. C. L. Cooney, *Science* **219**, 728 (1983).
32. M. L. Shuler, J. W. Pyne, G. A. Hallsby, *J. Am. Oil Chem. Soc.*, **61**, 1724 (1984); M. W. Glacken, R. J. Fleischaker, A. J. Sinskey, in *Biochemical Engineering III*, K. Venkatsubramanian, A. Constantinides, W. R. Vieth, Eds. (New York Academy of Sciences, New York, 1983), pp. 355–372; W. E. Goldstein, *ibid.*, pp. 394–408.
33. D. N. Bull, *Bio/Technology* **1**, 847 (1983); H. R. Lerner, D. Ben-Bassat, L. Reinhold, A. Poljokoff-Mayber, *Plant Physiol.* **61**, 213 (1978); J.

Feder and W. R. Tolbert, *Am. Biotechnol. Lab.* **3**(1), 24 (1985); P. Brodelius and K. Nilsson, *Eur. J. Appl. Microbiol. Biotechnol.* **17**, 275 (1983).
34. M. W. Fowler, in *Plant Biotechnology*, S. M. Mantell and H. S. Smith, Eds. (Cambridge Univ. Press, Cambridge, 1983), pp. 3–37; P. Hedman, *Am. Biotech. Lab.* **2** (No. 3), 29 (1984); O. Sahai and M. Knuth, *Biotechnol. Prog.* **1**, 1 (1985).
35. B. H. Kirsop, *Chem. Ind.* **7**, 218 (1981); J. W. Lee and A. Lopez, *CRC Crit. Rev. Food Sci. Nutr.* **21**, 289 (1984); K. M. Ulmer, *Science* **219**, 666 (1983).
36. B. Wolnak, in *Enzymes*, J. P. Danehy and B. Wolnak, Eds. (Dekker, New York, 1980), pp. 3–10; S. Hasegawa, U.S. Patent 4,447,456 (8 May 1984).
37. R. Aneja, *J. Am. Oil Chem. Soc.* **61**, 661 (1984); A. H. Rose, *Sci. Am.* **245**, 127 (September 1981); J. B. M. Rattray, *J. Am. Oil Chem. Soc.* **61**, 1701 (1984); D. L. Gierhart, U.S. Patents 4,485,172 and 4,485,173 (27 November 1984).
38. B. Jarvis and K. Paulus, *J. Chem. Techn. Biotechnol.* **32**, 233 (1982); L. R. Beuchat, *Food Technol.* **34** (No. 6), 65 (1984); D. Tuse, *CRC Crit. Rev. Food Sci. Nutr.* **19**, 273 (1983); S. Matz, *Sci. Am.* **251**, 123 (November 1984).
39. F. L. Davies and M. J. Casson, *J. Dairy Res.* **48**, 363 (1981); L. McKay, *Antonie van Leeuwenhoek*, **49**, 259 (1983); C. A. Batt and A. J. Sinskey, paper presented at the Symposium on the Importance of Lactic Acid Fermentation, Mexico City, December 1984; A. R. Huggins, *Food Technol.* **38** (No. 6), 41 (1984).
40. A. Kramer and B. A. Twigg, *Quality Control for the Food Industry* (AVI, Westport, Conn., 1970); R. D. Middlekauff, *Food Technol.* **38** (No. 10), 97 (1984); Y. Pomeranz and C. E. Meloan, *Food Analysis: Theory and Practice* (AVI, Westport, Conn., 1978).
41. R. L. Gatz, B. A. Young, T. J. Facklam, and D. A. Scantland, *Bio/Technology* **1**, 337 (1983).
42. H. J. Neujahr, in *Biotechnology and Genetic Engineering Reviews*, G. E. Russell, Ed., (Intercept, Newcastle-upon-Tyne, 1984), vol. 1, pp. 167–186; N. Smit and G. A. Rechnitz, *Biotechnol. Lett.* **6**, 209 (1984).
43. R. Dagani, *Chem. Eng. News* **62** (No. 46), 25 (1984).
44. E. L. Korwek, *Food Drug Cosmetic Law J.* **37**, 289 (1982); D. D. Jones, *Food Technol.* **39** (No. 6), 59 (1985).
45. H. Hemfort and W. Kohlstette, *Starch* **36**, 109 (1984); V. Wiesboden and H. Binder, in *Advances in Biochemical Engineering*, A. Fiechter, Ed. (Springer Verlag, Berlin, 1982), pp. 120–171.
46. W. A. Bough, *Process Biochem.* **11** (No. 1), 13 (1976); P. R. Austin, C. J. Brine, J. E. Castle, J. P. Zikakis, *Science* **212**, 749 (1981); S. Latlief and D. Knorr, *J. Food Sci.* **48**, 1587 (1983).
47. M. R. Kula, K. H. Kroner, H. Hustedt, in *Advances in Biochemical Engineering*, A. Fiechter, Ed. (Springer Verlag, Berlin, 1982), pp. 73–118.
48. E. Stahl and K. W. Quirin, *Naturwissenschaften* **71**, 181 (1984); L. G. Randall, *Separation Sci. Technol.* **17**, 1 (1982).
49. E. Stahl, E. Schütz, H. K. Mangold, *J. Agric. Food Chem.* **28**, 1153 (1980); J. P. Friedrich and E. H. Pryde, *J. Am. Oil Chem. Soc.* **61**, 223 (1984); H. J. Gährs, *ZFL Int. J. Food Technol. Food Process. Eng.* **35**, 302 (1984).
50. J. L. Dwyer, *Bio/Technology* **2**, 957 (1984).
51. P. D. Fullbrook, *J. Am. Oil Chem. Soc.* **60**, 476 (1983).
52. H. Ruttloff, J. Huber, F. Zicker, K. Mangold, *Industrielle Enzyme* (VEB, Leipzig, 1983).
53. W. Hartmeier, *Process. Biochem.* Feb. 40 (1984); B. Dixon, *Biotechnology* **2**, 594 (1984); D. Knorr, S. M. Miazga, R. A. Teutonico, *Food Technol.* **39**(10), 135 (1985).
54. C. V. Morr, *Food Technol.* **38** (No. 6), 39 (1984).
55. R. R. Zall, in *Food Processing Waste Management*, J. H. Green and A. Kramer, Eds. (AVI, Westport, Conn., 1979), pp. 175–201.
56. D. Knorr, *J. Food Technol.* **12**, 563 (1977); F. Holm and S. Eriksen, *ibid.* **15**, 71 (1980); D. Knorr, *Food Technol.* **37** (No. 2), 71 (1983); J. R. Rosenau, L. F. Whitney, J. R. Haight, *ibid.* **32** (No. 6), 37 (1978).
57. R. S. Tutunjian, in *Biochemical Engineering III*, K. Venkatsubramanian, A. Constantinides, W. R. Vieth, Eds. (New York Academy of Sciences, New York, 1983), pp. 238–253; P. Jelen, *Agric. Food Chem.* **27**, 658 (1979).
58. G. G. Birch, K. J. Parker, J. T. Worgan, *Food From Waste* (Applied Sciences, London, 1976); J. H. Green and A. Kramer, *Food Processing Waste Management* (AVI, Westport, Conn., 1980); M. W. M. Bewick, *Handbook of Organic Waste Conversion* (Van Nostrand Reinhold, New York, 1980); D. Knorr, in *Sustainable Food Systems*, D. Knorr, Ed. (AVI, Westport, Conn., 1983), pp. 249–78.
59. D. Knorr, *Food Technol.* **38** (No. 1), 85 (1984); D. Rodriquez-Sanchez and C. Rha, *J. Food Technol.* **16**, 469 (1981); K. D. Vorlop and J. Klein, *Biotechnol. Letters.* **3** (No. 1), 9 (1981).
60. S. Revah-Moiseev and A. Carroad, *Biotechnol. Bioeng.* **23**, 1067 (1981); I. G. Casio, R. A. Fisher, P. A. Carroad, *J. Food Sci.* **47**, 901 (1982); J. Zikakis, Ed., *Chitin, Chitosan and Related Enzymes* (Academic Press, Orlando, Fla., 1984); papers presented at the Third International Conference on Chitin/Chitosan, Senigellia, Italy, 1 to 4 April 1985; R. L. Rawin, *Chem. Eng. News* **62** (No. 20), 42 (1984).
61. H. Skogman, in *Food From Waste*, G. G. Birch, K. J. Parker, J. T. Worgan, Eds. (Applied Science, London, 1976), pp. 167–179; Z. G. Moulin and P. Galzi, in *Biotechnology and Genetic Engineering Reviews*, G. E. Russell, Ed. (Intercept, Newcastle-upon-Tyne, 1984), pp. 347–374.
62. D. L. Wise, Ed., *Fuel Gas Development* (CRC Press, Boca Raton, Fla., 1984); D. A. Stafford *et al.*, *Methane Production from Waste Organic Matter* (CRC Press, Boca Raton, Fla., 1984).
63. Institute of Food Technologists (IFT) Workshop on Research Needs, Arlington Heights, Ill., 11 to 14 November 1984; B. J. Liska and W. W. Marion, *Food Technol.* **39** (No. 6), 81 (1985).
64. We thank the participants of the biotechnology topic at the IFT workshop on research needs, especially A. J. Stevens. Valuable comments on drafts of this article were provided by S. P. Shoemaker, M. J. Haas, K. Venkat, P. M. Walsh, and D. Kukich.

19

Biotechnology in the American Pharmaceutical Industry: The Japanese Challenge

Mark D. Dibner

The pharmaceutical industry in the United States has enjoyed steady growth and stability throughout most of this century. In 1984 the industry posted greater profit margins than any other industry group (*1*). However, the industry is also undergoing great change, largely because of biotechnology. The use of living cells to produce commercial products is having a major impact on the drug industry, with the anticipation of there being new products, new processes, new entrants into the industry, and increased competition.

Pharmaceutical companies are now taking action to address the impact of biotechnology. How they incorporate the new technologies and how they face new competition may play a significant role in defining their financial success in the future. Of the new forms of competition confronting U.S. firms, the strongest is projected to be from Japan (*2, 3*). Current strategies used by U.S. and Japanese firms to incorporate biotechnology are the focus of this article.

Impacts of Biotechnology

Traditionally, pharmaceuticals have been manufactured through chemical synthesis or purification processes. Genetic engineering makes it possible to manufacture a host of new molecules with projected uses as therapeutic agents. Among those currently under development are the interferons, interleukins and other lymphokines, tissue and kidney plasminogen activators, and tumor necrosis factor. Other biotechnologically produced therapeutics, previously made by other methods, include

Science 229, 1230–1235 (20 September 1985)

human insulin, growth hormone, serum albumin, and clotting factor VIII. These proteins can be produced in abundance by genetic engineering techniques and may have fewer side effects than do proteins derived from nonhuman sources (*4, 5*). Moreover, new processes can be employed to produce vitamins, amino acids, steroids, antibiotics, enzymes, bioactive peptides, and many other molecules for potential use as drugs, as well as novel compounds for use as vaccines (*6, 7*). Products of hybridomas, such as the monoclonal antibodies, are already important for use in new, sensitive diagnostics. In addition, monoclonal antibodies with therapeutic uses are on the horizon. At present, insulin and human growth hormone are the only therapeutic entities developed by biotechnology approved for human use in the United States. In comparison, over 100 monoclonal antibodies have been approved for use in diagnostics. Many additional compounds are in various stages of development or testing, and should reach the marketplace within the next decade (*2, 6*).

Another impact of biotechnology will be in pharmaceutical manufacturing. In many instances, it should be possible to produce molecules with higher purity, and perhaps more cheaply, through cell growth and fermentation processes (*2, 5–8*). This will not be without cost; new production facilities must be built at an expense of up to $100 million each (*6*), and bioprocess engineering personnel must be trained or hired.

Most U.S. and foreign pharmaceutical companies are aware of the scientific and financial importance of biotechnology, and they are in the process of incorporating biotechnology skills into their programs and plans. In 1983 the worldwide market for pharmaceuticals was over $60 billion (*9*), and biotechnology may eventually affect the production of 20 percent of the current pharmaceutical products (*6*). In addition, sales of new drugs and diagnostics made possible by biotechnology are predicted to reach $10 billion within the next decade, and they should rise thereafter (*2*). With such large markets at stake, pharmaceutical companies in the United States are attempting to be strong competitors in biotechnology. But can these companies continue to enjoy profits and growth with new competition on the horizon?

Biotechnology Industry

Fermentation processes have long been in commercial use. In the early 1970's genetic engineering and hybridoma technologies were developed, primarily in academic laboratories (*2, 5*). Subsequently, many small biotechnology firms re formed, often by academic scientists, to commercialize advances made in basic research (*2, 5*). A 1985 compilation in *Genetic Engineering News* listed almost 300 companies with major biotechnology efforts, the vast majority of the companies being small and recently formed (*10*). More than half of these new companies are involved in pharmaceutical or diagnostic development (*2, 10*). More than 75 percent of these biotechnology firms worldwide have been formed in the United States, with an initial total investment of $2 billion to $2.5 billion (*3, 7, 10*). To provide a return on this investment, the new companies have entered into contract research, as well as the licensing of products they develop. Some are also attempting to directly market their products, but most do not have available the extensive resources necessary to take a pharmaceutical product through the regulatory process; nor do they have the necessary production or marketing expertise (*2, 11*). Thus, much of the effort of these small firms can be described as technology transfer to larger companies.

An analysis of strategies used by pharmaceutical companies to incorporate biotechnology yields three categories: (i) academic relationships, (ii) internal expansion, and (iii) agreements with biotechnology firms (*12*). All three strategies are used by many companies but the most frequently used one is the third—agreements with biotechnology firms (*2, 12*). Joint projects between large U.S. pharmaceutical companies and biotechnology firms (Table 1) should lead to the commercialization of a number of important products of biotechnology within the next decade (*2, 12*).

New Competition

The U.S. pharmaceutical industry is comprised mostly of large, established companies that have been responsible for the development and introduction of many major drugs. As a result of biotechnology, there should be major changes in the composition of the industry in the form of new competition.

One major source of competition will be large, nonpharmaceutical firms that have planned a future in pharmaceuticals

Table 1. Joint projects between large U.S. pharmaceutical companies and biotechnology firms. Interactions announced between 1981 and 1984, selected from database (*12*). Two Swiss-based companies, Hoffmann–La Roche and Sandoz, that have major facilities in the United States are also included.

Biotechnology firm	Pharmaceutical company	Products involved
Biogen	Merck	Hepatitis B vaccine
	Schering-Plough	Interferon
	SmithKline	Anticlotting factor
Centocor	Abbott	Cancer diagnostics
	Hoffmann–La Roche	Monoclonal antibodies for cancer treatment
	Warner-Lambert	Hepatitis B diagnostics
Collaborative Research	Sandoz	Kidney plasminogen activator
Genentech	Hoffmann–La Roche	Interferons
	Lilly	Insulin
	Baxter Travenol	Diagnostics
Genetic Systems	Syntex	Diagnostics
Genetics Institute	Baxter Travenol	Factor VIII
Genex	Bristol-Myers	Interferons
Hybritech	Baxter Travenol	Monoclonal antibodies for bacterial infection
Molecular Genetics	Lederle	Herpes simplex vaccine

through biotechnology programs. Companies such as Monsanto and Du Pont, which currently have only modest sales related to pharmaceuticals, have announced major expansion of their drug and diagnostic research efforts with emphasis on biotechnology. Table 2 demonstrates that, in addition to drug companies, there are large, nonpharmaceutical companies that are also buying equity positions in biotechnology firms. The new technologies also are important for future products in the chemical industry; nine of the ten largest U.S. chemical companies have annonced their involvement in biotechnology for the development of agrichemicals and pharmaceuticals (*13*). Large companies with current emphasis on chemicals, foods, textiles, and other goods could become prominent contenders in the pharmaceutical industry of the future.

A second source of competition will be from the biotechnology firms themselves. Although most do not have the capital to bring a drug to market, many of these firms do have the ability to market diagnostics. Through financial incentives, such as research and development limited partnerships (RDLP's), biotechnology firms may obtain the capital to directly market therapeutics in the future (*2, 14*). In each year from 1982 to 1984, a total of $166 million to $199 million was raised by five to nine firms to form RDLP's that involve the development of specific biotechnology-based products (*14*).

Foreign Competition

The strongest source of competition in biotechnology for U.S. companies appears likely to come from abroad. Whereas the United States has no significant federal program for the coordination of biotechnology efforts, the governments of Great Britain, France, West Germany, and Japan have mounted major programs to foster the growth of domestic biotechnology (*2*). The country predicted to have the greatest impact on the commercialization of biotechnology is Japan (*2, 3*). Although the new biotechnologies have been largely developed in the United States, the Japanese are expected to soon take the lead in commercialization of these technologies (*2, 3*). A large part of their success will be based on products first developed in the United States. An analysis of Japan's incorporation of biotechnology may help in understanding this process.

Strategies Used by the Japanese

Historically, the Japanese pharmaceutical industry has been quite different than that in the United States, as is indicated in Table 3. Japanese drug companies are considerably smaller than their U.S. counterparts; there are 11 U.S. drug companies with annual pharmaceutical sales (1983) of over $1 billion, but only one Japanese company, Takeda, has reached that mark (*15*). In the past, U.S. companies were responsible for the introduction of twice as many new drugs as Japanese firms (*2, 9*). In recent years, however, Japanese pharmaceutical companies have spent almost 50 percent more on research and development (as a proportion of sales) than U.S. companies (*1, 9*). As a result, Japanese companies introduced 70 percent more new drugs between 1981 and 1983 than did U.S. companies (*2*). Given the total amount spent on research and development, the Japanese have been at least six times more productive (as mea-

Table 2. Equity purchased in firms with a major focus on biotechnology. Equity purchases selected from database (*12*).

Large company (purchaser)	Biotechnology firm	Year
	Purchased by U.S. pharmaceutical companies	
Abbott	Amgen	1980
Baxter Travenol	Genetics Institute	1982
Becton Dickenson	Applied Biosystems	1984
Bristol-Myers	Genetic Systems*	1985
Johnson & Johnson	Enzo Biochem	1982
Lederle	Molecular Genetics	1981
Lederle	Cytogen	1983
Lilly	Synergen	1984
Lilly	Hybritech*	1985
Schering-Plough	Biogen	1982
Schering-Plough	DNAX Ltd.*	1982
SmithKline	Beckman*	1982
Syntex	Genetic Systems	1982
	Purchased by other U.S. companies	
Dow	Collaborative Research	1981
Du Pont	New England Nuclear*	1981
Fluor	Genentech	1981
W. R. Grace	Amicon*	1983
Martin Marietta	Molecular Genetics	1982
Monsanto	Biogen	1980
Monsanto	Collagen Corporation	1980
	Purchased by Japanese companies	
Green Cross	Collaborative Research	1981
Mitsubishi	BioVec	1984

*Acquisition. Each nonacquisition purchase involved an average of $8 million.

sured by number of drugs introduced) per dollar spent on research (*16*). A recent estimate from the Japanese Bio-Industry Development Center predicts $60.7 billion in biotechnology-related Japanese sales by the year 2000, with $12.8 billion coming from pharmaceuticals (*17*), confirming Japan's emphasis on biotechnology.

Japan is considered a world leader in fermentation technology (*3, 5, 17*). This is of key importance to commercial success in biotechnology, but basic research is also necessary to supply new products for companies to market. Recent Japanese progress in biotechnology has resulted from coordinated efforts by the government, individual companies, and academic laboratories. Moreover, Japanese companies have received considerable foreign help, mostly from the United States, in filling certain gaps in basic research and development.

Japanese government programs in biotechnology have emerged from three sources: the Science and Technology Agency (STA), the Ministry of International Trade and Industry (MITI), and the Ministry of Agriculture, Forestry and Fisheries (MAFF) (*2, 18*). The total gov-

Table 3. Comparison of U.S. and Japanese pharmaceutical industries and involvement in biotechnology. All 1983 data, except as noted. [Sources: (*1*, *2*, *9*, *15*)]

Data category	United States	Japan
Population (millions)	234.5	119.2
Gross national product	$3.3 trillion	$1.2 trillion
Domestic pharmaceutical market (world rank)	$21.3 billion (1)	$13.4 billion (2)
Number of pharmaceutical companies with sales over $1 billion*	11	1
Total pharmaceutical sales of ten largest pharmaceutical companies†	$16.7 billion	$6 billion
Pharmaceutical sales as percent of total sales‡	50.1	74.1
Number of new pharmaceutical products introduced:		
1961–1980	353	155
1981–1983	24	41
R&D expenditures as percent of sales‡	6.8	9.2
Scientists and engineers in industrial R&D§:		
Total number	573,900	272,000
Percent of work force	0.58	0.50
Government-funded research in biotechnology:		
Total	$520 million	$60 million
Percent of basic research	>98	<50
Targets of funding in biotechnology	Basic research	Basic research, scale-up, industrial projects, government laboratory facilities, manufacturing technology

*Pharmaceutical sales only. †Total world pharmaceutical sales in 1983 were approximately $60 billion. ‡Average of top ten companies. §All industries, 1977 data.

ernment support for biotechnology, $50 million to $60 million in 1984, is only about one-tenth of that spent by the U.S. government (Table 3) (*2*, *18*), but Japanese funding is much more focused on specific projects. For example, MITI, in a 10-year strategic program beginning in 1981, has targeted next-generation technologies to foster scale-up techniques, aimed at assisting in the commercialization of biotechnology (*2*). The STA is also funding applied research, such as the development of bioreactors (*2*). The latest announced budgets of STA,

MAFF, and MITI emphasize national centers related to biotechnology research, including the development of cell line and gene banks (*18*). Very little of the Japanese government's support for biotechnology is for basic research (*2*). In contrast, the U.S. government's support of biotechnology is almost ten times in magnitude, but support of applied research makes up only 1 to 2 percent of this total, with far less specificity than in Japan (Table 3) (*2*).

Another emphasis in Japan is to foster cooperation between companies and between industry and academia. There are more than a dozen joint ventures on record involving two or more Japanese companies that are aimed at developing therapeutics through research in biotechnology (*2, 19*). Similar cooperation between large U.S. companies does not (or cannot) exist (*2*).

In order to further foster cooperation between Japanese companies, a trade association, tentatively called the Society for Advanced Pharmaceutical Research, was formed in 1985 with 31 member companies and the support of Japan's Ministry of Health and Welfare (*19*). A trade group, the Industrial Biotechnology Association, exists in the United States with 46 member companies, but is not supported by the federal government (*20*).

Because government funding in Japan is focused on applied research, Japanese companies are also in the process of expanding in-house expertise in basic research and development in biotechnology. Many companies have announced the expansion of research facilities, such as Sankyo's new $53-million biotechnology laboratory to be completed by 1986 (*21*). The availability of personnel to staff basic research laboratories in Japan has been a problem, primarily owing to a paucity of university programs in molecular genetics (*2, 16*). To fill the need for researchers, some Japanese companies have begun in-house training programs, while others have sent employees abroad to be trained or have hired foreign researchers (*2*).

These methods are apparently successful; the number of in-house scientists involved in basic research in biotechnology has increased more than fivefold in the past 3 years (*16*). Japan has an ample supply of fermentation process engineers, which is important for the commercialization of biotechnology (*2, 16, 22*). In contrast, in the United States the supply of basic researchers for biotechnology has not been as much of a problem, although it is expected that there will be a shortage of scientists trained in bioprocess engineering (*2, 16, 22*).

In Japan, little venture capital has been available and very few biotechnology firms have been formed (*2, 16, 23*). With in-house incorporation of biotechnology still at an early stage, Japanese companies have turned to U.S. biotechnology firms for basic research assistance. Table 4 identifies a number of these contractual agreements, some of which involve the same U.S. biotechnology firms and the same products as covered by agreements with U.S. pharmaceutical companies (see also Table 1). In most cases the contractual agreement gives the Japanese company licensing rights to a product developed in the United States, often with marketing rights being limited to Japan and other Asian countries (*2, 16, 24*). In only a few instances have Japanese companies purchased equity in biotechnology firms in the United States (Table 2).

Table 4. Contractual agreements between U.S. biotechnology firms and major Japanese companies. Agreements announced between 1981 and 1984, selected from database (*12*). Abbreviations: HSA, human serum albumin; IL-2, interleukin-2; MAb, monoclonal antibodies.

Biotechnology firm	Japanese company	Products involved
Biogen	Shionogi	Interferon, IL-2, HSA
	Fujisawa	Tissue plasminogen activator
	Green Cross	Hepatitis B vaccine
	Suntory	Tumor necrosis factor
	Teijin	Factor VIII
Centocor	Toray	Hepatitis diagnostics
Collaborative Research	Green Cross	Urokinase, interferon
Genentech	Mitsubishi	Tissue plasminogen activator
	Toray	Interferons
Genetic Systems	Daiichi	Diagnostics for blood disorders
Genetics Institute	Chugi	Human erythropoietin
Genex	Yamanouchi	Fibrinolytic agent
	Green Cross	HSA
	Yoshitomi	IL-2
	Mitsui Toatsu	Urokinase
	Mitsubishi Chem.	HSA
Hybritech	Teijin	MAb's for cancer treatment
	Green Cross	Immunoglobulins
	Toyo Soda	MAb diagnostics

To assess the extent of both domestic and foreign involvement with U.S. biotechnology firms, I have created a database of recorded pharmaceutical-related interactions between these firms and large companies (*12*). This database includes 72 joint or contractual interactions between U.S. biotechnology firms and U.S. companies from 1981 to 1985. Additionally, the database includes 43 such interactions between U.S. biotechnology firms and Japanese companies, 60 percent as numerous as those with U.S. companies (*12*). It is clear that Japanese companies, like large U.S. companies, are relying on contractual agreements with small U.S. firms to provide basic research or newly developed products (*2, 16, 25*).

Some U.S. biotechnology firms have set up facilities in Japan to coordinate joint research with Japanese companies. For example, Genentech Ltd. was set up in Japan in 1982 to serve as a liaison between Genentech in the United States and multiple Japanese firms with which it does business (*26*).

As has occurred in the United States, Japanese companies not previously involved in pharmaceuticals have started biotechnology programs that could lead to pharmaceutical products. A recent listing noted 10 major Japanese chemical companies, 15 food processing companies, and 4 textile companies that have biotechnology-related pharmaceutical projects (*2*). Most of these companies have contractual agreements with U.S. biotechnology firms (*12*).

In increasing numbers, large U.S.

companies are forming joint subsidiaries with Japanese companies for the commercialization of pharmaceuticals, as evidenced by such company names as Merck-Banyu, Pfizer Taito, Nippon Upjohn and Du Pont–Sankyo (*15, 27*). Other joint agreements, not involving the formation of subsidiaries, have been recorded with U.S. pharmaceutical companies, such as Shionogi's agreements with Lilly and Merck (*12, 24*). In addition to U.S. companies entering Japan, at least one Japanese pharmaceutical company is setting up a biotechnology facility in the United States. Otsuka, Japan's fourth largest pharmaceutical manufacturer, is building a $7-million biotechnology research facility in Maryland, to be completed in 1985 (*28*). Other Japanese pharmaceutical companies already have or plan manufacturing facilities in the United States; they include Takeda, Japan's largest pharmaceutical company, which has a plan for a vitamin-production facility in North Carolina.

United States and Japan: Comparison of Strategies

In comparing the incorporation of biotechnology into the U.S. and Japanese pharmaceutical industries, some similarities are apparent. Companies in both countries have expanding in-house biotechnology efforts, but are also relying on the new biotechnology firms, primarily U.S. ones, to gain access to basic research in biotechnology and to obtain products for commercialization. This method appears successful; many initial products of these collaborations are in the pipeline for approval by the Food and Drug Administration (*29*). For example, human insulin, the first therapeutic developed by biotechnology, was evolved by Lilly in collaboration with Genentech; and alpha interferon, to be marketed by Schering-Plough, was developed by Biogen (*4, 6*).

In both countries there is new pharmaceutical industry competition attributable to biotechnology. Nonpharmaceutical companies, such as Exxon, Corning, W. R. Grace, Monsanto, Martin Marietta, and Du Pont, have planned major programs in biotechnology (*2, 12*). Giant Japanese corporations, such as Mitsubishi Chemical, Ajinomoto, Suntory, Kirin Brewery, and Asahi Chemical Industry also are in the process of developing pharmaceutical products employing biotechnology (*2, 12*). It is thus likely that the composition of the pharmaceutical industry in both countries will change radically over the next decade.

Where Japan and the United States differ is in the type and amount of government support for the development of biotechnology. In Japan there is a clear effort by government to enhance the future commercial success of the pharmaceutical industry by assisting in the development of biotechnology. Although this support is administered by a few different agencies and is small in size (by U.S. standards), it is viewed both externally (*2, 25*) and internally (*16*) as a single cohesive effort with a high potential for success. The companies involved must create their own basic research and development programs; government assistance is at the next level, helping to foster commercialization of products, manufacturing, and generic support, such as gene banks (*18*). In the United States, federal support for biotechnology is ten times greater in magnitude but is almost exclusively targeted at basic research. Although support of basic research programs in biotechnology should be continued and expanded to ensure maintained leadership in basic research, support for more applied areas is also needed (*2, 16*).

Another contrast between the two countries is in the availability of basic researchers in biotechnology and bioprocess engineering. There was a reported shortage in the United States of basic researchers trained in genetic engineering, but this problem appears to have abated (*2, 30*). Due to strong academic programs in this and related areas, the availability of basic researchers should continue to be sufficient (*2*). However, a paucity of academic programs in bioprocess engineering continues (*2*). As more companies generate products of biotechnology for scale-up, it is expected that there will be a severe shortage of personnel trained in production technologies, which may hamper commercial success (*2*). Japan has the opposite problem—an adequate supply of fermentation engineers but too few basic researchers with training in molecular genetics (*16*). This is another reason why Japanese companies have been borrowing U.S. basic research, but are predicted to outpace the United States in commercialization (*2, 3*).

Outlook

In January 1984 the U.S. Congress Office of Technology Assessment (OTA) published a 612-page analysis on commercial biotechnology (*2*). The report noted the importance of biotechnology both for its basic scientific benefit and for its potential commercial development. In assessing the competitive position for the United States, the OTA report stated the following (*2*, p. 7):

> Japan is likely to be the leading competitor of the United States for two reasons. First, Japanese companies in a broad range of industrial sectors have extensive experience in bioprocess technology. Japan does not have superior bioprocess technology, but it does have relatively more industrial experience using old biotechnology, more established bioprocessing plants, and more bioprocess engineers than the United States. Second, the Japanese Government has targeted biotechnology as a key technology of the future, is funding its commercial development, and is coordinating interactions among representatives from industry, universities, and government.

When the focus of analysis is narrowed to the pharmaceutical industry, it can also be concluded that the Japanese have the potential to be a leading competitor. An important factor in their success has been the borrowing of basic biotechnological research by Japanese companies from U.S. biotechnology firms. Although biotechnology licensed by U.S. firms to Japanese companies generally involves marketing rights in Japan or Asia (*2*), the Japanese market for pharmaceuticals is the second largest in the world. When added to other Asian markets, it becomes two-thirds the size of the North American or European markets (*9*). U.S. pharmaceutical companies have gained 40 percent of their revenues from foreign sales, and the loss of a foreign market may represent lost income (*9*).

In addition to basic biotechnology borrowed from the United States, Japan has been simultaneously building its own strength in this field. There are more and more frequent reports of new developments in basic biotechnology and discoveries of new drugs from Japanese industrial laboratories (Table 3) (*12*). It is thus possible that Japan's predicted future strength in pharmaceutical biotechnology will come both from internal developments and strategic government programs (*16*).

This is not to imply that with Japanese strength in biotechnology will come U.S. weakness in this area. As stated earlier, pharmaceutical and other companies in the United States are expanding their efforts in biotechnology and are nearing their goals of bringing new therapeutics and diagnostics to market. However, an analysis of Japanese strategies may help to understand how U.S. industry can optimize this process. In addition, U.S. industry will be strengthened if the U.S. government makes the commercialization of biotechnology a high priority and funds specific academic and other programs leading to that goal (*2*). As stated in the OTA report (*2*): "The United States may compete very favorably with Japan if it can direct more attention to research problems associated with the scaling-up of bioprocesses for production."

In addition, government activities that enhance cooperation between companies, decrease regulation, or provide centers to assist in biotechnology would help meet this goal (*2, 6, 31*). However, in the period since the OTA report was made public, no broad program of support to strengthen the U.S. position in biotechnology has been announced by the federal government.

Steps in the Right Direction

A few recent developments should prove useful to the future development of biotechnology in the United States. The first is the opening of biotechnology centers to assist in the transfer of biotechnology expertise from academia to industry. Two of these centers are at Pennsylvania State University and in Research Triangle Park, North Carolina. The Penn State Biotechnology Institute has planned research and educational facilities and will allow member companies access to "application-oriented research" and to a pilot production facility for assistance in scale-up (*32*).

The North Carolina Biotechnology Center currently receives $2.5 million in annual funding from state, federal, and industrial sources. The center funds specific programs, such as its Monoclonal Lymphocyte Technology Center, which involves academic research at the University of North Carolina and Duke University, the participation of industry, and funding by the National Science Foundation. The five industrial members agree on priorities for directed research to be funded by specific grants to participating laboratories. Although still in its infancy, the Monoclonal Lymphocyte Technology Center is fostering cooperation between companies in a university environment that probably would not have otherwise occurred (*33*).

The Center for Advanced Research in Biotechnology (CARB), to be built in Gaithersburg, Maryland, will combine federal, state, county, and university efforts (*34*). With CARB, the National Bureau of Standards will add its analytical expertise to molecular biology expertise from the University of Maryland. A CARB research facility to be completed in 1986 will house 130 scientists. In addition to basic and applied research in biotechnology, CARB will provide services to industry, including analytical measurements and molecular modeling on a supercomputer, and it will make available basic tools for research in biotechnology (*34*).

Lastly, the National Science Foundation has announced a $20-million, 5-year grant to establish the Center on Biotechnology Process Engineering at the Massachusetts Institute of Technology (*35*). This is part of a new program to advance engineering research with industrial applications. It is likely that industry will also provide support for this center (*35*). Although not part of a broad program to develop biotechnology, the creation of this center, CARB, and other biotechnology centers will greatly assist U.S. companies in developing the skills necessary to face competition.

Conclusions

The current situation involving biotechnology in the U.S. pharmaceutical industry and its competition can be summarized as follows: Many if not most of the significant advances in biotechnology over the past decade have occurred in academic laboratories in the United States. A large portion of this research has been funded by the U.S. government, currently at an annual rate of more than $500 million. As the commercial potential of the products of biotechnology has become apparent, these products and many of the researchers have been transferred to a new industry—the biotechnology industry. This fledgling industry has been formed with over $2 billion of venture capital, mostly from the United States.

A problem occurs with return on investment. The new biotechnology firms need income to remain in business, and, for the most part, they are not presently able to market their products directly. However, they can sell their research capabilities and rights to products.

The U.S. pharmaceutical industry will benefit greatly from these discoveries, but it does not have exclusive access. On the one hand, many other U.S. companies see this as an opportunity for future profits and a means to enter the pharmaceutical industry. On the other hand, many Japanese companies, poised with enormous fermentation expertise to commercialize the products of biotechnology, see this as an opportunity to buy products of basic research while developing basic research capabilities of their own. The investment to develop biotechnology has been made primarily in the United States, yet the commercial success of this research will be shared with companies from other countries, especially Japan. The dollars invested by the United States should lead to U.S. jobs and increased revenue for U.S. corporations, but some of that return will be lost. And Japan, with U.S. help, is predicted to soon become the leading competitor in this field.

It is not a question of whether biotechnology research should be supported; the benefits to mankind are large. However, the type of support can clearly affect the outcome. In addition to the continued support of basic research, a U.S. federal program needs to be strategically designed to foster academic, government, and industry programs that would lead to maximizing the commercial success of biotechnology within the United States. If this occurs, then the predicted future leader in both scientific advances and commercialization could be the United States. Without relation to defense, it is unlikely that the transfer of basic biotechnology will be highly regulated or curtailed in the future. However, it should be possible to borrow strategies from Japan for planning and coordinating the commercial success of technologies, present and future.

Biotechnology in the pharmaceutical industry is still at a very early stage. It may still be possible for U.S. industry to maximize the return on investment in basic research. New programs that emphasize shared knowledge and centralized facilities should assist industry in its ability to develop new therapeutics. Of key importance is the assistance provided by the federal government to optimize regulatory and financial environments and to furnish coordinated support for continued achievement in biotechnology research.

References and Notes

1. *Bus. Week* (corporate scoreboard issue) (22 March 1985), p. 24.
2. Office of Technology Assessment, *Commercial Biotechnology: An International Assessment* (Government Printing Office, Washington, D.C., 1984).
3. *Chem. Week* (16 December 1981), p. 38.
4. I. S. Johnson, *Science* **219**, 632 (1983).
5. W. Gilbert and A. Taunton-Rigby, *Res. Dev. Ind.* (June 1984), p. 176.
6. W. P. Patterson, *Ind. Week* (7 September 1981), p. 64.
7. *Chem. Week* (24 November 1982), p. 42.
8. J. Vane and P. Cuatrecasas, *Nature (London)* **312**, 303 (1984).
9. *SCRIP 1985 Yearbok* (PJB Publications, Surrey, England, 1984).
10. *Genet. Eng. News* ("Guide to Biotechnology Companies") (November–December 1985), p. 15.
11. D. Webber, *Chem. Eng. News* (18 June 1984), p. 10.
12. A database was created containing records of actions that involved pharmaceutical products and were taken by U.S. pharmaceutical companies, other U.S. companies, Japanese companies, and U.S. biotechnology firms between 1981 and 1985. The sources of information for these records were public information from the companies and firms involved, financial data, and news items. Sources most frequently used were *Genetic Engineering News, Biotechnology News, SCRIP World Pharmaceutical News, Chemical and Engineering News, Chemical Marketing Reporter, Biotechnology Newswatch, IMS Pharmaceutical Marketletter*, the *Wall Street Journal*, and *Chemical Week*. By sorting and selecting the 320 records in this database, it was possible to derive company involvement, company successes, and industry trends [M. D. Dibner, *Pharm. Exec.* (September 1985), p. 81. M. D. Dibner, *Trends in Pharmacol. Sci.* **6**, 343 (1985)].
13. *Chem. Week* (26 October 1983), p. 45.
14. D. Weld, *Genet. Eng. News* (July–August 1984), p. 28; *Biotech. News* 14 January 1985), p. 6.
15. *SCRIP Pharmaceutical Company League Tables 1983–1984* (PJB Publications, Surrey, England, 1984).
16. G. Gregory, *Chem. Econ. Eng. Rev.* (August 1984), p. 14.
17. *Biotech. Newswatch* (17 January 1985), p. 6.
18. *Ibid.* (4 March 1985), p. 7; *ibid.* (6 May 1985), p. 1; *Jpn. Chem. Week* (24 December 1984), p. 4.
19. *Biotech. Newswatch* (1 April 1985), p. 8.
20. Industrial Biotechnology Association, Rockville, Md., public literature.
21. *SCRIP World Pharm. News* (2 May 1984), p. 11; *Manuf. Chemist* (June 1984), p. 19.
22. N. Howard, *Dun's Bus. Month* (January 1982), p. 92.
23. *Eur. Chem. News* (19 November 1984), p. 21.
24. *Biotech. News* (17 April 1985), p. 4.
25. *The Engineer* (17 May 1984), p. 37; *IMS Pharm. Marketletter* (16 August 1982), p. 3.
26. *Chem. Eng.* (15 November 1982), p. 45.
27. *Jpn. Econ. J.* (16 November 1982), p. 16.
28. R. S. Johnson, *Genet. Eng. News* (March 1984), p. 3.
29. *Biotech. Newswatch* (21 January 1985), p. 8.
30. *Wall St. J.* (10 October 1984), p. 35.
31. *Biotechnol. Newswatch* (7 January 1985), p. 3.
32. Penn State Biotechnology Institute, public information.
33. North Carolina Biotechnology Center Newsletter (winter 1984); University-Industry Cooperative Research Center in Monoclonal Lymphocyte Technology, "Revised Prospectus" (January 1985).
34. R. S. Johnson, *Genet. Eng. News* (April 1984), p. 3.
35. C. Norman, *Science* **228**, 305 (1985).

36. This study was based in part on thesis research for an M.B.A. degree with emphasis in strategic planning. I thank N. Ackerman, W. K. Schmidt, F. Baldino, Jr., and L. G. Davis for their review of the manuscript and S. M. Vari for secretarial support.

Part VI

Virology

20

Nucleotide Sequence of Yellow Fever Virus: Implications for Flavivirus Gene Expression and Evolution

Charles M. Rice, Edith M. Lenches, Sean R. Eddy, Se Jung Shin, Rebecca L. Sheets, and James H. Strauss

The *Flavivirus* genus, family Flaviviridae, consists of a group of some 70 closely related human or veterinary pathogens causing many serious illnesses, including dengue fever, Japanese encephalitis, St. Louis encephalitis, Murray Valley encephalitis, tick-borne encephalitis, and yellow fever (*1*). Most flaviviruses are transmitted to vertebrate hosts by blood-sucking arthropods, mosquitoes or ticks, although some evidently lack an arthropod vector (*2*). Arthropod-transmitted flaviviruses replicate in the arthropod host as well as the vertebrate host. Human flavivirus diseases have diverse and complex pathologies and different viruses exhibit marked tissue tropisms. Many are neurotropic, causing encephalitic symptoms; others, such as the dengue group, replicate preferentially in host macrophages, whereas yellow fever is usually viscerotropic.

The disease known as yellow fever has been recognized for several hundred years (*3, 4*). Until the early 1900's recurrent epidemics occurred in the Caribbean area which caused great human suffering and had a profound influence on human activities in the area. From its focus in the Caribbean, epidemic yellow fever was spread by ship to ports as far north as Boston and as far east as England, where mortality rates in an epidemic could exceed 20 percent of those contracting the disease. Walter Reed and colleagues in pioneering studies in Cuba in 1900 demonstrated that yellow fever is transmitted by mosquitoes, and 2 years later showed that the disease agent is filterable (*5*). With the recognition that the mosquito *Aedes aegypti* is the vector for urban yellow fever, mosquito control measures rapidly led to the elimination

Science 229, 726–733 (23 August 1985)

of urban yellow fever. Subsequently, a safe and effective attenuated vaccine strain (17D) was developed by in vitro passage of the virulent Asibi strain in chicken embryo tissue (*6*). However, the virus persists in a sylvan cycle in the forests of South America and Africa, transmitted by numerous mosquito species including those of the genus *Haemagogus* in South America and of the genus *Aedes* in Africa. The vertebrate hosts in this cycle appear to be almost exclusively primates, demonstrating the limited natural host range of yellow fever. From the sylvan cycle periodic outbreaks in neighboring human populations have arisen on both continents. Furthermore, since *Aedes aegypti* is widespread in the world, a situation exacerbated by relaxation of mosquito abatement procedures in the Caribbean and elsewhere, the potential exists for future epidemics of urban yellow fever.

Previous studies have shown that flaviviruses contain single-stranded infectious RNA (thus defining them as plus-stranded RNA viruses in which the virion RNA serves as a messenger) encapsidated in a nucleocapsid possessing icosahedral symmetry and containing a single species of capsid protein [C, apparent mass of about 14 kilodaltons (kD)]. This in turn is surrounded by a lipid bilayer containing an envelope protein (E; about 50 to 60 kD) that is usually but not invariably glycosylated (*7*) and a second, nonglycosylated protein (M; about 8 kD) (*8, 9*). How the envelope is obtained is unclear, as budding flaviviruses are seldom identified in electron microscopic studies, although maturation does appear to occur in association with intracellular membranes (*9, 10*). Replication of flaviviruses in tissue culture is slow, with a long latent period, and only moderate titers of virus are produced. Host cell protein and RNA synthesis are shut off only poorly (vertebrate cells) or not at all (mosquito cells), making study of flavivirus replication and structure somewhat more difficult. Virus-specific protein synthesis appears to be associated with the rough endoplasmic reticulum, and RNA replication is localized in the perinuclear region (*11*). No subgenomic RNA has been detected in cells infected with flaviviruses, and it is believed that the genomic length RNA which is capped but not polyadenylated (*12, 13*) is the only messenger RNA (mRNA) species (*9, 12, 14*). This mRNA is translated into the three structural proteins and several nonstructural proteins. Translation of the flavivirus genome in vitro produces polypeptides related to the structural proteins (*15*) which, in the presence of appropriate membrane fractions, can be processed efficiently to yield C and E (*16*). Peptide mapping of in vitro translation products as well as selective incorporation of *N*-formylmethionine suggest that initiation in vitro occurs only with the capsid protein. Alternatively, studies on the in vivo translation of flavivirus Kunjin have been based on the use of pactamycin or high salt inhibition of translation initiation (*17*) or ultraviolet inactivation of translation (*18*) in an attempt to map the genome order of flavivirus proteins on the assumption that there is just a single site for initiation of translation. These experiments have led Westaway and collaborators to suggest that multiple independent translation initiation sites are used within flavivirus RNA, a situation not typically found with other eukaryotic mRNA's (*19*).

We now present the complete nucleotide sequence of the yellow fever

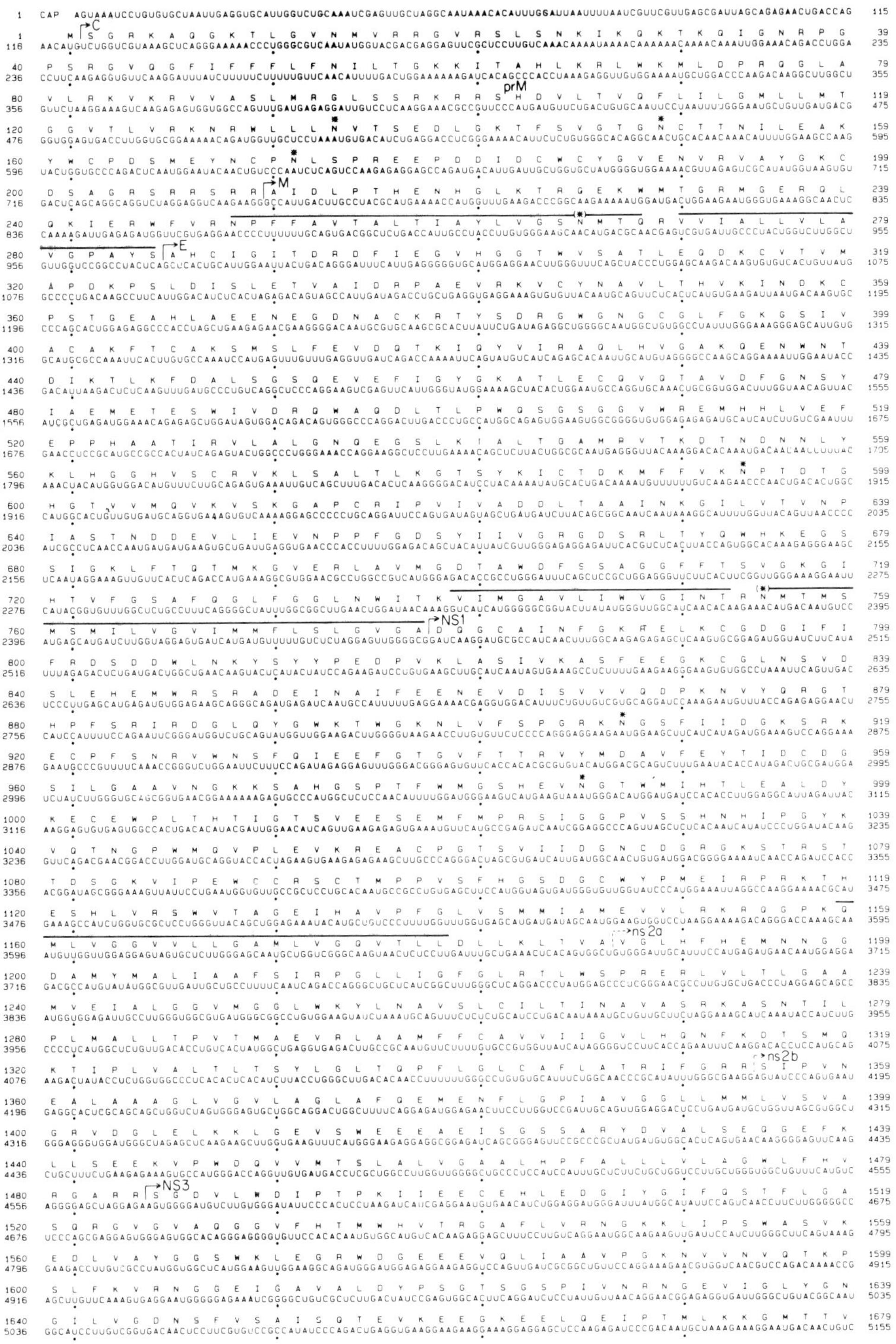

```
1680  L D F H P G A G K T R R F L P Q I L A E C A R R R L R T L V L A P T R V V L S E  1719
5156  CUUGAUUUUCAUCCUGGAGCUGGGAAGACAAGACGUUUCCUCCCACAGAUCUUGGCCGAGUGCGCACGGAGACGCUUGCGCACUCUUGUGUUGGCCCCCACCAGGGUUGUUCUUUCUGAA  5275

1720  M K E A F H G L D V K F H T Q A F S A H G S G R E V I D A M C H A T L T Y R M L  1759
5276  AUGAAGGAGGCUUUUCACGGCCUGGACGUGAAAUUCCACACACAGGCUUUUUCCGCUCACGGCAGCGGGAGAGAAGUCAUUGAUGCCAUGUGCCAUGCCACCCUAACUUACAGGAUGUUG  5395

1760  E P T R V V N W E V I I M D E A H F L D P A S I A A R G W A A H R A R A N E S A  1799
5396  GAACCAACUAGGGUUGUUAACUGGGAAGUGAUCAUUAUGGAUGAAGCCCAUUUUUUGGAUCCAGCUAGCAUAGCCGCUAGAGGUUGGGCAGCGCACAGAGCUAGGGCAAAUGAAAGUGCA  5515

1800  T I L M T A T P P G T S D E F P H S N G E I E D V Q T D I P S E P W N T G H D W  1839
5516  ACAAUCUUGAUGACAGCCACACCGCCUGGGACUAGUGAUGAAUUUCCACAUUCAAAUGGUGAAAUAGAAGAUGUUCAAACGGACAUACCCAGUGAGCCCUGGAACACAGGGCAUGACUGG  5635

1840  I L A D K R P T A W F L P S I R A A N V M A A S L R K A G K S V V V L N R K T F  1879
5636  AUCCUAGCUGACAAAAGGCCCACGGCAUGGUUCCUUCCAUCCAUCAGAGCUGCAAAUGUCAUGGCUGCCUCUUUGCGUAAGGCUGGAAAGAGUGUGGUGGUCCUGAACAGGAAAACCUUC  5755

1880  E R E Y P T I K Q K K P D F I L A T D I A E M G A N L C V E R V L D C R T A F K  1919
5756  GAGAGAGAAUACCCCACGAUAAAGCAGAAGAAACCUGACUUUAUAUUGGCCACUGACAUAGCUGAAAUGGGAGCCAACCUUUGCGUGGAGCGAGUGCUGGAUUGCAGGACGGCUUUUAAG  5875

1920  P V L V D E G R K V A I K G P L R I S A S S A A Q R R G R I G R N P N R D G D S  1959
5876  CCUGUGCUUGUGGAUGAAGGGAGGAAGGUGGCAAUAAAAGGCCCACUUCGUAUCUCCGCAUCCUCUGCUGCUCAAAGGAGGGGGCGCAUUGGGAGAAAUCCCAACAGAGAUGGAGACUCA  5995

1960  Y Y Y S E P T S E N N A H H V C W L E A S M L L D N M E V R G G M V A P L Y G V  1999
5996  UACUACUAUUCUGAGCCUACAAGUGAAAAUAAUGCCCACCACGUCUGCUGGUUGGAGGCCUCAAUGCUCUUGGACAACAUGGAGGUGAGGGGUGGAAUGGUCGCCCCACUCUAUGGCGUC  6115

2000  E G T K T P V S P G E M R L R D D Q R K V F R E L V R N C D L P V W L S W Q V A  2039
6116  GAAGGAACUAAAACACCAGUUUCCCCUGGUGAAAUGAGACUGAGGGAUGACCAGAGGAAAGUCUUCAGAGAACUAGUGAGGAAUUGUGACCUGCCCGUUUGGCUUUCGUGGCAAGUGGCC  6235

2040  K A G L K T N D R K W C F E G P E E H E I L N D S G E T V K C R A P G G A K K P  2079
6236  AAGGCUGGUUUGAAGACGAAUGAUCGUAAGUGGUGUUUUGAAGGCCCUGAGGAACAUGAGAUCUUGAAUGACAGCGGUGAAACAGUGAAGUGCAGGGCUCCUGGAGGAGCAAAGAAGCCU  6355
                                                                                     →ns4a
2080  L R P R W C D E R V S S D Q S A L S E F I K F A E G R R | G A A E V L V V L S E L  2119
6356  CUGCGCCCAAGGUGGUGUGAUGAAAGGGUGUCAUCUGACCAGAGUGCGCUGUCUGAAUUUAUUAAGUUUGCUGAAGGUAGGAGGGGAGCUGCUGAAGUGCUAGUUGUGCUGAGUGAACUC  6475

2120  P D F L A K K G G E A M D T I S V F L H S E E G S R A Y R N A L S M M P E A M T  2159
6476  CCUGAUUUCCUGGCUAAAAAAGGUGGAGAGGCAAUGGAUACCAUCAGUGUGUUCCUCCACUCUGAGGAAGGCUCUAGGGCUUACCGCAAUGCACUAUCAAUGAUGCCUGAGGCAAUGACA  6595

2160  I V M L F I L A G L L T S G M V I F F M S P K G I S R M S M A M G T M A G C G Y  2199
6596  AUAGUCAUGCUGUUUAUACUGGCUGGACUACUGACAUCGGGAAUGGUCAUCUUUUUCAUGUCUCCCAAAGGCAUCAGUAGAAUGUCUAUGGCGAUGGGCACAAUGGCCGGCUGUGGAUAU  6715

2200  L M F L G G V K P T H I S Y V M L I F F V L M V V V I P E P G Q Q R S I Q D N Q  2239
6716  CUCAUGUUCCUUGGAGGCGUCAAACCCACUCACAUCUCCUAUGUCAUGCUCAUAUUCUUUGUCCUGAUGGUGGUUGUGAUCCCCGAGCCAGGGCAACAAAGGUCCAUCCAAGACAACCAA  6835

2240  V A Y L I I G I L T L V S A V A A N E L G M L E K T K E D L F G K K N L I P S S  2279
6836  GUGGCAUACCUCAUUAUUGGCAUCCUGACGCUGGUUUCAGCGGUGGCAGCCAACGAGCUAGGCAUGCUGGAGAAAACCAAAGAGGACCUCUUUGGGAAGAAGAACUUAAUUCCAUCUAGU  6955

2280  A S P W S W P D L D L K P G A A W T V Y V G I V T M L S P M L H H W I K V E Y G  2319
6956  GCUUCACCCUGGAGUUGGCCGGAUCUUGACCUGAAGCCAGGAGCUGCCUGGACAGUGUACGUUGGCAUUGUUACAAUGCUCUCUCCAAUGUUGCACCACUGGAUCAAAGUCGAAUAUGGC  7075

2320  N L S L S G I A Q S A S V L S F M D K G I P F M K M N I S V I M L L V S G W N S  2359
7076  AACCUGUCUCUGUCUGGAAUAGCCCAGUCAGCCUCAGUCCUUUCUUUCAUGGACAAGGGGAUACCAUUCAUGAAGAUGAAUAUCUCGGUCAUAAUGCUGCUGGUCAGUGGCUGGAAUUCA  7195
                                                                                                        →ns4b
2360  I T V M P L L C G I G C A M L H W S L I L P G I K A Q Q S K L A Q R R | V F H G V  2399
7196  AUAACAGUGAUGCCUCUGCUCUGUGGCAUAGGGUGCGCCAUGCUCCACUGGUCUCUCAUUUUACCUGGAAUCAAAGCGCAGCAGUCAAAGCUUGCACAGAGAAGGGUGUUCCAUGGCGUU  7315

2400  A E N P V V D G N P T V D I E E A P E M P A L Y E K K L A L Y L L L A L S L A S  2439
7316  GCCGAGAACCCUGUGGUUGAUGGGAAUCCAACAGUUGACAUUGAGGAAGCUCCUGAAAUGCCUGCCCUUUAUGAGAAGAAACUGGCUCUAUAUCUCCUUCUUGCUCUCAGCCUAGCUUCU  7435

2440  V A M C R T P F S L A E G I V L A S A A L G P L I E G N T S L L W N G P M A V S  2479
7436  GUUGCCAUGUGCAGAACGCCCUUUUCAUUGGCUGAAGGCAUUGUCCUAGCAUCAGCUGCCUUAGGGCCGCUCAUAGAGGGAAACACCAGCCUUCUUUGGAAUGGACCCAUGGCUGUCUCC  7555
                                                                                 →NS5
2480  M T G V M R G N H Y A F V G V M Y N L W K M K T G R R | G S A N G K T L G E V W K  2519
7556  AUGACAGGAGUCAUGAGGGGGAAUCACUAUGCUUUUGUGGGAGUCAUGUACAAUCUAUGGAAGAUGAAAACUGGACGCCGGGGGAGCGCGAAUGGAAAAACUUUGGGUGAAGUCUGGAAG  7675

2520  R E L N L L D K R Q F E L Y K R T D I V E V D R D T A R R H L A E G K V D T G V  2559
7676  AGGGAACUGAAUCUGUUGGACAAGCGACAGUUUGAGUUGUAUAAAAGGACCGACAUUGUGGAGGUGGAUCGUGAUACGGCACGCAGGCAUUUGGCCGAAGGGAAGGUGGACACCGGGGUG  7795

2560  A V S R G T A K L R W F H E R G Y V K L E G R V I D L G C G R G G W C Y Y A A A  2599
7796  GCGGUCUCCAGGGGGACCGCAAAGUUAAGGUGGUUCCAUGAGCGUGGCUAUGUCAAGCUGGAAGGUAGGGUGAUUGACCUGGGGUGUGGCCGCGGAGGCUGGUGUUACUACGCUGCUGCG  7915

2600  Q K E V S G V K G F T L G R D G H E K P M N V Q S L G W N I I T F K D K T D I H  2639
7916  CAAAAGGAAGUGAGUGGGGUCAAAGGAUUUACUCUUGGAAGAGACGGCCAUGAGAAACCCAUGAAUGUGCAAAGUCUGGGAUGGAACAUCAUCACCUUCAAGGACAAAACUGAUAUCCAC  8035

2640  R L E P V K C D T L L C D I G E S S S S S V T E G E R T V R V L D T V E K W L A  2679
8036  CGCCUAGAACCAGUGAAAUGUGACACCCUUUUGUGUGACAUUGGAGAGUCAUCAUCGUCAUCGGUCACAGAGGGGGAAAGGACCGUGAGAGUUCUUGAUACUGUAGAAAAAUGGCUGGCU  8155

2680  C G V D N F C V K V L A P Y M P D V L E K L E L L Q R R F G G T V I R N P L S R  2719
8156  UGUGGGGUUGACAACUUCUGUGUGAAGGUGUUAGCUCCAUACAUGCCAGAUGUUCUCGAGAAACUGGAAUUGCUCCAAAGGAGGUUUGGCGGAACAGUGAUCAGGAACCCUCUCUCCAGG  8275

2720  N S T H E M Y Y V S G A R S N V T F T V N Q T S R L L M R R M R R P T G K V T L  2759
8276  AAUUCCACUCAUGAAAUGUACUACGUGUCUGGAGCCCGCAGCAAUGUCACAUUUACUGUGAACCAAACAUCCCGCCUCCUGAUGAGGAGAAUGAGGCGUCCAACUGGAAAAGUGACCCUG  8395

2760  E A D V I L P I G T R S V E T D K G P L D K E A I E E R V E R I K S E Y M T S W  2799
8396  GAGGCUGACGUCAUCCUCCCAAUUGGGACACGCAGUGUUGAGACAGACAAGGGACCCCUGGACAAAGAGGCCAUAGAAGAAAGGGUUGAGAGGAUAAAAUCUGAGUACAUGACCUCUUGG  8515

2800  F Y D N D N P Y R T W H Y C G S Y V T K T S G S A A S M V N G V I K I L T Y P W  2839
8516  UUUUAUGACAAUGACAACCCCUACAGGACCUGGCACUACUGUGGCUCCUAUGUCACAAAAACCUCAGGAAGUGCGGCGAGCAUGGUAAAUGGUGUUAUUAAAAUUCUGACAUAUCCAUGG  8635

2840  D R I E E V T R M A M T D T T P F G Q Q R V F K E K V D T R A K D P P A G T R K  2879
8636  GACAGGAUAGAGGAGGUCACAAGAAUGGCAAUGACUGACACAACCCCUUUUGGACAGCAAAGAGUGUUUAAAGAAAAAGUUGACACCAGAGCAAAGGAUCCACCAGCGGGAACUAGGAAG  8755

2880  I M K V V N R W L F R H L A R E K N P R L C T K E E F I A K V R S H A A I G A Y  2919
8756  AUCAUGAAAGUUGUCAACAGGUGGCUGUUCCGCCACCUGGCCAGAGAAAAGAACCCCAGACUGUGCACAAAGGAAGAAUUUAUUGCAAAAGUCCGAAGUCAUGCAGCCAUUGGAGCUUAC  8875

2920  L E E Q E Q W K T A N E A V Q D P K F W E L V D E E R K L H Q Q G R C R T C V Y  2959
8876  CUGGAAGAACAAGAACAGUGGAAGACUGCCAAUGAGGCUGUCCAAGACCCAAAGUUCUGGGAACUGGUGGAUGAAGAAAGGAAGCUGCACCAACAAGGCAGGUGUCGGACUUGUGUGUAC  8995

2960  N M M G K R E K K L S E F G K A K G S R A I W Y M W L G A R Y L E F E A L G F L  2999
8996  AACAUGAUGGGAAAAAGAGAGAAGAAGCUGUCAGAGUUUGGGAAAGCAAAGGGAAGCCGUGCCAUAUGGUAUAUGUGGCUGGGAGCGCGGUAUCUUGAGUUUGAGGCCCUGGGAUUCCUG  9115

3000  N E D H W A S R E N S G G G V E G I G L Q Y L G Y V I R D L A A M D G G G [F Y A  3039
9116  AAUGAGGACCAUUGGGCUUCCAGGGAAAACUCAGGAGGAGGAGUGGAAGGCAUUGGCUUACAAUACCUAGGAUAUGUGAUCAGAGACCUGGCUGCAAUGGAUGGUGGUGGAUUCUACGCG  9235

3040  D D T A G W D T R I T E A D L D D E Q E I L N Y M S P H H K K L A Q A V M E M T  3079
9236  GAUGACACCGCUGGAUGGGACACGCGCAUCACAGAGGCAGACCUUGAUGAUGAACAGGAGAUCUUGAACUACAUGAGCCCACAUCACAAAAAACUGGCACAAGCAGUGAUGGAAAUGACA  9355

3080  Y K N K V V K V L R P A P G G K A Y M D V I S R R D Q R G S G Q V V T Y A L N T  3119
9356  UACAAGAACAAAGUGGUGAAAGUGUUGAGACCAGCCCCAGGAGGGAAAGCCUACAUGGAUGUCAUAAGUCGACGAGACCAGAGAGGAUCCGGGCAGGUAGUGACUUAUGCUCUGAACACC  9475

3120  I T N L K V Q L I R M A E A E M V I H H Q H V Q D C D E S V L T R L E A W L T E  3159
9476  AUCACCAACUUGAAAGUCCAAUUGAUCAGAAUGGCAGAAGCAGAGAUGGUGAUACAUCACCAACAUGUUCAAGAUUGUGAUGAAUCAGUUCUGACCAGGCUGGAGGCAUGGCUCACUGAG  9595
```

3160 H G C D R L K R M A V S G D D C V V R P I D D R F G L A L S H L N A M S K V R K 3199
9596 CACGGAUGUGACAGACUGAAGAGGAUGGCGGUGAGUGGAGACGACUGUGUGGUCCGGCCCAUCGAUGACAGGUUCGGCCUGGCCCUGUCCCAUCUCAACGCCAUGUCCAAGGUUAGAAAG 9715

3200 D I S E W Q P S K G W N D W E N V P F C S H H F H E L Q L K D G R R I V V P C R 3239
9716 GACAUAUCUGAAUGGCAGCCAUCAAAAGGGUGGAAUGAUUGGGAGAAUGUGCCCUUCUGUUCCCACCACUUCCAUGAACUACAGCUGAAGGAUGGCAGGAGGAUUGUGGUGCCUUGCCGA 9835

3240 E Q D E L I G R G R V S P G N G W M I K E T A C L S K A Y A N M W S L M Y F H K 3279
9836 GAACAGGACGAGCUCAUUGGGAGAGGAAGGGUGUCUCCAGGAAACGGCUGGAUGAUCAAGGAAACAGCUUGCCUCAGCAAAGCCUAUGCCAACAUGUGGUCACUGAUGUAUUUUCACAAA 9955

3280 R D M R L L S L A V S S A V P T S W V P Q G R T T W S I H G K G E W M T T E D M 3319
9956 AGGGACAUGAGGCUACUGUCAUUGGCUGUUUCCUCAGCUGUUCCCACCUCAUGGGUUCCACAAGGACGCACAACAUGGUCGAUUCAUGGGAAAGGGGAGUGGAUGACCACGGAAGACAUG 10075

3320 L E V W N R V W I T N N P H M Q D K T M V K K W R D V P Y L T K R Q D K L C G S 3359
10076 CUUGAGGUGUGGAACAGAGUAUGGAUAACCAACAACCCACACAUGCAGGACAAGACAAUGGUGAAAAAAUGGAGAGAUGUCCCUUAUCUAACCAAGAGACAAGACAAGCUGUGCGGAUCA 10195

3360 L I G M T N R A T W A S H I H L V I H R I R T L I G Q E K Y T D Y L T V M D R Y 3399
10196 CUGAUUGGAAUGACCAAUAGGGCCACCUGGGCCUCCCACAUCCAUUUAGUCAUCCAUCGUAUCCGAACGCUGAUUGGACAGGAGAAAUACACUGACUACCUAACAGUCAUGGACAGGUAU 10315

3400 S V D A D L Q L G E L I 3411
10316 UCUGUGGAUGCUGACCUGCAACUGGGUGAGCUUAUCUGAAACACCAUCUAACAGGAAUAACCGGGAUACAAACCACGGGUGGAGAACCGGACUCCCCACAACCUGAAACCGGGAUAUAAA 10435

10436 CCACGGCUGGAGAACCGGGCUCCGCACUUAAAAUGAAACAGAAACCGGGAUAAAAACUACGGAUGGAGAACCGGACUCCACACAUUGAGACAGAAGAAGUUGUCAGCCCAGAACCCCACA 10555

10556 CGAGUUUUGCCACUGCUAAGCUGUGAGGCAGUGCAGGCUGGGACAGCCGACCUCCAGGUUGCGAAAAACCUGGUUUCUGGGACCUCCCACCCCAGAGUAAAAAGAACGGAGCCUCCGCUA 10675

10676 CCACCCUCCCACGUGGUGGUAGAAAGACGGGGUCUAGAGGUUAGAGGAGACCCUCCAGGGAACAAAUAGUGGGACCAUAUUGACGCCAGGGAAAGACCGGAGUCGUUCUCUGCUUUUCCU 10795

10796 CCAGAGGUCUGUGAGCACAGUUUGCUCAAGAAUAAGCAGACCUUUGGAUGACAAACACAAAACCACU 10862

Fig. 1 (pages 283–285). Entire sequence of the genome of yellow fever virus. Yellow fever virus, 17D vaccine strain, was obtained from the American Type Culture Collection. This sample represents in vitro passage 234 of the line originated by Theiler and colleagues who started with the virulent Asibi strain (*6*). After plaque purification in Vero cells and amplification in BHK cells, the virus was grown in SW13 monolayers (*50*) and purified by polyethylene glycol precipitation, in glycerol-tartrate gradients. The purified virus was diluted with aqueous buffer and sedimented in the ultracentrifuge; the RNA was isolated by phenol extraction (*51*). Briefly, single-stranded cDNA was synthesized with avian myeloblastosis virus reverse transcriptase using degraded calf thymus DNA for priming (*47*). Second strand synthesis was carried out essentially as previously described (*52*). After methylation of the Eco RI sites with Eco RI methylase, phosphorylated Eco RI linkers were added with T4 DNA ligase. Following complete digestion with Eco RI, the double-stranded cDNA was sized on an agarose gel and selected size fractions were inserted into the Eco RI site of a plasmid vector derived from pBR322. Colonies containing yellow fever–specific inserts were selected by colony hybridization and were characterized by restriction mapping to obtain clones which represented most of the yellow fever genome. Clones containing the 3′ end of the genome were constructed by poly(A)-tailing (polyadenylation) the genomic RNA with *Escherichia coli* poly(A) polymerase followed by synthesis of double-stranded cDNA with an oligo(dT) primer. Addition of the poly(A) tract was relatively inefficient but after digestion of the double-stranded cDNA with Bgl I, 3′-terminal Bgl I fragments were selectively cloned with a plasmid vector derived from cloned yellow fever DNA (*51*). Clones containing the 5′ end of the genome were constructed by primer extension followed by oligo(dC) tailing with terminal deoxynucleotidyl transferase and oligo(dG) primed second strand synthesis. The entire sequence was obtained by chemical sequencing of both strands of the DNA (*53*). In addition, sequence was obtained throughout from at least two clones. Wherever the sequence differed between two clones (due presumably to heterogeneity in the RNA population or errors introduced during cloning), a third and occasionally a fourth clone was sequenced in this area, and the preferred nucleotide is reported here. Nucleotides are numbered from the 5′ terminus. Amino acids are numbered from the first methionine in the polyprotein sequence. The beginning of each protein is labeled (see Table 1 and text for nomenclature); tentative assignments are indicated by dashed arrows. Putative hydrophobic membrane-associated segments in the structural region are overlined. Potential *N*-linked glycosylation sites are denoted by an asterisk. The region of NS5 homologous to other RNA viruses (see text) is enclosed by brackets and the conserved Gly-Asp-Asp sequence is boxed. Repeated nucleotide sequences are underlined. Closely spaced in phase stop codons that terminate the long open reading frame are boxed. The single letter abbreviations for the amino acid residues are: A, alanine; C, cysteine; D, aspartic acid; E, glutamic acid; F, phenylalanine; G, glycine; H, histidine; I, isoleucine; K, lysine; L, leucine; M, methionine; N, asparagine; P, proline; Q, glutamine; R, arginine; S, serine; T, threonine; V, valine; W, tryptophan; Y, tyrosine.

genome determined from complementary DNA (cDNA) clones of the 17D vaccine strain. Together with recent NH_2-terminal sequence analysis of both structural (*20*) and some nonstructural yellow fever proteins, the amino acid sequences of the encoded proteins have been deduced and a preliminary picture of flavivirus gene organization and expression has begun to emerge.

Sequence of yellow fever RNA. The complete sequence of yellow fever RNA is shown in Fig. 1. The 5′- and 3′-terminal sequences presented were derived from several independent clones, are homologous to the 5′ and 3′ termini of West Nile flavivirus genomic RNA (*21*) (see below), and thus probably reflect the extreme ends of the yellow fever genome. Given these assumptions, the RNA genome is 10,862 nucleotides in length and has a mass of 3.75×10^6 daltons (expressed as the sodium form). Previous reports have shown that flavivirus genomic RNA contains a type 1 cap at the 5′ terminus but lacks a polyadenylate tract at the 3′ terminus (*12, 13*). The base composition of the RNA is 27.3 percent A, 23.0 percent U, 28.4 percent G, and 21.3 percent C.

It is striking that the RNA contains an extremely long open reading frame, which spans virtually the entire length of the genome. This open reading frame, beginning from the first AUG triplet, is 10,233 nucleotides in length, terminating with a single opal codon (UGA), and could encode a polypeptide of 380,763 daltons, leaving 5′- and 3′-noncoding regions of 118 and 511 nucleotides, respectively. Examination of the remaining five possible reading frames (two in the virion RNA and three in the complementary RNA) reveals multiple stop codons in every case, with the longest possible other open reading frame being 804 nucleotides (in the complementary strand). Thus there is no reason to expect that any protein is translated from yellow fever RNA other than the polyprotein encoded by the long open reading frame shown in Fig. 1.

The structural proteins of yellow fever virus. The start points of the three yellow fever virus structural proteins (C, M, and E) have been positioned within the translated RNA sequence from NH_2-terminal amino acid sequences obtained for the structural proteins isolated from yellow fever virions (*20*) (Fig. 1). The capsid protein is the first protein found in the long open reading frame and begins one residue past the first methionine. Thus, in agreement with in vitro translation data from the flavivirus genomic RNA's of tick-borne encephalitis virus, West Nile virus, and Kunjin virus (*15, 16*), the translation of the yellow fever genome initiates with the capsid protein, and the NH_2-terminal methionine is removed during maturation of the protein (*20*). The capsid protein may be released from the precursor polyprotein by cleavage at or just past a series of basic amino acids (Figs. 1 and 2). From this deduced amino acid sequence, the capsid protein is quite basic containing about 25 percent lysine and arginine distributed throughout the protein. The capsid protein of tick-borne encephalitis virus contains a similar proportion of basic amino acids (*22*). Since the capsid protein forms complexes with the RNA, its highly basic character probably acts to neutralize some of the RNA charges in such a compact structure. There is also a hydrophobic stretch of 16 uncharged amino acids beginning with residue 42 from the NH_2 terminus (see Fig. 3), which is conserved among flaviviruses (*23*) and may be involved in

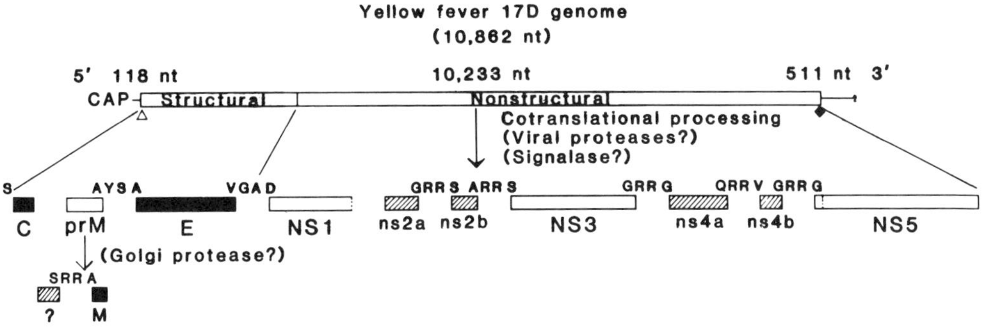

Fig. 2. Organization and processing of proteins encoded by the yellow fever genome. Untranslated regions are shown as single lines and the translated region as an open box. The open triangle is the initiation codon (AUG); the solid diamond the termination codon (UGA). The protein nomenclature is described in Table 1 and (*35*). The single letter amino acid code is used for sequences flanking assigned cleavage sites (solid lines). Two other potential cleavage sites are shown as dotted lines. Structural proteins, identified nonstructural proteins, and hypothesized nonstructural proteins (see text) are indicated by solid, open, and hatched boxes, respectively. Other potential cleavage sites have been found and are described in Table 1, footnote asterisk.

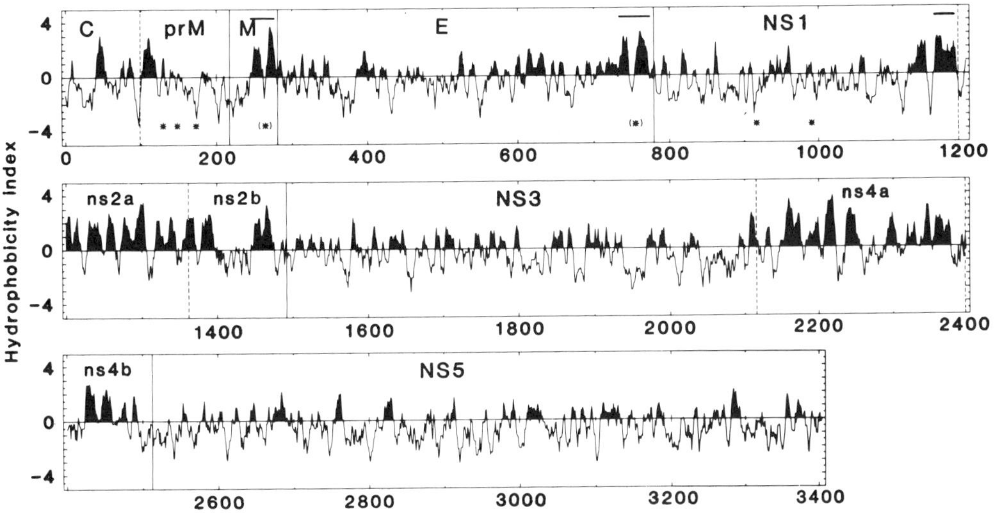

Fig. 3. Hydrophobicity plot of the yellow fever polyprotein sequence. The program of Kyte and Doolittle (*54*) with a search length of seven amino acids was used. Cleavage sites localized by NH_2-terminal protein sequence are indicated by solid vertical lines; putative cleavage sites are indicated by dotted vertical lines. The protein nomenclature is described in Table 1 and (*35*). The degree of hydrophobicity increases with distance above the horizontal line; hydrophilicity increases with distance below the horizontal line. Potential *N*-linked glycosylation sites are denoted by asterisks and putative membrane-associated anchors are indicated by solid bars.

protein-protein or specific protein-RNA interactions (or both) which assemble the nucleocapsid and lead to acquisition of the lipoprotein envelope by the capsid.

The start point of the virion M protein is also shown in Fig. 1. This protein contains a charged NH_2-terminal domain and two long uncharged stretches at its COOH terminus; these two stretches are separated by a single basic residue (Figs. 1 and 3) and could act as membrane spanning anchors similar to those observed in many virus envelope proteins. Protein M has not been identified in infected cells and is postulated to be derived from a precursor glycoprotein which we call prM (Table 1), which is also called by others GP23, GP19, or NV2 (*8, 24*). The sequence data support this hypothesis. A possible start point of prM, as deduced by limited homology with the NH_2-terminal sequence of the flavivirus St. Louis encephalitis NV2 (*20*) and homology in this region with Murray Valley encephalitis virus (*23*), follows the capsid protein; prM may begin with an uncharged stretch of amino acids which could function as an NH_2-terminal signal sequence for its cotranslational insertion into the endoplasmic reticulum (Fig. 3). After this hydrophobic domain, which may or may not be removed by signalase, the prM sequence contains three possible glycosylation sites of the type Asn-X-Ser/Thr. The NH_2 terminus of M (*20*) follows the sequence Arg-Ser-Arg-Arg in prM, indicating that the cleavage to produce M may be effected by the same enzyme that cleaves a number of viral envelope precursors at the sequence Arg-X-Arg/Lys-Arg and that has been postulated to be a host protease localized in the Golgi apparatus or Golgi-derived vesicles (*25*), perhaps similar to the cathepsins (*26*). As a result of this cleavage, which apparently occurs late during virus maturation and release, an 11.4-kD (not including carbohydrate) glycopeptide would be removed leaving the nonglycosylated M protein embedded in the virion membrane. Trace quantities of small virus-specific glycoproteins have been detected in cytoplasmic extracts (*27, 28, 29*), but whether the glycopeptide fragment remains cell-associated and is rapidly degraded or is released into the extracellular medium is unknown.

The E protein follows M. The NH_2 terminus of E is charged, and the more hydrophobic COOH-terminal domain of M (or its precursor, prM) may function as the signal sequence for the translocation of E across the rough endoplasmic reticulum. The protein E contains two sites of the form Asn-X-Ser/Thr which could serve as carbohydrate attachment sites, and both glycosylated and nonglycosylated forms have been detected in infected cells (*7, 27, 30*). The COOH-terminal domain of E contains uncharged stretches that could serve as a transmembrane anchor. Cleavage between M and E occurs after a Ser residue, and could be catalyzed by a host protease such as signalase. Since the COOH terminus of the mature M protein has not been determined, a small peptide, analogous to the 6 kD protein of alphaviruses (*25, 31*) could be produced during maturation of M and E. However, the apparent size of the M protein agrees well with the predicted molecular weight if cleavage occurs after the Ser at position 285.

This model for translation and processing of structural proteins and the features mentioned above predict that most of the

E protein and some of the M protein should be exposed on the mature virion surface, and therefore sensitive to digestion by appropriate proteases. Protease digestion of purified tick-borne encephalitis virus (*32*) and also yellow fever virus (*29*) support this hypothesis. Thus, the M protein (or prM) of flaviviruses is an integral membrane protein and may interact specifically with both the E protein as well as the capsid protein-RNA complex during virus assembly.

The nonstructural proteins. In addition to prM, at least four and as many as 12 nonstructural proteins have been described in flavivirus-infected cells (*9, 28, 33, 34*). Some or all of these proteins must be active in the replication of the viral RNA. The start points of the three largest nonstructural proteins (NV3, NV4, and NV5 by the old nomenclature) (*35*) have been located by NH_2-terminal amino acid sequence analysis (*36*). As previously suggested by peptide mapping of the corresponding nonstructural proteins from other flaviviruses (*9, 15, 34*), the sequence data show that these proteins map to nonoverlapping segments in the yellow fever virus nonstructural region (Figs. 1 and 2).

In an attempt to simplify the description of flavivirus encoded nonstructural polypeptides, in particular the smaller proteins, we suggest a modified nomenclature (*35*) (Table 1) based on the linear order of these proteins in the yellow fever virus genome to complement designations based on their apparent molecular weights (*37*). In taking this approach we assume that members of Flaviviridae will have similar genome organization and express homologous proteins from homologous regions of their genomes. This assumption has been partially verified by an extensive sequence comparison of yellow fever virus with another member of the flavivirus genus, Murray Valley encephalitis virus (*23*).

Several features of the yellow fever virus nonstructural region are apparent from the localization of NS1, NS3, and NS5 (formerly NV3, NV4, and NV5).

Table 1. Flavivirus polypeptides.

Protein nomenclature (*35*)		NH_2-terminal cleavage site*	M_r†	M_{pred}‡	Glycosylated?	Comments
Proposed	Old					
			Structural region			
C	V2 (NV1½)	M ↓ S	13,000 to 16,000	11,320	No	Nucleocapsid protein
prM	(NV2) (NV2½)	?	19,000 to 23,000	20,925	Yes	Precursor to M
M	V1	SRR ↓ A	8,000 to 8,500	8,526	No	Virion envelope protein
E	V3	AYS ↓ A	51,000 to 60,000	53,712	Both forms§	Major virion envelope protein
			Nonstructural region			
NS1	NV3	VGA ↓ D	44,000 to 49,000	45,869	Yes	Soluble complement-fixing antigen
ns2a	(NV2½) (NV2)	(TVA ↓ V)	16,000 to 21,000	18,086	No	Hydrophobic; function unknown
ns2b	(NV1½)	(GRR ↓ S)	12,000 to 15,000	13,823	No	Hydrophobic; function unknown
NS3	NV4	ARR ↓ S	67,000 to 76,000	69,319	No	Replicase component ?
ns4a	(NVX) (NV2½)	(GRR ↓ G)	24,000 to 32,000	31,196	No	Hydrophobic; function unknown
ns4b	(NV1)	(QRR ↓ V)	10,000 to 11,000	12,159	No	Hydrophobic; function unknown
NS5	NV5	GRR ↓ G	91,000 to 98,000	104,079	No	Replicase component ?

*Cleavage sites predicted by NH_2-terminal protein sequence data (*20, 36*). Tentative sites (indicated by parenthesis) are based on homology with confirmed cleavage sites and the sizes of yellow fever–specific polypeptides observed in infected cells (*27, 29*). Alternative cleavage sites in the nonstructural region occur after residue 1946 (Gln-Arg-Arg ↓ Gly), residue 2548 (Ala-Arg-Arg ↓ His), residue 2707 (Gln-Arg-Arg ↓ Phe), and residue 3104 (Ser-Arg-Arg ↓ Asp). †Range of flavivirus protein sizes estimated from acrylamide gel electrophoresis. Some of these proteins have not yet been identified for all flaviviruses thus far examined [for comparative analyses see (*33, 34*)]. In particular, definitive comparisons between NV3, NV2, NV2½, NVX, and NV1½ are difficult because of the complexity of flavivirus protein patterns in the $8,000 < M_r < 45,000$ size range. Alternative pathways of posttranslational cleavage may be used by different viruses for the production of the smaller nonstructural polypeptides. However, given the relatively consistent pattern of structural and larger nonstructural proteins, it seems likely that the small nonstructural proteins are also conserved, but their apparent migration on acrylamide gels and labeling efficiency are influenced by differences in amino acid composition. [For more complete discussion of this subject see (*49*).] ‡Polypeptide molecular weights calculated according to the cleavage sites shown in Fig. 1, with the C-prM cleavage site between residues 101 and 102. §Both glycosylated and nonglycosylated forms of E have been identified for yellow fever virus (*27*) and Kunjin virus (*7, 30*).

First, NS1 immediately follows the putative transmembrane segment of the E protein. It should be noted that NS1 is glycosylated (*27*), and monoclonal antibodies against NS1 are capable of mediating complement-dependent lysis of yellow fever virus–infected cells, suggesting its presence at the plasma membrane (*38*). Thus, the COOH-terminal uncharged hydrophobic sequence of E could function as a signal sequence for translocation of NS1 across the endoplasmic reticulum. NS1 contains two sites of the type Asn-X-Ser/Thr which could serve as glycosylation sites. The probable COOH terminus of NS1 from estimates of molecular weight could contain a hydrophobic sequence for anchoring the protein in the membrane (Fig. 3). Thus the three glycoproteins of yellow fever virus, prM, E, and NS1, are adjacent to one another in the genome and are possibly inserted into the membrane one after another during synthesis. The sequence data support the hypothesis that each has the usual membrane protein topology of an NH_2 terminus outside and a COOH-terminal hydrophobic anchor. However, additional experiments are required to rigorously establish their orientation with respect to the lipid bilayer and exact COOH termini. The function of NS1 is unknown, but it could be involved in virus assembly rather than RNA replication. In this regard, it is of interest that NS1 has been shown to be the soluble complement-fixing antigen for dengue 2 (*28*) and suggestive evidence exists for a comparable role of NS1 in yellow fever virus infection (*8, 27*). Thus, this protein may exist in alternative membrane-associated and soluble forms, perhaps because of the presence or absence of the COOH-terminal hydrophobic domain.

NS3 begins at residue 1485 in the polyprotein sequence and is produced by cleavage at the site Gly-Ala-Arg-Arg ↓ Ser; the NH_2-terminus of NS5 has been tentatively identified as residue 2507 after cleavage at Thr-Gly-Arg-Arg ↓ Gly. Since no host proteases with this specificity (which are active in the cytosol) have been characterized and animal viruses often encode proteases active in the processing of their cytoplasmic polyprotein precursors, yellow fever virus may encode a protease that cleaves after two Arg residues (or two basic residues) surrounded by amino acids with short side chains, often Gly (Table 1 and footnote asterisk).

These assignments leave two regions in the polyprotein for which polypeptide products have not yet been identified. Assuming that other nonstructural proteins will be produced from these regions by the same protease responsible for NH_2-terminal cleavage of NS3 and NS5, we have scanned the remaining sequences for additional cleavage sites. Estimates of molecular weight (*27*) have positioned the COOH terminus of NS1 near residue 1187. The next potential cleavage sequence, Gly-Arg-Arg ↓ Ser, at residue 1355 would produce two small nonstructural polypeptides of approximately 18 kD (ns2a) and 14 kD (ns2b) located between NS1 and NS3 (Fig. 2 and Table 1). Both of these polypeptides would be extremely hydrophobic (Fig. 3) with ns2b containing a short internal charged domain. The putative cleavage at the sequence Glu-Gly-Arg-Arg ↓ Gly (residue 2108) would produce a polypeptide whose calculated mass agrees well with the observed size of NS3 on polyacrylamide gels (*27, 29*). Between this site and the NH_2 terminus of NS5 a single potential cleavage site (Ala-Gln-

Arg-Arg ↓ Val) is found preceding residue 2395. Cleavage here would result in two methionine-rich, hydrophobic polypeptides of 31 kD (ns4a) and 12 kD (ns4b) (see Figs. 2 and 3 and Table 1). Polypeptides of these approximate sizes (10, 14, 18, and 30 kD) do exist in yellow fever–infected cells, but definitive mapping of these polypeptides as well as other minor species await additional NH_2-terminal sequence data. Similarly in the absence of COOH-terminal sequence data we cannot be sure of the exact terminal residues. Some heterogeneity in flavivirus polypeptides may result from variable exopeptidase digestion of the COOH-terminal residues or alternative internal cleavages. The predicted size of NS5, if the protein encompasses the remainder of the open reading frame, agrees well with its observed size (*27*).

Implications for flavivirus replication. It has been suggested that flavivirus RNA is translated by multiple internal initiation events (*17, 18*) which would make flaviviruses atypical among eukaryotic viruses and eukaryotic genes. The presence of a single long open reading frame in yellow fever virus RNA, the fact that the final proteins found do not initiate with methionine but appear to arise from a consistent set of proteolytic cleavages, the gene order deduced from the pactamycin runoff experiments of Westaway (*17*), the in vitro translation data (*15, 16*), and recent evidence for polyprotein precursors (*39*) all support the view that translation of the flavivirus genome in vivo initiates with the capsid protein near the 5′ end of the genome and proceeds sequentially through the genome to produce one precursor polyprotein.

Cleavage of this precursor is rapid and occurs during translation so that the precursor is not seen in its entirety. The location and frequency of characteristic cleavage sites in this precursor suggest that processing involves both virus encoded and cellular organelle bound proteases. Although internal translation initiation cannot be formally excluded, the 5′ terminal location of the structural genes and the 3′ terminal replicase genes implies that the relative amounts of structural and nonstructural gene products could also be regulated by premature termination as well as by nonuniform rates of translation (*40*) or differential stability of the final products. A potential secondary structure in yellow fever RNA just past the structural protein genes could possibly be active in the former mechanism. It is unclear why gene mapping experiments with ultraviolet light to inactivate translation (*18*) or high salt to synchronize initiation of translation (*17*) suggest multiple independent sites of initiation and do not allow prediction of the correct gene order. Possible explanations are that ribosomes might have slow transit velocities in some areas, due to RNA secondary structures or the presence of rare codons (*40*), or that it might be necessary to translate a functional protease to produce the final products.

Several features potentially important in RNA replication or packaging (or both) can be identified in the genomic sequence. First, the extreme 5′- and 3′-terminal sequences are homologous to those found for another flavivirus, West Nile virus (*21*) (Fig. 4), and the complement of the 5′-terminal sequence [equivalent to the 3′ terminus of the (−) strand] is related to the 3′-terminal sequence of the (+) strand. This suggests that the

YFV

```
(+)Strand 5'  A G U A A A U C C U G U G U G C · · · · A A A C A C A A A A C C A C U  3'
(-)Strand 3'  U C A U U U A G G A C A C A C G · · · · U U U G U G U U U U G G U G A  5'
```

WN

```
(+)Strand 5'  A G U A G U U C G C C U G U G U G · · · · · · A A C A C A G G A U C U  3'
(-)Strand 3'  U C A U C A A G C G G A C A C A C · · · · · · U U G U G U C C U A G A  5'
```

Fig. 4. Nucleotide homology between yellow fever virus (YFV) and West Nile virus (WN) [WN data are from (*21*)] at the 3′ termini of the genomic (+) strand and complementary (−) strand RNA's. Nucleotide identities in the 3′-terminal sequences of (+) and (−) strands are circled; those which are homologous between yellow fever and West Nile RNA's are underlined [(−) strand] or overlined [(+) strand].

viral replicase may have similar recognition sites for (+) and (−) strand synthesis. In addition, a stable secondary structure ($\Delta G = -40$ to 45 kcal) can be formed from the 3′-terminal 87 nucleotides of the yellow fever genomic RNA (Fig. 5). This may be involved in RNA replication as well as encapsidation, and if conserved among flaviviruses could explain the observation that many flavivirus RNA's are poor substrates for 3′-terminal enzymatic modification including ligation and addition of poly(A) (polyadenylate). Similar transfer RNA–like secondary structures and conserved sequences have been identified at the 3′ end of many plant viral RNA's (*41*); in addition to serving as substrates for aminoacylation both in vivo and in vitro (*42*) they are important for initiation of (−) strand RNA synthesis (*43*). Last, the 3′-untranslated region contains a set of three closely spaced repeated sequences (underlined in Fig. 1) (located between nucleotides 10,374 and 10,520) each approximately 40 nucleotides long with an average of six changes between them in pairwise comparisons. The significance of these repeats in flavivirus replication is unknown.

Evolution of flaviviruses. It is becoming clear that the flaviviruses deserve their recent reclassification as a family separate from the alphaviruses. Although the mature virions are morphologically similar to alphaviruses in that they have a single-stranded RNA (+) sense genome encapsidated in an icosahedral nucleocapsid and surrounded by a lipid bilayer containing virus-specified polypeptides, they differ markedly in genome organization and replication strategy (*44*). The location of the genes encoding the structural proteins at the 5′ end of the genome, the single long reading frame, and the lack of a subgenomic message are all characteristics shared with picornaviruses rather than togaviruses.

In order to understand the evolutionary role of flaviviruses and their relation to other RNA viruses we have searched for homologies within the putative polymerase genes of various plant and animal viruses. Significant homologies have been found between alphaviruses and plant viruses (*45*) and less extensive homologies between picornaviruses and alphaviruses (*46*). Kamer and Argos (*46*) have aligned the polymerase gene of poliovirus with those of several viruses

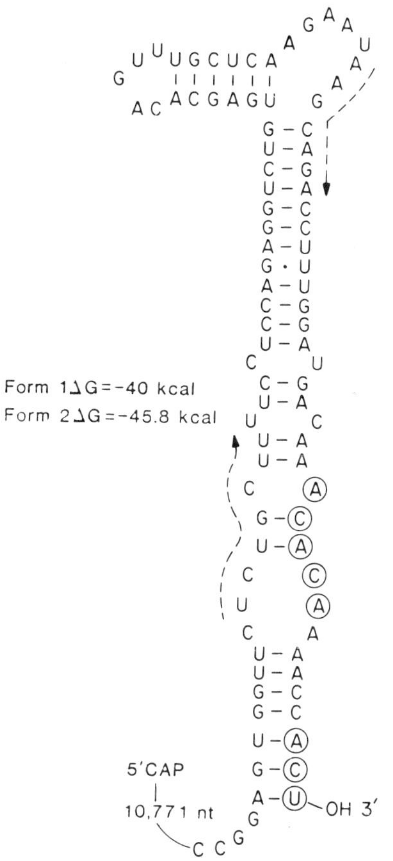

Fig. 5. Possible secondary structures at the 3′ terminus of yellow fever virus genomic RNA. Circled nucleotides are shared with the 3′ terminus of the yellow fever (−) strand (see Fig. 4). ΔG values were calculated according to Tinoco *et al.* (*55*). A more stable conformation than the one shown (form 1) can be formed if the two overlined sequences are base paired (form 2).

including alfalfa mosaic virus, bromegrass mosaic virus, tobacco mosaic virus, Sindbis virus, foot and mouth disease virus, encephalomyocarditis virus, and cowpea mosaic virus. The amino acid sequence of yellow fever virus NS5 between residues 3037 and 3181 can also be aligned with this collection of diverse RNA viruses (Fig. 1). These homologous regions are convincing but short and probably represent conserved functional domains for particular RNA-dependent polymerase functions. It is interesting to speculate on the origin of this diverse group of viruses. Whether they arose from one or a few protoviruses (perhaps insect viruses) and have radiated to their current divergent hosts or whether the viruses have repeatedly cannibalized their hosts, obtaining their replicases from eukaryotic cellular functions cannot be resolved at present. However, one possible measure of host adaptation or origin of viral genes from host functions is the CG doublet frequency in the RNA. Insects, insect viruses, and alphaviruses (insect-borne with vertebrate hosts) have the expected CG doublet frequency predicted from their base compositions (*47*), whereas vertebrate DNA (*48*), viruses with exclusively vertebrate hosts, and yellow fever virus have low CG doublet frequencies (2.4 percent CG found in yellow fever compared to 6.1 percent predicted from the base composition). Given the rapid evolution of RNA genomes, it is unlikely that this difference applies directly to the question of evolutionary origin of alphaviruses and flaviviruses but rather reflects alternative strategies of adaptation to their arthropod and vertebrate hosts in ways which are not currently understood.

Comparative studies with other flaviviruses should help to define areas of commonality of function in the nonstructural proteins, to localize biologically important antigenic epitopes on the structural polypeptides (and NS1) and to ascertain whether certain features of the yellow fever sequence (like the putative secondary structure at the extreme 3′ terminus and repeated nucleotide sequences) are functionally significant landmarks conserved among flaviviruses. In addition, the construction of cDNA clones designed for expression of functional virus gene products or production of infectious virus should provide useful new approaches for studying flavivirus molecular biology and pathogenesis as well as for development of flavivirus vaccines.

References and Notes

1. R. E. Shope, in *The Togaviruses*, R. W. Schlesinger, Ed. (Academic Press, New York, 1980), pp. 47–82.
2. R. W. Chamberlain, *ibid.*, pp. 175–227.
3. G. K. Strode, Ed., *Yellow Fever* (McGraw-Hill, New York, 1951).
4. W. G. Downs, *Yale J. Biol. Med.* **55**, 179 (1982).
5. W. Reed, *Med. Rec.* **60**, 201 (1901); ——— and J. Carroll, *Am Med.* **3**, 301 (1902); C. Norman, *Science* **223**, 1370 (1984).
6. M. Theiler and H. H. Smith, *J. Exp. Med.* **65**, 767 (1937); *ibid.*, p. 787.
7. P. J. Wright, *J. Gen. Virol.* **59**, 29 (1982).
8. P. K. Russell, W. E. Brandt, J. M. Dalrymple, in *The Togavirus*, R. W. Schlesinger, Ed. (Academic Press, New York, 1980), pp. 503–529.
9. E. G. Westaway, *ibid.*, pp. 531–581.
10. R. W. Boulton and E. G. Westaway, *Virology* **69**, 416 (1976); F. A. Murphy, in *The Togaviruses*, R. W. Schlesinger, Ed. (Academic Press, New York, 1980), pp. 241–316.
11. M. L. Ng, J. S. Pedersen, B. H. Toh, E. G. Westaway, *Arch. Virol.* **78**, 177 (1983).
12. G. Wengler, G. Wengler, H. J. Gross, *Virology* **89**, 423 (1978).
13. G. R. Cleaves and D. T. Dubin, *ibid.* **96**, 159 (1979); V. Deubel *et al.*, *Ann. Virol. (Inst. Pasteur)* **134E**, 581 (1983).
14. R. W. Boulton and E. G. Westaway, *Arch. Virol.* **55**, 201 (1977); C. W. Naeve and D. W. Trent, *J. Virol.* **25**, 535 (1978).
15. Y. V. Svitkin *et al.*, *FEBS Lett.* **96**, 211 (1978); G. Wengler, M. Beato, G. Wengler, *Virology* **96**, 516 (1979); Y. V. Svitkin *et al.*, *ibid.* **110**, 26 (1981); R. P. Monckton and E. G. Westaway, *J. Gen. Virol.* **63**, 227 (1982).
16. Y. V. Sitkin *et al.*, *Virology* **135**, 536 (1984).
17. E. G. Westaway, *ibid.* **80**, 320 (1977).
18. ———, G. Speight, L. Endo, *Virus Res.* **1**, 333 (1984).
19. M. Kozak, *Microbiol. Rev.* **47**, 1 (1983); we note a possible exception in the case of infectious pancreatic necrosis A RNA [P. P. C. Mertens and P. Dobos, *Nature (London)* **297**, 243 (1982)].
20. J. R. Bell *et al.*, *Virology* **143**, 224 (1985).
21. G. Wengler and G. Wengler, *ibid.* **113**, 544 (1981).
22. U. Boege, F. X. Heinz, G. Wengler, C. Kunz, *ibid.* **126**, 651 (1983).
23. L. Dalgarno, D. Trent, J. H. Strauss, C. M. Rice, in preparation.
24. D. Shapiro, W. E. Brandt, P. K. Russell, *Virology* **50**, 906 (1972).
25. H. Garoff, A. M. Frischauf, K. Simons, H. Lehrach, H. Delius, *Nature (London)* **288**, 235 (1980); C. M. Rice and J. H. Strauss, *Proc. Natl. Acad. Sci. U.S.A.* **78**, 2062 (1981).
26. K. Takio, T. Towatari, N. Katunuma, D. C. Teller, K. Titani, *Proc. Natl. Acad. Sci. U.S.A.* **80**, 3666 (1983).
27. J. J. Schlesinger, M. W. Brandriss, T. P. Monath, *Virology* **125**, 8 (1983).
28. G. W. Smith and P. J. Wright, *J. Gen. Virol.* **66**, 559 (1985).
29. J. Pata and C. M. Rice, unpublished data.
30. P. J. Wright, H. M. Warr, E. G. Westaway, *Virology* **109**, 418 (1981).
31. W. J. Welch and B. M. Sefton, *J. Virol.* **29**, 1186 (1979).
32. F. X. Heinz and C. Kunz, *Arch. Virol.* **60**, 207 (1979); F. Heinz, personal communication.
33. E. G. Westaway, *Virology* **51**, 454 (1973); ———, J. H. McKimm, L. G. McLeod, *Arch. Virol.* **53**, 305 (1977).
34. F. X. Heinz and C. Kunz, *J. Gen. Virol.* **62**, 271 (1982).
35. We propose an alternative nomenclature for flavivirus nonstructural proteins; our proposal is based on the yellow fever virus gene order determined by nucleic acid and protein sequence analysis (Fig. 1 and Table 1). The large nonstructural proteins (formerly NV3, NV4, and NV5) have been mapped and are numbered in order of appearance in the genome ($5' \rightarrow 3'$) with an upper case NS designation. We hypothesize that the remaining coding sequences in the nonstructural region encode several small flavivirus intracellular proteins (formerly NV1, NV1½, NV2, NV2½, and NVX), which are designated by a lower-case ns. Tentative identities with previously described flavivirus proteins are indicated by parentheses in Table 1. For alternative nomenclature see the text and (*37*). Minor virus-specific protein species that have been detected include two small glycoproteins (M_r ~13,000 and 17,000) (*27, 28*), and NV4½ (apparently related to NV4; perhaps equivalent to ns2b + NS3) (*5*).
36. J. Pata, J. Schlesinger, R. Aebersold, D. Teplow, S. Kent, J. H. Strauss, C. M. Rice, unpublished data.
37. E. G. Westaway *et al.*, *Intervirology* **14**, 114 (1980).
38. J. Schlesinger, personal communication.
39. G. Cleaves, personal communication.
40. R. Grantham, C. Gautier, M. Gouy, M. Jacobzone, R. Mercier, *Nucleic Acids Res.* **9**, 43 (1981).
41. T. C. Hall, in *International Review of Cytology*, (Academic Press, New York, 1979), vol. 60, pp. 1–26.
42. L. S. Loesch-Fries and T. C. Hall, *Nature (London)* **298**, 771 (1982); T. C. Hall, D. S. Shih, P. Kaesberg, *Biochem. J.* **129**, 969 (1972).
43. P. Ahlquist, J. J. Bujarski, P. Kaesberg, T. C. Hall, *Plant Mol. Biol.* **3**, 37 (1984).
44. E. G. Strauss and J. H. Strauss, *Curr. Top. Microbiol. Immunol.* **105**, 1 (1983).

45. J. Haseloff, J. Haseloff, P. Goelet, D. Zimmern, P. Ahlquist, R. Dasgupta, P. Kalsberg, *Proc. Natl. Acad. Sci. U.S.A.* **81**, 4358 (1984); P. Ahlquist *et al.*, *J. Virol.* **53**, 536 (1985).
46. G. Kamer and P. Argos, *Nucleic Acids Res.* **12**, 7269 (1984).
47. C. M. Rice and J. H. Strauss, *J. Mol. Biol.* **150**, 315 (1981).
48. G. J. Russell, P. M. B. Walker, R. A. Elton, J. H. Subak-Sharpe, *ibid.* **108**, 1 (1976); A. P. Bird, *Nucleic Acids Res.* **8**, 1499 (1980).
49. C. M. Rice, E. G. Strauss, J. H. Strauss, in *The Togaviruses and Flaviviruses*, S. Schlesinger and M. Schlesinger, Eds. (Plenum, New York, in press).
50. Obtained from Dr. Dennis Trent, Centers for Disease Control, Fort Collins, Colorado.
51. C. M. Rice, L. Dalgarno, D. W. Trent, J. H. Strauss; details of cloning and sequencing procedures are in preparation.
52. H. Okayama and P. Berg, *Mol. Cell. Biol.* **2**, 161 (1982).
53. A. M. Maxam and W. Gilbert, *Methods Enzymol.* **65**, 499 (1980).
54. J. Kyte and R. F. Doolittle, *J. Mol. Biol.* **157**, 105 (1982).
55. I. Tinoco *et al.*, *Nature (London)* **246**, 40 (1973).
56. We thank E. G. Strauss, L. Dalgarno, and C. Chang for helpful discussions and our many colleagues for critical comments on the manuscript; L. Hood and T. Hunkapiller for the use of their computer facilities; and C. S. Hahn for help with RNA secondary structure analysis. We also thank G. Cleaves, J. J. Schlesinger, and F. X. Heinz for allowing us to quote their unpublished results. Supported in part by grants AI 20612 and AI 10793 from NIH and by grant PCM 83-16856 from NSF.

17 May 1985; accepted 7 July 1985

21

Three-Dimensional Structure of Poliovirus at 2.9 Å Resolution

J.M. Hogle, M. Chow, and D.J. Filman

In 1908, Landsteiner and Popper identified poliovirus as the etiological agent of poliomyelitis (*1*). Since then poliovirus has played a unique role in biology and medicine. Poliovirus was one of the first viruses to be grown in cultured cells (*2*), an accomplishment that led to the development of a killed virus and a live attenuated vaccine (*3*), to the physical and chemical characterization of the virion and its assembly intermediates (*4*), and to characterization of the virus life cycle within the infected cell (*5*).

The genome structures and the RNA sequences of several strains have been determined (*6*). Upon transfection into animal cells, full-length complementary DNA (cDNA) copies of the viral genome yield infectious virus, permitting the genetic analysis and manipulation of the virus (*7, 8*). Recently, monoclonal antibodies (*9*) and synthetic peptide antigens (*10–12*) have been used to map the antigenic sites of the virion onto the amino acid sequences of the capsid proteins. As a result of these and other studies, poliovirus has become one of the best characterized viral pathogens.

Poliovirus is a member of the picornavirus family, which also includes foot and mouth disease virus, hepatitis A virus, the rhinoviruses, and the coxsackie viruses. Poliovirus is approximately 310 Å in diameter with a molecular mass of 8.5×10^6 daltons (*4*). The virion is composed of 60 copies each of four coat-protein subunits—VP1 (306 amino acids, ~ 33 kD), VP2 (272 amino acids, ~30 kD), VP3 (238 amino acids, ~26 kD), and VP4 (69 amino acids, ~7.5 kD), and a single-stranded, plus-sense RNA genome of approximately 7500 nucleo-

Science 229, 1358–1365 (27 September 1985)

tides (*6*). The RNA is polyadenylated at its 3′ end (*4–8*) and is covalently linked at the 5′ end to the 22-amino-acid protein VPg (*13*).

The RNA contains a single large open-reading frame from which a 220 kD polyprotein is synthesized (*5, 6*). All known viral proteins are generated from this polyprotein in a cascade of proteolytic cleavages by virally encoded proteases (*14*). The capsid protein sequences are located at the NH_2-terminus of the polyprotein in the order VP4-VP2-VP3-VP1 (*6*). Early in the cascade, the capsid precursor P1-1a (*15*) is cleaved from the polyprotein. Subsequent cleavage of the P1-1a protomer (*16*) to VP0-VP3-VP1 is coupled with virion assembly. The final step in virion maturation is the cleavage of VP0, yielding VP4 and VP2 (*4*).

The three-dimensional structures of several intact plant viruses have been determined by x-ray crystallographic methods, including tomato bushy stunt virus (TBSV) (*17*), southern bean mosaic virus (SBMV) (*18*), and satellite tobacco necrosis virus (STNV) (*19*). Analyses of these structures have provided insight into the architecture and assembly of small spherical viruses. These studies have also provided a background of methodology and structural principles for the crystallographic investigation of animal viruses, for which the biology has been better characterized. We report here the structure of the Mahoney strain of type 1 poliovirus as determined by x-ray crystallographic methods; this structure provides a three-dimensional context for studying the relationship between the structure of an animal virus and its biological properties.

Crystallization and structure determination. Seed stocks of poliovirus type 1 (Mahoney strain) were obtained from an isolated plaque produced by transfection of HeLa cells with an infectious cDNA clone of the viral genome (*7*). The virus was propagated in HeLa cells, purified by standard methods (*20*), and crystallized by microdialysis at neutral *p*H (*21*). Virus recovered from redissolved crystals is fully infectious (*22*). The crystals belong to the orthorhombic space group $P2_12_12$ with $a = 322.9$, $b = 358.0$, and $c = 380.2$ Å and contain one-half virus particle per asymmetric unit. The particle center is located on the crystallographic twofold axis at approximately $Z = 0.25$ (*21, 23*). Three dimensional x-ray diffraction data for native crystals and for a platinum derivative (prepared by soaking crystals in 5 m*M* K_2PtCl_4 for 48 hours) were collected by oscillation photography to 2.9 Å resolution. Because the crystals were extremely sensitive to x-irradiation, the data were collected at −15°C, with 25 percent ethylene glycol as a cryosolvent. The photographs were digitized on a 50-μm raster with an Optronics P-1000 scanner and integrated and processed with the use of locally developed software (*24*). These and all subsequent calculations described below were carried out on a VAX 11/750 computer with 8 megabytes of memory and a 500-megabyte disk. Statistics for the native and derivative data are shown in Table 1.

The orientation of the poliovirus particle about the crystallographic twofold axis was determined by evaluation of a one-dimensional icosahedrally locked rotation function, and the position of the particle along the twofold axis was estimated by inspection of native Patterson maps. Platinum substitution sites in the derivative crystals were determined by an icosahedrally locked Patterson vector search. One general and one fivefold

Table 1. Summary of the Mahoney native and platinum derivative data.

Items	Mahoney native	Pt derivative
Films (No.)*	120	37
Measurements (No.)	3,573,325	1,103,477
Unique *hkl* measured† (No.)	893,887	539,636
Selection criterion‡ (%)	50+	40+
Unique *hkl* used§ (No.)	736,926	310,063
R_{sym} versus mean‖	0.182	0.226
R_{sym} versus averaged whole¶	0.141	0.092

*The 0.5-degree oscillation photographs were taken on CEA reflex 25 film with Supper oscillation or with oscillation-precession cameras, with CuKα radiation from a Marconi GX-20 or GX-18 rotating anode generator with a 100-μm focus, and Franks mirror optics (*34*). Exposure times were 12 to 18 hours with the GX-20, and 6 to 8 hours with the GX-18. †Out of 950,000 unique reflections in the 2.9 Å sphere. ‡Partially observed reflections were corrected to their fully recorded equivalent by the method of Winkler, Schutt, and Harrison (*35*). Partials observed to an extent less than the indicated percentage were not used in constructing the unique data set. §A conservative cut based on the intensity and its standard deviation was used to exclude poorly measured reflections. ‖R_{sym} compares multiply measured individual corrected intensities (I_{hj}) with their sigma-weighted mean $<I_h>$.

$$R_{sym} = \frac{\sum_h \sum_j | I_{hj} - <I_h> |}{\sum_h \sum_j <I_h>}$$

¶Same as (‖), with $<I_h>$ computed with the use of fully recorded reflections only.

axial platinum site were found per icosahedral asymmetric unit. Single isomorphous replacement (SIR) centroid phases were applied to the native amplitudes between 50 and 5 Å resolution. Native phases were refined at 5 Å resolution by iterative electron density averaging about the icosahedral symmetry elements as described by Bricogne (*25*). Several times during this phase refinement, the particle position and orientation, and the platinum substitution sites in the particle were optimized, treating them as variables in a systematic search which maximized the electron density at the predicted heavy atom sites in unaveraged derivative difference Fourier maps. After the final application of SIR phases, phase refinement at 5 Å converged in eight cycles; but the progress of the refinement suggested that SIR phases would be of limited use at higher resolution.

The structure determination was extended from 5 Å to 2.9 Å resolution by the addition of 0.05 Å shells of structure factor amplitudes in each cycle of density averaging. The new terms were phased by inversion of the previous cycle's averaged electron density. Similar methods have been used for phase extension in the structure determinations of STNV (*19*) and hemocyanin (*26*).

Alpha carbon positions were determined for all four capsid proteins, and correlated with the known amino acid sequence, in 3.6 Å averaged electron density maps. Phase extension and refinement were subsequently continued to 2.9 Å resolution. The complete process of extending phases between 5 and 2.9 Å resolution required 67 cycles of density averaging, yielding a quadratic *R* value of 0.27 and a correlation coefficient of 0.84 (*27*). Representative sections of the final 2.9 Å electron density map are

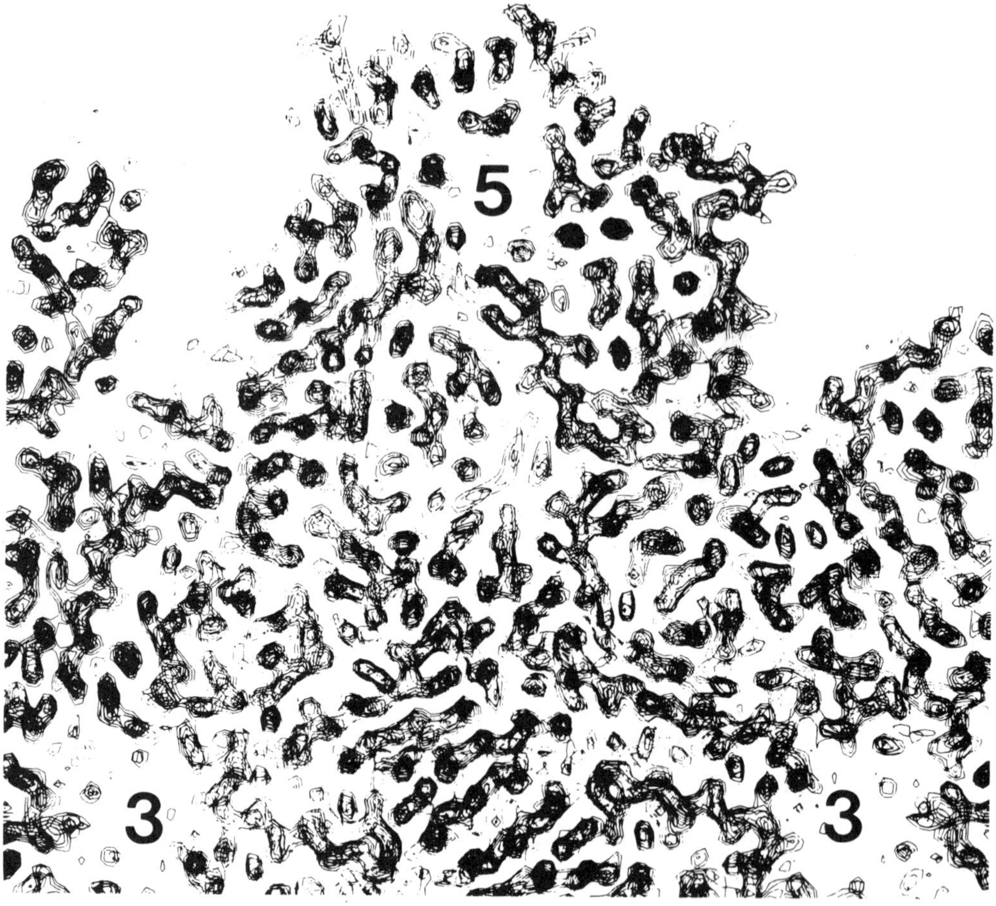

Fig. 1. Five representative sections of the final 2.9 Å averaged electron density map. The map encloses the icosahedral asymmetric unit of the poliovirion, sectioned perpendicular to the particle twofold axis. Sections from 120 to 124 Å are shown. Labels indicate the approximate positions of one fivefold and two threefold axes (which are skewed relative to the plane of the map). In this map, the connectivity of the main chain is unambiguous, the side chains are apparent, and there is partially resolved electron density for carbonyl oxygens. The quality of the map is also indicated by the level of noise at the protein-solvent interfaces. The lack of density at the top of the photograph is the result of solvent flattening.

shown in Fig. 1. Residue placement and side-chain identifications have been confirmed by inspection of this map. A full atomic model to fit the 2.9 Å map has not yet been completed.

Capsid protein subunits. The quality of the electron density map was such that it was possible to locate and identify almost all of the residues in all four capsid protein subunits. However, no electron density was present for residues 1 to 20 at the NH_2-terminus of VP1, residues 1 to 7 at the NH_2-terminus of VP2, residues 1 to 12 at the NH_2-terminus of VP4 (all of which are in the interior of the particle) and residues 234 to 238 at the

COOH-terminus of VP3 (on the outside surface of the virion). In addition, there was no significant unassigned density that could correspond to the viral RNA or to VPg. The lack of density for the NH_2-terminal residues and the RNA implies that the interior of the virion is spatially disordered with respect to the icosahedrally symmetric protein shell.

The structures of VP1, VP2, and VP3 (Plate IV) are remarkably similar to one another and surprisingly similar to the coat proteins of TBSV, SBMV, and STNV. Each of these proteins consists of a "core," or common structural motif, with variable "elaborations." The cores are topologically identical. Each core is composed of an eight-stranded antiparallel beta barrel with two flanking helices. Four strands (B, I, D, and G in Fig. 2a) make up a large twisted beta sheet which forms the front and bottom surfaces of the barrels (Fig. 2, b to d). The remaining four strands (C, H, E, and F) make up a smaller, flatter beta sheet which forms the back surface of the barrel. The strands comprising the front and back surfaces are joined at one end by four short loops, giving the barrel the shape of a triangular wedge. In poliovirus, each of the three coat proteins has different NH_2- and COOH-terminal extensions, and includes a different set of internal insertions. The largest internal insertion in VP1 (residues 207 to 237) is the loop connecting beta strands G and H. The largest insertion in VP2 (residues 127 to 185) connects beta strand E with the radial helix on the back surface of the barrel. The most significant excursion in VP3 (residues 53 to 69) connects the NH_2-terminal strand to beta strand B. The NH_2-terminal strands of VP1, VP2, and VP3 fold beneath the barrels, while the internal insertions and COOH-terminal additions are located on the top surface of the barrel.

In contrast to the compact structures found in the other three coat proteins, VP4 has a more extended conformation. The only significant compact structure in VP4 is a short two-stranded antiparallel beta sheet near its NH_2-terminus. VP4 is similar in its position and conformation to the NH_2-terminal strands of VP1 and VP3, and thus it appears to function as the (detached) NH_2-terminal extension of VP2 rather than as an independent capsid protein.

Protomer subunit. The position and orientation of the four capsid proteins on the icosahedral surface of the virion are shown in Plate V. The closed end of the barrel of VP1 is located near the fivefold axis. The closed ends of the VP2 and VP3 barrels alternate around the particle threefold axis. The subunits shown are those that have the most extensive intersubunit interactions, and thus they are likely to be the ones that are derived by cleavage of the same P1-1a protomer. On the outer surface of the particle, the COOH-terminus of VP1 wraps over VP3, while residues 207 to 237 and 271 to 295 of VP1 flank the outer surface of VP2 (Plate V, center). On the inner surface, the interactions are more extensive. The NH_2-terminal strand of VP3 wraps around the base of the VP1 barrel, while the NH_2-terminal strand of VP1 makes extensive contacts with the inner surface of the VP3 barrel (Plate V, bottom).

Outer surface of the virion. The continuous protein shell of the poliovirion extends from approximately 110 to 140 Å radius (Plate VI a). The exterior surface of the virion (Plate VI, b and c) is dominated by two sets of prominent radial extensions: A ribbed peak (extending to 165 Å radius) is formed at the fivefold

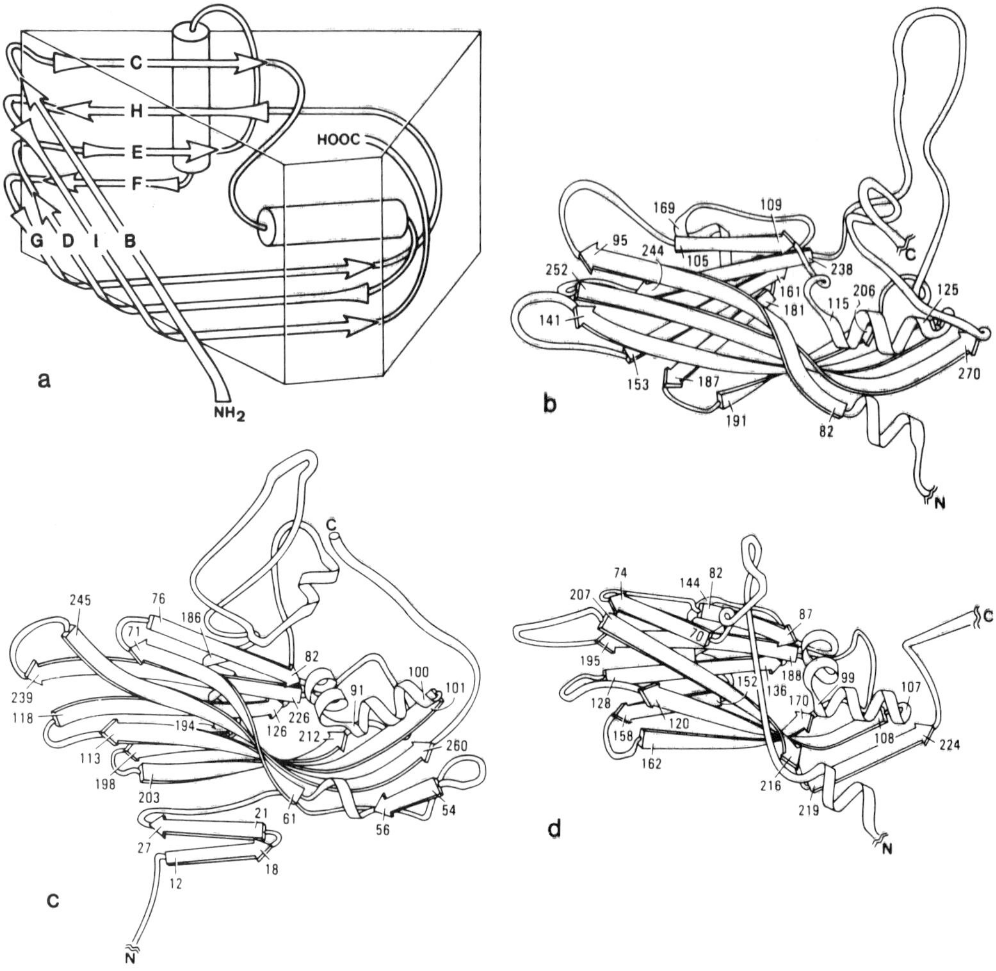

Fig. 2. Schematic representation of the poliovirus capsid proteins. (a) Simplified diagram showing the topology of the structurally conserved "core." Beta strand designations are consistent with those used to describe the capsid proteins of the plant viruses (*38*). Ribbon diagrams show (b) VP1, (c) VP2, and (d) VP3. The NH_2- and COOH-terminal extensions of VP1 and the NH_2-terminal extension of VP3 have been truncated for clarity.

axes by residues from VP1, and a broad plateau is formed at the threefold axes by residues from VP2 and VP3. The plateau extends to 150 Å radius at its center, and is ringed by promontories. The peaks at the fivefold axes are surrounded by broad valleys, while adjacent plateaus are separated by saddle surfaces at the particle twofold axes. Thus, the exterior surface of the poliovirus particle is approximated by the geometric construction shown in Plate VII a. This construction appears in the other figures to aid in orientation.

VP1, VP2, and VP3 occupy sites in the poliovirus capsid that are analogous to those of the A, C, and B subunits, respectively, in the $T = 3$ plant viruses. However, the external surface of poliovirus differs significantly from those of

SBMV and of the S domain of TBSV. In the $T = 3$ plant viruses, the upper surfaces of the cores of the three capsid proteins in one icosahedral asymmetric unit, and the three in a twofold related icosahedral asymmetric unit are coplanar, forming the face of a rhombic triacontahredron (Plate VII b). The greater relief of the poliovirion surface can be ascribed to differences in the packing and orientation of the capsid protein subunits.

A pronounced tilt of the closed end of the VP1 barrel outward along the fivefold axis is responsible for the prominent fivefold peaks (Plate VIII, a and b). The upper surface of the VP1 barrel forms the slope of the peak, and the uppermost three loops of the barrel (residues 96 to 104, 245 to 251, and 142 to 152) are exposed at the summit. The top loop (96 to 104) twists out from the surface of the subunit so that it is particularly well exposed, giving the fivefold peak a ribbed appearance. Indeed, this loop contains the only trypsin-sensitive site in the Sabit strain of type 1, and the Sabin and Leon strains of type 3 poliovirus (*28*).

The outward tilt of the closed ends of the VP2 and VP3 barrels along the threefold axes is less pronounced. As a result, only the outermost two loops (residues 72 to 75 and 240 to 244 in VP2, and residues 75 to 81 and 196 to 206 in VP3) are exposed. The plateau at the threefold axis is broadened by two sets of outward projections. The smaller projection, which extends to a radius of 155 Å, is formed by the insertion in the B strand of VP3 (residues 53 to 69). The larger projection, which extends to 165 Å radius, is formed by the internal insertion in VP2 (residues 127 to 185), and is supported at its base by the COOH-terminal strand of VP2 (residues 261 to 272), and by residues 207 to 237 and 271 to 295 from VP1.

Subunit packing. In the $T = 3$ plant viruses, the three subunits in each icosahedral asymmetric unit (as defined in Plate VII b) are related by a local threefold axis. Within the asymmetric unit, the pairwise contacts between the A and B, B and C, and C and A subunits are nearly identical. In addition, the contact between fivefold related A subunits is nearly identical to one of the two distinct types of interaction betweeen B and C subunits at the threefold axes (*17, 18*). These similarities reflect the constraints imposed by quai-equivalent packing of chemically identical subunits (*29*).

In poliovirus, VP1, VP2, and VP3 are chemically distinct, so the constraints are relaxed. The relaxation of constraints permits qualitative differences in the "tilts" of the subunits (as described above), causing significant differences in the contacts between subunits. Two specific examples serve to demonstrate these differences. (i) The "tangential" helices of VP1 and VP3 (shown in the front of Fig. 2a) are packed against one another at approximately a 45-degree angle. In contrast, the analogous packing of helices between twofold related molecules of VP2 resembles that in the $T = 3$ plant viruses, wherein the helix axes are nearly parallel. (ii) In VP2, VP3, and in the A, B, and C subunits of the $T = 3$ plant viruses, the loop that connects beta strand E to the "radial" helix (shown in the back of Fig. 2a) lies in the interface between subunits within the triangular icosahedral asymmetric unit defined in Plate VII b. In contrast, the analogous loop of VP1 does not lie in the VP1–VP3 interface. As a result of the pronounced tilt of VP1, the loop is entirely exposed on the surface of the virion. These differences in subunit contacts indicate that the packing of VP1 cores around the fivefold axes is dissimilar from the packing of VP2 and VP3 around the threefold axes. They also demonstrate that, unlike the $T = 3$ plant viruses, there is no satisfactory local threefold axis relating the subunits within an icosahedral asymmetric unit.

Inner surface of virion. The interior surface of the capsid is formed by the bottom surfaces of the beta barrels of VP1, VP2, and VP3, by VP4, and by the NH_2-terminal strands of VP1, VP2, and VP3. The NH_2-termini form an extensive network (Plate VIII c) which links fivefold related protomers together to form pentamers; twofold related pentamers are also linked by this network, though less strongly. The most striking feature of the inner surface is the interaction between the NH_2-termini of VP3 and VP4 about the fivefold axis (Plate VIII, c and d). The NH_2-termini of five subunits of VP3 intertwine about the fivefold axis forming a five-stranded tube of extremely twisted parallel beta structure. This twisted tube is flanked on its lower surface by five two-stranded antiparallel beta sheets from the NH_2-termini of VP4. A structure analogous to the twisted tube, described as a "beta annulus," occurs at the threefold axes of the $T = 3$ plant viruses (*17, 18*).

Implications for assembly. The extensive network of interactions between fivefold related protomers is consistent with earlier proposals that the pentameric association of VP0, VP3, and VP1 is an intermediate in the assembly of poliovirions (*4*). In particular, the twisted tube formed around fivefold axes by the NH_2-termini of VP3 is a structure that can form only upon pentamer assembly. This structure may direct pentamer formation and is expected to contribute substantially to the stability of the pentamers once formed. The role of the NH_2-terminal network in poliovirus differs from that of an analogous network in the $T = 3$ plant viruses, wherein the "arms" which are linked at the threefold "beta annuli" form a $T = 1$ scaffold which determines local curvature and thereby directs assembly of the $T = 3$ capsid (*30*).

The cleavage of VP0, yielding VP4 and VP2, is believed to be the final step in virion maturation, occurring after capsid assembly and the encapsidation of RNA (*4*). The COOH-terminus of VP4 and the NH_2-terminus of VP2 are close to one another in the interior of the mature virion. Unless there is a substantial rearrangement of the capsid structure after the cleavage of VP0, the cleavage site is inaccessible to external proteases. Consequently, the cleavage of VP0 may be autocatalytic. The internal position of VP4 also suggests that the loss of this protein upon inactivation of the virus at elevated temperatures (55° to 60°C) or upon cell attachment (*4*) signifies a substantial rearrangement of the virion structure.

One remarkable discovery resulting from the structure determination is that the NH_2- and COOH-termini generated by proteolytic processing of P1-1a to VP0, VP3, and VP1 are spatially distant, and indeed, are located on opposite surfaces of the pentamer. The separation of these termini and the extensive interaction of the NH_2-terminal strands of VP3 about the fivefold axes are most consistent with a proposed assembly pathway (*4*) in which the cleavage of P1-1a occurs prior to pentamer formation. Processing of P1-1a after pentamer formation would require far more elaborate conformational changes. In either case, substantial structural rearrangements must occur subsequent to proteolytic cleavage. This confirms the proposal (*4*) that posttranslational processing plays a significant role in controlling virion assembly.

Sequence differences between strains and serotypes. Sequence differences between naturally occurring strains of influenza virus have been used to identify several distinct sites on the hemagglutinin (*31*) and neuraminidase (*32*) which are responsible for antigenic drift. The success of this approach may be attributed to the availability of several strains which were derived from one another in a known temporal sequence under immune selection. In poliovirus, however, the evolutionary relationships between the three serotypes are unknown, and the role of immune selection in the emergence of new field isolates is unclear. As a result, specific sequence differences in the existing strains cannot be ascribed to immune selection with any confidence.

Differences in the amino acid sequences of the capsid proteins of type 1 Mahoney and type 3 Leon strains were mapped onto the Mahoney structure. The sequence differences were distributed widely over the inner and outer surfaces of the capsid proteins, but were conspicuously absent from the cores of the subunits. This distribution shows that, notwithstanding the selective pressures for change, the inherent constraints imposed by protein folding and assembly exert a strong selective pressure for sequence conservation.

Location of neutralizing antigenic determinants. A more direct procedure for determining antigenic sites is the use of neutralizing antiviral monoclonal antibodies to select for mutant viruses resistant to neutralization (*9*). The locations of sequence changes that confer such resistance are shown in Plate VIII e. The mutation sites map onto the major external surface features of the virion. These "escape mutations" can be grouped on the basis of proximity into four discrete clusters (Plate VIII, e and f). Cluster 1 (colored yellow) contains residues from the loops of VP1 which form the fivefold peak, as well as nearby residues from the loop (162 to 178) that lies along the upper surface of the VP1 barrel. Cluster 2 (blue) includes residues from the internal insertion in VP3 and from the top loop of the barrel in VP2. Cluster 3 (green) is formed by residues from the internal insertion in VP1, the internal insertion in VP2, and the COOH-terminus of VP2. The grouping of these three segments into one antigenic cluster by proximity is confirmed by the observation that a mutation in the insertion of either VP1 or VP2 conferred resistance to the same neutralizing monoclonal antibody (*33*). Cluster 4 (red) contains residues from a loop (271 to 295) in the COOH-terminal strand of VP1. Although this site has been identified as a distinct cluster, it is sufficiently close to clusters 2 and 3 that it might become continuous with either one when a larger number of viral mutants are identified. One unexplained observation is that most of the escape mutations in the Mahoney and Sabin strains of type 1 poliovirus occur in clusters 2 and 3, whereas nearly all of the escape mutations in the Leon and Sabin strains of type 3 poliovirus are located in cluster 1, in residues 90 to 103 of VP1. Nonetheless, all of the mutation sites occur in exposed loops on the surface of the virion which are readily accessible to antibody binding. Thus, it is unnecessary to propose that any of these mutations allow the virus to escape neutralization by acting at a site which is distant from the antibody binding site.

Synthetic peptide immunogens have also been used to identify the antigenic sites of poliovirus (*10–12*). A number of these peptides have been shown to induce antibodies that bind or neutralize virions (or both); or alternatively, to prime animals for a neutralizing response to a subsequent subimmunizing dose of virions. With the exception of several peptides from VP1, the immunogenic peptides correspond to sequences that are located on the outer surface of the virion, within the clusters of antibody-induced mutation sites. One exception is the peptide corresponding to residues 113 to 121 of VP1 (*12*). The NH_2-terminal portion of this peptide is partially exposed and lies close to cluster 4, so that it may represent an extension of this cluster. The other exceptions are peptides from the NH_2-terminal strand of VP1 or from the bottom loop of VP1 (residues 182 to 201), all of which are deeply buried in the interior of the virion. Two of these peptides (61 to 80 and 182 to 201) induce a significant (1:150 to 1:200) neutralizing response in both rabbits and rats (*10*). The ability of these buried peptides to induce an antiviral response may prove to be important in understanding the mechanism of neutralization.

In conclusion, the crystal structure of poliovirus has yielded important insights into the structure and assembly of small spherical viruses. In addition, the structure furnishes a three-dimensional framework for studying properties that are especially relevant to animal viruses. The mapping of monoclonal release mutants and of peptides that induce neutralizing antibodies to virus has produced a detailed description of the distribution of antigenic sites on an intact virus. This information is expected to lead to an understanding of the mechanism of neutralization. We also anticipate that the structure will be useful for the design and interpretation of experiments in other areas of biological and medical importance, such as the role of capsid structure in host- and tissue-specificity, and in the attenuation of neurovirulence. Thus, the poliovirus structure provides an excellent opportunity to study viral pathogenesis at the molecular level.

References and Notes

1. K. Landsteiner and E. Popper, *Wien. Klin. Wochenshr.* **21**, 1830 (1908).
2. J. F. Enders, T. H. Weller, F. C. Robbins, *Science* **109**, 85 (1949).
3. J. Salk and D. Salk, *ibid.* **195**, 837 (1977); A. B. Sabin and C. R. Boulger, *J. Biol. Stand.* **1**, 115 (1973).
4. As reviewed by: R. R. Rueckert, in *Comprehensive Virology*, H. Fraenkel-Conrat and R. Wagner, Eds. (Plenum, New York, 1976), vol. 6, pp. 131–213; J. R. Putnak and B. A. Phillips, *Microbiol. Rev.* **45**, 287 (1981).
5. As reviewed by L. Levintow, in *Comprehensive Virology*, H. Fraenkel-Conrat and R. Wagner, Eds. (Plenum, New York, 1974), vol. 2, pp. 109–171.
6. N. Kitamura *et al.*, *Nature (London)* **291**, 547 (1981); V. Racaniello and D. Baltimore, *Proc. Natl. Acad. Sci. U.S.A.* **78**, 4887 (1981); G. Stanway *et al.*, *Nucleic Acids Res.* **11**, 5629 (1983); G. Stanway *et al.*, *Proc. Natl. Acad. Sci. U.S.A.* **81**, 1539 (1984); H. Toyoda *et al.*, *J. Mol. Biol.* **174**, 561 (1984).
7. V. Racaniello and D. Baltimore, *Science* **214**, 916 (1981).
8. B. L. Semler, A. J. Dorner, E. Wimmer, *Nucleic Acids Res.* **12**, 5123 (1984); T. Omata *et al.*, *Gene* **32**, 1 (1984).
9. E. A. Emini, B. A. Jameson, A. J. Lewis, G. R. Larsen, E. A. Wimmer, *J. Virol.* **43**, 997 (1982); R. Crainic *et al.*, *Infect. Immun.* **41**, 1217 (1983); P. Minor *et al.*, *Nature (London)* **301**, 674 (1983); P. Minor *et al.*, *J. Gen. Virol.* **65**, 1159 (1985); M. Ferguson *et al.*, *Virology* **143**, 505 (1985); D. C. Diamond *et al.*, *Science* **229**, 1090 (1985).
10. M. Chow, R. Yabrov, J. Bittle, J. Hogle, D. Baltimore, *Proc. Natl. Acad. Sci. U.S.A.* **82**, 910 (1985).
11. E. A. Emini, B. A. Jameson, E. Wimmer, *Nature (London)* **304**, 699 (1983); *J. Virol.* **52**, 719 (1984); in *Modern Approaches to Vaccines*, R. M. Channock and R. A. Lerner, Eds. (Cold Spring Harbor Laboratory, Cold Spring Harbor, N.Y., 1984), pp. 65–76; M. Ferguson *et al.*, *Virology* **143**, 505 (1985).

12. B. A. Jameson, J. Bonin, M. G. Murray, E. Wimmer, O. Kew, in *Vaccines 85*, R. Lerner, R. M. Channock, F. Brown, Eds. (Cold Spring Harbor Laboratory, Cold Spring Harbor, N.Y., 1985), pp. 191–198.
13. E. Wimmer, *Cell* **28**, 199 (1982).
14. M. A. Pallansch *et al.*, *J. Virol.* **49**, 873 (1984).
15. R. R. Rueckert and E. Wimmer, *ibid.* **50**, 957 (1984).
16. The term "protomer" will refer to a single molecule of P1–1a, or to a set of proteins derived from it by further proteolytic processing (VP0-VP3-VP1 or VP4-VP2-VP3-VP1).
17. S. C. Harrison, A. J. Olson, C. E. Schutt, F. K. Winkler, G. Bricogne, *Nature (London)* **276**, 368 (1978).
18. C. Abad-Zapatero *et al.*, *ibid.* **286**, 33 (1980).
19. L. Liljas *et al.*, *J. Mol. Biol.* **159**, 93 (1982).
20. D. Baltimore, M. Girard, J. E. Darnell, *Virology* **29**, 179 (1966); M. H. Baron and D. Baltimore, *Cell* **28**, 395 (1982).
21. J. M. Hogle, *J. Mol. Biol.* **160**, 663 (1982).
22. J. P. Icenogle and J. M. Hogle, unpublished data.
23. J. T. Finch and A. Klug, *Nature (London)* **183**, 1709 (1959).
24. D. J. Filman and J. M. Hogle, in preparation.
25. G. Bricogne, *Acta Crystallogr.* **A30**, 395 (1974); *ibid.* **A32**, 832 (1976).
26. W. P. J. Gaykema *et al.*, *Nature (London)* **309**, 23 (1984).
27. The quadratic *R* value and the linear correlation coefficient are measures of the agreement between two sets of structure factor amplitudes: those observed and those calculated by Fourier inversion of the averaged electron density map. The square of *R* is equal to the sum of the squares of the unweighted differences, divided by the sum of the observed reflection intensities.
28. C. E. Fricks, J. P. Icenogle, J. M. Hogle, *J. Virol.* **54**, 856 (1985).
29. D. L. D. Caspar and A. Klug, *Cold Spring Harbor Symp. Quant. Biol.* **27**, 1 (1962).
30. S. C. Harrison, *Biophysics J.* **32**, 139 (1980).
31. D. C. Wiley, I. A. Wilson, J. J. Skehel, *Nature (London)* **289**, 373 (1981).
32. P. M. Colman, J. N. Varghese and W. G. Laver, *ibid.* **305**, 41 (1983).
33. P. Minor, personal communication.
34. S. C. Harrison, *J. Appl. Cryst.* **1**, 84 (1968).
35. F. K. Winkler, C. E. Schutt, S. C. Harrison, *Acta Crystallogr.* **A35**, 901 (1979).
36. Color figures were produced by Dan Bloch using the graphics language GRAMPS [T. J. O'Donnell and A. J. Olson, *Comput. Graphics* **15**, 133 (1981)]. The molecular modeling program GRANNY [M. L. Connolly and A. J. Olson, *Comput. Chem.* **9**, 1 (1985)], and the molecular surface program MS [M. L. Connolly, *Science* **221**, 709 (1983)].
37. Space-filling representations were produced by A. J. Olson, using the molecular surface programs AMS [M. L. Connolly, *J. Appl. Crystallogr.* **16**, 548 (1983)] and RAMS [M. L. Connolly, *J. Mol. Graph.* **3**, 19 (1985)].
38. M. G. Rossmann *et al.*, *J. Mol. Biol.* **165**, 711 (1983).
39. Supported by NIH grant AI-20566 (J.M.H.), in part by a grant from the Rockefeller Foundation and by NIH grant AI-22346 to D. Baltimore (Whitehead Institute for Biomedical Research), and by NIH training grants NS-07078 and NS-12428 (D.J.F.). Preliminary crystallographic studies that led to this work were supported by NIH grant CA-13202 to S. C. Harrison. We thank R. Lerner, D. Baltimore, and S. C. Harrison for interest and support throughout various phases of this work; D. Bloch for assistance in the computer graphics, E. Getzoff for helping to prepare the ribbon diagrams in Fig. 2. A. J. Olson for preparing Plate VI, a and b, and J. Icenogle, C. Fricks, M. Oldstone, and I. Wilson for comments and discussions. We also thank P. Minor for sharing his data on monoclonal release mutants prior to publication. This is publication No. MB4075 of the Research Institute of Scripps Clinic.

13 August 1985; accepted 30 August 1985

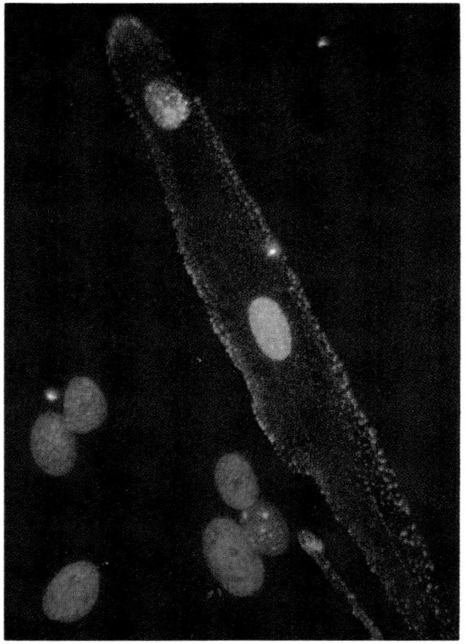

Plate I, Chapter 10. Activation of human muscle gene for 5.1H11 in heterokaryons. Live cells were incubated with monoclonal antibody to 5.1H11 followed by biotin conjugated to an antimouse antibody and avidin conjugated to Texas red. Heterokaryons are shown in fluorescence microscopy at two different wavelengths. The binucleate heterokaryon containing one punctate mouse muscle nucleus and one uniformly stained human hepatocyte nucleus expresses the antigen, which is uniformly distributed on the cell surface. The trinucleate heterokaryon (lower center) has not activated the gene for 5.1H11.

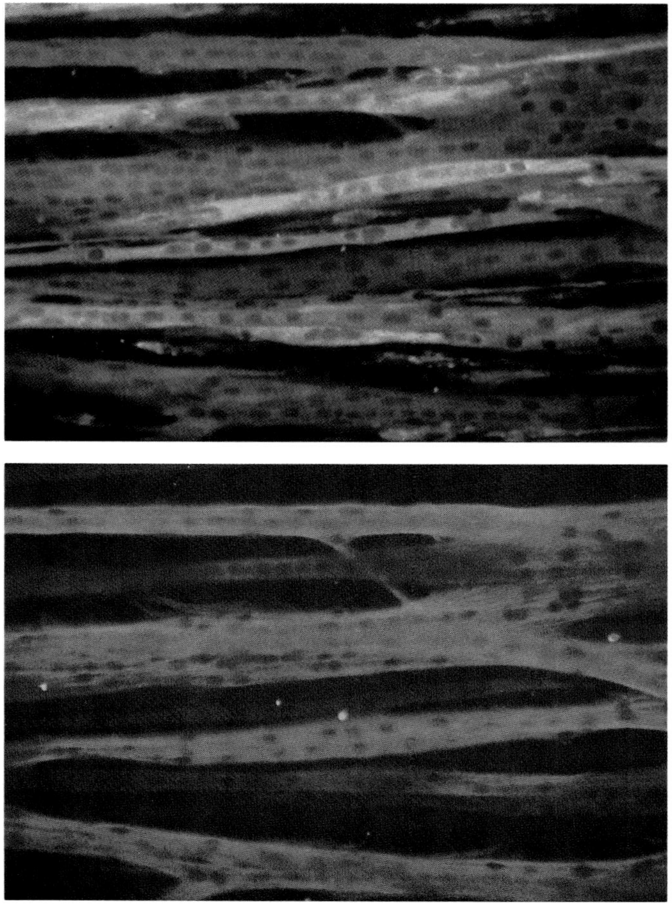

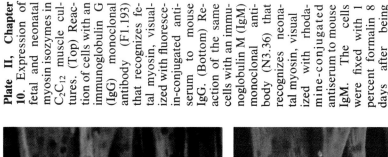

Plate II, Chapter 10. Expression of fetal and neonatal myosin isozymes in C_2C_{12} muscle cultures. (Top) Reaction of cells with an immunoglobulin G (IgG) monoclonal antibody (F1.193) that recognizes fetal myosin, visualized with fluorescein-conjugated antiserum to mouse IgG. (Bottom) Reaction of the same cells with an immunoglobulin M (IgM) monoclonal antibody (N3.36) that recognizes neonatal myosin, visual ized with rhodamine-conjugated antiserum to mouse IgM. The cells were fixed with 1 percent formalin 8 days after being transferred to differentiation medium (2 percent horse serum in Dulbecco's modified Eagle's medium).

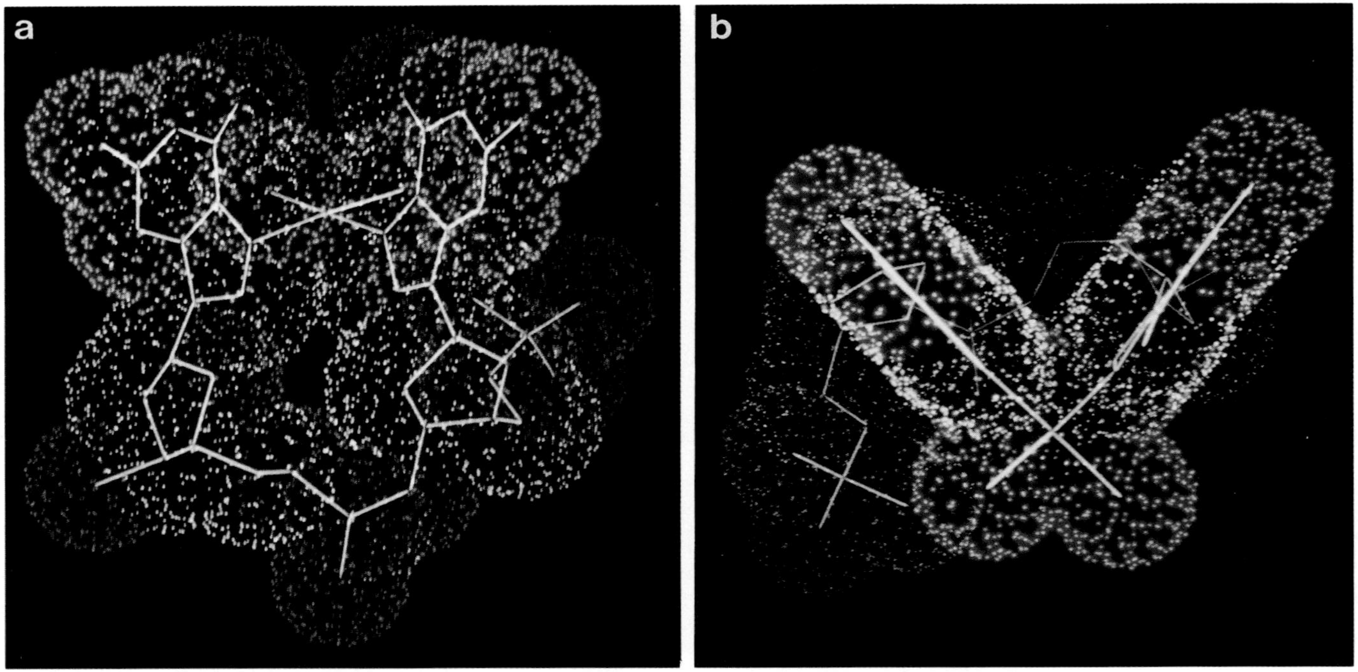

Plate III (a and b), Chapter 13. Two views of the van der Waals spheres of the structure in Fig. 2. The right-hand view clearly reveals the hydrogen bond formed between the terminal 5′-phosphate and coordinated ammine groups as interpenetrating van der Waals spheres.

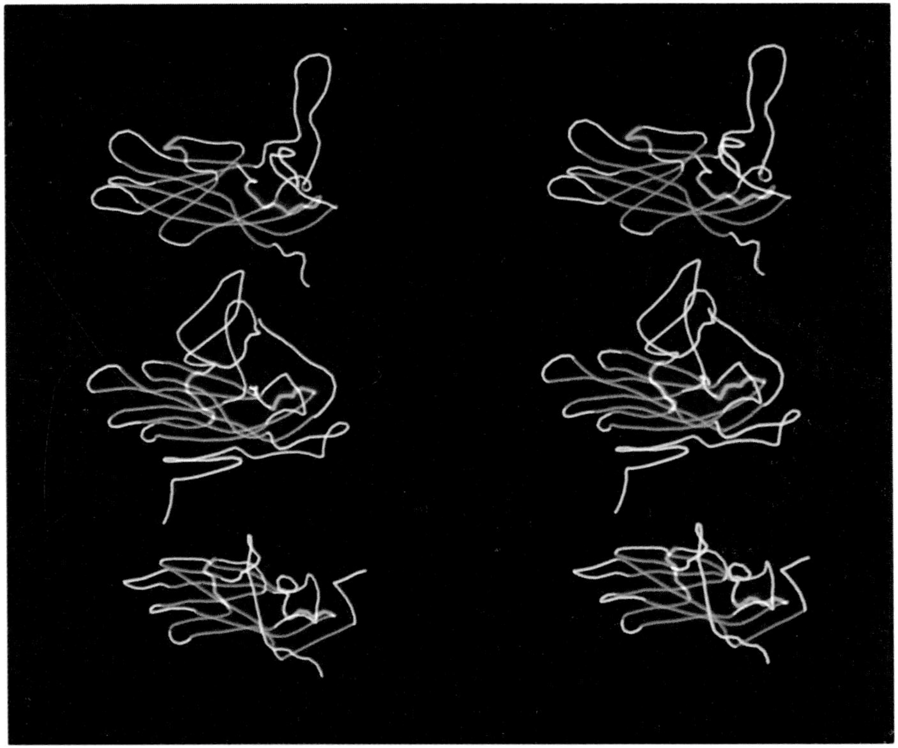

Plate IV, Chapter 21. Stereo picture of alpha carbon models of the poliovirus capsid proteins (*36*): VP1 (top), VP2 (middle), VP3 (bottom). Portions of the main chain corresponding to the eight-stranded antiparallel beta barrel are red; the conserved flanking helices are blue; variable internal loops and terminal extensions are yellow. The NH_2- and COOH-terminal extensions of VP1 and the NH_2-terminal extension of VP3 have been truncated for clarity.

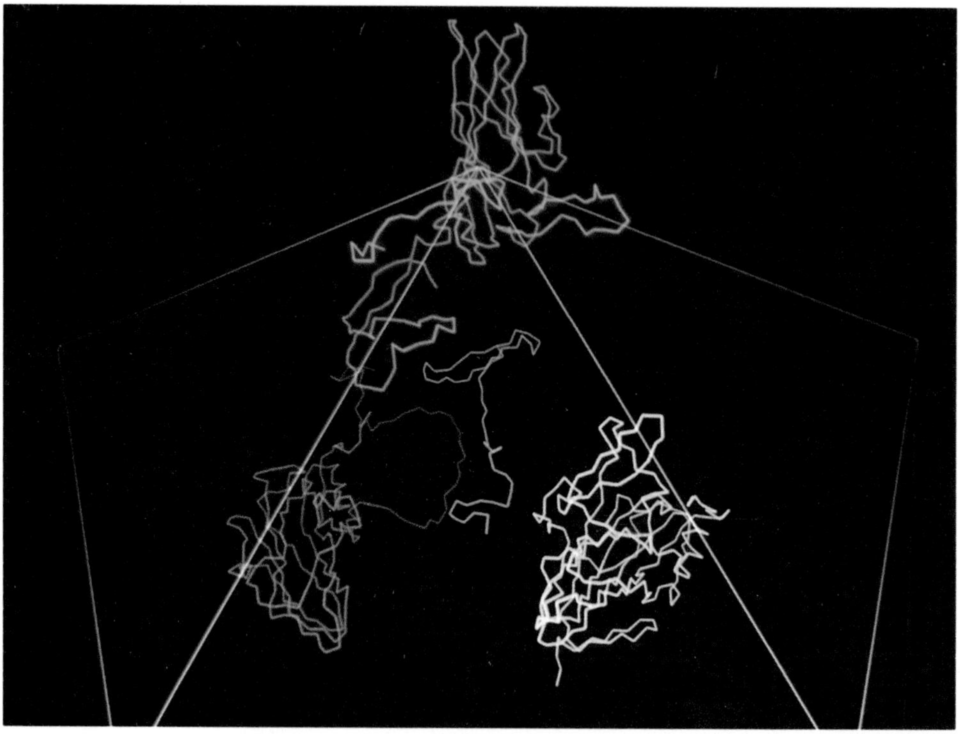

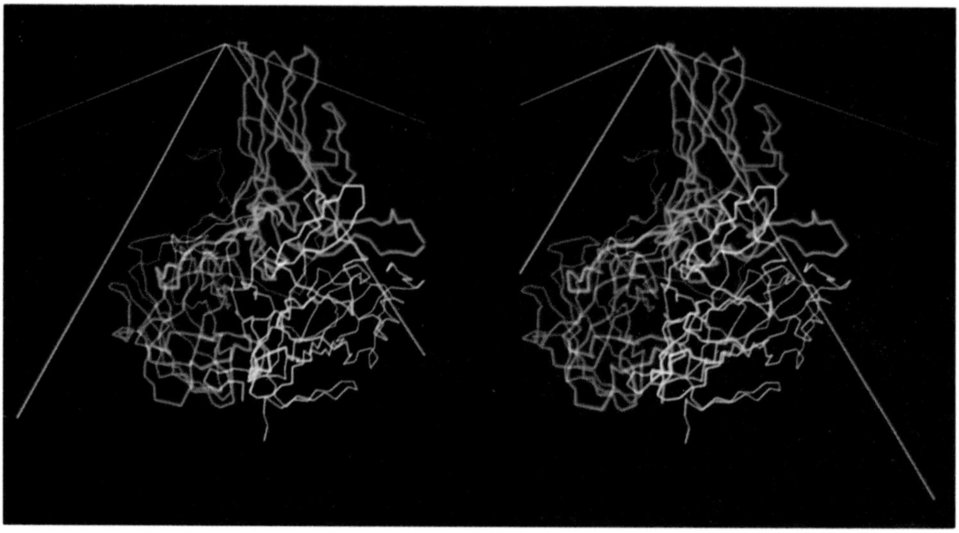

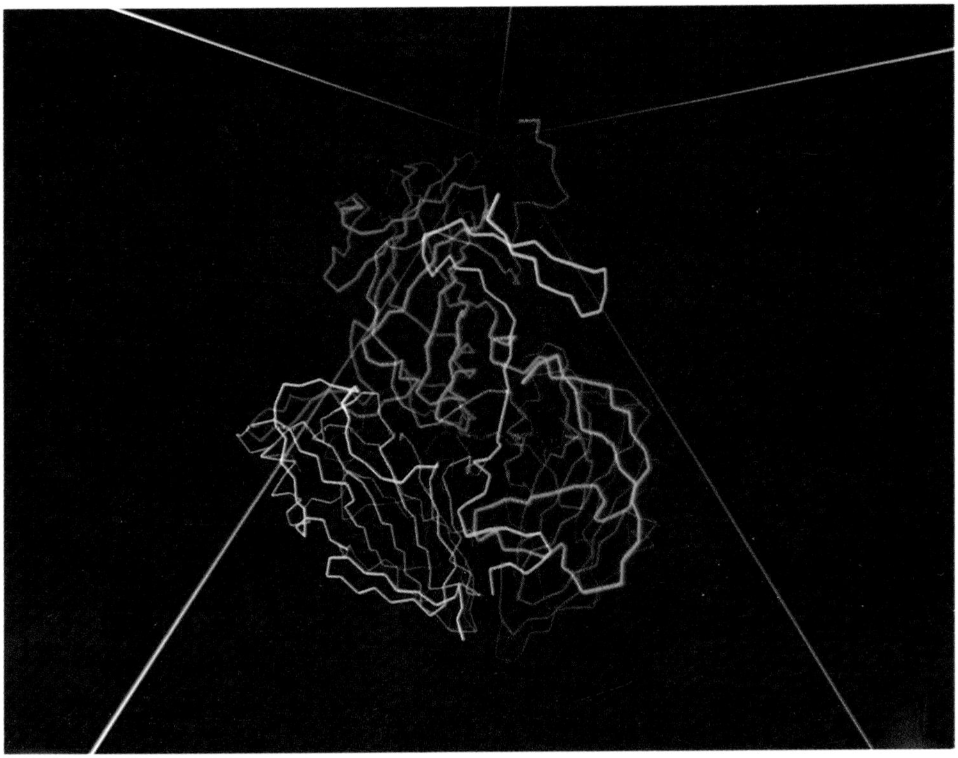

Plate V, Chapter 21. Alpha carbon models of the four poliovirus capsid proteins which constitute a protomer (derived by cleavage of a single molecule of the P1-1a precursor) (*36*). VP1 is blue, VP2 is yellow, VP3 is red, VP4 is green. Portions of the structure close to the view point are brighter than portions which are further away. The edges of an icosahedron are included to indicate the orientation of the protomer relative to the icosahedral symmetry elements of the virus particle. (Left, above) An "exploded" view of the capsid proteins, showing the relationships of VP4 and the terminal extensions of VP1 and VP3 to the major compact domains of the capsid proteins. (Left, below) Stereo picture of the protomer, viewed from the outside of the virion. VP1 is adjacent to the fivefold axis, VP2 and VP3 alternate around particle threefold axes. (Above) View of the protomer from the inside of the virion. This view is heavily depth-cued to emphasize VP4 and the NH_2-terminal extensions of VP1 and VP3.

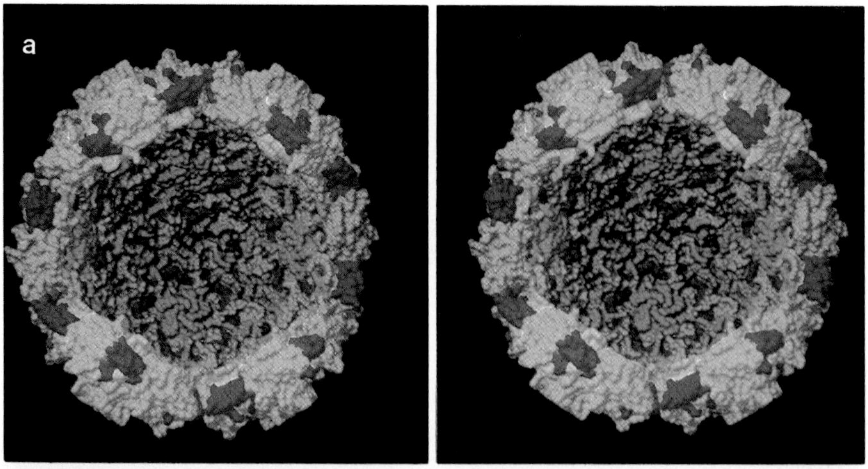

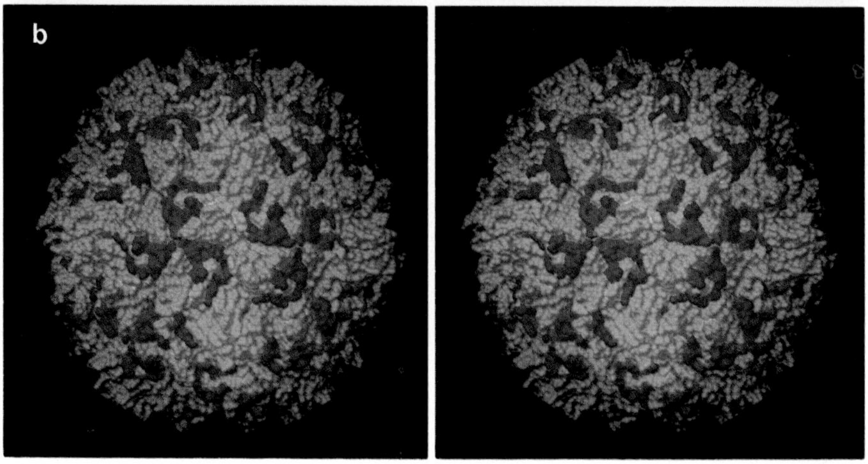

Plate VI, Chapter 21. Major features on the surface of poliovirus. VP1 is blue, VP2 is yellow, VP3 is red, VP4 is green. (a) Stereo space-filling representation of the interior surface (*37*). Four pentamers have been removed from the capsid sheel. The thickness of the shell and the volume of the cavity occupied by RNA are apparent. (b) Stereo space-filling representation of the exterior surface (*37*). (c) Packing of eight protomers on the exterior surface of poliovirus (*36*). The symmetry elements of the particle are indicated by the edges of an icosahedron and a dodecahedron (as described in the legend of Plate VII). Portions of the structure close to the view point are brighter than portions which are further away. In (b) and (c), note the ribbed peak formed by VP1; the broad plateau ringed by promontories formed by VP2 and VP3; the broad valley surrounding the fivefold peak; and the saddle surface across the twofold axis.

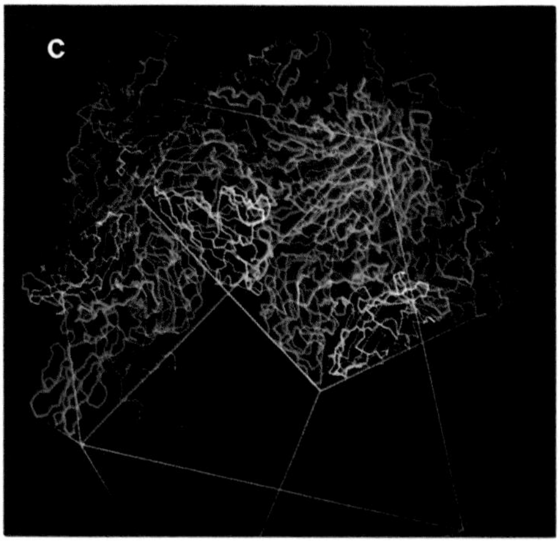

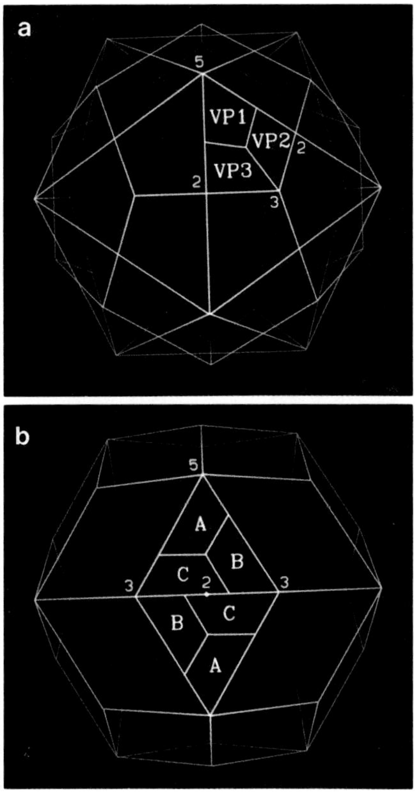

Plate VII, Chapter 21. Geometric representations of the exterior surface features of poliovirus and of the $T = 3$ plant viruses. (a) In this construction, the edges of an icosahedron and a dodecahedron intersect along twofold axes. The resulting surface resembles poliovirus in that it has somewhat larger prominences at its fivefold axes than at its threefold axes. Labels indicate the positions of VP1, VP2, and VP3 from the same protomer. Each protomer occupies an icosahedral asymmetric unit bounded by one fivefold, one threefold, and two twofold axes. (b) A rhombic triacontahedron: a simplified representation of the surface of southern bean mosaic virus (*18*), and tomato bushy stunt virus (*17*). Labels indicate the positions of the six capsid subunits that are contained in two icosahedral asymmetric units. The A, C, and B subunits of the $T = 3$ plant viruses lie in positions analogous to those of the poliovirus subunits VP1, VP2, and VP3, respectively.

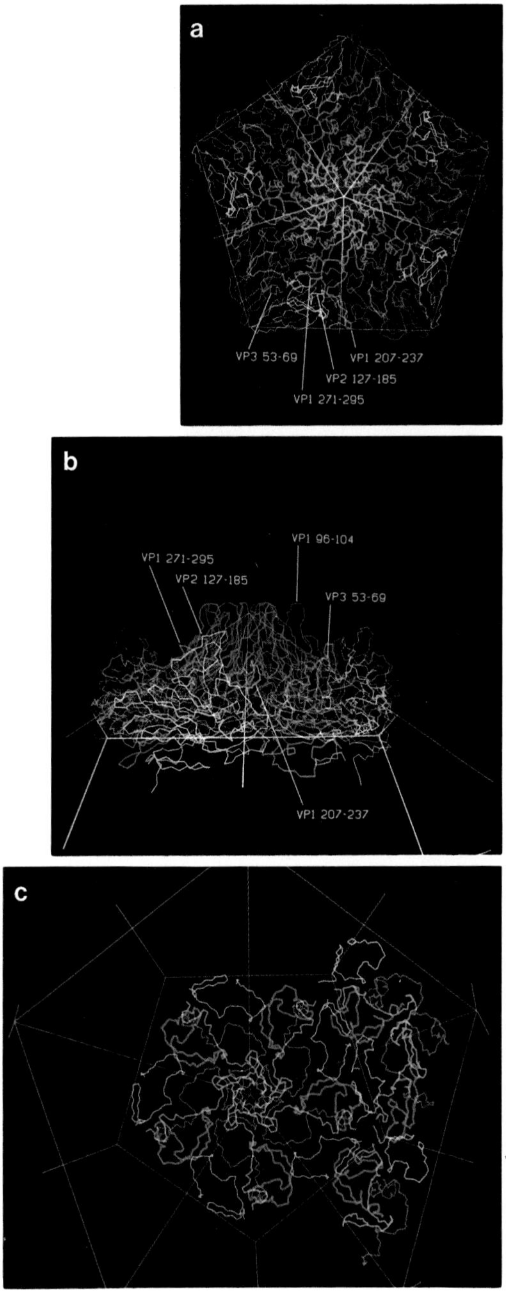

Plate VIII, Chapter 21. Selected views of portions of the poliovirus structure (*36*). VP1 is blue, VP2 is yellow, VP3 is red, VP4 is green. Portions of the structure close to the view point are brighter than (*continued*)

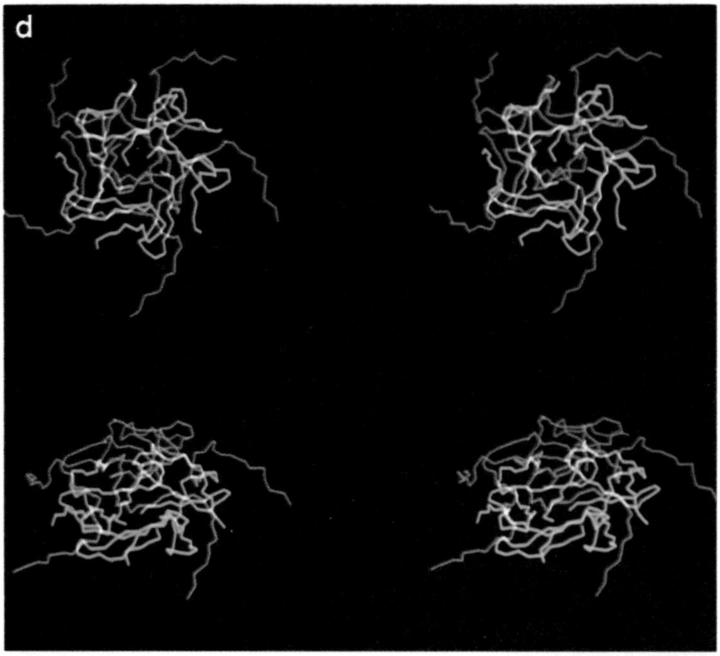

Plate VIII, cont. portions which are further away. The edges of an icosahedreon and a dodecahedron (as described in Plate VII) are included as an orientation aid. (a) The outer surface of five VP1-VP2-VP3-VP4 protomers, viewed along the fivefold axis. Loops discussed in the text are labeled by residue number. (b) Side view of the same pentamer shown in (a). (c) Network formed by VP4 and the NH_2-terminal extensions of VP1, VP2, and VP3, viewed from the inside of the virus particle. The remainder of the capsid protein molecules have been omitted for clarity. Strands from eight protomers are depicted. (d) Stereo pictures showing the structure formed around the fivefold axis by the interaction of five copies each of VP4 and the NH_2-termini of VP3. The structure is located on the interior surface of the capsid. Top: the view along the fivefold axis looking toward the particle center. Bottom: side view. (e) Dot surfaces indicate the positions of monoclonal release mutations on the virion outer surface. (f) Dot surfaces have been color-coded to indicate clusters of monoclonal release mutations grouped by proximity. Clusters 1, 2, 3, and 4 (as defined in the text) are colored yellow, blue, green, and red, respectively. Dot surfaces in (e) and (f) were computed using a 3 A probe sphere and the alpha carbon model. Only residues corresponding to monoclonal release mutations are highlighted.

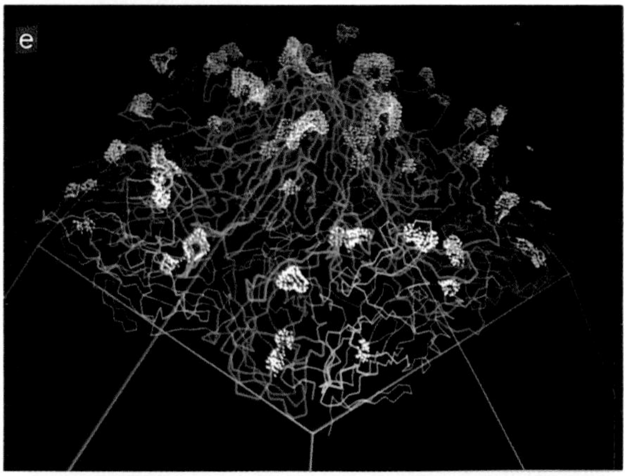

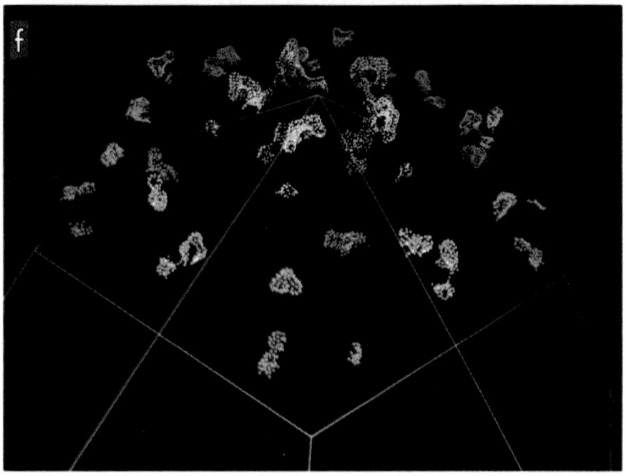

Part VII

Plant Sciences

22

Arabidopsis thaliana *and Plant Molecular Genetics*

Elliot M. Meyerowitz and Robert E. Pruitt

It is worth understanding the molecular genetics of plants not only because such understanding has practical value in the improvement of crops but also because studies of plants offer an opportunity to gain insight into various basic life processes unique to plants. Plants do not use the same hormones as animals; nor do they use hormones in the same way that animals use them. Most or all of the cells in a plant may both produce and respond to plant hormones at some time. Plants respond to stress differently than animals. Plants also respond to light in diverse and subtle ways, with photosynthesis being only one of the responses of plants to light that are not found in animals. Even the basic developmental processes of plants have features that distinguish them from those of animals. Among these features are the absence of cell migration in plant development and the fact that each flowering plant has certain parts (the meristems) that remain embryonic and can produce adult organs, including germ cells, throughout the life of the plant. Further, individual differentiated cells taken from the vegetative parts of plants can dedifferentiate and regenerate to form entire, fertile plants.

Although it is sensible and necessary to study crop plants for purposes of crop improvement, the crop plants now used for basic classical and molecular genetic studies have disadvantages for some of the types of experimentation used in this work. Classical genetics depends on researchers being able to raise many successive generations of organisms in large numbers. Typical crop plants have generation times of several months, and they require a great deal of field space

Science 229, 1214–1218 (20 September 1985)

for growth in large numbers. The genetics of some of these plants is also made more difficult by polyploidy or allopolyploidy. The ease with which recombinant DNA work can be done with any organism depends in part on the size of the organism's nuclear genome; the smaller the genome, the less work is required to screen recombinant DNA libraries and thus to isolate any particular gene. The genomes of the plants presently used for recombinant DNA work are large, being similar in size to those found in mammals (Table 1). In addition, the plants generally used for recombinant DNA work contain large amounts of dispersed repetitive DNA (*1*), which makes it very difficult or impossible to perform such procedures as genomic blot analysis with genomic clones and chromosome walking. *Arabidopsis thaliana* provides a system without these specific disadvantages, and it offers the possibility of doing some types of molecular genetic experiments more rapidly, more easily, and at less expense than other plants currently permit.

Classical and Biochemical Genetics

Arabidopsis is a flowering plant that has been used in classical genetic work for over 40 years (*2*) and that has an extensive genetic and ecological literature. The plant is a member of the mustard family, along with some more familiar plants, such as cabbages and radishes. It is a harmless weed of no food or other economic value. Nonetheless, the literature on the plant indicates that for several reasons, *Arabidopsis* may be of considerable value in molecular genetic research. The plant is well suited to classical genetic work: it has a generation time of only 5 weeks; it can produce more than 10,000 seeds per plant; and it is of such small size that dozens can be grown in a small pot, and tens of thousands can be grown in a small room (*3*). It requires only moist soil and fluorescent light for rapid growth. The small flowers contain both anthers and pistils, and the plant typically self-fertilizes. Self-fertilization allows new mutations to be made homozygous with minimal effort. When desired, cross-fertilization can be simply and rapidly effected, thereby making possible the crosses necessary for genetic mapping and for the production of multiple mutant stocks. Mutagenesis can be performed by soaking seeds in chemical mutagens such as ethyl methanesulfonate or by irradiation of seeds soaked in water. The mutagenized seeds are planted, grown to maturity, and allowed to self-fertilize, thereby producing seeds that are homozygous for new mutations.

Various mutations have been identified, including visible mutations useful as markers in genetic mapping, with phenotypes affecting every part of the plant. These include mutations affecting the wax coat of the epidermal cells (*eceriferum* mutants) (*4*) and the trichomes normally found on leaves and stems (*distorted trichomes* and *glabra* mutants) (*5*), as well as those mutations causing more easily visible effects, such as changes in flower morphology (*agamous, apetala* and *pistillata* mutants) (*6*) and growth habit (*erecta* and *compacta* mutants) (*6, 7*). Other mutations affect the embryonic development of *Arabidopsis* and lead to embryonic lethality; the stage at which developmental arrest occurs depends on the specific gene mutated (*8*). Many leaf-color mutants exist, including two that give variegated plants.

Plants homozygous for the immutans mutation have leaves that are variegated white and green; the degree of this variegation depends both on the particular allele of the immutans mutation and on the intensity of the light falling on the plant at the time of a given leaf develops. This recessive mutation is inherited from both white and green sectors (*9*). Plants homozygous for the chloroplast mutator mutation also have leaves that bear white and green sectors. In this case the white sectors are attributable to the failure of the chloroplasts to develop normally. Although plants heterozygous for the chloroplast mutator mutation do not produce aberrant chloroplasts, chloroplasts that heterozygous plants maternally inherit from homozygous chloroplast mutator plants continue to be of abnormal structure. In fact, from these outcrossed chloroplast mutator strains it is possible to isolate stable homoplastidic lines of *Arabidopsis* in which all the chloroplasts have the same abnormal appearance. It has been proposed that the nuclear chloroplast mutator mutation acts by causing errors during chloroplast DNA replication (*10*).

Biochemical mutants of various types have also been isolated. Mutations in several genes give rise to plants that for growth require thiamine or the thiamine precursors 2,5-dimethyl-4-aminopyrimidine or 4-methyl-5-hydroxyethylthiazole (*11, 12*). Somerville and Ogren have characterized biochemical mutations that affect photorespiration and photosynthesis. These include nuclear mutations that result in the absence of activity of chloroplast glutamate synthase, phosphoglycolate phosphatase, mitochondrial glycine decarboxylase, mitochondrial serine transhydroxymethylase, and serine-glyoxylate aminotransferase (*13*). Mutations that affect the activity of nitrate reductase and of alcohol dehydrogenase have also been characterized (*14*).

In addition, mutations have been obtained that appear to disrupt the normal action of several of the plant growth regulators. Koornneef and his collaborators have recovered mutations that cause absence of gibberellins or of abscisic acid, as well as mutations that confer resistance to high levels of exogenous abscisic acid. The mutations that affect gibberellins fall into five complementation groups. Three of these contain alleles that, when homozygous, prevent germination unless the seeds are treated with exogenous gibberellins (*15*). A mutant strain lacking abscisic acid has been used by Karssen *et al.* to examine the relative effects of maternal and embryo-derived abscisic acid on seed dormancy (*16*). Mutations that result in resistance to high levels of a synthetic auxin have also been recovered, and some alleles of these mutations produce plants with agravitropic roots (*17*). One of these mutations is dominant and produces a plant with a dwarf rosette and inflorescence when heterozygous, but is lethal when homozygous.

More than 75 of the known mutations have been assembled into a genetic linkage map by Koornneef *et al.* (*6*). This genetic map consists of five linkage groups, in concordance with the haploid chromosome number of five. The location of the centromeres on the genetic map has been determined by the use of strains that are trisomic for one arm of a chromosome (*6, 18*). *Arabidopsis* stocks containing many of the mutations on the genetic map in combinations suitable for mapping specific chromosome regions, together with many strains collected from the wild, are available from a central international seed collection (*19*).

In addition to growing them in soil, it is possible to grow whole *Arabidopsis* plants on sterile, biochemically defined media; both solid and liquid media have been used (*11, 20*). *Arabidopsis* cells have also been grown in tissue culture, and plants have been regenerated from such cells (*21*). Some types of biochemical mutations have been isolated by selection of cultured cells; it may also be possible to use biochemical selections to assay transformation of cells in culture.

Molecular Genetics: Genome Size and Organization

Three lines of evidence indicate that *Arabidopsis* has a small genome. Microspectrophotometry of Feulgen-stained nuclei gives an estimate of 0.2 pg, or approximately 2×10^8 base pairs (bp), for the quantity of DNA in the *Arabidopsis* haploid genome. This is the smallest published size for an angiosperm genome (*22*). A second line of evidence indicating that *Arabidopsis* has a small genome is based on the proportionality of genome size and nuclear volume. This relationship implies that *Arabidopsis* has a haploid genome size of roughly 10^9 bp (*23*). These estimates indicate that the *Arabidopsis* genome is small enough to greatly simplify the task of screening recombinant DNA libraries to isolate genes.

We have performed a more detailed analysis of the *Arabidopsis* genome to verify its small size and to determine the fraction and nature of repetitive sequences present. Our initial study was a reassociation analysis of DNA extracted from whole plants (*24*). This gave an estimated genome size of 7×10^7 bp, a size only five times larger than that of the yeast genome (*25*) and much smaller than that of other flowering plants (Table 1). Our analysis also showed that the whole-plant DNA contains 10 to 14 percent very rapidly reannealing sequences (either highly repeated sequences or inverted repeat sequences) and 23 to 27 percent middle repetitive sequences. By us-

Table 1. Haploid genome size in various flowering plants and the number of lambda clones that must be screened to have a 99 percent chance of isolating a single-copy sequence from these genomes. The genome sizes are calculated from kinetic complexity measurements (*38*). The library sizes are calculated assuming a random nuclear DNA library with an average clone insert length of 20 kb; 4.6 genome equivalents must be screened for a 99 percent probability of isolating any individual unique sequence.

Plant	Haploid genome size (in kilobase pairs)	Number of lambda clones in complete library
Arabidopsis	70,000	16,000
Mung bean	470,000	110,000
Cotton	780,000	180,000
Tobacco	1,600,000	370,000
Soybean	1,800,000	440,000
Pea	4,500,000	1,000,000
Wheat	5,900,000	1,400,000

ing a labeled tracer containing cloned chloroplast sequences, we demonstrated that almost all the middle repetitive sequences are from the chloroplast genome—and thus that *Arabidopsis* has far less nuclear repetitive DNA than other angiosperms (*1*) (Table 2). Leutwiler *et al.* have also examined the degree of cytosine methylation in *Arabidopsis* and have found that only 4.6 percent of the cytosines derived from whole-plant DNA are present as 5-methylcytosine; this is the lowest value known for a flowering plant (*24*).

More recently, we have used recombinant DNA techniques to examine random segments of genomic DNA (*26*). We have subjected randomly selected recombinant lambda phage to a variety of experimental manipulations to determine if they contain unique or repetitive sequences (or both) and to characterize the nature of the repetitive sequences that are contained in the clones. These experiments confirm and extend our previous analysis. A majority of the 50 clones analyzed contain only unique sequences, and most of the clones that contain repetitive sequences are derived either from the chloroplast genome (four clones) or from the nuclear DNA that codes for the large ribosomal RNA's (eight clones). There are approximately 570 copies of the ribosomal DNA per haploid genome, each ribosomal DNA repeat unit is about 10 kilobase pairs (kb) in length, and the repeats are largely arranged in tandem array. In addition, there are two clones that appear to contain duplicated sequences, one clone that contains a low copy number repeat unit that is apparently conserved in all copies (this may represent a fragment of mitochondrial DNA), and three clones that contain both repeated sequences and unique sequences interspersed with each other. These results indicate that the *Arabidopsis* nuclear genome consists of predominantly unique sequences and that most of the nuclear repetitive DNA is ribosomal DNA. The experiments show that much of the nuclear DNA of *Arabidopsis* is organized as extremely long blocks (averaging 120 kb) of unique sequences.

These two sets of experiments demon-

Table 2. Average size of single-copy DNA sequences interspersed with repetitive sequences and total amount of repetitive DNA in various plant genomes. The data for *Arabidopsis* are from (*24*) for amount of repetitive DNA and from (*26*) for size of unique DNA stretches; the unique sequence size is from measurements of random cloned fragments. The data for the other plants are from reassociation analysis (*38*).

Plant	Average size in predominant class of single-copy sequences (in kilobase pairs)	Amount of repetitive DNA in haploid genome (in kilobase pairs)
Arabidopsis	120	18,000
Mung bean	>6.7	160,000
Cotton	1.8	310,000
Tobacco	1.4	1,200,000
Soybean	<3	1,100,000
Pea	0.3	3,800,000
Wheat	1	4,400,000

strate that *Arabidopsis* has a remarkably small and simple genome. This fact, taken together with the results from all of the previous work on *Arabidopsis*, indicates that the plant does represent a very good model system for basic research in plant molecular biology.

Molecular Genetics: Specific Genes

We have started to do additional work in studying individual genes from *Arabidopsis* to provide material for analysis of the mechanisms by which gene expression is regulated by both developmental and environmental stimuli. So far, we have examined three different genes or gene families, and this work has led us to two generalizations that emphasize the utility of *Arabidopsis* for experiments in molecular cloning. The first generalization is that it is possible to cross-hybridize genes cloned from a wide variety of flowering plants with the homologous gene or genes from *Arabidopsis*. This perhaps is not surprising because fossil evidence indicates that the first major radiation of the angiosperms took place only 1×10^8 to 1.5×10^8 years ago. Our second generalization is that proteins that are encoded by multiple genes or large gene families in other plants are encoded by single genes or small gene families in *Arabidopsis*.

The first gene we examined was the one that encodes the large seed storage protein (*27*). We cloned this gene from a recombinant library containing genomic DNA by using a complementary DNA clone encoding the 12*S* seed storage protein of *Brassica napus* (*28*). In *Arabidopsis* this protein is encoded by a single gene, which indicates that the heterogeneity observed in other storage protein gene families is not required in this plant. Further, the fact that there is only one gene coding for this protein makes it clear that the tissue- and time-specific regulation of the protein is attributable to the activity of only one gene and also that experiments to understand this regulated gene expression need to deal only with this single sequence and not with a large family of similar genes as in other plants.

A second series of experiments involved the cloning of the genes encoding the light-induced chlorophyll a/b binding protein of *Arabidopsis* (*29*). Chlorophyll a/b binding protein, or light-harvesting chlorophyll protein, is a nuclear-coded component of the light-harvesting antenna complex of the chloroplast thylakoid membranes. These genes were isolated from an *Arabidopsis* genomic library by using as a probe a genomic clone from the tiny aquatic monocot *Lemna gibba* (*30*). There are three genes encoding this protein in *Arabidopsis*, and they are all located in a small gene cluster of less than 6000 bp. The three genes of the *Arabidopsis* chlorophyll a/b binding protein family contrast with the much larger number in the homologous families in other plants. One example is *Petunia*; the thorough studies of Dunsmuir *et al.* have shown that there may be as many as 16 or more members of this gene family, divided into at least five major subclasses (*31*).

Perhaps even more surprising is the fact that it has been determined by DNA sequencing that all the *Arabidopsis* genes encode an identical product after the transit peptides of the three proteins are removed. As with the seed storage protein, therefore, there does not appear to be any protein heterogeneity for the chlorophyll a/b binding protein encoded in the genome of *Arabidopsis*. Any model for this gene family in *Arabidopsis* must recognize that the purpose of the multiple genes is not to provide a series of variant proteins of slightly different function.

G. An (*32*) has taken advantage of the small number of *Arabidopsis* chlorophyll a/b binding protein genes; because there are only three copies, it is possible in a short series of experiments to test all of the upstream regulatory regions of the genes for function and light inducibility. An has fused two of these regions to bacterial chloramphenicol acetyltransferase genes and has used *Agrobacterium* Ti-plasmid constructs to introduce the fusion genes to tissue culture cells of tobacco. In some instances each of the two *Arabidopsis* DNA fragments has conferred light-inducibility on the bacterial gene.

One final example of a specific *Arabidopsis* gene that has been cloned and characterized is the gene that codes for alcohol dehydrogenase (Adh) (*33*). A classical genetic study of alcohol dehydrogenase allozymes had already demonstrated that there is only a single alcohol dehydrogenase gene in *Arabidopsis*, in contrast to the two or three such genes in many other plants (*34*). The *Arabidopsis* gene was isolated from a recombinant library by cross-hybridization with a labeled maize Adh 1 gene fragment. The gene appears to be single-copy, in confirmation of the earlier genetic results; DNA sequencing shows that the *Arabidopsis* gene shares more than 70 percent nucleic acid homology with the maize Adh 1 gene. The proteins coded by the *Arabidopsis* gene and the maize Adh 1 gene have slightly more than 80 percent identity of their amino acids. Further, the structure of the *Arabidopsis* gene and that of the maize gene are strikingly similar. Both the maize Adh 1 and Adh 2 genes contain nine intervening sequences at identical positions (*35*). The *Arabidopsis* gene has six intervening sequences; the positions of all six are coincident with the corresponding positions of six of the maize introns.

The existing evidence, therefore, leads to two conclusions. The first is that *Arabidopsis* genes will cross-hybridize with homologous genes from other angiosperms, both monocots and dicots; the second is that genes found in large gene families in other plants can exist either in small families or in single copy in *Arabidopsis*. The practical importance of the first conclusion is that genes of interest can be simply cloned from the extraordinarily small *Arabidopsis* genome and then used as probes for the isolation of the homologous genes from plants of economic value. The importance of the second conclusion is that genes found in gene families can be more easily and more thoroughly studied in *Arabidopsis* than in other angiosperms.

The Future of *Arabidopsis* Research

For *Arabidopsis* to be truly useful as a tool for molecular genetic research, two additional techniques must be developed. First, it must become possible to clone genes about which no more is known than their mutant phenotype. Second, it must become possible to take cloned genes that have been modified in vitro and introduce them back into the plant to assay their in vivo function. Development of the second of these techniques is currently being pursued by many laboratories using various methods. *Arabidopsis* is known to be susceptible to infection by *Agrobacterium tumefaciens*, and it is known that Ti-plasmid strains of *Agrobacterium* cause typical tumors on *Arabidopsis* (*36*). It will very likely prove possible to introduce cloned sequences into the *Arabidopsis* genome by using the same methods currently used in the transformation of other plants. If a method can be found that works efficiently enough, the small genome of *Arabidopsis* may be used to advantage in isolation of genes by so-called shotgun transformation.

Gene isolation would involve transformation of mutant plants or plant cells with a complete random recombinant library derived from wild-type DNA cloned in a Ti-plasmid-based vector, followed by assay of individual transformed cells or plants for complementation of the mutant phenotype by the expression of the introduced DNA. Using a cosmid vector with a capacity of 20 to 25 kb, it would require approximately 3000 transformation events to introduce one genome equivalent of wild-type DNA into a set of mutant plants or cells and 14,000 transformation events to introduce the 4.6 genomes necessary to have a 99 percent chance of recovering any specific gene. Even with an efficient technique, these are large numbers of transformation events unless the gene of interest has a phenotype that can be selected directly.

An alternative approach would be to use a procedure of successive isolations of overlapping cloned segments starting from a known cloned genetic location and proceeding to any nearby genetic locus. The *Arabidopsis* genome is unique among those of flowering plants in its suitability for this sort of process, owing both to its small size and to the near absence of dispersed repeated sequences. No other flowering plant is known to have a genome with few enough of these sequences, spaced far enough from each other, to allow such a procedure to be practical (Table 2).

For a chromosome walking procedure to be used, it is necessary to have starting points that are known to be located relatively close to the gene of interest. To provide these starting points, we are currently involved in the production of a genetic map using restriction fragment length polymorphisms (RFLP's) (*37*). It is easy to find RFLP's in different wild-type strains of *Arabidopsis*. We have tested a number of different strains and have selected the Niederzenz strain (first collected in Niederzenz, West Germany) and the Landsberg strain (first collected in Landsberg, East Germany) as the parental strains.

Landsberg was chosen because most of the mutations used in construction of the published genetic maps of *Arabidopsis* are in the Landsberg genetic background. Niederzenz was selected as the second parental strain because it grows as quickly as Landsberg (many other wild strains do not) and because it shows many RFLP's when compared to Landsberg. We have crossed these two parental strains and allowed the F_1 progeny to self-fertilize to form F_2 seed. F_2 plants were grown and allowed to self, and F_3 seeds were collected from each plant. These pools of F_3 seeds have been grown and DNA has been prepared from the resulting plants. This DNA contains the alleles present in the F_2 plant from which the F_3 plants descended. By genome blot analysis of the DNA, it will be possible to determine the genotype of each of the F_2 plants at each polymorphic locus. From these data, linkage distances can be calculated just as for any other genetic markers. By including visible markers in the original cross and performing subsequent crosses with other visible markers, it will be possible to align the RFLP map with the published genetic map of visible mutations. The clones used as probes on the RFLP genome blots can then serve as starting points for successive clone isolations.

To tell when the desired gene has been cloned, it will be necessary to transform cloned DNA segments into the plant and to assay them for the ability to complement mutations in the gene. Because only a small number of clones close to the gene of interest need to be tested by introduction into the plant genome, this procedure will not require the high frequency of transformation demanded by the first general gene isolation method described.

The ability to transform *Arabidopsis* not only has the potential to permit cloning of any gene having a known mutant phenotype but will also make possible the type of detailed analysis of these genes that is now being performed with genes of yeast, *Drosophila*, and mice. Among the genes that may be cloned and analyzed are those affecting particular enzyme activities, those with specific effects on the ability of *Arabidopsis* to synthesize and respond to plant hormones, and those in which mutations give specific developmental abnormalities. Our hope is that *Arabidopsis* will soon join the other organisms for which a combined genetic and molecular approach has led to both fundamental and practical scientific advances.

References and Notes

1. R. Flavell, *Annu. Rev. Plant Physiol.* **31**, 569 (1980).
2. F. Laibach, *Bot. Arch.* **44**, 439 (1943); *Naturwissenschaften* **31**, 246 (1943); E. Reinholz, *ibid.* **34**, 26 (1947).
3. G. P. Redei, *Annu. Rev. Genet.* **9**, 111 (1975).
4. L. M. W. Dellaert, J. Y. P. Van Es, M. Koornneef, Arabidopsis *Inf. Serv.* **16**, 10 (1979).
5. W. J. Feenstra, *ibid.* **15**, 35 (1978); S. Lee-Chen and L. M. Steinitz-Sears, *Can. J. Genet. Cytol.* **9**, 381 (1967); M. Koornneef, L. W. M. Dellaert, J. H. van der Veen, *Mutat. Res.* **93**, 109 (1982).
6. M. Koornneef *et al.*, *J. Hered.* **74**, 265 (1983).
7. G. P. Redei, *Z. Vererbungsl.* **93**, 164 (1962).
8. D. W. Meinke and I. M. Sussex, *Dev. Biol.* **72**, 50 (1979); D. W. Meinke, *Theor. Appl. Genet.* **69**, 543 (1985).
9. G. P. Redei, *Genetics* **56**, 431 (1967); G. Robbelen, *Planta (Berlin)* **80**, 237 (1968).
10. G. P. Redei, *Mutat. Res.* **18**, 149 (1973).
11. J. Langridge, *Nature (London)* **176**, 260 (1955).
12. S. L. Li and G. P. Redei, *Biochem. Genet.* **3**, 163 (1969).
13. C. R. Somerville and W. L. Ogren, *Nature (London)* **286**, 257 (1980); *ibid.* **280**, 833 (1979); *Biochem. J.* **202**, 373 (1982); *Plant Physiol.* **67**, 666 (1981); *Proc. Natl. Acad. Sci. U.S.A.* **77**, 2684 (1980).
14. F. J. Braaksma and W. J. Feenstra, *Theor. Appl. Genet.* **64**, 83 (1982); M. Jacobs and D. Schwartz, Arabidopsis *Inf. Serv.* **17**, 88 (1980).
15. M. Koornneef and J. H. van der Veen, *Theor. Appl. Genet.* **58**, 257 (1980); M. Koornneef, M. L. Jorna, D. L. C. Brinkhorst–van der Swan, C. M. Karssen, *ibid.* **61**, 385 (1982); M. Koornneef, G. Reuling, C. M. Karssen, *Physiol. Plant.* **61**, 377 (1984).
16. C. M. Karssen, D. L. C. Brinkhorst–van der Swan, A. E. Breekland, M. Koornneef, *Planta* **157**, 158 (1983).
17. J. I. Mirza, G. M. Olsen, T.-H. Iversen, E. P. Maher, *Physiol. Plant.* **60**, 516 (1984); G. M. Olsen, J. I. Mirza, E. P. Maher, T.-H. Iversen, *ibid.*, p. 523.
18. M. Koornneef and J. H. van der Veen, *Genetica* **61**, 41 (1983).
19. The seed collection is maintained by A. R. Kranz, Botanisches Institut, J. W. Goethe-Universität, Frankfurt am Main 11, West Germany. Kranz also edits and distributes Arabidopsis *Information Service*, an annual newsletter that includes original contributions, reviews, reports of new mutations, and lists of available stocks.
20. G. P. Redei and C. M. Perry, Arabidopsis *Inf. Serv.* **8**, 34 (1971); N. Goto, *ibid.* **19**, 55 (1982).
21. I. Negrutiu, M. Jacobs, W. de Greef, *Z. Pflanzenphysiol.* **90**, 363 (1978).
22. M. D. Bennett and J. B. Smith, *Proc. R. Soc. London Ser. B* **274**, 227 (1976).
23. A. H. Sparrow, H. J. Price, A. G. Underbrink, *Brookhaven Symp. Biol.* **223**, 451 (1972).
24. L. S. Leutwiler, B. R. Hough-Evans, E. M. Meyerowitz, *Mol. Gen. Genet.* **194**, 15 (1984).
25. G. D. Lauer, T. M. Roberts, L. C. Klotz, *J. Mol. Biol.* **114**, 507 (1977).
26. R. E. Pruitt and E. M. Meyerowitz, unpublished manuscript.
27. R. E. Pruitt, D. Ruff, E. M. Meyerowitz, in preparation.
28. A. E. Simon, K. M. Tenbarge, S. R. Scofield, R. R. Finkelstein, M. L. Crouch, unpublished manuscript.
29. L. S. Leutwiler, E. M. Meyerowitz, E. M. Tobin, in preparation.
30. W. J. Stiekema, C. F. Wimpee, J. Silverthorne, E. M. Tobin, *Plant Physiol.* **72**, 717 (1983).

31. P. Dunsmuir, S. M. Smith, J. Bedbrook, *J. Mol. Appl. Genet.* **2**, 285 (1983).
32. G. An, personal communication.
33. C. Chang and E. M. Meyerowitz, in preparation.
34. R. Dolferus and M. Jacobs, *Biochem. Genet.* **22**, 817 (1984).
35. E. S. Dennis *et al.*, *Nucleic Acids Res.* **12**, 3983 (1984); E. S. Dennis, M. M. Sachs, W. L. Gerlach, E. J. Finnegan, W. J. Peacock, *ibid.* **13**, 727 (1985).
36. M. Aerts, M. Jacobs, J.-P. Hernalsteens, M. van Montagu, J. Schell, *Plant Sci. Lett.* **17**, 43 (1979).
37. D. Botstein, R. L. White, M. Skolnick, R. W. Davis, *Am. J. Hum. Genet.* **32**, 314 (1980).
38. The sources of the genome measurements are *Arabidopsis*: (*24*); mung bean: M. G. Murray, J. D. Palmer, W. F. Thompson, *Biochemistry* **18**, 5259 (1979); cotton: V. Walbot, L. S. Dure III, *J. Mol. Biol.* **101**, 503 (1976); tobacco: J. L. Zimmerman and R. B. Goldberg, *Chromosoma (Berlin)* **59**, 227 (1977); soybean: R. B. Goldberg, *Biochem. Genet.* **16**, 45 (1978); pea: M. G. Murray, R. E. Cuellar, W. F. Thompson, *Biochemistry* **17**, 5781 (1978); wheat: D. B. Smith and R. B. Flavell, *Chromosoma (Berlin)* **50**, 223 (1975) and R. B. Flavell and D. B. Smith, *Heredity* **37**, 231 (1976).
39. We thank the other members of the laboratory of E.M.M. for comments on the manuscript. Our *Arabidopsis* work is supported by NSF grant PCM-8408504 to E.M.M.

23

Safety Concerns and Genetic Engineering in Agriculture

Winston J. Brill

Federal agencies are considering various regulations to protect the public from environmental and health problems that might arise from the release of genetically engineered organisms. Concern has been expressed because several agricultural practices, such as the widespread use of DDT in past decades (*1*), have caused serious problems that were unintended and unexpected. Also, movement of weeds and insect pests into new environments has created problems that have become difficult to control. Examples include kudzu, hydrilla, the gypsy moth, and the Japanese beetle. Because of these experiences, it is necessary to consider the potential effects of releasing organisms containing genes from related and unrelated genera. This article will focus on the safety issues involved in using genetically engineered plants and microorganisms (bacteria and fungi) to benefit agriculture. Other applications to which the same principles should hold with respect to safety issues include the use of genetically engineered organisms for mining, waste treatment, and detoxifying chemical spills.

The economic and environmental benefits expected to accrue from agricultural use of recombinant organisms are great (*2*) and should be considered in relation to the potential risks. By splicing foreign genes into plant chromosomes it may be possible to create plants resistant to a wide array of pests. The hope and expectation is that they will lead to decreased use of chemical fungicides and insecticides, many of which are toxic to man. Recombinant DNA techniques may be used to develop plants that utilize fertilizers more efficiently, thereby minimizing fertilizer runoff into streams

Science 227, 381–384 (25 January 1985)

and lakes. In many crop species a relatively narrow base of germplasm is being used to develop varieties. There is concern that this has created genetic vulnerability to disease (*3*). Genetic engineering can be used to introduce new genes and thereby increase genetic variability for the future. The time it takes to develop new plant varieties should be greatly decreased by this new technology.

Genetically engineered bacteria and fungi also have potential value. For example, *Rhizobium* strains isolated from many locations around the world are being applied to soils in large numbers so that legumes can produce high yields without needing expensive nitrogenous fertilizers. Several approaches are being considered to increase legume yields with genetically engineered *Rhizobium* (*4*). Other microbes, such as mycorrhizae, *Pseudomonas*, and *Frankia* (*5*), are also promising candidates for use in agriculture, and there is a good chance that the value of these organisms can be increased through recombinant DNA technology as well as traditional mutation and recombination techniques. As in traditional agriculture, the value of the new plants and microbes can be assessed only after they have been tested under a variety of field conditions. This article will discuss ways to predict the safety level of an organism that has received several foreign genes.

Of particular concern in the introduction of new organisms is the potential to self-perpetuate and spread. For the purpose of this discussion, however, a problem plant that gets no farther than the next field is not defined as a serious problem. Nor is a microbe that unexpectedly kills plants that it was sprayed on but does not damage plants in a neighboring field.

Plants

Plants have been crossed (traditional "genetic engineering") by man for centuries. New variants resulting from such breeding have not caused serious problems. Most of our high-yielding crops, productive forest trees, popular ornamentals, and garden plants have been derived through breeding programs. Some crosses include those that would not occur without man's intervention, such as crosses between high-yielding Midwestern corn and its putative wild ancestor, teosinte (*6*). Species that do not readily cross-pollinate have been crossed, without recombinant DNA technology, by many scientists around the world. As an example, cultivated oats have been crossed with wild species to increase the protein concentration of seeds and to introduce resistance to diseases (*7*). Protoplast fusion between cells of plants that normally are unable to cross have yielded new variants (*8*). Also, plants obtained by mutation have frequently been grown in experimental fields with the hope of detecting useful new phenotypes. These experiments produce novel plants and, with the exception of mutated plants, the progeny are the result of uncontrolled recombination of tens of thousands of genes. The exact properties of progeny from most of these crosses are impossible to predict. Breeders have never taken and do not now take special precautions in testing these plants in the field because they know from experience that these extensive mixings have not produced serious problems. If we compare plants derived from breeding programs with those derived through genetic engineering, it is clear that, in the latter case, the addition of a few characterized genes to the plant results in properties that are relatively easy to predict.

One ecological concern is the inadvertent release of a new weed that will be difficult to control. However, the long and diverse experience of breeders and plant geneticists indicates that genetic crosses among nonweedy plants will not result in a serious problem (*9*). From our growing understanding of the genetic and biochemical basis of competition by weeds, it is obvious that many genes must interact appropriately for the plant to display the undersirable properties of a weed (e.g. efficient seed dispersal, long seed viability, rapid growth in an environment not normally favorable to other plants) (*10*). It is possible that hundreds or thousands of specific interacting genes are necessary for a plant to be a weed that will cause problems that approach the magnitude of those of kudzu. Thus the chance is exceedingly small that a cross between nonweeds will yield a problem weed. Most commercial field tests with genetically engineered plants will involve cultivated crops that have been specifically bred for high yield under intensive agricultural practices. As crops are bred for characteristics favorable to agriculture, competitive properties are weakened. Such crops, if left unattended, are not capable of competing well with other plants (*11*). Addition of characterized foreign genes to these crops should not produce an undesirable weed. Obviously, if weedy species are to be purposely genetically engineered, both the weed and the recombinant derivative need to be considered in light of potential environmental damage. Precautions currently used by scientists who purposely plant problem weeds (or crops that readily cross with problem weeds) should be sufficient for those who would plant weeds with incorporated foreign genes (*12*).

Genetic changes in weeds through man's activity but not involving genetic engineering occurred prior to application of recombinant DNA technology. In recent decades the use of chemical herbicides has caused uncharacterized genetic changes by which weeds have become herbicide-resistant (*13*). Problems can be overcome merely by using a herbicide to which the weed is not resistant, thus removing the environmental pressure for maintaining the resistance genes. A similar situation is found with insecticide application to plants, whereby insecticide-resistant insects may arise (*14*). Therefore, uncharacterized genetic changes causing problems with undesirable organisms have already been generated through traditional practices (*15*). Some of these changes probably are due to a single-gene modification. It would be extremely difficult (if not impossible) to purposely mutate a plant now considered safe to become a serious problem weed. It should be even more difficult to derive such a weed through acquisition of characterized foreign genes.

It can be anticipated that problems encountered in traditional breeding programs will occur in plant genetic engineering. For instance, certain popular corn hybrids were especially susceptible to the fungus *Helminthosporium* (*16*). This resulted in the corn leaf blight that destroyed a large portion of the U.S. corn crop in 1970. Breeders prepare for this type of situation, however. They were ready to quickly replace the susceptible variety with ones resistant to corn leaf blight. Field tests, therefore, are necessary to assess the threat of pathogens and to check for undesirable characteristics of new varieties, whether they are products of traditional breeding or genetic engineering.

There is a very small chance that plants resulting from genetic engineering with uncharacterized genes may produce a toxic secondary metabolite or protein toxin. For this reason, animal feeding experiments might be desirable before an edible crop is introduced commercially. Even through traditional breeding, however, toxin production can be a concern, especially when exotic plants are used in the breeding program. Several plants currently marketed, including rhubarb, cotton, and castor bean, contain toxins and therefore need to be carefully processed. Another example is a cultivar of potato that was removed from market shelves because, under certain stress conditions, it produced potentially hazardous levels of glycoalkaloids (*17*). Plant toxins are natural products, whether polypeptides or secondary metabolites, and should be rapidly degraded and not accumulate in the soil or water supply.

One reason critics urge caution over the release of genetically engineered plants is experience with problem plant species such as kudzu. This plant has been extremely difficult to control since its introduction, from the Orient, to the southern United States (*18*). The problems were not caused by changes in the genetic makeup of the plant, however, but rather by its introduction into a new environment. Each plant species evolved over eons to be competitive, which is why it exists naturally in at least one environment. In that natural environment a variety of factors, such as other plants, pests, and weather, keeps the population in check. Most U.S.–grown crops were introduced from other countries (*19*), and the Department of Agriculture maintains large collections of wild members of our cultivated species to improve our crops (*20*). These collections are not normally maintained under strict quarantine.

Microorganisms

From the early years of this century certain microbes were grown in large volumes and, in many cases, became the foundation of new industries. Examples include the production of antibiotics, solvents, vitamins, amino acids, *Rhizobium*, *Azotobacter*, *Bacillus*, and yeast. In a number of industries mutated organisms have been utilized (*21*). These organisms have not caused nor will they create environmental or health problems that are difficult to control. This is not surprising considering microbial behavior in the environment. Every time rotted food is discarded in the woods, streams, or fields a culture of millions or billions of uncharacterized microbes is added to the environment. A rotted tree stump contains billions of lignin-degrading fungal cells, which are readily transported by people, animals, and insects that come in contact with the stump. No one is concerned that such uncharacterized organisms will cause difficult-to-control problems. During this century many cultures of bacteria and fungi (inoculants) have been added to soils or plants in the environment in the hope of finding useful applications, as in oil and chemical waste removal (*22*), plant residue decomposition (*23*), plant pest protection (*24*), and plant growth stimulation (*25*). No substantiated damage of any significance has been caused through such practices with fungi and bacteria not considered dangerous to our health or the environment.

Known pathogens have been used in field studies. For example, microorganisms that are weed pathogens have been used in experiments to control weeds (*26*). Precautions are required in these types of experiments; however, it is extremely unlikely that addition of characterized genes to such pathogens would increase their potential to cause serious problems.

There is no reason to think that a bacterium or fungus that is not known to damage the environment will cause environmental problems after it has obtained several characterized foreign genes. It is also extremely unlikely that a dangerous organism in the soil (for example, *Clostridium tetani*) will become more of a problem after acquiring these new genes from the introduced organism. Certainly, microorganisms intentionally and unintentionally added to the environment have naturally exchanged genes with other microorganisms. Such organisms have moved through wind and water and with man to distant places (*27*). Without man's intervention microbes are continually mutating, sharing, and rearranging genes through such agents as transposons, viruses, and plasmids.

Randomly introduced microorganisms generally are unable to predominate in new habitats because preexisting organisms already have evolved to successfully compete for those niches. In most cases a microbe in nature grows far more slowly than it does in laboratory cultures; thus the newly introduced organism will probably have a difficult time surviving and an even more difficult time significantly increasing and maintaining its population, whether it is genetically engineered or not. The extra burden to the organism carrying new genes should decrease its ability to compete and persist.

What is the chance that a harmless microorganism can become a pathogen after it has been genetically engineered to be agriculturally useful? Studies with pathogens have demonstrated that many specific genes with interacting activities (usually not all genetically linked to each other) are required for a microbe to cause disease, persist outside the host, and be transferred to subsequent hosts (*28*). Most of these studies involved animal pathogens, but it is becoming apparent that the same is true for plant pathogens (*29*). The chance that one could accidentally convert a microbe that normally is nonpathogenic to a problem pathogen through introduction of characterized foreign genes seems very small. That an organism has obtained genes involved in pathogenesis does not necessarily mean that the recipient will become a problem pathogen even if it damages a host in the laboratory. To become a serious problem (as defined earlier), it has to maintain the genes, be able to spread from host to host, and retain the genes during times when no host is available. Appreciation of this should minimize concern over natural dynamic exchange of genes among uncharacterized microbes in the field. There are many plant pathogens that can naturally exchange genes with *Escherichia coli*, but we do not see *E. coli* strains becoming pathogenic to plants.

One can argue that special problems may arise because of the very high concentration of recombinant organisms applied to crops. As mentioned earlier, high concentrations of microbes have been purposely added to crops for decades. In many cases such microbes can naturally exchange genes with pathogens. Certainly these inoculants have come in contact with pathogens, but no problems have been reported.

Examples are known from current practices not involving genetic engineering in which acquisition of a single gene or a mutation in a microbe causes ecological and health problems. Applications of certain herbicides or pesticides to soils enrich the soil for microbes *(30)* that degrade the chemical, resulting in the need to apply more of the chemical in subsequent years. Another example is acquisition of antibiotic resistance genes that have caused major medical problems. These problems arose not by man's ability to genetically manipulate organisms but rather by introducing new chemicals to the environment. The problems can be reversed by eliminating application of such chemicals. In fact, many current genetic engineering experiments are focused on projects expected to decrease the use of some industrially produced chemicals *(31)*.

Need for Field Tests

Experience has shown that it is important to test the degree of toxicity of each newly synthesized chemical before it is used internally or added to large areas of land. Even if a new chemical is only a slightly modified analog of a known safe chemical, the degree of safety cannot be extrapolated from that safe chemical. In fact, analogs of normal metabolites can be most dangerous. By comparison, minor modifications obtained by breeding safe plants or mutating safe microbes do not yield progeny that become serious problems. Minor modifications are expected from genetic engineering; agronomic problems that may arise can be assessed only by field testing. To allay concerns about the safety of a recombinant organism, it would be useful to follow testing protocols before the organism is generally released. However, the task of designing relevant tests for most situations seems to be enormous, if achievable at all *(32)*. How will a greenhouse test show that a corn line resulting from a standard genetic cross will not become a problem weed? If a bacterium increases corn yield in the greenhouse, how will researchers know, without field testing, that it will not harm the following season's crop? Tests aimed toward predicting the level of microbe persistence in a field could be very difficult and not relevant *(33)*. Because different soils, soil treatments, and weather conditions can dramatically alter the growth rate, population, and persistence of a microbe, greenhouse or growth chamber experiments have little relevance to field results. It is easy to imagine that one type of pathogen can damage a plant at a density of ten cells per gram of soil whereas another pathogen may initiate plant disease only at a million cells per gram. Current field testing practices seem to be the best guide to predicting safety.

Certain microorganisms and plants have been introduced in the environment without need for regulation. Such organisms containing recombinant DNA should not be of concern unless the organisms or introduced genes have obvious potential problems (for example, *Clostridium botulinum* and the botulinus toxin gene) that require special precautions. It is unlikely, however, that such experiments would be proposed for field testing. Because of the complex interaction of genes required for an organism to cause a serious disease problem or major environmental disruption, it would be extremely difficult to purposely engineer an organism now considered safe to an organism that would be a significant problem. A program that aims to utilize, in agriculture, a plant, bacterium, or fungus considered to be safe but with several foreign genes will have essentially no chance of accidentally producing an organism that would create an out-of-control problem. The chance of a problem resulting from genetic engineering should be viewed in perspective and compared to known problems caused by currently accepted genetic and chemical practices, such as breeding and the use of chemical pesticides. It may be valuable for one or more laboratories to test, under appropriate containment, several worst-case scenarios, as was done with *E. coli* (*34*). Such a test might utilize an *E. coli* strain genetically engineered to contain pectolytic enzymes required for soft-rot disease of plants (*35*). Regulations governing release of genetically engineered organisms should be based on scientific experience and informed debate of the issues.

To summarize, traditional agricultural practices continually improve useful crops and microbes by taking advantage of new genetic modifications. In almost all cases the exact nature of these modifications is unknown. There has not been any special concern about the new variants. By comparison, genetic engineering will make well-characterized and specific modifications. Thus there does not seem to be any reason to expect greater problems arising from recombinant organisms in agriculture than from organisms produced through traditional practices.

References and Notes

1. F. Moriarty, *Pollutants and Animals* (Allen & Unwin, London, 1975), p. 140.
2. Office of Technology Assessment, *Impacts of Applied Genetics: Micro-organisms, Plants, and Animals* (Government Printing Office, Washington, D.C., 1981); K. A. Barton and W. J. Brill, *Science* **219**, 671 (1983); R. L. Phillips, in *Genetic Engineering of Plants*, T. Kosuge, C. P. Meredith, A. Hollaender, Eds. (Plenum, New York, 1983), pp. 453–465; National Academy of Sciences, *Genetic Engineering of Plants* (National Academy Press, Washington, D.C., 1984).
3. A. L. Hooker, *Crop Sci.*, **17**, 689 (1977); G. F. Sprague, *Annu. Rev. Phytopathol.* **18**, 147 (1980).
4. D. A. Phillips, *Annu. Rev. Plant Physiol.* **31**, 29 (1980); J. M. Lyons, R. C. Valentine, D. A. Phillips, D. W. Rains, R. C. Huffaker, Eds., *Genetic Engineering of Symbiotic Nitrogen Fixation* (Plenum, New York, 1981); R. A. Haugland, M. A. Cantrell, J. S. Beaty, F. J. Hanus, S. A. Russell, H. J. Evans, *J. Bacteriol.* **159**, 1006 (1984).
5. W. J. Brill, *Sci. Am.* **245**, 198 (September 1981).
6. H. G. Wilkes, *Science* **177**, 1071 (1972); H. H. Iltis, *ibid.* **222**, 886 (1983).
7. P. M. Lyrene and H. L. Shands, *Crop Sci.* **15**, 398 (1975).
8. G. Melchers, M. D. Sacristan, S. A. Holder, *Carlsberg Res. Commun.* **43**, 203 (1978); J. B. Power, S. F. Berry, J. V. Chapman, E. C. Cocking, *Theor. Appl. Genet.* **57**, 1 (1980); D. A. Evans, *Biotechnology* **1**, 253 (1983).
9. There are instances of crop plants either mutating to become weedy species or crossing with related wild species. Examples, such as shattercane, are unusual, and the crops (sorghum in the case of shattercane) themselves have many of the properties required for weediness. Sorghum itself is an introduced species. Generation of weeds through agricultural practices can be controlled through chemicals integrated with proper rotation and cultivation techniques.
10. J. R. Harlan, *Crops and Man* (American Society of Agronomy, Madison, Wis., 1975), pp. 85–104.
11. J. B. Lawes, *Agric. Stud. Gaz.* **7**, 65 (1895).
12. D. L. Klingman and J. R. Coulson, *Plant Dis.* **66**, 1205 (1982).
13. J. D. Bandeen, G. R. Stephenson, E. R. Cowett, in *Herbicide Resistance in Plants*, H. M. LeBaron and J. Gressel, Eds. (Wiley-Interscience, New York, 1982), pp. 9–27.
14. F. W. Papp, *Annu. Rev. Entomol.* **21**, 179 (1976).
15. E. Anderson, in *Man's Role in Changing the Face of the Earth*, W. L. Thomas, Ed. (Univ. of Chicago Press, Chicago, 1956), pp. 763–777.

16. L. A. Tatum, *Science* **171**, 1113 (1971); R. M. Atlas, *Crit. Rev. Microbiol.* **5**, 371 (1977).
17. S. L. Sinden, L. L. Sanford, R. E. Webb, *Am. Potato J.* **61**, 141 (1984).
18. J. H. Miller and B. Edwards, *South. J. Appl. For.* **7**, 165 (1983).
19. W. R. Courtenay, in *Wildlife and America*, H. D. Brokaw, Ed. (Council on Environmental Quality, Washington, D.C., 1978), pp. 237–252.
20. G. F. Sprague, *Annu. Rev. Phytopathol.* **18**, 147 (1980).
21. S. Abe, in *The Microbial Production of Amino Acids*, Y. Yamada, S. Kinoshita, T. Tsunoda, K. Aida, Eds. (Wiley, New York, 1972), pp. 39–66.
22. *Biotechnol. Newswatch* **4**, 7 (1984).
23. R. P. Wolkowski and K. A. Kelling, *Agron. J.* **76**, 189 (1984).
24. J. M. Cullen, P. F. Kable, M. Catt, *Nature (London)* **244**, 462 (1973); J. T. Daniel, G. E. Templeton, R. J. Smith, Jr., W. T. Fox, *Weed Sci.* **21**, 303 (1973); K. E. Conway, *Phytopathology* **66**, 914 (1976); L. W. Moore and G. Warren, *Annu. Rev. Phytopathol.* **17**, 163 (1979); A. Klausner, *Bio/Technology* **2**, 408 (1984).
25. M. E. Brown, S. K. Burlingham, R. M. Jackson, *Plant Soil* **20**, 194 (1964); P. R. Merriman, R. D. Price, J. F. Kollmorgen, T. Piggott, E. H. Ridge, *Aust. J. Agric. Res.* **25**, 219 (1974); J. W. Kloepper and M. N. Schroth, *Phytopathology* **71**, 590 (1981); Y. Kapulnik, S. Sarig, I. Nur, Y. Okon, K. Kigel, Y. Henis, *Exp. Agric.* **17**, 179 (1981); M. Datta, S. Banik, R. K. Gupta, *Plant Soil* **69**, 365 (1982); R. H. Howeler, L. F. Cadavid, E. Burkhardt, *ibid.* **69**, 327 (1982); S. E. Lindow, D. C. Arny, C. D. Upper, *Phytopathology* **73**, 1097(1983).
26. S. Hason, *PANS* (*Pest Artic. News Summ.*) **20**, 437 (1974).
27. J. C. Zadoks, *Neth. J. Plant Pathol.* **73**, 61 (1967); J. D. Anderson and C. S. Cox, in *Airborne Microbes*, P. H. Gregory and J. L. Monteith, Eds. (Cambridge Univ. Press, Cambridge, 1967), pp. 203–226.
28. P. J. Sansonetti, T. L. Hale, G. J. Dammin, C. Kapfer, H. H. Collins, Jr., S. B. Formal, *Infect. Immun.* **39**, 1392 (1983); N. Bauman, *Bio/Technology* **2**, 749 (1984).
29. E. W. Nester and T. Kosuge, *Annu. Rev. Microbiol.* **35**, 531 (1981).
30. T. Obrigawitch, A. R. Martin, F. W. Roeth, *Weed Sci.* **31**, 187 (1983).
31. *What Is Biotechnology?* (Industrial Biotechnology Association, Rockville, Md., 1984).
32. D. Simberloff, in *Biotic Crises in Ecological and Evolutionary Time*, M. H. Nitechki, Ed. (Academic Press, New York, 1981), pp. 53–81.
33. M. H. Dughri and P. J. Bottomley, *Appl. Env. Microbiol.* **46**, 1207 (1983); W. R. Ellis, G. E. Ham, E. L. Schmidt, *Agron. J.* **76**, 573 (1984).
34. M. A. Israel, H. W. Chan, W. P. Rowe, M. A. Martin, *Science* **203**, 883 (1979).
35. N. T. Keen, D. Dahlbeck, B. Staskawicz, W. Belser, *J. Bacteriol.* **159**, 825 (1984).

Part VIII

Behavior and Sensory Phenomena

24

The Cellular Basis of Hearing: The Biophysics of Hair Cells

A.J. Hudspeth

With over a million essential moving parts, the auditory receptor organ, or cochlea, is the most complex mechanical apparatus in the human body. There has been for decades a good understanding of how sounds are conducted to the inner ear, and there are excellent descriptions of the signals that move from the cochlea to the brain along the eighth cranial nerve. The daunting complexity of the cochlea, however, has delayed our understanding of how this pea-sized organ produces electrical signals in response to acoustical stimulation. Within the last decade, the availability of tools for the study of the ear's receptor cells in vitro has made possible a detailed biophysical description of how hearing occurs at a cellular level. At the same time, experimental data from intact animals have revealed a new level of complexity in the cochlea's operation.

The hair cell is the receptor cell of the auditory, vestibular, and related sensory systems. It performs an essential duty of these systems, transduction, the rendering of sensory inputs into electrical signals. The hair cell carries out this task with subtlety, making use of a variety of mechanical, hydrodynamic, and electrical strategies to measure signals with great sensitivity and remarkable frequency discrimination.

Hair Cells

Hair cells are epithelial cells that originate not from the neural tube that forms the brain but from the surface ectoderm of the vertebrate embryo. The cells maintain their epithelial character in the mature organs of the ear (Fig. 1); they

Science 230, 745–752 (15 November 1985)

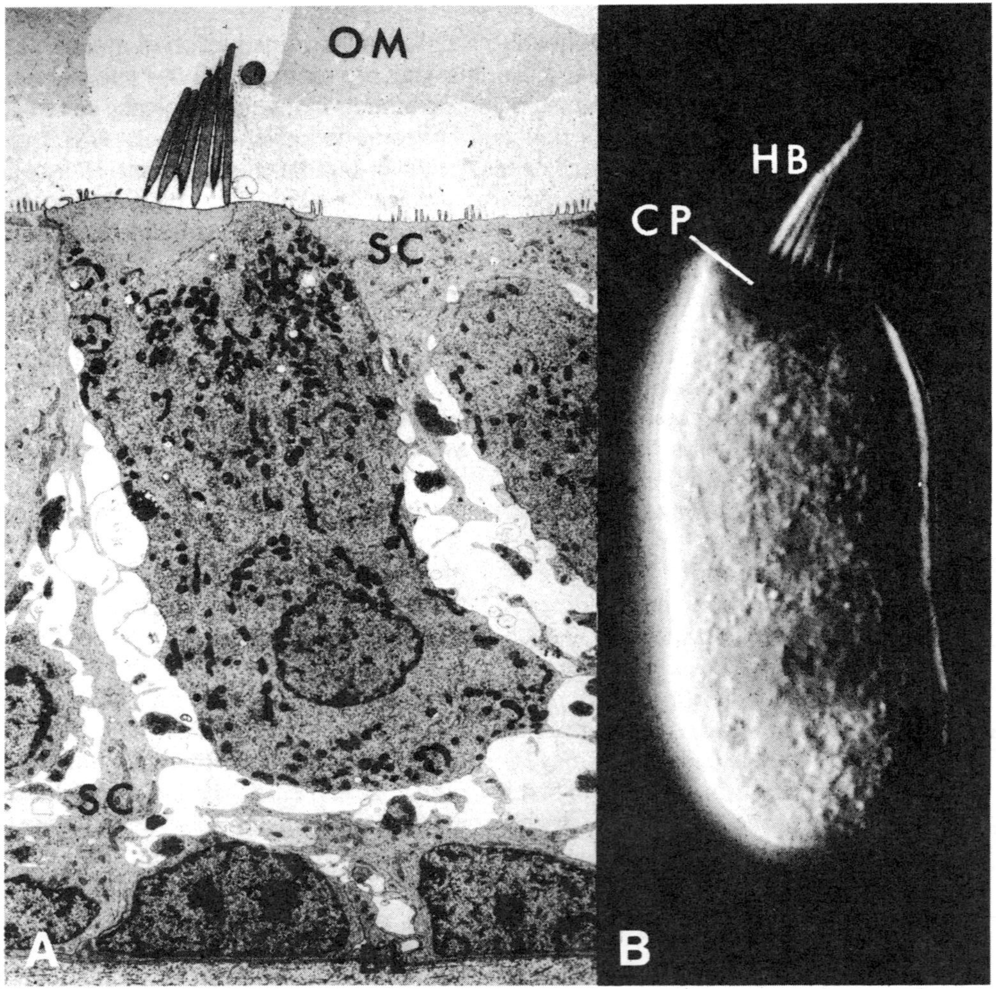

Fig. 1. (A) A transmission electron micrograph of a typical hair cell in its normal environment. The hair cells and adjacent supporting cells (SC) from the bullfrog's sacculus form an epithelial sheet resting upon the basal lamina (BL). The hair bundle extends from the apical surface of the hair cell to contact the otolithic membrane (OM), an accessory structure that conveys mechanical stimuli to the bundle. (B) A light micrograph of a living hair cell enzymatically isolated from the sacculus. The cell body is cylindrical; afferent and efferent synaptic contacts occur basal to the nucleus (N). The hair bundle (HB) protrudes from the fibrous cuticular plate (CP) at the cell's apex (magnification ×2300; corrected for shrinkage during histological processing).

are constituents of a sheet of cells, joined to one another by tight and intermediate junctions and resting upon a basal lamina. Although they lack axons and dendrites, the characteristic processes of neurons, hair cells make synapses onto afferent nerve fibers of the eighth cranial nerve and also receive efferent synaptic contacts from axons originating in the brainstem.

The unique structural feature of the hair cell is the hair bundle, an elegant assemblage of microscopic processes protruding from the cell's top or apical surface (Figs. 1 and 2). Each of these processes, which are termed stereocilia, consists of a straight rod of fasciculated actin filaments surrounded by a tube of membrane. Because the microfilaments are extensively cross-bridged (*1*), each stereocilium behaves as a rigid rod. When mechanically disturbed, it remains relatively straight along its length but pivots about a flexible basal insertion (*2*).

Hair bundles occur in a variety of sizes and shapes (*3*); the functional significance of this diversity of form will be considered below. They range in length from 0.8 μm in the bat's cochlea to about 50 μm in certain vestibular organs. Their constituent processes, termed stereocilia, number from 30 in the cochleas of some lizards to more than 300 in the chick's ear; stereociliary diameters range from 0.1 μm in the cochleas of mammals to 1 μm in those of lizards. Certain features of the hair bundle are universal, however (*4*); stereocilia uniformly lie in a hexagonal array and always increase in length from one edge of the hair bundle to that opposite.

Despite the variety of stimuli to which hair cells respond in various organs, including sound in the cochlea, linear acceleration and ground-borne vibration in the utriculus and sacculus, angular acceleration in the semicircular canals, and water motion in lateral-line organs, every hair cell is evidently sensitive to the same proximal stimulus. The hair cell is a mechanoreceptor; it produces an electrical signal, or receptor potential, in response to mechanical stimulation of its hair bundle.

Hair bundles are deflected in two distinct manners. In many organs, the bundles protrude unencumbered into the surrounding fluid. When the fluid moves in response to sound, the bundles are bent by the force of viscous drag, thereby initiating a response. Under these circumstances, the physical properties of the hair bundles are of paramount importance in determining which stimuli are effective. The freestanding hair bundles of lizards (Fig. 2D) exemplify this structural pattern; the inner hair cells of the mammalian cochlea (Fig. 2C) are probably stimulated in a similiar manner. In other organs, each hair bundle is ligated at or near the distal tip of its tall edge to an extracellular accessory structure. Stimuli are then initially transmitted to the accessory structure—a tectorial membrane, otolithic membrane, or cupula—that in turn deflects the attached hair bundles. Although the hair bundles may exert a substantial influence on the activities of such organs, the properties of the accessory structures undoubtedly play important roles in determining the organs' sensitivities. Hair bundles of the sacculus (Figs. 1A and 2, A and B) and the outer hair cells of the mammalian cochlea typify this arrangement.

Because naturally occurring sounds are generally not pure tones, auditory receptor organs not only must detect acoustical stimuli but also must analyze them into constituent frequencies as a

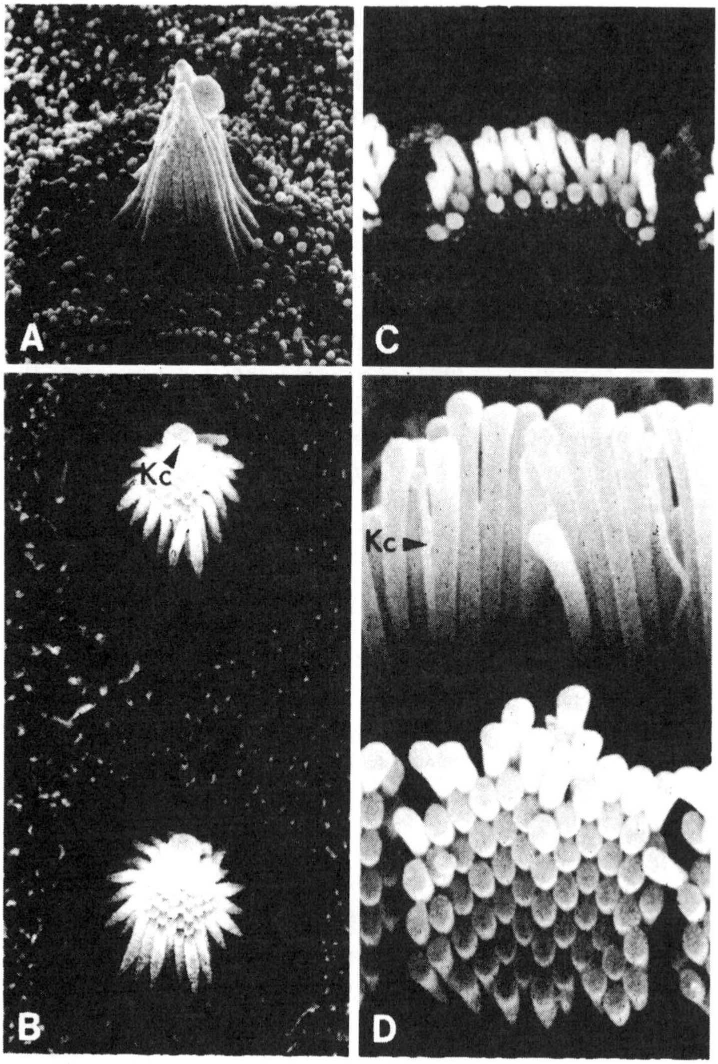

Fig. 2. A variety of hair bundles as demonstrated by scanning electron microscopy. (A) The general structural features of the hair bundle in a lateral view of a bundle from the bullfrog's sacculus. The bundle comprises about 60 hypertrophied microvilli, called stereocilia, whose lengths increase progressively from the bundle's left to its right edge. A single true cilium, the kinocilium, occurs at the tall edge of the bundle; its distal tip, which is normally attached to the otolithic membrane that provides mechanical input to the bundle, has a bulbous swelling. (B) A top view of two hair cells from the same organ, revealing the regular packing of the stereocilia and the eccentric location of the kinocilium (Kc). (C) The bundle of an inner hair cell from the cat. Here the stereocilia form a palisade rather than a compact cluster. (D) Top view of a hair bundle from the fence lizard (*Sceloporus jarrovi*) with very large stereocilia in a compact hexagonal array. Two other bundles, seen *en face* in the background, display kinocilia (Kc) at their long edges. The four micrographs are shown at the same magnification (×4300) to illustrate the morphological diversity of hair bundles from various organs. The largest stereocilia of the lizard's cells are over 30 μm long and 0.9 μm in diameter, whereas the cat's hair bundle is just 2 μm tall and its smallest stereocilia measure only 0.1 μm across.

first step in the stimulus-recognition process. Accordingly, hair cells in many organs are tuned; that is, they are selective for particular frequencies of mechanical stimulation. The restricted responsiveness of hair cells to specific frequencies also serves a second purpose; by limiting its sensitivity to a narrow band of frequencies, a cell can reject noise components outside this range. Frequency tuning is therefore not only a means of analyzing complex signals but also a strategy for noise reduction.

Before considering how hair cells translate mechanical inputs into electrical outputs, I shall describe how stimuli reach the cells and how hair bundles influence cellular responses. I shall consider stimulation and frequency tuning in organs representative of both the freestanding and the assisted forms of stimulation, the primitive cochlea of the lizard and the complex mammalian cochlea.

Mechanical Tuning at the Cellular Level

The observation that various organs of the inner ear possess a wide variety of hair bundles, but that each bundle in a given part of an organ tends to be stereotyped in form, suggests that variations in hair-bundle morphology are of some functional significance. The role of hair bundles in determining a cell's responsiveness is most clearly illustrated by hair cells with freestanding bundles that do not contact any accessory structure. The activity of such hair cells must be dominated by the physical properties of the hair bundles: their dimensions, especially their lengths, their masses, and their hydrodynamic drag in the surrounding fluid.

Within the cochleas of many lizards, hair bundles vary severalfold in length in a systematic manner. In the alligator lizard, for example, the bundles are shortest at the basal end of the organ; they progressively grow longer toward the organ's center until they are nearly threefold as long as at the base (*5*). Electrophysiological investigations indicate that the frequencies to which hair cells are most sensitive, their characteristic frequencies, vary inversely with the lengths of the hair bundles; cells with short bundles are tuned to frequencies near 4 kHz, whereas cells with the longest bundles are tuned to frequencies near 1 kHz (*6*).

When the lizard's cochlea is stimulated at various frequencies and observed microscopically under stroboscopic illumination, it is apparent that hair bundles of different heights do not move equivalently (*7*). Low-frequency stimuli extensively deflect the long, massive bundles at the organ's center, while the short, comparatively stiff bundles remain relatively undisturbed (Fig. 3, B and C). High-frequency stimuli, conversely, bend the short hair bundles to the exclusion of long ones. Although the hydrodynamic behavior of a hair bundle is quite complex, mathematical models suggest that the observed tuning may be accounted for by the bundle's physical properties (*8*).

Mechanical Tuning in the Mammalian Cochlea

While the structure of the mammal's cochlea is extraordinarily complex, this organ operates in a manner that is fundamentally simple. Sound, which consists of a pattern of pressure changes at the eardrum, is mechanically conducted through the chain of bones within the middle ear. The last of these three bones, the stapes, is mounted like a piston in contact with fluid within the cochlea. As the stapes moves back and forth in response to stimulation, it transmits pressure changes into the cochlear fluids.

Although it superficially resembles a snail and derives its name from the Greek word for that mollusk, the cochlea actually consists of three fluid-filled chambers. These tubes are separated

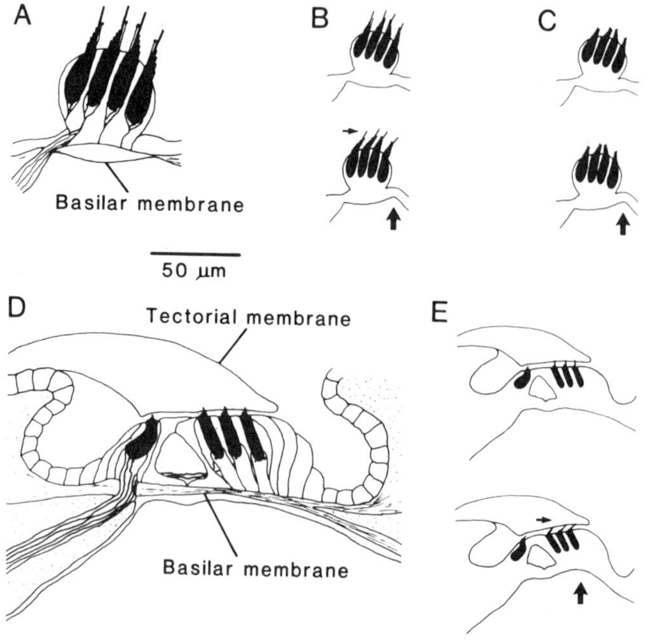

Fig. 3. Structures and operations of representative cochleas. Hair cells are shown solid black. (A) A schematic cross-section through the cochlea of the alligator lizard (*Gerrhonotus multicarinatus*), which contains the organ's longest freestanding hair bundles. (B) When the lizard's ear is stimulated by low-frequency sound, differences in fluid pressure across the basilar membrane cause it to move vertically. As the hair cells upon the basilar membrane sway laterally, the inertia and hydrodynamic resistance of the hair bundles cause them to be bent; this deflection in turn excites the hair cells. (C) Because of their lesser inertia and drag, the short hair bundles at the organ's opposite end do not respond to low-frequency stimulation. (D) The organ of Corti of the cat, shown schematically. The organ possesses four rows of hair cells whose hair bundles are surmounted by a gelatinous tectorial membrane. (E) When sound pressure deviates the basilar membrane, the pivoting of the organ of Corti causes a shear between it and the tectorial membrane. The hair bundles of the three rows of outer hair cells are inserted into the tectorial membrane and are deflected directly by this shear. The bundles of the inner hair cells are probably not attached to the tectorial membrane, but it is thought that they are bent by fluid moving between the tectorial membrane and the underlying hair cells.

from one another by two elastic partitions and are helically coiled, one atop another, about a common axis. When the stapes compresses the fluid within one chamber, the basilar membrane, one of the partitions between cochlear chambers, is deflected. Even when stimulated with as simple a sound as a pure tone, the basilar membrane moves in a complex manner. Because the dimensions and mechanical properties of the membrane vary from its base to its apex, it does not act like a homogeneous string on a plucked musical instrument. Instead, the basilar membrane develops a traveling wave in a region along its length that depends upon the stimulus frequency: low frequencies, down to 20 Hz in humans, excite motions near the apex of the cochlea, whereas high frequencies, up to 20 kHz in humans, deflect the basal parts of the partition.

On the human basilar membrane rests the organ of Corti, a 34-mm-long helical structure containing, among other cell types, some 16,000 hair cells in four parallel rows (Fig. 3D). Each hair cell has about 100 stereocilia, so there are over a million of these receptive organelles in an ear. Because each frequency moves a specific zone of the basilar membrane with the greatest amplitude, it follows that any given tone influences a particular group of hair cells most strongly. One of the cochlea's principal virtues ensues from this arrangement:

the basilar membrane functions as a spectral analyzer, decomposing a complex sound—the human voice, for example—into its pure tonal constituents. It is left for the hair cells that receive information about a particular tonal input somehow to transduce the mechanical motions of the basilar membrane into electrical signals that are suitable for analysis by the nervous system.

When driven up and down by acoustical stimuli, the basilar membrane carries with it hair cells of the organ of Corti. The distal ends of the hair bundles of the outer hair cells are attached to the tectorial membrane, a gelatinous shelf of protein that spirals along the entire extent of the organ of Corti. As each hair cell moves, a shear develops between its top surface and the lower surface of the tectorial membrane; this shear bends the hair bundles that extend across the intervening space (Fig. 3E).

Having seen what stimuli reach representative hair cells, we may formulate a number of specific questions about the operation of this cell type. How does the hair cell convert a mechanical input into an electrical output? How can the ear reliably measure mechanical stimuli whose average amplitudes at threshold approximate the diameter of a hydrogen atom? How can one explain a sensory system that can measure events that occur 20 times more frequently than nerve fibers can signal? How does a hair cell contribute to the frequency selectivity of the internal ear?

Mechanoelectrical Transduction

Although it has been possible to learn a considerable amount about the operation of hair cells in the mammalian cochlea, a more nearly complete picture of the transduction process has emerged from in vitro investigations of hair cells from lower vertebrates. One valuable experimental preparation is that of the bullfrog's sacculus, an organ containing large hair cells of typical structure and great hardiness during physiological recording. I shall discuss below some of the details about transduction that have emerged from study of the sacculus and then assimilate them in a general model for transduction by vertebrate hair cells.

When a saccular hair cell is stimulated by deflecting the tip of its hair bundle with a glass probe, the membrane potential deviates from its resting value of approximately −60 mV. Stimuli directed toward the tall edge of the hair bundle—the edge at which there also occurs a single true cilium, the kinocilium—produce depolarizations of up to 20 mV. Motion in the opposite direction elicits hyperpolarizations about a quarter as large (*9*).

If the bundle is deflected at right angles to its plane of morphological symmetry, the membrane potential does not change at all (*10*). The hair cell's responsiveness is highly directional. It can decompose an arbitrarily directed stimulus to its hair bundle into two components and respond in a graded fashion to the component along its axis of symmetry while disregarding the perpendicular component altogether.

As is the case in neurons, the electrical signals in the hair cell originate from the flow of ionic currents across the membrane. This may be demonstrated by voltage-clamping a hair cell so that its transmembrane current is measured directly (Fig. 4, A and B). When the hair bundle is moved in the positive direction (toward its taller edge) the membrane conductance increases; that is, the membrane becomes more permeable to posi-

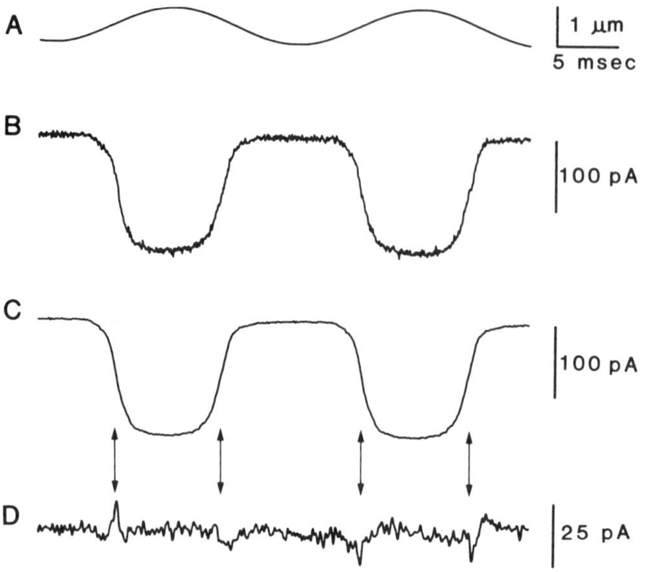

Fig. 4. Transduction current and the associated noise. When the hair bundle's distal tip is moved back and forth sinusoidally (A), a voltage-clamp amplifier measures the flow of transduction current [in picoamperes (pA)] into the cell (B). Positive stimuli [upward deflections in (A)] open transduction channels, allowing an inward flow of positive ions [downward deflections in (B)]. The trace in (B) is noisy; the level of noise, which is in excess of that due to instrumentation, reflects the random gating of transduction channels. When the responses to 136 repetitions of the same stimulus are averaged (C), the transduction-associated noise largely vanishes. Subtraction of the mean record in (C) from an individual trace such as that in (B) demonstrates, at a higher gain, the noise due to the random opening and closing of transduction channels (D). The noise is minimal when most channels are closed and the mean current is most positive and also when most channels are open and the mean current is most negative. The greatest noise occurs when half the channels are open (arrows). Analysis of the data from this cell indicates the presence of 160 transduction channels, each with a conductance of 12 pS at 10°C.

tively charged ions (*9*). This behavior suggests the existence of transduction channels, transmembrane pores through which ions can pass when the hair bundle is appropriately stimulated.

It is possible to use electrophysiological techniques to estimate how many transduction channels occur in each hair cell. The basis of this measurement is shown in Fig. 4. When a cell's hair bundle is moved back and forth sinusoidally with a stimulating probe, transduction current flows into the cell in a cyclical pattern. Because each of the transduction channels in the membrane operates independently of the others, however, each cycle of the response is not exactly the same as any other. Each time the stimulus is repeated, a slightly different pattern of activation occurs, resulting in fluctuations (noise) in the response. This noise may be quantified and analyzed on the basis of particular models for transduction. If each channel has the same ionic conductance when open and zero conductance when closed, the variance, or square of the measured noise, is related parabolically to the mean transduction current (*11*). Measurement of both the average transduction current and the noise from trace to trace leads to estimates for the number and conductance of the individual channels (*12*).

In the frog's sacculus, there are only about 280 transduction channels per hair cell, fewer than half a dozen for each stereocilium (*13*). This makes the isolation and biochemical characterization of transduction channels a rather bleak prospect, for there seem to be a millionfold fewer transduction molecules per hair cell than there are, for example, rhodopsin molecules in each rod photoreceptor.

Despite the paucity of transduction channels, it is possible to obtain some information about their chemical nature by examining their ability to "catalyze" the movement of ions across the membrane. Voltage-clamp measurements after substitution of various ions in the fluid bathing the hair bundle indicate which ions can traverse the transduction channel. The cation normally present in the highest concentration in the fluid, K^+, readily passes throught the channel; so do the other alkali ions: Li^+, Na^+, Rb^+, and Cs^+. Divalent cations, such as Ca^{2+}, Sr^{2+}, and Ba^{2+}, are still more permeant (*14*, *15*). Even small organic cations, such as choline, tetramethylammonium, and tris(hydroxymethyl)aminomethane, can carry transduction current. The fact that such ionic species can traverse the transduction channel, while slightly larger ones cannot, suggests that the bore of the channel is about 0.7 nm (*14*). The channel's preference for cations over anions indicates that the amino-acid residues exposed on its inner surface bear negative charges. Finally, the relatively poor discrimination among cations implies that there is not a close fit between the ions and the groups lining the channel. The ions are probably moving in hydrated form through a relatively large pore. Variance measurements indicate that the hair cell's transduction channel has a conductance of 17 picosiemens (pS) at room temperature (*13*), a value in line with the pore's meager ionic selectivity and hydrated nature.

Because the chemical identity of the transduction channel is not yet known, there is at present no way of localizing the channel by biochemical means, for example through the use of antibodies. The flow of ionic current into transduction channels, however, provides a means of inferring with moderate spatial resolution where they occur. As ionic current streams toward the transduction channels, its flow across the resistance afforded by the saline solution around the hair bundle produces minute potential changes. Although these variations in voltage are less than 15 μV in amplitude, they can be detected and their spatial distribution mapped with suitable electrodes (*16*). Rather surprisingly, the strongest electrical signal, that indicative of the site of transduction, does not occur at the bases of stereocilia, where they bend during mechanical stimulation. Instead, the maximum response is found at the top of the hair bundle, a result suggesting that the transduction channels occur at or near the distal tips of the stereocilia.

A striking feature of the transduction process of vertebrate hair cells is its rapidity. Humans can detect stimuli at frequencies as great as 20 kHz, and bats and whales can hear sounds of five- to tenfold higher frequency. Channel opening in hair cells is so rapid that it is difficult to resolve at room temperature. At an experimental temperature of 28°C, channels begin to carry current within a few microseconds of the application of a stimulus (*17*). The response, moreover, becomes faster as the amplitude of stimulation increases; saturating stimuli evoke responses that peak within 100 μsec. The dependence of response kinetics on stimulus amplitude is intriguing, for such behavior suggests that mechanical stimuli somehow affect the rate constants for the reactions of channel opening and closing.

A Model for Mechanoelectrical Transduction

How are the hair cell's transduction channels gated; that is, how are their opening and closing controlled by mechanical stimuli? The localization of transduction to the hair bundle's top, together with transduction's being too rapid to permit the intervention of second messengers, suggests that some interaction among the stereocilia initiates the response. The kinetic evidence led to the suggestion that displacement of the stereocilia exerts tensile force upon a linkage connected to each transduction channel (*17*, *18*). Since that model was put forward, anatomical experiments have supported it by demonstrating a morphological candidate for the hypothetical linkage. When hair cells are appropriately prepared for electron microscopy, the tip of each stereocilium displays a fine filament linking it to the flank of the adjacent longer stereocilium along the hair bundle's axis of symmetry (*19*). This filament is admirably positioned to report shear between adjacent stereocilia. Its alignment along the cell's axis of symmetry also explains the directional sensitivity of transduction in hair cells. Although the kinetic model does not depend on a particular site for the transduction linkage, it is tempting to believe that the linkage has now been observed.

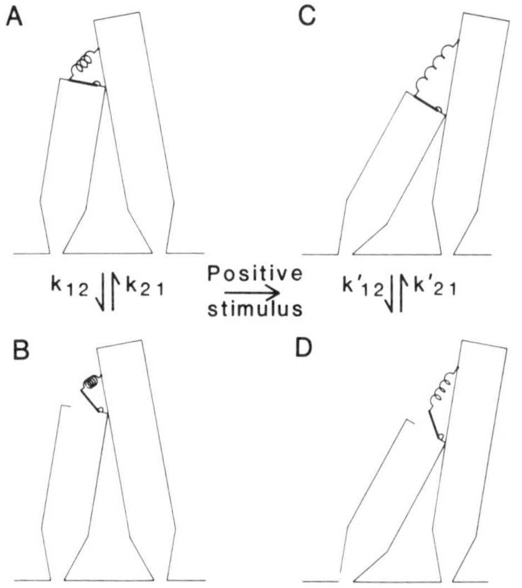

Fig. 5. A model for mechanoelectrical transduction by hair cells. At any instant, each transduction channel at a stereocilium's tip may be either closed (A) or open (B). The relative values of the rate constants for channel opening and closing, k_{12} and k_{21}, respectively, determine the fraction of the transduction channels open in the undisturbed steady state. When the hair bundle is deflected with a positive stimulus (C and D), the values of the rate constants are altered; the opening rate constant (k'_{12}) is larger and the closing rate constant (k'_{21}) smaller than the original values. The new steady-state transduction current is therefore greater, and the cell is depolarized. Pushing the hair bundle in the opposite direction has a contrary effect on the rate constants, culminating in a hyperpolarizing response.

The essential features of the model are shown in Fig. 5. At or near the tip of each stereocilium lie one to a few transduction channels, each equipped with a gate that regulates the flow of ionic current into the cell. When the hair bundle is in its resting position, each gate swings back and forth between its closed and open configurations (Fig. 5, A and B, respectively) under the influence of a constant thermal buffeting by surrounding molecules. Depending on the strength of its intrinsic "door-closing" spring, each channel will, on average, spend a certain fraction of its time open in the absence of stimulation; experiments indicate that this fraction is about 20 percent.

When a stimulus pushes the tip of the hair bundle in the positive direction, the stereocilia in successive ranks slide along one another. The transduction linkage connecting each channel to the adjacent stereocilium is therefore presumably stretched. A given channel continues to fluctuate between its closed and open states (Fig. 5, C and D, respectively). Now, however, the additional force exerted upon the channel's gate by the elongated transduction linkage causes the channel to spend more of its time open. As a result, a larger average current flows through each channel, and the cell is depolarized.

Although somewhat abstract, the model includes the requisite properties for a description of transduction by hair cells. Because the channels rattle between the open and closed states with little in the way of an intervening energy barrier, they are capable of responding to an imposed stimulus with the great rapidity characteristic of auditory transduction. The kinetic data indicate that the interstate barrier is 0.8 kcal/mol in the bullfrog (*17*); in the absence of stimulation, a channel opens and closes over a thousand times a second. Because the response is probabilistic, it is capable, when averaged over a sufficiently long period, of providing information about stimuli of the atomic dimensions associated with the auditory threshold. The model is in good quantitative agreement with some experimental measurements (*20*). The steady-state probability of a channel's being open, for example, nicely fits the predicted Boltzmann relationship (*13*). Although detailed experiments remain to be done on single cells, the model also qualitatively predicts the hair cell's response kinetics (*17*).

Absolute Sensitivity of the Transduction Process

The primary function of any sensory receptor cell is the detection of some particular form of stimulus energy. It is of considerable importance to the organism that a receptor be capable of reliably detecting very low levels of stimulation. It is desirable, in fact, for a receptor's sensitivity to be limited only by physical constraints on the stimulus rather than by inefficiency or noise in the receptor itself. Rod photoreceptors, for example, are able to detect, with a modest error rate, the capture of individual photons. These detectors could not be much more efficient. Their absolute sensitivity is limited by the quantum nature of light. How well does a hair cell perform in this regard? In the absence of a stimulus of quantum nature, what limits the sensitivity of this receptor?

The rod photoreceptor provides a revealing comparison with the hair cell in regard to absolute sensitivity. The initial step in phototransduction, the light-induced isomerization of retinal, proceeds through an activated intermediate whose free energy exceeds that of the starting material by about 30 kcal/mol. A photon imparts a free-energy change of about 57 kcal/mol upon absorption, an energy so much greater than that of the activated intermediate that isomerization is highly efficient. Of equal importance, however, is the relationship between isomerization energy and thermal energy. At room temperature, the average thermal energy imparted to each degree of freedom in a rhodopsin molecule is only 0.3 kcal/mol. This is so much less than the energy of the activated complex that spontaneous isomerizations are very rare; in agreement with calculations, they occur every millenium per molecule. For a photoreceptor, even one with over a billion rhodopsin molecules, spontaneous signals are therefore rather infrequent, occurring only once a minute (*21*).

The hair cell is less fortunate, for the stimulus energies with which this receptor cell must reckon are far smaller. When one estimates the energy imparted to the auditory or the vestibular system at the human behavioral threshold, in fact, it appears that the energy supplied to each hair cell is of the same order of magnitude as that of thermal motion (*22*). Although the point has not been demonstrated experimentally at the cellular level, it appears that the sensitivity limit in hair cells is set by Brownian motion—we may be able to hear everything "louder" than the molecular motions within the ear (*23*).

As a result of the nature of acoustical stimuli, the hair cell is probably able to do somewhat better than one might think possible at detecting faint signals. Most natural sounds have durations that are many times as long as the periods of their constituent sinusoidal frequencies. Because a hair cell attending to a particular frequency generally receives many cycles of the stimulus, it can in effect average its input. Random noise, in this case due to the transduction of Brownian motion, has a diminishing effect as a hair cell averages its input over an increasing period of time. It was suggested recently that the limit on auditory sensitivity is actually set by quantum indeterminancy (*24*); it remains to be seen whether this provocative notion can be substantiated.

The photoreceptor, then, has evolved for the reliable detection of a single, brief, high-energy input in a low-noise environment. The hair cell, by contrast, specializes in the measurement of prolonged input of low energy compared with the background noise.

Electrical Tuning

The traveling-wave phenomenon doubtlessly constitutes the principal means by which the mammalian cochlea discriminates among stimulus frequencies, whereas the mechanical properties of free-standing hair bundles dominate the responsiveness of some simpler cochleas. These mechanical strategies, however, are not the only means by which tuning may be accomplished. In the internal ears of some lower vertebrates, substantial tuning occurs after mechanoelectrical transduction has taken place, through the action of an electrical resonance in the hair cell's membrane.

When a hair cell of the turtle's cochlea or the frog's sacculus is stimulated acoustically or mechanically, it responds best at some particular frequency, its characteristic frequency (*25, 26*). If the cell is instead stimulated by injection of a constant-current pulse across its membrane, the membrane potential does not settle exponentially, as is the case with most neurons and other cells, but instead manifests a damped, sinusoidal oscillation at the cell's characteristic frequency (Fig. 6A).

How does this membrane resonance arise? My research group has investigated the problem in hair cells of the bullfrog's sacculus, an organ dedicated to

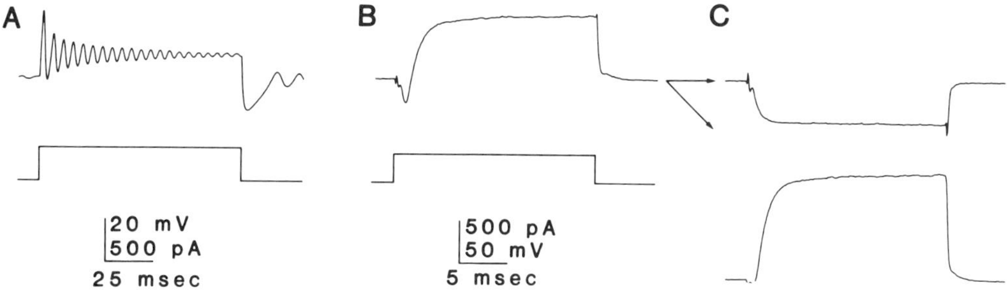

Fig. 6. Electrical resonance and its ionic basis in the bullfrog's saccular hair cell. When a single cell is stimulated by the intracellular injection of current (A, lower trace), the membrane potential displays a damped, sinusoidal resonance at a frequency of 203 Hz (A, upper trace). This is the frequency at which the cell would be most sensitive upon mechanical stimulation of its hair bundle. When the same cell is depolarized from −85 mV to −40 mV under voltage-clamp conditions (B, lower trace), the ionic current that flows across its membrane is biphasic (B, upper trace); there is an early, inward component (downward deflection) followed, after a delay, by a larger, outward component (upward deflection). If the same cell is exposed to 10 m*M* tetraethylammonium ion, the late component is blocked, unmasking a voltage-sensitive Ca^{2+} current that persists throughout the depolarizing stimulus (C, upper trace). Subtraction of the Ca^{2+} current from the total current isolates the Ca^{2+}-sensitive K^+ current (C, lower trace).

the detection of ground-borne vibration at frequencies below 300 Hz (*27*). Hair cells from this preparation may be enzymatically isolated so that whole-cell recordings may be made from them with gigaohm-seal microelectrodes (*26*). When a voltage-clamped cell is depolarized, the resulting transmembrane current is biphasic (Fig. 6B). The initial component represents the flow of positive current into the cell, and the larger, later component indicates the delayed flow of positive charge out of the cell. Substitution of various ions in the experimental solution demonstrates that the early, inward current is carried by Ca^{2+} (Fig. 6C). The second, delayed component of the response is carried by K^+. Unlike the Ca^{2+} channels, or for that matter the K^+ channels associated with the action potential in nerve fibers, the K^+ channels involved in resonant tuning are relatively insensitive to membrane potential. The gating of the K^+ channels is instead primarily influenced by the intracellular concentration of Ca^{2+}. Calcium-sensitive K^+ channels have been demonstrated in a variety of neuronal cell types, where they characteristically serve to modulate electrical excitability (*28*). In the hair cell, these channels have been pressed into service in a more rapid and more complex process.

The way electrical resonance occurs can be readily visualized in qualitative terms (Fig. 7). Stimulation of the hair bundle depolarizes the cell directly, by opening transduction channels, and through the regenerative activation of Ca^{2+} channels. The accumulation of Ca^{2+} intracellularly, however, opens Ca^{2+}-sensitive K^+ channels that act to repolarize the membrane. The resonance arises from the interplay of inward Ca^{2+} currents and outward K^+ currents. The process somewhat resembles the production of action potentials by the se-

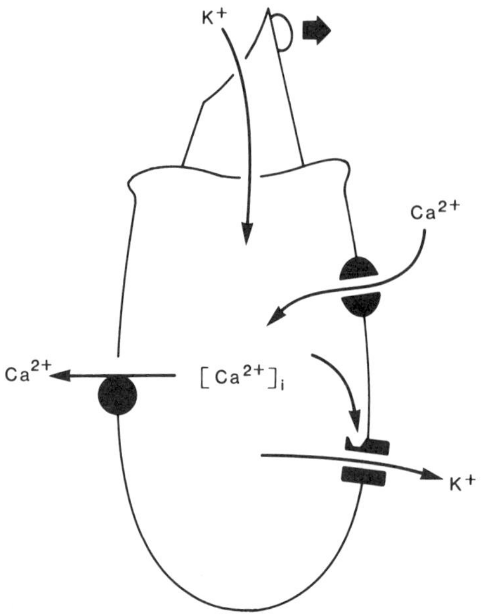

Fig. 7. A model for electrical resonance in hair cells. When the hair bundle is deflected, transduction channels open and positive ions, largely K^+ in vivo, enter the cell. The depolarization evoked by this transduction current activates voltage-sensitive Ca^{2+} channels. As Ca^{2+} ions flow into the cell, they augment the depolarization. At the same time, however, the influx of Ca^{2+} raises the intracellular concentration of this ion, $[Ca^{2+}]_i$, especially the local concentration just beneath the surface membrane. The high Ca^{2+} concentration brings into play the Ca^{2+}-sensitive K^+ channels. As K^+ exits through these pores, it begins to repolarize the membrane, thereby diminishing the activation of Ca^{2+} channels. The fluid bathing the apical surface of a hair cell characteristically has a much higher K^+ concentration than that contacting the basolateral cellular surface. As a consequence, K^+ can both enter and leave the cell passively. By the time the membrane potential is somewhat more negative than its steady-state value, the intracellular Ca^{2+} concentration is reduced by the sequestering of the ion within organelles and by its extrusion through ion pumps. As the Ca^{2+}-sensitive K^+ channels close, the cell returns to approximately its initial condition, and another cycle of the electrical resonance commences.

quential activation of Na^+ and K^+ channels, with the striking difference that an action potential is of a stereotyped amplitude, whereas the resonance in hair cells is graded with the intensity of the signal that initiates it.

At least in hair cells of the frog, our understanding of the electrical resonance is rather complete. The various ionic currents may be modeled on the basis of their behavior during voltage-clamp experiments on whole cells or on excised patches of membrane. Solutions of the resultant equations (*29*) closely resemble actual experimental records (*30*). A tantalizing problem remains, however: how do individual hair cells become tuned to the proper frequencies? During each cell's development, it somehow acquires an electrical resonance that makes the cell most sensitive at a particular frequency of stimulation. One can imagine many ways in which this could come about: by regulation of the number of Ca^{2+} or K^+ channels, by adjustment of the kinetics of channel gating, by changes in membrane capacitance, by regulation of the cytoplasmic Ca^{2+}-buffering capacity, or by control of the rate at which intracellular Ca^{2+} is removed. There are as yet insufficient recordings, however, to reveal which of these or other possible parameters sets the tuning frequency. Still further in the future lies an understanding how this tuning is established, whether it is wholly programmed or is adjusted on the basis of experience.

Conclusion

Despite the considerable recent growth in our understanding of the auditory system, its performance continues to exceed our ability to account for it, even taking into account the possibilities of mechanical and electrical tuning at the cellular level. It appeared for a time that the remarkable performance of the cochlea could be explained by positing a "second filter" interposed between initial frequency analysis on the basilar membrane and transduction by hair cells. More refined measurements, however, recently demonstrated that sharp frequency discrimination with high sensitivity occurs equally in basilar-membrane motion (*31*) and in hair-cell receptor potentials (*32*). This implies that the mechanical performance of the basilar membrane somehow exceeds the predictions from simple hydrodynamic modeling. Models can be made to agree reasonably well with the data, however, by incorporating into them negative damping of the basilar membrane (*33*)—by assuming, in other words, that under some circumstances the organ of Corti can supply energy to the basilar membrane's motion rather than passively dissipate it.

During the period when the sharp frequency tuning of basilar-membrane motion was being measured, other unexpected evidence pointed toward the existence of an active mechanical process in the cochlea. When Kemp (*34*) recorded the pressure near the eardrum with sensitive instrumentation, he found that presenting a click stimulus to some human ears evoked the transitory emission of one or more bursts of pure tones. This evoked auditory emission is not simply an acoustical echo: the tones emerge much later than anticipated for the passive conduction of sound through the cochlea and back to the recording apparatus. Both the frequency tuning of hair cells and the mechanical impedance of the ear are affected by stimulating the efferent nerve supply to the cochlea (*35*). Because the efferent fibers largely terminate on outer hair cells, these cells are presently considered the most likely site of mechanical changes and of emissions. Even more striking manifestations of mechanical activity within the cochlea are ears that spontaneously emit pure tones (*36*). The existence of an animal whose ear produces 59-dB sound (*37*) is a potent argument that cochlear mechanical activity is more than a modeler's abstraction.

How can the evidence for active mechanical processes in the cochlea be reconciled with the cochlea's high sensitivity and sharp frequency selectivity? The widespread belief at present is that outer hair cells are capable of some form of motility and that, when appropriately stimulated, they move in a manner that amplifies the incoming mechanical stimulus. According to this model, the basilar membrane's motion is unexpectedly sensitive and frequency-selective because it is governed not only by passive hydrodynamics but also by an active contribution from the organ of Corti (*38*). This mechanical activity would, of course, supply the negative damping of cochlea models. The suggested process, a form of positive feedback, might be mediated through electrical resonance (*39*). Excessive gain in such feedback presumably produces the oscillations of evoked and spontaneous auditory emissions.

A central goal of current auditory research is the elucidation of the cellular and molecular bases for the active process in the organ of Corti. If the present models are correct, the contribution of this active process must be made on a cycle-by-cycle basis. In other words, the activity must occur every few microseconds or tens of microseconds to facilitate high-frequency hearing. Two active processes have been observed in vitro. Outer hair cells from the organ of Corti are capable of reversible contractions, but to date these have been demonstrated to occur on a time scale of seconds to minutes, not of microseconds (*40*). The hair bundles of the turtle's cochlea display oscillatory motions in phase with their electrical resonance; as yet, however, the observations extend to a frequency of only 171 Hz (*41*). Because the mechanism of motility is unknown in both instances, it remains to be seen whether either type of movement can account for sharp, high-frequency tuning and, therefore, whether a process that explains cochlear activity is now in hand.

References and Notes

1. D. J. DeRosier, L. G. Tilney, E. Egelman, *Nature (London)* **287**, 291 (1980).
2. Å. Flock, B. Flock, E. Murray, *Acta Otolaryngol.* **83**, 85 (1977).
3. M. R. Miller, *J. Morphol.* **156**, 381 (1978); V. Bruns and M. Goldbach, *Anat. Embryol.* **161**, 51 (1980); L. G. Tilney and J. C. Saunders, *J. Cell Biol.* **96**, 807 (1983).
4. A. J. Hudspeth, *Annu. Rev. Neurosci.* **6**, 187 (1983).
5. M. J. Mulroy, *Brain Behav. Evol.* **10**, 69 (1974).
6. T. Holton and T. F. Weiss, *J. Physiol. (London)* **345**, 241 (1983).
7. L. S. Frishkopf and D. J. DeRosier, *Hear. Res.* **12**, 393 (1983); T. Holton and A. J. Hudspeth, *Science* **222**, 508 (1983).
8. T. F. Weiss and R. Leong, *Hear. Res.*, in press.
9. A. J. Hudspeth and D. P. Corey, *Proc. Natl. Acad. Sci. U.S.A.* **74**, 2407 (1977).
10. S. L. Shotwell, R. Jacobs, A. J. Hudspeth, *Ann. N.Y. Acad. Sci.* **374**, 1 (1981).
11. F. J. Sigworth, *J. Physiol. (London)* **307**, 97 (1980).
12. By the two-conductance-state model, the mean current through N transduction channels is:
$$\bar{I} = N\gamma V_m p_o$$
in which p_o is the probability of any channel's being open. γ is the single-channel conductance, and V_m is the membrane potential. The variance in the transduction current is:
$$\sigma^2 = N\gamma^2 V_m^2 p_o(1 - p_o)$$
Eliminating p_o from these formulas yields:
$$\sigma^2 = \gamma V_m \bar{I} - \bar{I}^2/N$$
Fitting experimental data to this parabolic relation allows estimation of N and γ.
13. T. Holton and A. J. Hudspeth, *Soc. Neurosci. Abstr.* **10**, 10 (1984); *J. Physiol. (London)*, in press.
14. D. P. Corey and A. J. Hudspeth, *Nature (London)* **281**, 675 (1979).
15. H. Ohmori, *J. Physiol. (London)* **359**, 189 (1985).
16. A. J. Hudspeth, *J. Neurosci.* **2**, 1 (1982).
17. D. P. Corey and A. J. Hudspeth, *ibid.* **3**, 962 (1983).
18. A. J. Hudspeth, in *Contemporary Sensory Neurobiology*, M . J. Correia and A. A. Perachio, Eds. (Liss, New York, 1985), pp. 193–205.
19. J. O. Pickles, S. D. Comis, M. P. Osborne, *Hear. Res.* **15**, 103 (1984); D. N. Furness and C. M. Hackney, *ibid.* **18**, 177 (1985).
20. The rate constants for channel opening and closing are:
$$k_{12} = (kT/h)e^{-[G_{12} + \delta(G_1 - Zx)]/RT}$$
$$k_{21} = (kT/h)e^{-[G_{12} - (1 - \delta)(G_1 - Zx)]/RT}$$
G_{12} is the height of the energy barrier between the closed and open states and G_1 is their intrinsic free-energy difference, represented in Fig. 5 by the "door-closing" spring on the transduction channel's gate. When the hair bundle's tip is deflected by a distance x, the transduction linkage is stretched by a fraction δ of this displacement; the resultant increment in the free-energy barrier between the channel's two states is proportional to the sensitivity Z, which is approximately 6 kcal/mol-μm. The other parameters have their usual thermodynamic meanings: k and h are, respectively, Boltzmann's and Planck's constants, R the ideal-gas constant, and T the absolute temperature. When the hair bundle is abruptly displaced from one position to another, the transduction current relaxes from one steady-state value to another with a time constant τ, given by:
$$\tau = \frac{1}{k_{12} + k_{21}}$$
In the steady-state condition, the rates of channel opening and closing are equal; the probability p_o that a given channel is open at any instant, or the fraction of the channels that are open, is:
$$p_o = \frac{1}{1 + (k_{21}/k_{12})} = \frac{1}{1 + e^{(G_1 - Zx)/RT}}$$
In agreement with experimental results (*13*), plotting p_o against x produces a symmetrical, saturating, sigmoidal curve.
21. K.-W. Yau, G. Matthews, D. A. Baylor, *Nature (London)* **279**, 806 (1979).
22. H. deVries, *Physica* **14**, 48 (1948); *Acta Otolaryngol.* **37**, 218 (1949).
23. G. G. Harris, *J. Acoust. Soc. Am.* **44**, 176 (1968).
24. W. Bialek and A. Schweitzer, *Phys. Rev. Lett.* **54**, 725 (1985).
25. A. C. Crawford and R. Fettiplace, *J. Physiol. (London)* **312**, 377 (1981); J. F. Ashmore, *Nature (London)* **304**, 536 (1983).
26. R. S. Lewis and A. J. Hudspeth, *Nature (London)* **304**, 538 (1983).
27. E. R. Lewis, R. A. Baird, E. L. Leverenz, H. Koyama, *Science* **215**, 1641 (1982).

28. R. W. Meech, *Annu. Rev. Biophys. Bioeng.* **7**, 1 (1978).
29. The Ca^{2+} current fits a third-order exponential relaxation:

$$i_{Ca^{2+}} = (m^3 \cdot \bar{g}_{Ca^{2+}})(V_m - E_{Ca^{2+}})$$

in which $\bar{g}_{Ca^{2+}}$ is the maximal conductance of the Ca^{2+} channels, $E_{Ca^{2+}}$ the associated reversal potential, and m an activation parameter whose rate of change depends exponentially upon the membrane potential, V_m. It is thought that the Ca^{2+} accumulates in a buffered, submembrane volume from which it is removed by a first-order process:

$$d[Ca^{2+}]_i/dt = \alpha i_{Ca^{2+}} - \beta[Ca^{2+}]_i$$

in which α and β are constants. The simplest model that explains the kinetic behavior and the sensitivity of the K^+ channel to Ca^{2+} includes three closed states for the channels, corresponding to the binding of zero, one, or two Ca^{2+} ions, and two open states, corresponding to the binding of two or three ions (*42*). The solution of the differential equations for Ca^{2+}- and K^+-channel gating, together with the simultaneous equations for leakage and capacitive currents, reconstructs the observed electrical resonance.
30. R. S. Lewis, *Soc. Neurosci. Abstr.* **10**, 11 (1984).
31. S. M. Khanna and D. G. B. Leonard, *Science* **215**, 305 (1982); P. M. Sellick, R. Patuzzi, B. M. Johnstone, *Hear. Res.* **10**, 93 (1983).
32. I. J. Russell and P. M. Sellick, *J. Physiol. (London)* **284**, 261 (1978).
33. S. T. Neely and D. O. Kim, *Hear. Res.* **9**, 123 (1983).
34. D. T. Kemp, *J. Acoust. Soc. Am.* **64**, 1386 (1978).
35. D. C. Mountain, *Science* **210**, 71 (1980); M. C. Brown and A. L. Nuttall, *J. Physiol. (London)* **354**, 625 (1984).
36. P. M. Zurek, *J. Acoust. Soc. Am.* **69**, 514 (1981).
37. M. A. Ruggero, B. Kramek, N. C. Rich, *Hear. Res.* **13**, 293 (1984).
38. T. Gold, *Proc. R. Soc. London Ser. B* **135**, 492 (1948).
39. T. F. Weiss, *Hear. Res.* **7**, 353 (1982).
40. W. E. Brownell, C. R. Bader, D. Bertrand, Y. de Ribaupierre, *Science* **227**, 194 (1985); H. P. Zenner, U. Zimmerman, U. Schmitt, *Hear. Res.* **18**, 127 (1985).
41. A. C. Crawford and R. Fettiplace, *J. Physiol. (London)* **364**, 359 (1985).
42. K. L. Magleby and B. S. Pallotta, *ibid.* **344**, 585 (1983).
43. The research presented from the author's laboratory was conducted in association with D. P. Corey, T. Holton, R. S. Lewis, L. Robles, and S. L. Shotwell and was supported by NIH grants NS13154, NS20429, and NS22389 and by the System Development Foundation. I thank R. Jacobs, M. Miller, P. Leake, J. Howard, L. Robles, and M. P. Stryker for their assistance in preparing this article.

25

The Sociogenesis of Insect Colonies

Edward O. Wilson

Together with flight and metamorphosis, colonial life was one of the landmark events in the evolution of the insects and evidently served as a source of their ecological success. Preliminary studies indicate that approximately one-third of the entire animal biomass of the Amazonian terra firme rain forest may be composed of ants and termites, with each hectare of soil containing in excess of 8 million ants and 1 million termites (*1, 2*). On the Ivory Coast savanna the density of ants is 20 million per hectare, with one species, *Camponotus acvapimensis*, alone accounting for 2 million (*3*). Such African habitats are often visited by driver ants (*Dorylus* spp.), single colonies of which occasionally contain more than 20 million workers (*4*). And the driver ant case is far from the ultimate. A "supercolony" of the ant *Formica yessensis* on the Ishikari Coast of Hokkaido was reported to be composed of 306 million workers and 1,080,000 queens living in 45,000 interconnected nests across a territory of 2.7 square kilometers (*5*).

The environmental impact of these insects is correspondingly great. In most terrestrial habitats ants are among the leading predators of insects and other small invertebrates (*3, 6, 7*), and leafcutter ants (*Atta* spp.) are species for species the principal herbivores and most destructive insect pests of Central and South America (*8*). *Pogonomyrmex* and other harvester ants compete effectively with mammals for seeds in deserts of the southwestern United States (*9*). Other ants move approximately the same amount of soil as earthworms in the woodlands of New England, and they surpass them in tropical forests. Both are exceeded in turn by termites, which also

Science 228, 1489–1495 (28 June 1985). This article is based on the lecture of the 1984 Tyler Prize for Environmental Achievement, delivered at the University of Southern California on 24 May 1984.

break down a large part of the vegetable litter and diffuse the products through the humus (*10, 11*).

The Reasons for Success

In general, the most abundant social insects are the evolutionarily more advanced groups of ants and termites, in other words, those with the highest percentage of derived traits in anatomy and physiology as well as the more populous and complexly organized societies (*6, 12, 13*). What is the real origin of this competitive advantage in the environment as a whole? At the risk of oversimplification, it can be said that entomologists have come to recognize three qualities as being most important. First, coördinated groups conduct parallel as opposed to serial operations and hence make fewer mistakes, especially when labor is divided among specialists. If different cadres of workers in an ant colony simultaneously forage for food, feed the queen, and remove her eggs to a safe place, they are more likely as a whole to complete the operation than if they perform the steps in repeated sequences in the manner of solitary insects (*13*). Second, groups can concentrate more energy and force at critical points than can single competitors, using sheer numbers to construct nests in otherwise daunting terrain, as well as to defend the young, and to retrieve food more effectively. Finally, there is caste: in ways that vary among species, the food supply is stabilized by the use of larvae and special adult forms to store reserves in the form of fat bodies and nutrient liquids held in the crop, while defense, nest construction, foraging, and other tasks are mostly accomplished by specialists (*14*).

The aim of much of contemporary research on social insects is to identify more fully the mechanisms by which colony members differentiate into castes and divide labor—and to understand why certain combinations of mechanisms have produced more successful products than others. The larger hope is that more general and exact principles of biological organization will be revealed by the meshing of comparable information from developmental biology and sociobiology. The definitive process at the level of the organism is morphogenesis, the set of procedures by which individual cells or cell populations undergo changes in shape or position incident to organismic development (*15*). The definitive process at the level of the colony is sociogenesis, the procedures by which individuals undergo changes in caste, behavior, and physical location incident to colonial development. The question of interest for general biology is the nature of the similarities between morphogenesis and sociogenesis.

The study of social insects is by necessity both a reductionistic and holistic enterprise. The behavior of the colony can be understood only if the programs and positional effects of the individual members are teased apart, ultimately at the physiological level. But this information makes full sense only when the patterns of colonial behavior of each species are examined as potential idiosyncratic adaptations to the natural environment in which the species lives. At both levels social insects offer great advantages over ordinary organisms for the study of biological organization. Although no higher organism can be readily dissected into its constituent parts for study and then reassembled, this is not the case for the insect colony. The colony can be fragmented into any conceivable combination of sets of its members, manipulated experimentally, and reconstituted at the end of the day, unharmed and ready for replicate treatment at a later time. The technique is used for analysis of optimization in social organization as follows: the colony is modified by changing caste ratios, as though it were a mutant. The performance of this "pseudomutant" is then compared with that of the natural colony and other modified versions. The same colony can be turned repetitively into pseudomutants in random sequences on different days, eliminating the variance that would otherwise be due to between-colony differences (*16*). At a still higher level of explanation, that of the ecosystem, the large numbers of species of various kinds of social insects (more than 1000 each in the ant genera *Camponotus* and *Pheidole* alone) give a panoramic view of the evolution of colonial patterns and make correlative analysis of adaptation more feasible.

Principles of Sociogenesis

In all species of social insects thus far tudied, caste differences among colony members have proved to be principally or exclusively phenotypic rather than genetic. The environmental factors in each instance belong to one or more of the following six categories: larval nutrition (which is especially important in ants); inhibition caused by pheromones or other stimuli from particular castes (the key factor in many kinds of termites); egg size and hence quantity of nutrients available to the embryo; winter chilling; temperature during development; and age of the queen (*6, 17, 18*). Phenotypic caste determination is similar to restriction during cell differentiation. That is, the growing individual reaches one or more decision points at which it loses some of its potential, and this diminution continues progressively until it reaches the final decision point, where it is determined to the caste it will occupy as an adult. For example, in the ant genus *Pheidole* the restriction to either the queen line or worker line occurs in the egg; then larvae in the worker line become committed to development as either minor or major workers in the fourth and final instar. The cues affecting these two decisions, which include nutrition, winter chilling of queens, and inhibitory pheromones, are mediated to the developing tissue by juvenile hormone (*19, 20*).

The differentiation of the colony members into physical castes is supplemented in the great majority of social species by a regular progression on the part of most workers through different work roles during aging. In this way the individual belongs not only to one physical caste but to a sequence of temporal castes as it passes through its life-span. By far the most common sequence is for the worker to join in the care of the queen or immature stages shortly after it emerges into the adult stage, then to participate in nest building, and, finally, to forage outside the nest for food. Temporal castes are a derived trait in evolution, having become most clearly demarcated in species with the largest societies. They are typically weak or absent in anatomically primitive species with small colony populations (*6, 21*).

Although individual workers are flexible with respect to caste at the start of their personal development in the egg stage, the colony as a whole is rigidly limited to a single array of castes. Each species also has a particular size-frequency distribution of adult workers (*13, 22, 23*). Workers in the ant genus *Pheidole*, for example, are divided into two subcastes, the minors and the majors, by size and body proportions. Among ten species selected for their taxonomic diversity, the majors were found to range from 3 percent in *Pheidole distorta* to 25 percent in *Pheidole minutula* (*23*). A lesser amount of variation exists among colonies belonging to the same species, and recent work suggests indirectly that some of the variation is genetic. Seven colonies of *Pheidole dentata* raised under uniform laboratory conditions through three brood cycles maintained relatively constant major worker percentages, and these levels varied significantly among the colonies, from approximately 5 to 15 percent (*24*).

The size-frequency distribution can also persist through relatively long periods of geological time. A fragment of a colony of the extinct weaver ant *Oecophylla leakeyi* preserved intact from the African Miocene (the only fossil insect society collected to date) proved to have the same distinctive pattern as the two living species of the genus, *Oecophylla longinoda* and *Oecophylla smaragdina*. In particular, the frequency curve was sharply bimodal, with the major workers somewhat more numerous than the minors and with a small number of medias connecting the two moieties. The allometry, or disproportionate variation in body parts, is also similar between the extinct and living species (*25*).

These several lines of evidence have led to the hypothesis of adaptive demography (*13, 26*), which can be summarized as follows. The vast majority of insect, vertebrate, and other animal populations evolve primarily through selection at the level of the individual organism. As a consequence, survivorship curves and natality schedules are directly adaptive, whereas the age-frequency distribution of the population as a whole emerges as an epiphenomenon. In the advanced social insects, in contrast, selection occurs primarily at the level of the colony, with workers mostly or entirely eliminated from reproduction and colonies competing against one another as compact units. Colonies whose members possess the most effective age-frequency distribution are more likely to survive and to reproduce, regardless of the fate of individual colony members. It is generally believed that the workers will increase the replication of genes identical to their own by promoting the physical well-being of the colony, even if they sacrifice themselves to achieve this end. Hence the age-frequency distribution of the colony members is directly subject to natural selection. Survivorship and natality schedules are indirectly subject to natural selection, in the sense of being shaped according to the effect they have on the age-frequency distribution of the colony as a whole.

The adaptive demography hypothesis has begun to be tested by both correlative analysis and experimentation. For example, linear programming models predict that as a caste specializes, its members should decrease in proportion within the colony membership (*26*). This relation does hold among the species of *Pheidole* so far studied: the repertory size of the major caste is correlated significantly across species with the percentage of the majors in the worker force. Put another way, as the majors perform fewer tasks and devote more time proportionately to roles for which they are anatomically specialized, they become scarcer in the colony population (*23*).

And yet the major workers of *Pheidole* retain a remarkable flexibility. When the minor-major ratio was experimentally reduced to below 1:1 in three widely different species of the genus, the majors increased the number of kinds of acts they performed by as much as 4.5 times and their rate of activity 15 to 30 times. The change occurred within 1 hour of the ratio change and was reversed in comparably short time when the original ratio was restored. Thus the major workers were found to respond in a manner reminiscent of the genome of a somatic cell. Under normal circumstances most of their brain programs are silent: the active repertory is limited in a fashion appropriate to the tasks for which the majors are anatomically specialized. But when an emergency arises a much larger program is quickly summoned, the majors supply about 75 percent of the activity of the missing minors, and as a result the colony continues to feed and grow (*23*).

A second line of evidence of adaptive demography has been provided by studies of the leafcutter ant *Atta cephalotes*. New colonies of *Atta*, like those of most kinds of ants, are founded by single queens after the nuptial flights. These individuals dig a shaft into the ground, then eject a wad of symbiotic fungus from their mouths onto the ground and fertilize the hyphae with droplets of feces. During the next 6 weeks they rear the first brood of workers with reserves from their own bodies while bringing the small garden to flourishing condition. The queens have only enough ovarian yolk and other storage materials to rear one small group to maturity. In order for the colony to survive thereafter, the workers must range in size from a head width of 0.8 mm, which is small enough to culture the fungus, through 1.6 mm, which is just large enough to cut fresh leaves for the fungal substrate. It turns out that the first brood of workers possess a nearly uniform frequency distribution from 0.8 through 1.6 mm, which comes close to maximizing the number of individuals and at the same time achieves the minimum size range required to grow the fungus on which the colony depends (*27*).

As the leafcutter population expands afterward, the size-frequency distribution of the workers changes in dramatic fashion. The range is increased at both ends and the curve becomes strongly skewed toward the media and major worker classes (Fig. 1). An interesting question then arises: suppose that by some misadventure most of the population of a leafcutter colony were destroyed, reducing it to near the colony-founding state. Would the size-frequency distribution of new workers produced by the colony be characteristic of the beginning stage, or would it remain at the older stage? In other words, which is the more important in the ontogeny of the caste system, the size of the colony or its age? If age were more important, causing much of the available energy to be invested in workers larger than the minimum required to harvest leaves, the colony would be imperiled because of a shortage of the small gardener classes. The creation of just one new major worker, possessing a body weight 300 times that of a gardener worker, would bankrupt the already impoverished colony. In order to provide an answer, I selected four colonies 3 to 4 years old and with about 10,000 workers and reduced the population of each to 236, giving them an

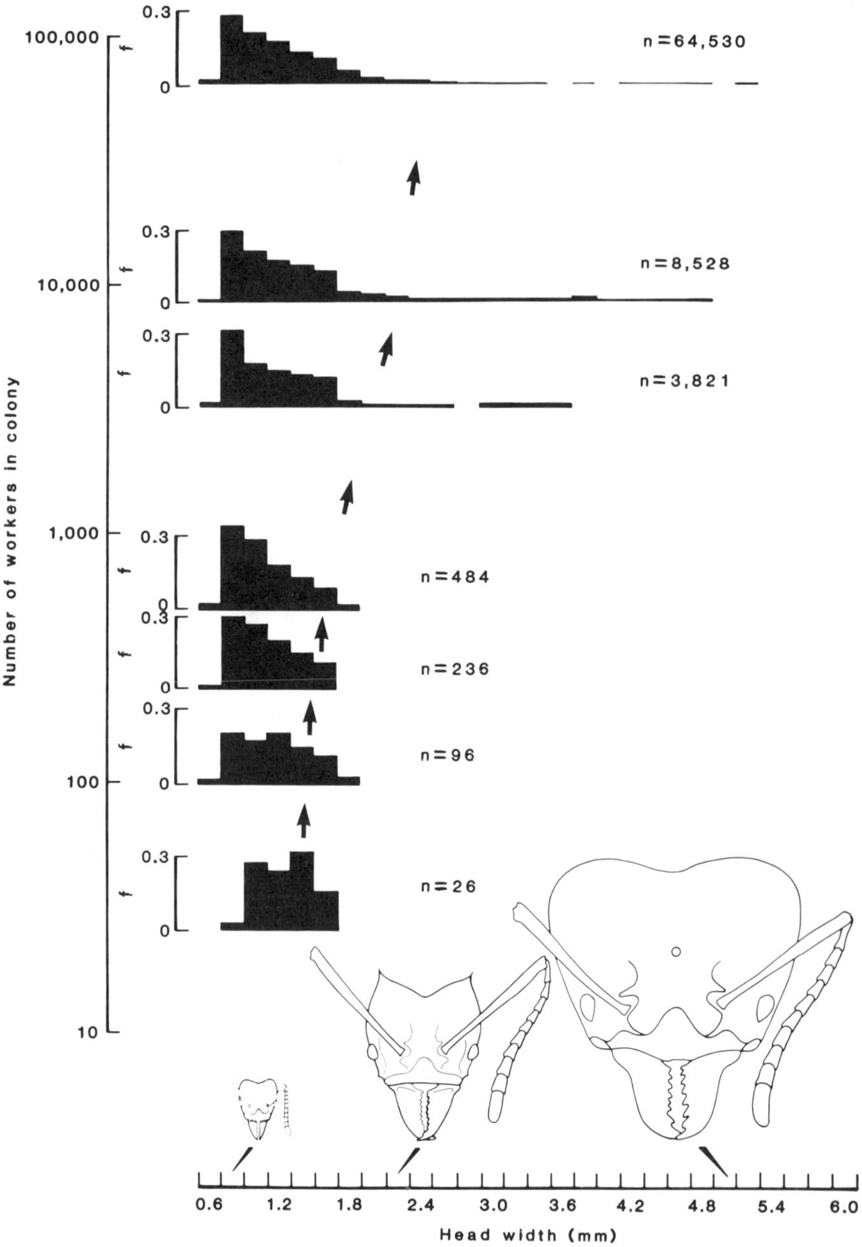

Fig. 1. The ontogeny of the caste system of the leafcutter ant *Atta cephalotes*, illustrated by seven representative colonies collected in the field or reared in the laboratory. The worker caste is differentiated into subcastes by continuous size variation associated with disproportionate growth in various body parts. The number of workers in each colony (*n*) is based on complete censuses; *f* is the frequency of individuals according to size class. The heads of three sizes of workers are shown in order to illustrate the disproportionate growth. Modified from Wilson (*27*).

artificially imposed juvenile size-frequency distribution. The worker pupae produced at the end of the first brood cycle possessed a size-frequency distribution like that of small, young colonies rather than larger, older ones. Thus colony size is more important than age, and "rejuvenated" colonies are prevented from extinguishing themselves through an incorrect investment of their resources (*27*).

Such programmed resiliency implies the existence of control mechanisms operating at the level of the colony during population growth. An increasing fraction of the research on social insects is now being directed at the discovery of such mechanisms. This work has begun to reveal a fascinating pattern of feedback loops, pacemakers, and positional effects.

An example of negative feedback is provided by the events leading to the fission of honeybee colonies. The queen secretes a "queen substance," *trans*-9-keto-2-decenoic acid, which under most circumstances inhibits the construction of royal cells by the workers and hence the rearing of new queens (*28*). However, in large, freely growing colonies this pheromone must be supplemented by a second substance, the footprint pheromone, which is secreted in relatively large amounts from glands in the fifth tarsal segment of the queen. When bee colonies become overcrowded, the queen is unable to walk along the bottom edges of the comb, where the royal cells are ordinarily built. As a result the inhibition fails in that zone, the cells are built, and the colony reproduces. With the population density now reduced to below threshold density, the queen is able to resume her inhibitory control (*29*).

Most such controls are negative and hence contribute to physiological stability and smooth growth cycles within the colony. What appear to be properties of positive feedback and explosive chain reactions nevertheless do occur during nest evacuation in a few species. When attacking fire ant workers press closely on nests of the ant *Pheidole dentata*, the defending minor workers start laying odor trails back into the brood area. This causes excited movement through the nest and further bouts of recruitment. At the height of this expanding activity the workers and queen suddenly scatter from the nest and seek individual cover. When the fire ants are then experimentally removed, the *Pheidole* adults return to the nest and reoccupy it (*30*).

The coordination of activity is still imperfectly understood. Although the typical insect society is not quite the "feminine monarchie" envisioned by early entomologists (*31*), it is also much more than a republic of specialists. According to the species, certain immature stages and castes function as pacemakers and coordinators of colony activity. Ant larvae are specially effective in initiating foraging and nest construction by the adult workers. In army ants (*Eciton*), the hatching of larvae triggers the monthly nomadic cycle during which the entire colony marches to a new location daily (*32*). But in the great majority of other species thus far studied it is the queen that provides the maximum regulation. In more primitive societies, such as those of bumblebees (*Bombus*) and paper wasps (*Polistes*), she physically dominates her daughters and other females occupying the nests, prevents them from laying eggs, and by these actions forces most into foraging and other nonreproductive tasks. Such influence can transcend simple displacement. For instance, the presence of the queen of *Polistes fuscatus*, probably a typical species at this evolutionary level, increases and synchronizes overall worker activity (*33*). In carpenter ants (*Camponotus*), the mother queen is the principal source of the nest odor (*34*). When she is removed, the workers, now in a more chaotic state, fall back on odor cues emanating from their own bodies (*35*).

Workers of social insects move to different positions with reference to the queen and brood according to their ages. This pattern is usually centrifugal: soon after the worker emerges from the pupa into the adult stage, it attends the queen and immature stages, then drifts toward the outer chambers to assist in nest construction, and finally devotes itself primarily to foraging outside the nest. The progression is accompanied by physiological change. The details vary greatly among species, and even among members of the same colony, but in general the ovaries reach maximum development early in adult life, along with fat bodies and exocrine glands devoted to nutritive exchange (*6, 17, 36–38*). Afterward these tissues regress more than enough to counterbalance the growth of exocrine glands associated with nest construction and foraging, so that the worker declines overall in weight. Mortality due to accidental causes increases sharply among workers when they commence foraging. But this attrition has far less effect on the size-and-age structure of the worker population than if individuals commenced foraging early in life, because the natural life-span is curtailed in any case past the onset of foraging by physiological senescence. In the best documented case, the honeybee worker born in early summer typically begins foraging at 2 to 3 weeks of adult life and dies from senescence by 10 weeks into this period (*39*).

The workers of advanced insect societies are not unlike cells that emigrate to new positions, transform into new types, and aggregate to form tissues and organs. With relatively small adjustments in response thresholds according to size and age, intricate new patterns are created at the level of the colony. In the fungus-growing termite *Macrotermes subhyalinus*, for example, 90 percent of the foragers are large major workers past 30 days of age. Younger major and minor workers accept the grass collected by these foragers, consume it, and pass the partly digested material out into the fungus comb. Workers of various castes older than 30 days eat the fungus comb and produce the final feces (*40*). In the leafcutting ant *Atta sexdens* most of the fresh vegetation is gathered by workers of intermediate size (which, incidentally, achieve the highest net energetic yield of all the size groups). The material is then converted into new fungus substrate within the nest by an assembly-line operation that penetrates ever more deeply into the combs: successively smaller workers cut the leaves into tiny fragments, chew them into pulp, stick the processed lumps onto the growing combs, and transfer strands of fungi onto this newly prepared substrate. Finally, the smallest workers of all care for the proliferating fungus, virtually strand by strand (*16, 41*).

Such patterns are in fact much more intricate than a description of sequences alone indicates. In the ant *Pheidole dentata* and the honeybee *Apis mellifera* the tasks are broken into sets that are linked not by the similarity of the behaviors performed but by the proximity of the

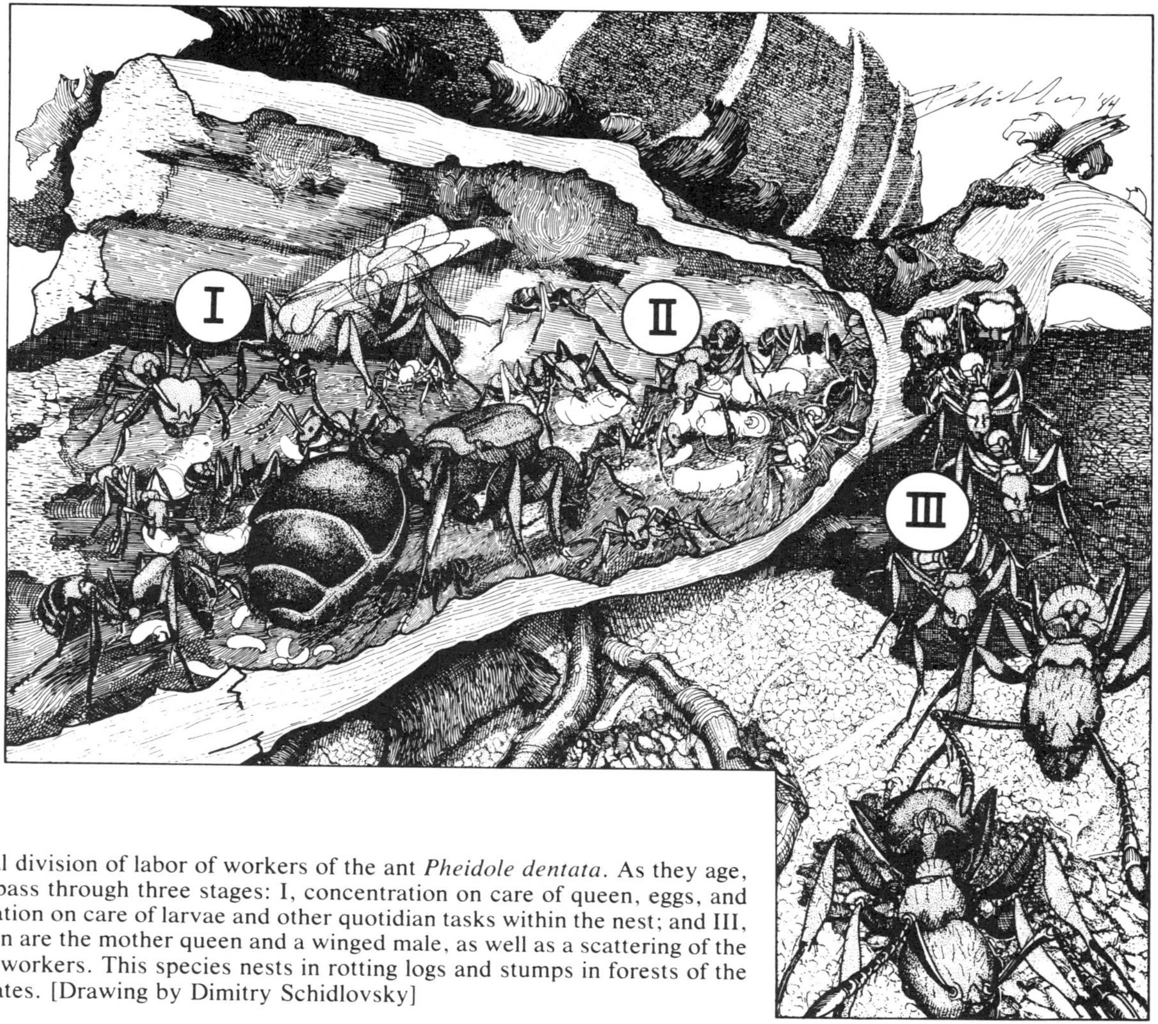

Fig. 2. The temporal division of labor of workers of the ant *Pheidole dentata*. As they age, the minor workers pass through three stages: I, concentration on care of queen, eggs, and pupae; II, concentration on care of larvae and other quotidian tasks within the nest; and III, foraging. Also shown are the mother queen and a winged male, as well as a scattering of the large-headed major workers. This species nests in rotting logs and stumps in forests of the southern United States. [Drawing by Dimitry Schidlovsky]

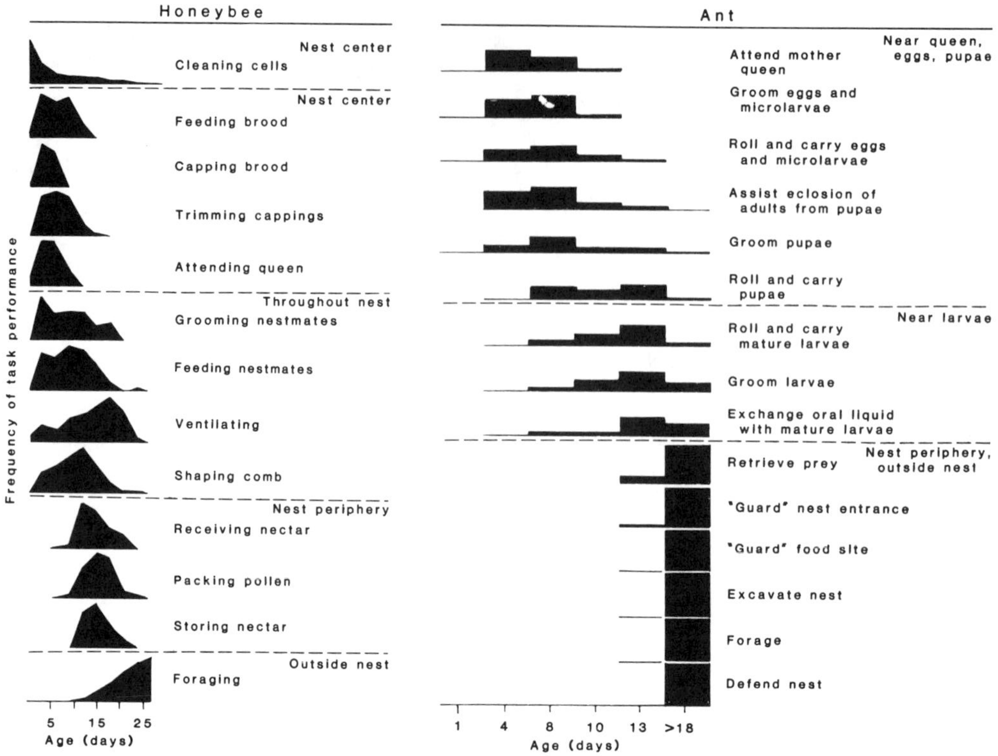

Fig. 3. The temporal division of labor, based on changes of behavior in the adult workers with aging, is shown in the ant *Pheidole dentata* and honeybee *Apis mellifera*: the insects shift from one linked set of tasks to another as they move their activities outward from the nest center (see Fig. 2). The similarities between the two species are convergent and believed to be adaptive. The sum of the frequencies in each histogram is 1.0. Adapted from Wilson (*54*) and Seeley (*55*).

objects to which they are directed, thus reducing the travel time and energy expenditure of the individual workers (Figs. 2 and 3). The similarities between the two patterns can only be due to convergent evolution, since ants and bees arose during Mesozoic times from widely different stocks of aculeate wasps (*42*).

The Imperfection of Insect Societies

Although insects as a whole originated at least 350 million years ago, higher social insects did not appear until the Jurassic Period, roughly 200 million years ago, and they began an extensive evolutionary radiation only in the late Cretaceous and early Tertiary Periods, about 100 million years later (*42*). Even then, advanced social organization originated in as few as 13 stocks, 12 within the aculeate Hymenoptera (ants, bees, and wasps) and one in the cockroach-like orthopteroids that produced the termites (*6*).

Two possible explanations for this evolutionary conservatism have emerged from more detailed studies of individual colony members. The first is that the small size of the insect brain and the heavy reliance of social forms on chemical signaling place inherent limits on the amount of information flow through the colonies. This circumstance leads to frequent near-chaotic states and the dependence on colony decision-making by *force majeure*, a statistical preponderance of certain actions over others that lead to a dynamic equilibrium rather than clean binary choices (*6, 13, 43, 44*). Thus when released from threshold concentrations of the queen inhibitory pheromones, some honeybee workers build royal cells while a smaller number of workers set out to dismantle them. The final result is an equilibrial number of cells sufficient for the rearing of new queens (*44*).

On the other hand, a few mechanisms are coming to light that sharpen the precision of mass response and bring it closer to binary action. Markl and Hölldobler (*45*) reported the existence of "modulatory communication" in ants, a form of signaling in one channel that alters the threshold of response in another. For example, when harvester ants of the genus *Novomessor* encounter large food objects they make sounds by scraping together specialized surfaces on the thin postpetiole and adjacent abdominal segment. This stridulation does not cause an overt behavioral change in nestmates but raises the probability that they will release short-range recruitment chemicals. The overall result is a speeding and tightening of the coordination process.

The second force inhibiting social evolution, at least in the case of hymenopterans, is the substantial conflict among individuals for reproductive privileges. Dominance rank orders, once thought to be confined to simply organized societies of halictine bees, bumblebees, and polistine wasps, as well as associations of queens of a few kinds of ants [*Nothomyrmecia, Myrmecocystus*, and *Eurhopalothrix* (*46*)] have also been discovered in the workers of some species of ants as well (*47*). West-Eberhard has argued that competition among workers is more pervasive among advanced societies than has been recognized and that selection at the level of the individual has consequently played a key role in the division of labor (*36, 48*). She explains the centrifugal pattern of temporal castes (Figs. 2 and 3) as the product of such selection. The individual worker, by staying close to the brood chambers while still young and while her personal reproductive value is highest, maximizes her potential to contribute personal offspring. But as death approaches and fertility declines because of senescence, the optimum strategy for contributing genes to the next generation is to enhance colony welfare through more dangerous occupations such as defense and foraging, thus producing more brothers and sisters as opposed to personal offspring. By this criterion, Porter and Jorgensen (*37*) were correct to call foraging harvester ants the "disposable" caste. Hölldobler (*49*) has recently described what may be the ultimate case: aging workers of the Australian tree ant (*Oecophylla smaragdina*) occupy special "barracks nests" around the periphery of the main nest area. They stand idle most of the time and are among the first defenders to enter combat during territorial battles with other tree ant colonies.

Individual selection appears likely to have inhibited the refinement of social behavior, especially in the earliest stages of the evolution. Indeed, there is evidence that species of the bee genus *Exoneurella*, trading production of siblings for the production of offspring, have returned from primitive sociality back to a more nearly solitary state (*50*). Yet there does appear to be a point of no return in the rise of sociality. When colonies become very complex, organized by an intricate caste system and highly coordinated group movements, the advantages of queenlike behavior on the part of workers is diminished and may even disappear. In a few advanced ant genera, such as *Pheidole* and *Solenopsis*, the workers no longer even possess ovaries (*51*).

The pattern emerging from comparative studies suggests that as reproductive competition has declined during the elaboration of sociogenesis, dominance interactions have been ritualized to serve as part of the communicative signals dividing labor. In the more complex societies of bees and wasps, overt aggression is replaced by queen pheromones, but the inhibition of the ovaries of the subordinates and their induction into worker roles remain essentially the same (*6, 14*). Also, traces of aggressive and subordinate interactions persist in ritual form. The workers of stingless bees either hurriedly withdraw from the area when the queen approaches, thus clearing a path for her, or else they mock-attack, then bow to her head, and finally swing to her side to become part of the retinue (*52*). Ritualized dominance interactions may also be important between sterile workers. Major workers of the ant *Pheidole pubiventris* turn away from minor workers when they encounter them around the brood, thus yielding most of the care of the immature forms to these smaller nestmates. This aversion neatly divides colony labor into several principal categories (*23*).

Although seldom acknowledged in the literature, regulatory mechanisms are often found lacking even when they are intuitively anticipated by the investigator. For example, the major workers of *Pheidole dentata* are specialized for response against fire ants and other members of the genus *Solenopsis*, but when colonies are stressed continually with these enemies the major-minor ratio remains the same.

In other words, there is no increase in the defense expenditure in the face of a major threat (*24*). Leafcutter workers with head widths from 1.8 through 2.2 mm are responsible for most of the foraging, but when members of this important caste are removed experimentally, the colonies fail to compensate for the loss by increasing representation of the size class in later broods. The result is a reduction in energetic efficiency through two brood cycles (*53*).

On the whole, insect societies display impressive degrees of complexity and integrity on the basis of what appear to be relatively few sociogenetic processes. The mechanisms that do exist, together with their strengths, precision, and phylogenetic distribution, constitute a subject in an early and exciting period of investigation. Of comparable importance are the expected mechanisms that do not exist, so that investigators are likely to pay closer attention to them than has been the case in the past. As the full pattern becomes clearer, it may be possible to compare sociogenesis with morphogenesis in a way that leads to a more satisfying general account of biological organization.

References and Notes

1. L. Beck, *Amazoniana* **3**, 69 (1971).
2. F. J. Fittbau and H. Klinge, *Biotropica* **5**, 2 (1973).
3. J. Levieux, in *The Biology of Social Insects*, M. D. Breed, C. D. Michener, H. E. Evans, Eds. (Westview, Boulder, 1982), pp. 48–51.
4. A. Raignier and J. Van Boven, *Ann. Mus. R. Congo Belg. Tervuren* **2**, 1 (1955).
5. S. Higashi and K. Yamauchi, *Jpn. J. Ecol.* **29**, 257 (1979).

6. E. O. Wilson, *The Insect Societies* (Harvard Univ. Press, Cambridge, Mass., 1971).
7. R. L. Jeanne, *Ecology* **60**, 1211 (1979).
8. N. A. Weber, *Gardening Ants: The Attines* (American Philosophical Society, Philadelphia, 1972); J. M. Cherrett, in *The Biology of Social Insects*, M. D. Breed, C. D. Michener, H. E. Evans, Eds. (Westview, Boulder, 1982), pp. 114–118.
9. D. Davidson, J. H. Brown, R. S. Inouye, *BioScience* **30**, 233 (1980).
10. W. H. Lyford, *Harv. For. Pap.* **7**, 1 (1963).
11. T. Abe, in *The Biology of Social Insects*, M. D. Breed, C. D. Michener, H. E. Evans, Eds. (Westview, Boulder, 1982), pp. 71–75.
12. E. O. Wilson, *Stud. Entomol.* **19**, 187 (1976).
13. G. F. Oster and E. O. Wilson, *Caste and Ecology in the Social Insects* (Princeton Univ. Press, Princeton, N.J., 1978); J. M. Herbers, *J. Theor. Biol.* **89**, 175 (1981).
14. The large literature on the advantages of social life, much of it based on experimental studies, is reviewed by Wilson (*6*) and investigators in (*13*), as well as by C. D. Michener, *The Social Behavior of the Bees: A Comparative Study* (Harvard Univ. Press, Cambridge, Mass., 1974); H. R. Hermann, Ed., *Social Insects* (Academic Press, New York, 1979–1982), vols. 1–4; T. Seeley and B. Heinrich, in *Insect Thermoregulation*, B. Heinrich, Ed. (Wiley, New York, 1981), pp. 159–234.
15. See, for example, N. K. Wessells, *Tissue Interaction and Development* (Benjamin, Menlo Park, Calif., 1977).
16. E. O. Wilson, *Behav. Ecol. Sociobiol.* **7**, 157 (1980).
17. M. V. Brian, in *Social Insects*, H. R. Hermann, Ed. (Academic Press, New York, 1979), vol. 1, pp. 121–222.
18. J. de Wilde and J. Beetsma, *Adv. Insect Physiol.* **16**, 167 (1982).
19. D. E. Wheeler and H. F. Nijhout, *J. Insect Physiol.* **30**, 127 (1984).
20. L. Passera and J.-P. Suzzoni, *Insectes Soc.* **26**, 343 (1979).
21. For example, the very primitive ant *Amblyopone pallipes* appears to lack temporal castes completely [J. F. A. Traniello, *Science* **202**, 770 (1978)].
22. M. I. Haverty, *Sociobiology* **2**, 199 (1977); in *The Biology of Social Insects*, M. D. Breed, C. D. Michener, H. E. Evans, Eds. (Westview, Boulder, 1982), p. 251.
23. E. O. Wilson, *Behav. Ecol. Sociobiol.* **16**, 89 (1984).
24. A. B. Johnston and E. O. Wilson, *Ann. Entomol. Soc. Am.* **78**, 8 (1985).
25. E. O. Wilson and R. W. Taylor, *Psyche* **71**, 93 (1964).
26. E. O. Wilson, *Am. Nat.* **102**, 41 (1968); J. M. Herbers, *Evolution* **34**, 575 (1980).
27. E. O. Wilson, *Behav. Ecol. Sociobiol.* **14**, 55 (1983).
28. C. G. Butler and R. K. Callow, *Proc. R. Entomol. Soc. London* **B43**, 62 (1968).
29. Y. Lenski and Y. Slabezki, *J. Insect Physiol.* **27**, 313 (1981).
30. E. O. Wilson, *Behav. Ecol. Sociobiol.* **1**, 63 (1976).
31. C. Butler, *The Feminine Monarchie* (Barnes, Oxford, 1609).
32. T. R. Schneirla, *Army Ants: A Study in Social Organization*, H. R. Topoff, Ed. (Freeman, San Francisco, 1971).
33. H. K. Reeve and G. J. Gamboa, *Behav. Ecol. Sociobiol.* **13**, 63 (1983).
34. N. F. Carlin and B. Hölldobler, *Science* **222**, 1027 (1983).
35. ———, personal communication.
36. M. J. West-Eberhard, in *Natural Selection and Social Behavior*, R. D. Alexander and D. W. Tinkle, Eds. (Chiron, New York, 1981), pp. 3–17.
37. S. D. Porter and C. D. Jorgensen, *Behav. Ecol. Sociobiol.* **9**, 247 (1981).
38. T. D. Seeley, *ibid.* **11**, 287 (1982).
39. M. Rockstein, *Ann. Entomol. Soc. Am.* **43**, 152 (1950); S. F. Sakagami and H. Fukuda, *Res. Popul. Ecol.* **10**, 127 (1968).
40. S. Badertscher, C. Gerber, and R. H. Leuthold, *Behav. Ecol. Sociobiol.* **12**, 115 (1983).
41. E. O. Wilson, *ibid.* **7**, 143 (1980).
42. F. M. Carpenter and H. R. Hermann, in *Social Insects*, H. R. Hermann, Ed. (Academic Press, New York, 1979), vol. 1, pp. 81–89.
43. P. Hogeweg and B. Hesper, *Behav. Ecol. Sociobiol.* **12**, 271 (1983).
44. D. H. Baird and T. D. Seeley, *ibid.* **13**, 221 (1983).
45. H. Markl and B. Hölldobler, *ibid.* **4**, 183 (1978).
46. S. H. Bartz and B. Hölldobler, *ibid.* **10**, 137 (1982); B. Hölldobler and R. W. Taylor, *Insectes Soc.* **30**, 384 (1983); E. O. Wilson, *ibid.* **30**, 408 (1985).
47. B. J. Cole, *Science* **212**, 83 (1981); N. Franks and E. Scovell, *Nature (London)* **304**, 724 (1983).
48. M. J. West-Eberhard, *Proc. Am. Philos. Soc.* **123**, 222 (1979).
49. B. Hölldobler, *Biotropica* **15**, 241 (1983).
50. C. D. Michener, *Kansas Univ. Sci. Bull.* **46**, 317 (1965).
51. E. O. Wilson, *J. Kansas Entomol. Soc.* **51**, 615 (1978).
52. S. F. Sakagami, in *Social Insects*, H. R. Hermann, Ed. (Academic Press, New York, 1982), vol. 3, pp. 361–423.
53. E. O. Wilson, *Behav. Ecol. Sociobiol.* **15**, 47 (1983).
54. ———, *ibid.* **1**, 141 (1976).
55. T. D. Seeley, *ibid.* **11**, 287 (1982).
56. I am grateful to D. M. Gordon, B. Hölldobler, T. Seeley, and D. Wheeler for critical readings of the manuscript. Supported by a series of grants from the National Science Foundation, the latest of which is BSR 81-19350.

26

Neurotrophic Factors

Hans Thoenen and David Edgar

The development and maintenance of function of the nervous system results from the concerted interaction of a great variety of genetic and epigenetic regulatory mechanisms. Transplantation and ablation experiments performed during this century have demonstrated that the survival of developing vertebrate neurons can be determined by their fields of innervation (*1*). So far, however, only one trophic factor has been shown to be responsible for this epigenetic determination of neuronal survival: the protein nerve growth factor (NGF) was demonstrated to be required for the survival of developing peripheral sympathetic and sensory neurons by showing that neutralization of endogenous NGF by antibodies to NGF (anti-NGF) resulted in the death of these neurons (*2*). More recently, numerous tissue culture experiments have been used to show that NGF is only one of a number of molecules able to maintain the survival of embryonic neurons in vitro, implying that such molecules might also function as trophic factors to support neuronal survival in vivo (*3*). The central thesis of this article is that in order to prove that putative trophic factors (detected by experiments in vitro) do have a physiological role, it is necessary to purify them to produce specific antibodies to them. Accordingly, the consequences of neutralization of the endogenous molecules in vivo—and hence their physiological role—can then be established.

Nerve Growth Factor

The detection of large amounts of NGF in the submandibular gland of the male mouse some 30 years ago was a

Science 229, 238–242 (19 July 1985)

prerequisite for its purification, necessary for the production of anti-NGF to delineate the neurotrophic actions of NGF in vivo (*2*). In addition, determination of the amino acid sequence of mouse NGF more than a decade ago (*4*) provided the information necessary for its recent molecular cloning (*5*). This work has now led to the elucidation of the structure of the NGF precursor and its genomic organization. Thus, the major part of the precursor sequence has been shown to be on the amino terminal side of β-NGF (the active subunit of the NGF molecule), whereas the carboxyl terminal arginine is followed only by two amino acids. The region coding for β-NGF represents about one-third of the total precursor messenger RNA (mRNA). Analysis of the organization of mouse and human genomes has shown that the NGF gene is present as a single copy, and that all the information for the β-NGF sequence is located in a single exon. Moreover, the amino acid sequence homology of mouse and human NGF is more than 90 percent, as deduced from the genomic DNA sequence (*5*).

A sensitive two-site enzyme immunoassay allowing determination of the NGF present in effector organs has only recently been developed, allowing a major gap in the NGF story to be closed (*6*). These investigations demonstrated a correlation between the density of sympathetic innervation and the levels of NGF in the corresponding peripheral target tissues; experiments with tissue culture have shown that the local concentration of NGF determines the extent of ramification of sympathetic nerve fibers in vitro (*7*), implying that the levels of NGF in target tissues may be responsible for the density of sympathetic innervation. Tissue culture experiments have also shown that target tissues can synthesize NGF in vitro (*8*), and recent work with nucleic acid probes to quantify the mRNA for NGF demonstrates that levels of NGF are correlated with the amounts of its mRNA (*9*). Thus, the rate of synthesis of NGF in target tissues is probably determined by regulation of production of its mRNA. It is not yet clear, however, which cells of the target tissues actually synthesize NGF.

Although we know more about NGF than any other neurotrophic factor, our information on the mechanism (or mechanisms) that regulate its synthesis and release is fragmentary. It seems that NGF synthesis in peripheral effector organs is not controlled by androgens, in contrast to its synthesis in the mouse submandibular gland (*6*). Furthermore, the release of NGF from the iris in organ cultures is not dependent on calcium influx, suggesting a constitutive release pathway (*8*). The marked increase in NGF after sympathetic or sensory denervation (or both) may be due to a loss of inhibition of NGF synthesis exerted by the innervating neurons or, alternatively, it may simply reflect the lack of removal of NGF by retrograde transport; ligation experiments have shown that endogenous NGF is transported retrogradely from the periphery to the corresponding neuronal cell bodies (*10*). The relatively high levels of NGF in sympathetic ganglia (*6*) do not, therefore, result from local synthesis [only very small amounts of NGF mRNA are detectable in the ganglia (*9*)], but from accumulation by retrograde axonal transport. Experiments where ^{125}I-labeled exogenous NGF has been retrogradely transported indicate that the NGF arriving at the cell body is intact, as shown by immunoprecipitation and sodium dodecyl sulfate (SDS)–polyacrylamide gel electrophoresis (*11*). Furthermore, the endogenous NGF arriving at the ganglionic neuronal cell bodies is biologically active (*6*).

There is compelling evidence that the receptor binding domain of the NGF molecule has remained highly conserved during evolution; the biological activities of mouse and bovine NGF are identical although they show little immunological cross-reactivity (*12*). Tryptophan residues, in particular residue 21, are essential for the biological activity of NGF (*13*). However, an unambiguous identification of the receptor binding domain has not yet been accomplished. A report that a peptide fragment consisting of residues 10 to 25 and 75 to 88, linked by a disulfide bridge (cysteines 15 and 80), is 100 times more active than native NGF (*14*) appears not to have been confirmed, and the corresponding synthetic fragment was reported to be inactive (*15*). Knowledge of the sequences of mouse and human NGF along with the sequences of NGF's from other sources should tell us which regions of the NGF molecule are most highly conserved. This information together with the possibility of producing such molecules by expression vectors and subjecting them to site-directed mutagenesis could lead to the resolution of the receptor binding domain of NGF.

The molecular mechanism of action of NGF on its target cells is still unresolved, although the kinetics of the interaction of NGF with its receptors have been established (*2*). The NGF receptors have been identified recently by affinity-labeling and partially purified by affinity chromatography (*16*). The binding of NGF induces a change in molecular weight of the receptor from 100,000 to 158,000 in PC12 cells. This change is accompanied by a decreased off rate for NGF, that is, a transformation from low- to high-affinity receptors (*17*); it also corresponds to a change in receptor extractability by Triton X-100 (*18*). Whether the detergent-resistant association of the high-affinity receptor with the cytoskeleton (*18*) reflects a mechanism whereby information might be transferred from the receptor via the cytoskeleton to regulatory sites inside the cell remains to be determined.

Second Messenger Question

The information available seems not to support postulated second messenger mechanisms. NGF does not act as its own second messenger after binding to the cell surface receptors by being transferred into the cytosol. When injected into the cytoplasm of PC12 cells (*19, 20*) or directly into the nuclear chromatin (*20*), NGF did not induce either fiber outgrowth (*19, 20*) or enzymes typical of NGF action via membrane receptors (*21*). Conversely, the injection of antibodies to NGF into the cytoplasm did not abolish the membrane-mediated effects of NGF (*19–21*). The possibility that proteolytic (nonantigenic) degradation products of NGF act as second messengers has been shown to be unlikely: inhibition of the rate of degradation of internalized NGF in PC12 cells did not interfere with its effects on enzyme induction or fiber outgrowth (*21*).

The role of calcium influx as a potential "second messenger" has been discounted in experiments with cultured sympathetic neurons of the newborn rat; in these experiments NGF-mediated induction of tyrosine hydroxylase (TH) was not inhibited by EGTA, calcium channel blockers, or calmodulin antagonists, although the antagonists did block TH induction resulting from high concentrations of potassium (*22*). Similarly, the possibility that cyclic adenosine monophosphate (cyclic AMP) acts as second messenger is unlikely because NGF-mediated selective enzyme induction in calf adrenal medullary cells is distinctly different from that initiated by cyclic AMP, with respect to both the regulation and the pattern of enzyme induction (*23*). The NGF-mediated induction of enzymes involved in the synthesis of adrenergic transmitter is restricted to tyrosine hydroxylase, dopamine β-hydroxylase and phenylethanolamine *N*-methyltransferase. In contrast, cyclic AMP induces not only these enzymes but also dopa decarboxylase. Moreover, the cyclic AMP–mediated enzyme induction can be blocked by the mRNA transcription inhibitor α-amanitin, whereas NGF-mediated enzyme induction is unaffected (*23*). The adenosine derivative 9-β-arabinofuranosyladenine does block enzyme induction by NGF (*23*), suggesting that the regulatory action of NGF may be at the level of mRNA processing, although other actions of this derivative, such as inhibition of transmethylation reactions (*24*), have yet to be discounted.

Closely related to the unresolved question of second messenger mechanisms are the "early effects" induced by the presentation of NGF to neurons that depend on it for survival. After chick sensory and sympathetic neurons are cultured for several hours in the absence of NGF, subsequent addition of NGF to the culture medium has been reported to result in a rapid activation of the sodium-potassium pump and subsequent restoration of sodium-dependent uptake mechanisms of the neurons such as those for glucose and uridine (*25*). However, such experiments cannot determine whether the activation of the sodium pump is the primary action of NGF, or whether it is a secondary effect resulting from restoration of general cell functions by other, still unknown, mechanisms; in the absence of NGF the neurons are dying.

Although the rapidity of the response of the sodium pump to NGF favored the assumption of a direct activation, NGF did not directly stimulate the sodium, potassium adenosine triphosphatase (ATPase) in membrane preparations from the same neurons (*26*). Thus these early effects most probably reflect a rapid but indirect restoration of general cell functions. Similarly, although the rapid activation of *N*-methylation of phospholipids by NGF may be causally related to fiber outgrowth (*27*), many other ligand-receptor interactions in other systems also result in the enhanced *N*-methylation of phospholipids (*28*). Thus, it is essential to establish whether other ligand-receptor interactions—such as tetanus toxin– or lectin-binding, which have no neurite outgrowth–promoting activity—activate *N*-methylation before the significance of this observation can be evaluated.

Not only NGF but also high potassium concentrations lead to the survival of chick sympathetic neurons, and both induce dephosphorylation of a 70-kilodalton protein (*29*). Again, however, it remains to be established whether this dephosphorylation of the 70-kilodalton molecule is causally related to survival, or if it is merely a correlated effect.

Purification of New Neurotrophic Molecules

A fundamental requirement for the purification and characterization of new neurotrophic factors is the availability of defined neuronal culture systems (*3*). These are necessary so that the activity of preparations containing putative trophic agents can be quantified from data on the ability to support neuronal survival. When such an assay based on the survival of cultured sensory neurons dissociated from embryonic chick dorsal root ganglia was used, a neurotrophic activity in mammalian brain could be detected and quantitatively assessed (*30*). A purification factor estimated to be more than 10^6 was necessary to achieve homogeneity of the active molecule, as judged by two-dimensional gel electrophoresis (*31*). Although this molecule exhibits some physicochemical properties (molecular weight, 12,300; isoelectric point, >10.1) similar to those of the monomer of β-NGF (molecular weight, 13,259; isoelectric point, 9.3), the immunological and biological properties are distinctly different from those of NGF. (i) There is no immunological cross-reactivity between NGF and the new brain factor (*31*). (ii) The two molecules act maximally on sensory neurons of different developmental ages (*30*). (iii) In contrast to NGF, the brain factor does not support the survival of sympathetic neurons but may support retinal neurons, which do not respond to NGF (*32*). Thus, the brain-derived putative neurotrophic factor seems to be able to exert a survival-promoting activity on neurons that have a projection in the central nervous system.

Manthorpe and co-workers have reported the purification of a neurotrophic factor, from embryonic chick eye tissue (CNTF) that supports the survival of cultured chick parasympathetic neurons (*33*). This molecule, with a molecular weight 20,600 and an isoelectric point of 5.0, did not cross-react with antibodies to mouse NGF and was also clearly different from the factor derived from mammalian brain. The eye-derived protein has a comparatively unspecific spectrum of action in that it supports the survival of both sympathetic and parasympathetic neurons (that do not respond to the brain factor), in addition to its ability to maintain sensory neurons for at least 24 hours in culture (*33*). Elucidation of the physiological roles of both of these putative neurotrophic proteins awaits the production of antibodies that could be used to determine their cellular location and to observe the consequences of their neutralization in vivo.

"Instructive" and Neurite Growth-Promoting Molecules

In addition to the identification and purification of neuronal survival factors, the purification of molecule(s) responsible for the induction of cholinergic properties in cultured adrenergic neurons of newborn rat is relatively far advanced (*34*). The main cholinergic-inducing activity found in heart cell–conditioned medium migrates as a molecule of 40 to 45 kilodaltons on SDS gel electrophoresis, and has been purified some 10,000-fold (*35*). It seems that the same molecule also increases choline acetyltransferase activity in spinal cholinergic neurons (*36*) and possibly also in chick parasympathetic neurons (*37*).

Progress is also being made in the identification and isolation of molecules that affect the morphological phenotypes of neurons rather than affecting their biochemical properties or survival. Kligman has reported that a soluble brain extract contains an activity that stimulates neurite outgrowth from cultured cerebral cortex neurons (*38*). This activity was purified and shown to be associated with a dimeric protein with subunits with a molecular weight of 37,000 upon reduction. Similarly, Davis and co-workers have partially purified a neurite-promoting activity of RN 22 schwannoma cells that acts by adhesion to polycationic culture substrates to stimulate the rate of neurite growth (*39*). They showed that their most pure preparation contains two proteins with apparent molecular weights 200,000 and 190,000 on SDS–polyacrylamide gel electrophoresis. It is not yet clear which of these proteins possess neurite outgrowth–promoting activity, although the larger cross-reacted with antibodies to the basement membrane protein laminin, which stimulates neurite outgrowth (see below). The antibodies to laminin failed, however, to block the neurite outgrowth–promoting activity of the schwannoma factor; but they could be used to immunoprecipitate it, and they do block the effect of laminin, which indicates that the molecules are not identical (*39*) even though there may be antigenic similarity between the factor and laminin.

Adopting a different approach, Gurney has shown that antibodies to the proteins secreted by denervated muscle can block neuronal sprouting in vivo (*40*). Thus, it appears that denervated muscle cells produce a sprouting factor, and an antiserum that blocks its activity recognizes a protein of apparent molecular weight 56,000. It will be interesting to see if this molecule, which is apparently responsible for neuronal sprouting in vivo, is also able to act as a neurite-promoting factor in vitro. Tissue culture experiments have indicated that this molecule may be able to support the survival or stimulate (or both) neurite outgrowth of spinal neurons from young (E4-5) chick embryos for short periods of time (24 hours) in vitro (*40*).

Influence of Extracellular Matrix Molecules on Neuronal Development

Not only the direction and rate of growth of neurites is dependent on the presence of appropriate substrates (*41*), but substrate-associated molecules can also affect neuronal differentiation (*42*) and can modulate the survival effect of neurotrophic factors (*43*). For example, fixed rat heart cells when used as culture substrates are able to induce cholinergic properties in sympathetic neurons, which otherwise would remain adrenergic (*42*). It is not yet established, however, what relation exists between the molecules released into the medium that also have this effect (*34–36*) and those present on the fixed heart cell membranes (*42*). Moreover, molecules of the extracellular matrix produced by embryonic chick heart cells modulate the NGF-dependent survival of sympathetic neurons (*43*), although they cannot promote survival themselves (*43, 44*). The maximal survival of chick sympathetic neurons resulting from supramaximal concentrations of NGF is 40 to 50 percent when the neurons are cultured on a polycationic substrate; however, when this substrate was first treated with heart cell–conditioned medium, virtually all of the neurons could be induced to survive by NGF. A distinct subpopulation of sympathetic neurons was subsequently shown to require the presence of the heart cell matrix deposited from the conditioned medium in order to survive. Neurons with adrenergic properties survived in response to NGF alone whereas those neurons present in chick sympathetic ganglia with nonadrenergic (presumably cholinergic) properties required both NGF and the heart cell matrix (*45*).

The neurite outgrowth–promoting effects of the substrate-attached materials from heart-conditioned medium are shared by those of conditioned medium produced by a rat schwannoma cell line RN 22 and various other conditioned media (*46*). These agents also potentiate the survival effects of NGF (and also potassium) on sympathetic neurons (*43, 46*), and the brain-derived growth factor on sensory neurons (*47*). From the experiments with conditioned media, however, it cannot be decided whether the potentiation of the survival effect and the neurite–promoting activity are due to the same or to different molecules. The fact that the basement membrane protein laminin has both a strong neurite promoting activity (*48, 49*) and can potentiate the survival effect of NGF (*49*) shows that these two properties can belong to the same molecule.

Laminin has proved to be an excellent model to analyze the mechanism of interaction of substrate molecules with cell membranes (*50*). Laminin can be cleaved proteolytically into fragments to which antibodies can be produced (*51*). Antibodies to parts of the three short arms of laminin in which the binding domains to tumor cells and hepatocytes reside did not block the fiber outgrowth promoting activity and the survival potentiating effects of laminin (*49*). However, antibodies to the globular domain at the end of the long arm of laminin abolished both the neurite outgrowth promoting activity and the enhancement of the survival effect of NGF (*49*). Previously, the only functional property ascribed to the globular domain at the end of the long arm of laminin was that it contains a heparin binding site (*51*). Thus, neurons may interact with laminin via the heparan sulfate of the neuronal membrane (*52*). Furthermore, the fact that the laminin molecule has two distinct binding sites for neurons and nonneuronal cells points to the possibility that laminin mediates intercellular interactions within the developing nervous system. In the adult, laminin is located in the basal laminae of peripheral nerves and is apparently synthesized by Schwann cells (*53*). Although such laminae are generally present on the outside of the Schwann cell-axon "unit," away from the axons, during axonal regeneration in injured peripheral nerves the axons have been seen to grow along the inside of the remaining basal laminae (*54*). This indicates that the laminin of the peripheral nervous system may be necessary for the ability of these neurons to regenerate their projections.

Future Developments

Even if the purification of a putative trophic factor has been accomplished, this does not guarantee the determination of its physiological significance. The small quantities available are not only inadequate for pharmacological studies but also hamper the production of antibodies: although the principle of monoclonal antibody production allows immunization with impure preparations, if the molecule is a poor antigen then it may prove exceedingly difficult to obtain suitable antibodies. Indeed, the production of antibodies to neither the brain factor nor CNTF has been reported. However, recent advances in peptide chemistry and molecular genetics offer a way around these problems (*55*). Thus, appropriate oligopeptide sequences from such molecules may be determined and subsequently synthesized in sufficient quantities to produce antibodies against the putative trophic factors. Furthermore the cloning of these molecules and subsequent production in prokaryotic or eukaryotic systems by expression vectors may be the only way to produce them in quantities large enough to test the possibility of their therapeutic applications.

That neurotrophic molecules might be useful for promoting regeneration can be deduced from the observation that the regeneration of lesioned adult sympathetic nerve fibers is enhanced by NGF and delayed by antibodies to NGF (*56*). The local application of neurotrophic and appropriate substrate molecules therefore may aid regeneration, although mere regrowth is no guarantee for success since the stimulation of regeneration has to be followed by the formation of the correct corrections (*57*). Whether the promotion of the regeneration of nerve fibers is followed by the "correct wiring" is not known.

The lack of regeneration in the central nervous system is a multifactorial problem including glial scar formation and insufficient production of neurotrophic molecules or matrix molecules. In cases of more general processes of degeneration or atrophy the local application of neurotrophic molecules does not seem to be appropriate. Better understanding of the physiology of these molecules may make it possible to influence pharmacologically their production and release. For example, an atrophy of central cholinergic neurons is found in Alzheimer's disease (*58*). Although these neurons do not appear to depend on endogenous NGF for their maintenance, they do respond to exogenous NGF with increased levels of the enzyme choline acetyltransferase (*59*). Thus, neurotrophic factors may eventually be used either by direct local administration or by pharmacological modifications of their synthesis in vivo.

References and Notes

1. M. Jacobson, *Developmental Neurobiology* (Plenum, New York, 1978), pp. 253–307.
2. R. Levi-Montalcini, *Harvey Lect.* **60**, 217 (1966); L. A. Greene and E. M. Shooter, *Annu. Rev. Neurosci.* **3**, 353 (1980); H. Thoenen and Y.-A. Barde, *Physiol. Rev.* **60**, 1284 (1980); B. Yankner and E. Shooter, *Annu. Rev. Biochem.* **51**, 845 (1982).
3. S. Varon and R. Adler, in *Adv. Cell. Neurobiol.* **2**, 115 (1981); D. K. Berg, in *Neuronal Development*, N. C. Spitzer, Ed. (Plenum, New York, 1982), p. 297; Y.-A. Barde, D. Edgar, H. Thoenen, *Annu. Rev. Physiol.* **45**, 601 (1983).
4. R. H. Angeletti and R. A. Bradshaw, *Proc. Natl. Acad. Sci. U.S.A.* **68**, 2417 (1971).
5. J. Scott *et al.*, *Nature (London)* **302**, 538 (1983); U. Francke *et al.*, *Science* **222**, 1248 (1983); A. Ullrich *et al.*, *Nature (London)* **303**, 821 (1983).
6. S. Korsching and H. Thoenen, *Proc. Natl. Acad. Sci. U.S.A.* **80**, 3513 (1983); H. Thoenen, Y.-A. Barde, S. Korsching, D. Edgar, *Cold Spring Harbor Symp. Quant. Biol.* **48**, 679 (1983).
7. R. Campenot, *Dev. Biol.* **93**, 1 (1982).
8. T. Ebendal, L. Olson, Å. Seiger, K.-O. Hedlung, *Nature (London)* **286**, 25 (1980); T. Ebendal, L. Olson, Å. Seiger, *Exp. Cell Res.* **148**, 311 (1983); E.-M. Barth, S. Korsching, H. Thoenen, *J. Cell Biol.* **99**, 939 (1984).
9. R. Heumann, S. Korsching, J. Scott, H. Thoenen, *EMBO J.* **3**, 3183 (1984); D. Shelton and L. F. Reichardt, *Soc. Neurosci. Abst.* **10**, 369 (1984).
10. S. Korsching and H. Thoenen, *Neurosci. Lett.* **39**, 1 (1983).
11. M. Schwab, R. Heumann, H. Thoenen, *Cold Spring Harbor Symp. Quant. Biol.* **46**, 125 (1982); M. E. Schwab and H. Thoenen, in *Handbook of Neurochemistry*, A. Lajtha, Ed. (Plenum, New York, 1983), p. 381.

12. G. P. Harper *et al.*, *Neuroscience* **8**, 375 (1983).
13. P. Cohen, A. Sutter, G. Landreth, A. Zimmermann, E. M. Shooter, *J. Biol. Chem.* **255**, 2949 (1980).
14. D. Mercanti, R. Butler, R. Revoltella, *Biochim. Biophys. Acta* **496**, 412 (1977).
15. M. W. Riemen, *Diss. Abst. Int.* **42**, 1023B (1981); L. Moroder, S. Romani, E. Wünsch, G. Harper, H. Thoenen, in *Peptides, Structure and Function*, V. J. Hruby and D. H. Rich, Eds. (Pierce Chemical, New York, 1983), p. 175.
16. J. Massagué, B. J. Guillette, M. P. Czech, C. J. Morgan, R. A. Bradshaw, *J. Biol. Chem.* **256**, 9419 (1981); P. Puma, S. E. Buxer, L. Watson, D. J. Kelleher, G. L. Johnson, *ibid.* **258**, 3370 (1983).
17. A. Sutter, M. Hosang, R. D. Vale, E. M. Shooter, in *Cellular and Molecular Biology of Neuronal Development*, I. B. Black, Ed. (Plenum, New York, 1984), p. 201.
18. J. Schechter and M. Bothwell, *Cell* **24**, 867 (1981).
19. R. Heumann, M. E. Schwab, H. Thoenen, *Nature (London)* **292**, 838 (1981).
20. P. Seeley, C. H. Keith, M. L. Shelanski, L. A. Greene, *J. Neurosci.* **3**, 1488 (1983).
21. R. Heumann, M. Schwab, R. Merkl, H. Thoenen, *ibid.*, in press.
22. F. Hefti, H. Gnahn, M. Schwab, H. Thoenen, *ibid.* **2**, 1554 (1982).
23. A. Acheson, K. Naujoks, H. Thoenen, *ibid.* **4**, 1771 (1984).
24. P. K. Chang, A. Guranowski, J. E. Segall, *Arch. Biochem. Biophys.* **207**, 175 (1981).
25. S. Skaper and S. Varon, *J. Neurochem.* **34**, 1654 (1980); *Exp. Cell Res.* **131**, 356 (1981); *Trend Biochem. Sci.* **8**, 22 (1983).
26. ______, *J. Neurosci. Res.* **6**, 133 (1981).
27. K. H. Pfenninger and M. P. Johnson, *Proc. Natl. Acad. Sci. U.S.A.* **78**, 7797 (1981); S. D. Skaper and S. Varon, *J. Neurochem.* **42**, 116 (1983).
28. J. Axelrod and F. Hirata, *Trends Pharmacol. Sci.* **4**, 156 (1982); J. M. Mato and S. Alemany, *Biochem. J.* **213**, 1 (1983).
29. S. B. Por and W. B. Huttner, *J. Biol. Chem.* **159**, 6526 (1984).
30. Y.-A. Barde, D. Edgar, H. Thoenen, *Proc. Natl. Acad. Sci. U.S.A.* **77**, 1199 (1980).
31. ______, *EMBO J.* **1**, 549 (1982).
32. J. E. Turner, Y.-A. Barde, M. E. Schwab, H. Thoenen, *Dev. Brain Res.* **6**, 77 (1983).
33. M. Manthorpe, G. Barbin, S. Varon, *J. Neurochem.*, in press.
34. P. Patterson, *Annu. Rev. Neurosci.* **1**, 1 (1978); M. Weber, *J. Biol. Chem.* **256**, 3447 (1981).
35. K. Fukada, *J. Neurochem. Suppl.* **41**, 589 (1983).
36. M. C. Giess and M. Weber, *J. Neurosci.* **4**, 1442 (1984).
37. R. Nishi and D. K. Berg, *ibid.* **1**, 505 (1981).
38. D. Kligman, *Brain Res.* **250**, 93 (1982); see also N. Micki, Y. Hayashi, H. Hagashida, *J. Neurochem.* **37**, 627 (1981); Y. Hayashi, H. Higashida, C. H. Kuo, N. Mike, *ibid.* **42**, 504 (1984); M. Coughlin and J. Kessler, *J. Neurosci. Res.* **8**, 289 (1982).
39. G. E. Davis, M. Manthorpe, S. Varon, *Soc. Neurosci. Abstr.* **19**, 36 (1984).
40. M. E. Gurney, *Nature (London)* **307**, 546 (1984); ______ and B. Apatoff, *Soc. Neurosci. Abstr.* **10**, 1051 (1984).
41. P. Letourneau, in *Neuronal Development*, N. C. Spitzer, Ed. (Plenum, New York, 1982), p. 213.
42. E. Hawrot, *Dev. Biol.* **50**, 541 (1980).
43. D. Edgar and H. Thoenen, *Dev. Brain Res.* **5**, 89 (1982).
44. F. Collins, *Proc. Natl. Acad. Sci. U.S.A.* **75**, 5210 (1978); R. Adler and S. Varon, *Brain Res.* **188**, 437 (1980).
45. H. Rohrer, H. Thoenen, D. Edgar, *Dev. Biol.* **99**, 34 (1983).
46. H. Adler, M. Manthorpe, S. D. Skaper, S. Varon, *Brain Res.* **206**, 129 (1981).
47. R. M. Lindsay, H. Thoenen, Y.-A. Barde, *Dev. Biol.* **112**, 319 (1985).
48. A. Baron van Evercooren *et al.*, *J. Neurosci. Res.* **8**, 179 (1982); S. Rogers, P. C. Letourneau, S. L. Palm, J. McCarthy, L. T. Furcht, *J. Cell Biol.* **98**, 212 (1983); M. Manthorpe, E. Engvall, E. Ruoslahti, F. M. Longo, G. E. Davis, S. Varon, *ibid.* **97**, 1882 (1983).
49. D. Edgar, R. Timpl, H. Thoenen, *EMBO J.* **3**, 1463 (1984).
50. N. C. Rao *et al.*, *J. Biol. Chem.* **257**, 9740 (1982); V. P. Terranova, C. N. Rao, T. Kalebic, M. K. Margulies, L. A. Liotta, *Proc. Natl. Acad. Sci. U.S.A.* **80**, 444 (1983); R. Timpl, S. Johansson, V. Van Delden, I. Oberbäumer, M. Höök, *J. Biol. Chem.* **258**, 8922 (1983).
51. U. Ott, E. Odermatt, J. Engel, H. Furthmayr, R. Timpl, *Eur. J. Biochem.* **123**, 63 (1982).
52. K. Greif and L. F. Reichardt, *J. Neurosci.* **2**, 843 (1982).
53. C. J. Cornbrooks, D. J. Carey, J. A. McDonald, R. Timpl, R. P. Bunge, *Proc. Natl. Acad. Sci. U.S.A.* **80**, 3850 (1983); S. Palm and L. T. Furcht, *J. Cell Biol.* **96**, 1218 (1983).
54. C. Ide, K. Tokyama, R. Yokota, T. Nitatori, S. Anodera, *Brain Res.* **288**, 61 (1983).
55. F. O. Schmitt, S. J. Bird, F. E. Bloom, *Molecular Genetic Neuroscience* (Raven, New York, 1982), p. 1.
56. B. Bjerre, A. Björklund, W. Mobley, *Z. Zellforsch. Mikrosk. Anat.* **146**, 15 (1973); B. Bjerre, A. Björklund, D. C. Edwards, *Cell Tissue Res.* **148**, 441 (1974); B. Bjerre, A. Björklund, W. Mobley, E. Rosengren, *Brain Res.* **94**, 263 (1975); B. Bjerre, L. Wiklund, D. C. Edwards, *ibid.* **92**, 257 (1975).
57. J. G. Nicholls, Ed., *Repair and Regeneration of the Nervous System*, (Springer Verlag, New York, 1982); G. Bray, M. Rasminsky, A. Aguayo, *Annu. Rev. Neurosci.* **4**, 127 (1981).
58. E. K. Perry and R. H. Perry, in *Metabolic Disorders of the Nervous System*, F. C. Rose, Ed. (Pitman, London, 1981), pp. 382–417; R. C. A. Pearson *et al.*, *Brain Res.* **289**, 375 (1983).
59. H. Gnahn, F. Hefti, R. Heumann, H. Thoenen, *Dev. Brain Res.* **9**, 45 (1983); P. Honegger and D. Lenoir, *ibid.* **3**, 229 (1983).

About the Authors

Jay A. Berzofsky is a senior investigator in the Metabolism Branch, National Cancer Institute, National Institutes of Health, Bethesda, Maryland 20205.

Helen M. Blau is in the Department of Pharmacology, Stanford University School of Medicine, Stanford, California 94305.

David T. Bonthron is a member of the Department of Pediatrics, Harvard Medical School, Boston, Massachusetts 02115, and also the Hematology Division, Children's Hospital and Dana-Farber Cancer Institute, Boston, Massachusetts 02115.

David Botstein is a professor in the Department of Biology, Massachusetts Institute of Technology, Cambridge, Massachusetts 02139.

Winston J. Brill is vice president of research and development, Agracetus, 8520 University Green, Middleton, Wisconsin 53562. He is also an adjunct professor of bacteriology, University of Wisconsin, Madison, Wisconsin 53706.

Michael S. Brown is a professor in the Department of Molecular Genetics, University of Texas Health Science Center at Dallas, Southwestern Medical School, Dallas, Texas 75235.

Gail A.P. Bruns is a member of the Department of Genetics, Harvard Medical School, Boston, Massachusetts 02115, and also the Genetics Division, Children's Hospital, Boston, Massachusetts 02115.

Kathryn Calame is an assistant professor in the Department of Biological Chemistry, UCLA School of Medicine, Los Angeles, California 90024.

Choy-Pik Chiu is in the Department of Pharmacology, Stanford University School of Medicine, Stanford, California 94305.

M. Chow is in the Department of Applied Biological Sciences, Massachusetts Institute of Technology, and the Whitehead Institute for Biomedical Research, Cambridge, Massachusetts 02139.

Adolfo J. de Bold is associated with the Department of Pathology, Queen's University and Hotel-Dieu Hospital, Kingston, Ontario, Canada K7L 3H6.

Mark D. Dibner is a neurobiologist in the Central Research and Development Department, E.I. du Pont de Nemours and Company, Experimental Station, E400, Wilmington, Delaware 19898. He is also an adjunct senior research fellow in the Management and Technology Program, Wharton School, University of Pennsylvania, Philadelphia, Pennsylvania 19104.

Timothy A. Donlon is a member of the Department of Genetics, Harvard Medical School, Boston, Massachusetts 02115, and also the Genetics Division, Children's Hospital, Boston, Massachusetts 02115.

Dennis Drayna, formerly at the Howard Hughes Medical Institute, University of Utah Medical Center, Salt Lake City, Utah 84132, is now in the Department of Molecular Biology, Genentech, South San Francisco, California 94080.

Margaret J. Duncan is a member of the Department of Molecular Genetics, Collaborative Research, Inc., Lexington, Massachusetts 02173.

Sean R. Eddy is a student at the California Institute of Technology, Pasadena, California 91125.

David Edgar is in the Department of Neurochemistry, Max-Planck-Institute for Psychiatry, D-8033 Martinsried/Munich, Federal Republic of Germany.

D.J. Filman is in the Department of Molecular Biology, Research Institute of Scripps Clinic, La Jolla, California 92037.

Dan Gibson is a Chaim Weizmann Postdoctoral Fellow in the Department of Chemistry, Massachusetts Institute of Technology, Cambridge, Massachusetts 02139.

David Ginsburg is at the Howard Hughes Medical Institute, and Department of Medicine, University of Michigan, Ann Arbor, Michigan 48109.

Joseph L. Goldstein is a professor in the Department of Molecular Genetics, University of Texas Health Science Center at Dallas, Southwestern Medical School, Dallas, Texas 75235.

Robert S. Goodenow is an assistant professor in the Department of Genetics and the Department of Microbiology and Immunology, University of California, Berkeley, California 94720.

Joan Goverman is a postdoctoral fellow in the Department of Biological Chemistry, UCLA School of Medicine, Los Angeles, California 90024.

Robert I. Handin is a member of the Department of Medicine, Harvard Medical School, Boston, Massachusetts 02115, and also the Hematology Division, Department of Medicine, Brigham and Women's Hospital, Boston, Massachusetts 02115.

Edna C. Hardeman is in the Department of Pharmacology, Stanford University School of Medicine, Stanford, California 94305.

Katherine Harding is in the Department of Biological Sciences, Fairchild Center, Columbia University, New York, New York 10027.

J.M. Hogle is in the Department of Molecular Biology, Research Institute of Scripps Clinic, La Jolla, California 92037.

A.J. Hudspeth is a professor of physiology and otolaryngology at the University of California School of Medicine, San Francisco, California 94143.

Frank J. Jenkins is a postdoctoral fellow at the Marjorie B. Kovler Viral Oncology Laboratories, University of Chicago, Chicago, Illinois 60637.

Dietrich Knorr is a professor in the Biotechnology Group, Department of Food Science, University of Delaware, Newark, Delaware 19716.

Samuel A. Latt is a member of the Department of Genetics, Harvard Medical School, Boston, Massachusetts, and also the Genetics Division, Children's Hospital, Boston, Massachusetts 02115.

Edith M. Lenches is a member of the Division of Biology, California Institute of Technology, Pasadena, California 91125.

Michael Levine is in the Department of Biological Sciences, Fairchild Center, Columbia University, New York, New York 10027.

Richard L. Linsk is a graduate student in the Department of Genetics, University of California, Berkeley, California 94720.

Stephen J. Lippard is a professor in the Department of Chemistry, Massachusetts Institute of Technology, Cambridge, Massachusetts 02139.

Harvey F. Lodish is a professor in the Whitehead Institute for Biomedical Research, Cambridge, Massachusetts 02142, and the Department of Biology, Massachusetts Institute of Technology, Cambridge, Massachusetts 02139.

William McGinnis is in the Department of Molecular Biophysics and Biochemistry, Yale University, New Haven, Connecticut 06520.

Mark Mercola is a graduate student at the Molecular Biology Institute, University of California, Los Angeles, California 90024.

Donald Metcalf is head of the Cancer Research Unit, Walter and Eliza Hall Institute of Medical Research, Post Office, Royal Melbourne Hospital 3050, Victoria, Australia.

Elliot M. Meyerowitz is an associate professor in the Division of Biology, California Institute of Technology, Pasadena, California 91125.

Steven C. Miller is in the Department of Pharmacology, Stanford University School of Medicine, Stanford, California 94305.

Carol Mirell is a postdoctoral fellow in the Department of Biological Chemistry, UCLA School of Medicine, Los Angeles, California 90024.

Donald T. Moir is a member of the Department of Molecular Genetics, Collaborative Research, Inc., Lexington, Massachusetts 02173.

Sherie L. Morrison is a professor of microbiology at both the Cancer Center and the Institute for Cancer Research, Columbia University College of Physicians and Surgeons, New York, New York 10032.

Stuart H. Orkin is a member of the Department of Pediatrics, Harvard Medical School, Boston, Massachusetts 02115, and also the Hematology Division, Children's Hospital and Dana-Farber Cancer Institute, Boston, Massachusetts 02115.

Grace K. Pavlath is in the Department of Pharmacology, Stanford University School of Medicine, Stanford, California 94305.

Robert E. Pruitt is a graduate student in the Division of Biology, California Institute of Technology, Pasadena, California 91125.

Charles M. Rice is a member of the Division of Biology, California Institute of Technology, Pasadena, California 91125.

Bernard Roizman is a Joseph Regenstein Distinguished Service Professor of Virology at the Marjorie B. Kovler Viral Oncology Laboratories, University of Chicago, Chicago, Illinois 60637.

David W. Russell is an assistant professor in the Department of Molecular Genetics, University of Texas Health Science Center at Dallas, Southwestern Medical School, Dallas, Texas 75235.

Rebecca L. Sheets is doing graduate work in the Department of Cellular, Viral, and Molecular Biology, University of Utah, University Medical Center, Salt Lake City, Utah 84132.

Suzanne E. Sherman is a graduate student in the Department of Chemistry, Massachusetts Institute of Technology, Cambridge, Massachusetts 02139.

Se Jung Shin is a student at the California Institute of Technology, Pasadena, California 91125.

David Shortle is an assistant professor in the Department of Biological Chemistry, The Johns Hopkins University School of Medicine, Baltimore, Maryland 21205

Laura Silberstein is in the Department of Pharmacology, Stanford University School of Medicine, Stanford, California 94305.

Anthony J. Sinskey is a professor in the Department of Applied Biological Sciences, Massachusetts Institute of Technology, Cambridge, Massachusetts 02139.

Robert A. Smith is a member of the Department of Molecular Genetics, Collaborative Research, Inc., Lexington, Massachusetts 02173.

James H. Strauss is a member of the Division of Biology, California Institute of Technology, Pasadena, California 91125.

Thomas C. Südhof is a postdoctoral fellow in the Department of Molecular Genetics, University of Texas Health Science Center at Dallas, Southwestern Medical School, Dallas, Texas 75235.

Hans Thoenen is in the Department of Neurochemistry, Max-Planck-Institute for Psychiatry, D-8033 Martinsried/Munich, Federal Republic of Germany.

Julie M. Vogel is a graduate student in the Department of Genetics, University of California, Berkeley, California 94720.

Andrew H.-J. Wang is a senior research scientist in the Department of Biology, Massachusetts Institute of Technology, Cambridge, Massachusetts 02139.

Cecelia Webster is in the Department of Pharmacology, Stanford University School of Medicine, Stanford, California 94305.

Steven G. Webster is in the Department of Pharmacology, Stanford University School of Medicine, Stanford, California 94305.

Cathy Wedeen is in the Department of Biological Sciences, Fairchild Center, Columbia University, New York, New York 10027.

Robert A. Weinberg is at the Whitehead Institute for Biomedical Research, Cambridge, Massachusetts 02142, and in the Department of Biology, Massachusetts Institute of Technology, Cambridge, Massachusetts 02139.

Ray White is associated with the Howard Hughes Medical Institute and the Department of Human Genetics, University of Utah Medical Center, Salt Lake City, Utah 84132.

William T. Wickner is a professor in the Molecular Biology Institute and Department of Biological Chemistry, University of California, Los Angeles, California 90024.

Edward O. Wilson is Frank B. Baird, Jr. Professor of Science and Curator in Entomology, Museum of Comparative Zoology, Harvard University, Cambridge, Massachusetts 02138.

Index